WORKING PAPERS

Astronomy and Astrophysics Panel Reports

ASTRONOMY AND ASTROPHYSICS SURVEY COMMITTEE
BOARD ON PHYSICS AND ASTRONOMY
COMMISSION ON PHYSICAL SCIENCES, MATHEMATICS, AND APPLICATIONS
NATIONAL RESEARCH COUNCIL

NATIONAL ACADEMY PRESS
Washington, D.C.   1991

This project was supported by the Department of Energy under Grant No. DE-FG05-89ER40421, the National Aeronautics and Space Administration and the National Science Foundation under Grant No. AST-8901685, the Naval Research Laboratory under Contract No. N00173-90-M-9744, and the Smithsonian Institution under Purchase Order No. SF0022430000.

Library of Congress Catalog Card Number 90-63608

International Standard Book Number 0-309-04383-2

Cover: Near-infrared image of the Milky Way. A new view of the Milky Way Galaxy obtained on April 17, 1990, by the Diffuse Infrared Background Explorer (DIRBE) on NASA's Cosmic Background Explorer (COBE) satellite. Courtesy of the COBE Science Working Group and NASA's Goddard Space Flight Center.

Additional copies of this document are available from:

National Academy Press
2101 Constitution Avenue, NW
Washington, DC 20418

S242

Printed in the United States of America

ASTRONOMY AND ASTROPHYSICS SURVEY COMMITTEE

Preface

This volume contains the working papers of the panels appointed by the Astronomy and Astrophysics Survey Committee. These papers were advisory to the survey committee and represent the opinions of the members of each panel in the context of their individual charges. They have not been edited by the survey committee, nor have they been edited or reviewed by the National Research Council.

The committee's full survey report is contained in a separately published document, *The Decade of Discovery in Astronomy and Astrophysics* (National Academy Press, Washington, D.C., 1991), issued simultaneously with this volume of the panels' working papers.

Selected by the committee, the chairs of the panels in turn helped the committee to select a broad and representative group of experts to advise in 15 areas of concern. The chairs were responsible, together with their panel members, for obtaining the views of a wide cross-section of the astronomy and astrophysics community and for preparing a paper on their discussions and findings. A member of the survey committee served as a vice-chair of each panel. In some cases, the panel chairs selected a core group to assume primary responsibility for writing the panel's paper; members of such core groups are designated by an asterisk in the list of panel members that precedes each paper.

The panel chairs presented their papers in oral and written form at the June and July 1990 meetings of the survey committee and were invited to participate with the committee in the initial attempts to generate a cohesive set of overall recommendations. The views of the participants were modified by the discussions that took place between the different advocates and experts. The committee based its final decisions and recommendations in significant part on the contents of the panel papers and on the discussions with the panel chairs.

Ten panels had charges that reflected specific scientific areas, eight of them based on wavelength region and two (those of the Planetary Astronomy Panel and Solar Astronomy Panel) on particular subdisciplines with special needs. The committee asked these ten science panels to identify the most important scientific goals in their respective areas, to prioritize the new initiatives needed to achieve these goals, to recommend proposals for technology development, to consider the possibilities for international collaboration, and to discuss any policy issues relevant to their charge. The Astronomy and Astrophysics Survey Committee served as an interdisciplinary panel to guarantee that scientific questions that did not fit conveniently into this organizational structure were handled appropriately on an ad hoc basis.

Four other panels were appointed to explore computing and data processing, policy opportunities, the benefits of astronomy to the nation, and the status of the profession. The working papers written on the first three topics were used by the committee as a basis for developing the chapters with corresponding subject matter (Chapters 5, 7, and 8, respectively) in the survey report. Data from the working paper titled "Status of the Profession" were used in preparing various chapters and Appendix B of the survey report and by other panels in preparing their papers. The Science Opportunities Panel, the fifteenth panel appointed by the committee, prepared a paper that the committee believed should be expanded and published separately as a popular book accessible to as large an audience as possible. An abbreviated and adapted version of this panel's paper appears as Chapter 2 of the survey report.

Members of the panels consulted widely with their colleagues to solicit advice and to inform other members of the astronomical community of the main issues facing the committee. Each panel held an open meeting at a session of the American Astronomical Society, and most of the panels held sessions at other professional gatherings, as well as at astronomical centers at different places in the United States. Each panel discussed with the relevant federal agency personnel the problems and issues of its particular area. These interactions with agency personnel provided valuable background to the discussions, although the panels were careful to preserve the independence and confidentiality of the National Research Council deliberative process.

The Astronomy and Astrophysics Survey Committee believes that the material in the panel papers is of general interest and may be of special use to students and research scientists in astronomy and astrophysics, as well as to university and governmental administrators.

John Bahcall
Chair
Astronomy and Astrophysics Survey Committee

Contents

Panel Reports

RADIO ASTRONOMY PANEL

KENNETH I. KELLERMANN, National Radio Astronomy Observatory, *Chair*
DAVID HEESCHEN, National Radio Astronomy Observatory, *Vice-Chair*

DONALD C. BACKER, University of California, Berkeley
MARSHALL H. COHEN,* California Institute of Technology
MICHAEL DAVIS, National Astronomy and Ionosphere Center
IMKE DE PATER, University of California, Berkeley
DAVID DE YOUNG, National Optical Astronomy Observatories
GEORGE A. DULK,* University of Colorado, Boulder
J.R. FISHER, National Radio Astronomy Observatory
W. MILLER GOSS, National Radio Astronomy Observatory
MARTHA P. HAYNES,* Cornell University
CARL E. HEILES, University of California, Berkeley
WILLIAM M. IRVINE, University of Massachusetts, Amherst
KENNETH J. JOHNSTON,* Naval Research Laboratory
JAMES MORAN, Harvard-Smithsonian Center for Astrophysics
STEVEN J. OSTRO, Jet Propulsion Laboratory
PATRICK PALMER,* University of Chicago
THOMAS G. PHILLIPS, California Institute of Technology
ALAN E.E. ROGERS, Haystack Observatory
NICHOLAS Z. SCOVILLE, California Institute of Technology
PHILIP M. SOLOMON, State University of New York, Stony Brook
JILL C. TARTER, NASA Ames Research Center
JOSEPH H. TAYLOR, JR.,* Princeton University
PATRICK THADDEUS,* Harvard-Smithsonian Center for Astrophysics
JUAN M. USON, National Radio Astronomy Observatory
WILLIAM JOHN WELCH,* University of California, Berkeley
ROBERT W. WILSON,* AT&T Bell Laboratories

Radio Astronomy

EXECUTIVE SUMMARY

Since Karl Jansky's first observations in 1932, improvements in technology have increased the sensitivity of radio telescopes by an average of about two orders of magnitude per decade, improved the angular resolution of radio images from tens of degrees to better than a thousandth of an arcsecond, and extended the short wavelength limit of radio astronomy from meter to millimeter and sub-millimeter wavelengths. The radio telescope is now the instrument of choice for high resolution and high fidelity images of many types of celestial objects.

During the past decade the unique facilities at the national radio observatories have made possible dramatic discoveries ranging from fundamental physics and cosmology to the spectacular radar imaging of asteroids. At the same time, pioneering observations made at millimeter and sub-millimeter wavelengths have provided the best picture yet of the spiral structure of our Galaxy, and have led to a much better understanding of the structure, dynamics, and chemistry of star-forming regions. New radio techniques have been developed to measure distances throughout the Universe. These methods are already leading to reevaluations of the size scale of the Galaxy and the Universe. Other advances in high resolution imaging, signal processing, and millimeter and sub-millimeter spectroscopy have opened many other new opportunities for radio astronomy in the 1990's. Unfortunately, however, the funding for radio astronomy has not been able to keep pace with the growth of the science.

Over the past ten to fifteen years, important radio telescopes have been closed, and there has been minimal new capital investment in existing national facilities to upgrade them to the state of the art, or even to maintain them and replace obsolete instrumentation. Of particular concern are the deteriorating state of the VLA—the world's premier radio telescope—the inadequate support for the newly developed fields of millimeter and sub-millimeter radio astronomy, and the decrease in the number and level of research grants to individual scientists.

As we enter the decade of the 1990's radio astronomy looks forward to the timely completion of the Very Long Baseline Array (VLBA), the Green Bank Telescope (GBT), the Arecibo upgrading project, the Arizona-German Sub-Millimeter Telescope (SMT), and the Smithsonian Sub-Millimeter Wavelength Array (SMA). Additional funds will be needed for operating these new facilities. At the same time, it is important to exploit the dramatic technical developments of the 1980's and to start now on the design and construction of facilities that will provide powerful new research opportunities during the decade following the 1990's.

The Radio Astronomy Panel recommends as the highest priority for new instrumentation for radio astronomy the construction of a Millimeter Wavelength Array (MMA) with a collecting area about 2000 square meters, receivers operating in all atmospheric windows in the range of 30 to 350 GHz, angular resolution better than 0.1" at the shortest wavelengths, and versatile high resolution spectroscopic capability.

The Millimeter Array will make possible the study of a wide variety of objects in the solar system, star formation and evolution, stellar nucleosynthesis, chemical and physical structure of the interstellar medium in the Milky Way as well as in distant galaxies, and the structure and evolution of the Universe. The sensitivity, angular resolution, speed, and image quality of the MMA will each exceed that of any existing millimeter wave instrument in the world by more than an order of magnitude.

Due to the fact that the MMA will not be complete before late in the decade, it is essential that adequate support be provided in the interim to the millimeter and sub-millimeter telescopes currently in operation. These instruments will advance the science and technology in this field during the next decade and train the young scientists who will use the MMA when it goes into operation. The existing university-based millimeter interferometers will play a particularly important role because they have begun and will continue to develop the scientific and technical program leading to the MMA. They will also provide a vital source of student and postdoctoral training in millimeter interferometry.

The Radio Astronomy Panel also recommends, in order of priority, the following new moderate scale instruments:

The construction of a filled aperture Large Millimeter Wavelength Radio Telescope.

The expansion of the VLA to cover the range of resolution intermediate between the current VLA and the VLBA, and to greatly enhance the imaging power of both the VLA and the VLBA.

The deployment in space of a 25-m class radio telescope, in collaboration with an international group of partners in Europe, Japan, and the USSR, to operate as a Very Long Baseline Interferometer (VLBI) element in space.

The Radio Astronomy Panel recognizes the need for a continuing opportunity for initiating new small-scale projects. Although the Panel fully expects that new ideas will be continually developed over the next decade, we have identified the following initiatives as being particularly meritorious at this time:

A Large Southern Radio Telescope in Brazil to be constructed and operated by an international consortium for research in atmospheric sciences, radio, and radar astronomy in the southern skies.

The construction of a small radio telescope especially designed to detect spatial fluctuations in the cosmic background radiation (CBR) at levels of one part in a million.

The participation in the Soviet and Japanese space VLBI missions planned for the mid 1990's.

The establishment of small research groups at universities to develop advanced instrumentation and carry out observational programs to search for extraterrestrial intelligence (SETI).

The development of a frequency agile, image-forming radio telescope for solar research.

The construction of a Fast All Sky Telescope to survey the sky for variable radio sources.

The Panel has identified the following areas of technological research which have particularly great potential to enhance the power of existing and future radio telescopes: a) the continued development of receiver technology for millimeter and sub-millimeter wavelengths; b) the development of broad bandwidth recording systems and data links for VLBI; and c) strengthening of efforts toward the protection from radio frequency interference (RFI) to ground, space, and lunar based radio telescopes, together with the development of effective techniques to suppress or eliminate the effects of RFI on radio astronomy observations.

The Panel also recognizes the opportunity for the development of major new capabilities that will be possible beyond the year 2000, and recommends that an orderly program begin during the 1990's directed toward the development of low frequency radio astronomy techniques on the ground and in space, ultimately leading to the establishment of a low frequency, high resolution radio astronomy telescope on the moon.

Introduction

Radio astronomy began just before the second World War and matured in the 1950's, mostly through the pioneering efforts of scientists with backgrounds in radio science, electrical engineering, or wartime radar. Their work led to remarkable discoveries in the 1950's and 1960's, including radio galaxies, quasars, pulsars, radio bursts from the Sun and Jupiter, giant molecular clouds, interstellar masers, and the cosmic microwave background. The radio observations also led toward much better understanding of a number of other astrophysical topics, including the nature of planetary atmospheres, surfaces, and spin-orbit resonances, the physical conditions in star-forming regions, the importance of galactic nuclei, the gas content of circumstellar shells and interstellar space, and conditions in the most distant parts of the Universe corresponding to epochs shortly after its creation.

In the 1970's, radio astronomers undertook an ambitious radio telescope construction program to exploit these new astrophysical areas, as well as the vigorous development of the specialized technologies needed for such fruitful new techniques as very long baseline interferometry, millimeter wavelength spectroscopy, and fast data acquisition and signal processing for pulsar and planetary radar studies.

The techniques of radio astronomy have continued to develop rapidly during the 1980's. Specialized hardware and algorithms have been developed for aperture synthesis imaging, with angular resolution and image quality unequaled by any other technique, and for making detailed measurements of the weak periodic signals from pulsars. Lessons learned in long baseline interferometry experiments led to the construction of the transcontinental Very Long Baseline Array, with antenna elements located from Hawaii to the Caribbean. At the same time millimeter and sub-millimeter techniques have been developed and exploited in this nearly unexplored region of the electromagnetic spectrum. But, for more than a decade, NSF funding of ground-based astronomy has been inadequate to keep pace with the growth of the science. This has serious consequences which now threaten the health of all of astronomy in the United States. Radio astronomy, which depends on the NSF for nearly all of its support, is in a particularly critical situation.

The lack of adequate funds for the support of individual scientists, for the operation, maintenance, and upgrading of existing radio telescopes to the state of the art, and for instrumentation and computing resources is the most important problem facing radio astronomy.

As we enter the decade of the 1990's, opportunities for new research initiatives will depend on the timely completion of the VLBA, the GBT, the Arecibo upgrading project, the Arizona-German Sub-Millimeter Telescope, and the Smithsonian Sub-Millimeter Wavelength Array. Additional funds will be needed for operating these new instruments. At the same time, it is important to exploit the dramatic technical developments of the 1980's and to start now on the construction of radio astronomy facilities which will provide powerful new research opportunities during the decade following the 1990's.

Scientific Opportunities

The history of radio astronomy has been characterized by the discovery of a wide range of fundamentally new phenomena and objects that have revolutionized our understanding of the Universe. Radio galaxies, quasars, pulsars, molecular masers, and solar radio bursts were serendipitous discoveries resulting from the use of powerful new technologies. Other new phenomena, such as gravitational lenses, neutron stars, and the microwave background radiation, were discussed prior to their discovery, but theoretical considerations played little role in their actual discovery.

Even among the more traditional cosmic bodies, such as stars, planets, and the Sun, radio observations have opened up a whole new domain of previously unknown phenomena. Planetary radio and radar observations first revealed the retrograde rotation of Venus and the unexpected rotation of Mercury. Other unexpected solar system discoveries include the excessive temperature of the Sun's corona, the high surface temperature of Venus likely the result of a runaway greenhouse effect, the high temperature of the outer planets apparently due to internal heat sources, the Van Allen Belts around Jupiter, and the spectacular low frequency bursts caused by violent electromagnetic activity in the atmospheres of Jupiter and the Sun.

For many years the analytic power of radio telescopes suffered from two major limitations: poor angular resolution and the inability to measure distances. But, during the decade of the 1980's, this situation has dramatically changed.

Because of the long wavelengths involved, it was thought for a long time that the angular resolution of radio telescopes must be severely limited compared with that of optical or infrared telescopes. In fact, the reverse is true; the long wavelength radio waves pass relatively unaffected through the terrestrial atmosphere while optical telescopes are limited by "seeing." Also, because the precision needed to build diffraction limited instruments at radio wavelengths is not as demanding as at optical wavelengths, radio telescopes may have essentially unlimited resolution. Sophisticated new techniques for analyzing radio interferometer data effectively eliminate any effects of image distortion from the atmosphere to give radio images with extraordinary image quality and angular resolution better than one thousandth of an arcsecond. This is several orders of magnitude better than available by any other technique on the ground or in space.

Radio distance measurements are now able to reach beyond the local flow to give fundamentally new determinations of the size of the Galaxy, the Hubble Constant, and the size of the Universe itself. These techniques, some of which are completely independent of evolutionary effects or the usual hierarchical arguments, include: the direct trigonometric parallax of pulsars and other galactic objects; statistical parallax measurements of H_2O masers; the time delay of OH emission in late type stars; VLBI measurements of supernovae expansion velocities; HI and CO spectroscopic redshifts; the Tully-Fisher Relation; VLBI observations of superluminal component motions; gravitational lensing; and the Sunyaev-Zeldovich effect.

Millimeter and Sub-Millimeter Wavelength Astronomy

Millimeter wave astronomy has opened up new opportunities to study the evolution of stars, galaxies, and the Universe itself. The chemistry and composition of the interstellar medium, the earliest stages of star formation, and the internal kinematics of luminous galaxies are uniquely revealed at millimeter wavelengths. Array-type radio telescopes for millimeter and sub-millimeter wavelengths, built with recently developed technology and exploiting powerful new imaging techniques, will provide tremendous improvements in sensitivity and resolution in these spectral bands.

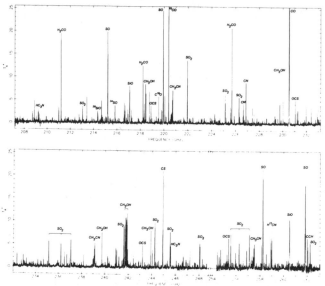

FIGURE 1 The millimeter wavelength spectrum of the Orion Molecular Cloud (OMC-1) showing more than one thousand lines identified with about thirty different molecular species. High resolution images of the chemical and isotopic distributions map the gradients of temperature and density, as well as the kinematics, and give insight into the process of how these clouds collapse to form stars. (Photo courtesy of T. G. Phillips, California Institute of Technology)

New stars are continually being born in giant clouds containing millions of solar masses of molecular gas. Studies of carbon monoxide made at 2.6 mm wavelength have led to the determination of the size, mass, and location of hundreds of molecular clouds throughout the Galaxy, and have provided the best picture yet of the spiral structure of the Milky Way. The study of isotope abundances in molecular clouds gives evidence for the survival of interstellar molecular material in primitive solar system objects, and allows the study of conditions relevant to the origin of the solar system, and perhaps, life itself. In the most luminous galaxies and quasars, the molecular gas appears to play a pivotal role in promoting energetic starbursts and possibly fueling active galactic nuclei.

Millimeter wavelength observations of the gaseous envelopes around very old stars give insight into their morphology, dynamics, nucleosynthesis and chemical abundance. High resolution millimeter wavelength images of outflowing envelopes of old giant stars show that they contain shells of gas containing molecules which must have been produced in a remarkably short time of a few thousand years. Improved sensitivity and resolution at millimeter and sub-millimeter wavelengths has also led to much better understanding of the structure, dynamics, and chemistry of star-forming regions, the detection of interstellar polyatomic organic molecules, and to the unexpected discovery of gaseous outflows from young stars.

Millimeter and sub-millimeter wavelength observations are particularly critical to our understanding of galaxies, because these wavelengths penetrate the dust obscuring the galactic cores at other wavelengths and allow the large-scale gas and dust distributions and their relationship to global star formation to be determined. Carbon monoxide has now been detected in hundreds of galaxies, and imaged in dozens. The data reveal galaxies with central disks, rings, bars, strong nuclear concentrations, and prominent spiral arms. Molecular gas is found to be concentrated primarily in the inner parts of spiral galaxies, especially those that are very luminous in the infrared. The recent detection of CO in several quasars serves as a prominent indication of the future potential of extragalactic molecular astronomy.

Meter to Hectometer Wavelength Astronomy

During the past decade several unexpected discoveries have led to a resurgence of interest in radio astronomy at long wavelengths. Surprisingly, strong meter wavelength recombination lines have been found in the interstellar medium throughout the galaxy. A prominent meter wavelength continuum source led to the discovery of the first millisecond pulsar. The variability of Cassiopeia A at meter wavelengths is difficult to explain within the context of any conventional understanding. Solar radio bursts due to electron streams and shock waves have been observed and need to be imaged with high angular resolution, particularly in the nearly unexplored hectometer wavelength band where the radiation originates in the region of solar wind acceleration. Planetary radio observations at long wavelengths have also resulted in the recognition of a new coherent emission mechanism, known as cyclotron maser radiation, which provides an elegant explanation for the extraordinarily bright (up to 10^{16} K) circularly polarized radiation seen in the Earth's auroral zones, from Jupiter and the other giant planets, from the Sun, and from a variety of stars. An important challenge for meter wave radio astronomy during the next decade will be the attempts to detect highly redshifted primordial "pancake" clouds of neutral hydrogen.

The Sun, Stars, Pulsars, Interstellar Masers, and Extrasolar planets

Millisecond and binary pulsars, formed in the complicated evolution of an interacting pair of stars, have taught us important lessons about the last stages of stellar evolution in close binary systems. Pulsars will continue to be extremely productive tools for probing a wide range of phenomena in gravitational physics, cosmology, astrometry, time-keeping metrology, and nuclear and plasma physics. The upgraded Arecibo antenna and the Green Bank Telescope, together with sophisticated new signal processing and data acquisition systems, will provide unprecedented sensitivity and flexibility for pulsar studies of all kinds.

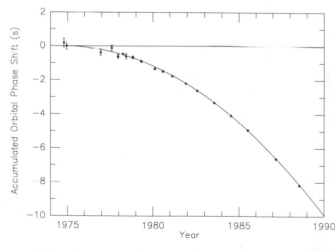

FIGURE 2 The emission of gravitational radiation by the binary pulsar PSR 1913+16 leads to an increasing change in the orbit compared with a hypothetical system whose orbital period remains constant. The observations agree to better than one percent with the change predicted from general relativity and provide *the only experimental evidence for the existence of gravitational radiation.* (Photo courtesy of J. H. Taylor, Princeton University)

Molecular maser clouds are found surrounding newly formed as well as very old stars. Measurements of Zeeman splitting of OH and H_2O maser lines determine the magnetic field strength which has been important in understanding the energy balance and kinematics of the molecular clouds. Astrometric VLBI measurements made with the extraordinary precision of 10 microarcseconds per year have made it possible to track the motions of H_2O masers in the envelopes of young stellar objects, and to determine directly their distances. Extension of this technique to space VLBI offers the promise of the direct measurement of distances to nearby galaxies and the recalibration of the distance scale of the Universe.

The high sensitivity of the Very Large Array and the Arecibo telescope has also made possible the detection and imaging of radio emission from a variety of stars. Radio emission with a thermal spectrum has been identified with stellar winds which transfer mass between the components in binary star systems, while non-thermal emission is associated with a wide range of phenomena including short-lived flares up to a million times more intense than those seen on the sun. The VLA has identified the locations where high energy electrons are accelerated and confined during solar flares, and has revealed a remarkable correlation between radio brightness and the magnetic field structure of the chromosphere and corona. The Millimeter Array and the added resolution of the expanded VLA will be particularly important in imaging the radio emission from stars of every spectral type and luminosity.

Interest in the existence of planets around other stars and their possible consequences has never been higher. Astrometric detection of dark stellar companions may be possible with the VLBA. The formation of planetary systems around individual stars is a fundamental problem that is best studied at millimeter wavelengths where the dynamics and chemistry of the dust and gas surrounding newly formed stars can be directly observed. The millimeter and sub-millimeter arrays and the Large Millimeter Telescope will be extremely powerful tools for probing preplanetary circumstellar disks. In addition, the Search for Extraterrestrial Intelligence, SETI, continues to fascinate the layman as well as scientists. SETI provides a powerful intellectual and technical challenge, and will be expanded during the 1990's with powerful new instrumentation and techniques that will greatly extend the horizons of the search.

The Planets, Asteroids, and Comets

With the detection of Pluto, thermal radio emission has now been observed from all of the planets, several of their satellites, and from a number of asteroids and comets. Millimeter interferometry of ammonia in the atmosphere of the giant planets and of carbon dioxide in the atmospheres of Venus and Mars offer the possibility to directly observe the diurnal, latitude, and seasonal variations of atmospheric temperature and molecular abundance. Variable sulfur dioxide emission has been observed at millimeter wavelengths on Io, probably as a result of volcanic activity. Millimeter observations of hydrogen cyanide, a cometary parent molecule, provides direct information on the kinematics in cometary coma as well as on its chemical composition.

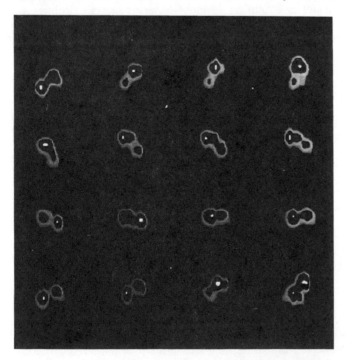

Figure 3 Radar images of asteroid 1989 PB made at the Arecibo Observatory near the time of closest approach of 2.5 million miles. The dumbbell-shaped asteroid is between one and two kilometers across and rotates with a period of about four hours. These images have an effective resolution of better than ten milliarcseconds. (Photo courtesy of S. Ostro, Caltech Jet Propulsion Laboratory)

Worldwide VLBI observations have been used to track the two Soviet VEGA balloons to give information on the circulation of winds in the atmosphere of Venus. Radar observations during the past decade have yielded the first direct detection of a cometary nucleus, the discovery of large-particle clouds associated with comets IRAS-Araki-Alcock and Halley, the extremely irregular, non-convex shapes of some near-Earth asteroids, and the first direct evidence that the rings of Saturn contain centimeter or larger sized particles.

Radio Galaxies, Quasars, and Cosmology

Radio observations continue to play a key role in understanding galaxies, quasars, and active galactic nuclei (AGN's), and have changed our understanding of cosmology in a fundamental way. Observations of neutral hydrogen gas in thousands of galaxies have revealed the existence of structures with dimensions of at least 50-100 Mpc. These results establish important boundary conditions on the evolution of large-scale structures in the Universe, and have been used for dynamical studies of the mass distribution within galaxies, placing lower limits on the amount of "dark matter" that they contain. Extragalactic neutral hydrogen and carbon monoxide surveys will continue to be extremely productive, especially with the upgraded capabilities of the Arecibo telescope, the VLA, the new Green Bank Telescope, and the proposed new millimeter radio telescopes. The observation of highly redshifted atomic and molecular gas provides information about conditions in galaxy disks at early epochs. Together with optical redshift surveys, these data provide an invaluable pool of cosmological information to investigate the formation, evolution, and large-scale distribution of galaxies, and to address the question of whether the Universe is open or closed. Gravitationally focused images of distant quasars give us a new technique for studying the mass distribution in galaxies as well as a new and potentially important method of determining the size and age of the Universe.

Probably the most important discovery in cosmology in modern times is the radio detection of the cosmic microwave background radiation. Except for the effect of our motion through the Universe, the background radiation is found to be remarkably smooth to within a few parts per hundred thousand. This simple experimental fact provides one of the most stringent constraints on models of the early Universe, and particularly on the enigmatic process of galaxy formation. Testing the isotropy of the cosmic microwave background at the one-part-per-million level is now experimentally feasible and needs to be vigorously pursued. This type of experiment will continue to be one of the observational cornerstones of cosmology.

Long-standing problems still remain in understanding the source of energy in quasars and active galactic nuclei, and the conversion of energy into the relativistic plasma which generates the observed synchrotron radiation. VLA observations during the past decade have revealed jets, filaments, and hot spots in both extragalactic radio sources and in the center of our own Galaxy. These complex structures reflect the wealth of detail in the radio emitting plasma and the important role played by magnetic fields.

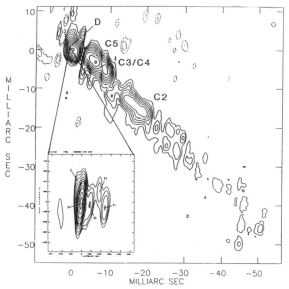

FIGURE 4 Six centimeter VLBI image of the core of the quasar 3C 273 made from a 14 element "world-array." The image has an angular resolution of 1 by 2 milliarcseconds. Repeated observations at this and at shorter wavelengths show the ejection of superluminal components which emerge from the core and appear to move along a fixed curve path with similar speeds of about one milliarcsecond per year. This corresponds to an apparent transverse linear velocity of about ten times the speed of light, and is referred to as *superluminal motion*.

The insert shows the core component observed at 3 mm wavelength with an angular resolution about ten times better than at 6 cm. Millimeter interferometry gives the highest resolution images ever obtained in astronomy. The individual components are less than half a light year in extent. (Photo courtesy of L. Baath, Onsala Space Observatory, Sweden)

VLBI observations have concentrated on the small but incredibly energetic cores of quasars and active galactic nuclei where the relativistic plasma is accelerated and focused into narrow jets which flow with apparent superluminal motion toward the extended radio lobes located hundreds of thousands of light years away. Superluminal motion is thought to be due to bulk relativistic motion of the radiating plasma nearly along the line of sight. An important consequence of the relativistic motion is that the synchrotron radiation is beamed along the direction of motion, so that the apparent radio luminosity of quasars and active galactic nuclei is very dependent on the orientation of the beam and in favorable cases may be enhanced by orders of magnitude. It is not clear how important the effects of relativistic beaming are in other parts of the electromagnetic spectrum, but the correlation of time variability and the continuity of the spectra suggest that the apparent optical, IR, and X-ray luminosity of quasars and active galactic nuclei may also be enhanced by this phenomenon. But the beaming models are difficult to reconcile in detail with the observations, and attempts to establish unified geometric models have been only partially successful. The increased resolution, sensitivity, and dynamic range expected from the expanded VLA, the VLBA, and space VLBI experiments will provide greatly enhanced capabilities for attacking these problems.

Challenges For Radio Astronomy in the 1990's

Many astrophysical puzzles are not yet solved, and almost certainly some presently "known" answers are wrong. The apparent neutral hydrogen links between the distant quasars and nearby bright galaxies, the apparent anisotropy and anomalous nature of the counts of strong radio sources, the absence of expected relativistic effects in the angular size-redshift distribution of quasars, and the apparent quantization of quasar and galaxy redshifts, are all difficult to understand within the framework of conventional cosmology and astrophysics.

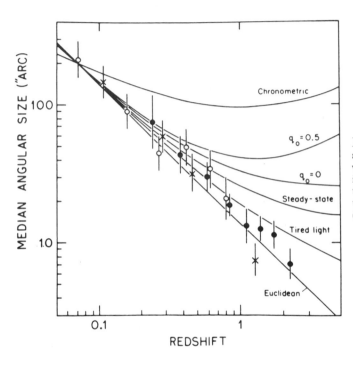

FIGURE 5   Angular size - redshift relation for different samples of radio galaxies and quasars compared with various world models. The only simple model consistent with the data is the static Euclidean model. Friedmann models require that the evolution of linear size with cosmic epoch just compensate the geometric effects to reproduce the Euclidean relationship. (Photo courtesy of V. Kapahi, Tata Institute, Bangalore, India)

By its very nature, basic scientific research addresses questions that lie at the boundaries between the known and unknown. If an answer to a scientific question is predictable with any degree of confidence, the question is probably not very close to this boundary! For this reason, it is difficult and probably even inappropriate to speculate on the most important scientific advances during the coming decade, even if the discussion is in rather general terms and the time scale is the relatively near future.

Too great a reliance on detailed planning may limit truly innovative thinking. We note that many of the radio astronomy highlights of the 1980's—millisecond pulsars and the detection of gravitational radiation damping, the extreme isotropy of the cosmic background radiation, the ordered clumpiness of the distribution of galaxies, bi-polar outflows from very young stars, gravitational lensing, and the high-dynamic-range mapping capability of the Very Large Array—were unexpected developments and largely unforeseen before their discovery. Earlier, radio galaxies, AGN's, quasars, pulsars, radio bursts from the Sun and Jupiter, the high surface and atmospheric temperatures of the planets, giant molecular clouds, interstellar molecular masers, and the cosmic background radiation itself were initially discovered as a result of the drive to exploit emerging new technology. Considerations of specific scientific issues had little impact on these major discoveries which now dominate much of our astrophysical thinking. As a result of these discoveries, radio astronomy has probably generated more new problems and questions than it has solved old problems, and has shown not only the inadequacy of our understanding even a few decades ago but, more importantly, the inadequacy of the questions we were asking. In an experimental discipline like radio astronomy, progress depends on the availability of the most advanced technology used by talented people with access to the best possible opportunities for training.

Recommendations for New Facilities

The Radio Astronomy Panel recommends as the highest priority for new construction a Millimeter Wave Array with sub-arcsecond resolution, comparable to that of the VLA and having good image quality, a sensitivity adequate to study faint continuum and line emission, and a flexible spectroscopic capability in all of the millimeter wavelength windows between 30 GHz and 350 GHz.   Cost: $115 M.

The Panel also identifies the following moderate sized projects, in order of priority, as being complementary to the Millimeter Array and important to the continued development of radio astronomy during the decade of the 1990's.

A Large Millimeter Radio Telescope Working to at Least 230 GHz[1]	$ 15 M
Expansion of the VLA	$ 33 M
A VLBI Antenna in Space[2]	$ 200 M

The Panel recognizes the need for a continuing opportunity to develop small new instruments and programs in response to newly developed discoveries, techniques, or theoretical ideas. The following small new initiatives at university facilities and national laboratories have been identified as being particularly important and timely.

A Large Southern Radio Telescope to be Built and Operated in Brazil in Collaboration With an International Consortium of Partners[3]	$ 10 M
A Dedicated Cosmic Background Imager	$ 10 M
RADIOASTRON and VSOP Space VLBI Missions[4]	$ 10 M
Establishment of University-Based SETI Research Programs	$ 5 M
A Fast All Sky Telescope	$ 10 M
A Solar Radio Telescope	$0.4 M

The Panel also recognizes the importance of developing long-range plans and instrumentation needed for new facilities in the beginning of the 21st century, including:

The identification of technological innovations leading to the development of new instrumentation for radio astronomy, including receiver technology for millimeter and sub-millimeter wavelengths, broadband recording systems for VLBI, advanced computing facilities and algorithms for imaging and pulsar analysis, and the strengthening of efforts to control radio frequency interference.

Radio telescopes in space for observations at sub-millimeter wavelengths

An astrophysical observatory in Antarctica with large millimeter and sub-millimeter radio telescopes

A low frequency radio telescope on the far side of the moon

---

[1] Federal share representing about half of the total cost of project.
[2] Approximate US share of proposed international mission
[3] US share of approximately $ 100 M project
[4] Cost of US participation in Soviet and Japanese space VLBI missions

The Millimeter Array

Large radio telescopes may require a decade to design and construct. In order to ensure the continued preeminence of American radio astronomy into the next decade, it is important to begin now the work leading to the next generation of radio telescopes.

The highest priority of the Radio Panel for new instrumentation is for the construction of a Millimeter Wavelength Array with sensitivity, resolution, image quality, and speed adequate to investigate the wide range of astrophysical phenomena that are best studied at millimeter wavelengths.

FIGURE 6 Artists conception of the Millimeter Array showing the antennas in the 250 meter configuration. The road for the 900 meter configuration is shown as well as the compact configuration in which the antennas are arranged to simulate the response of a single 70 meter diameter antenna. The 3 km outer road is coincident to the 900 meter road in this inner part of the array. (Photo courtesy of the National Radio Astronomy Observatory—Associated Universities, Inc.)

Dramatic advances in technology have caused an explosive growth of millimeter and sub-millimeter wave astronomy. The high spectral resolution provided by heterodyne spectroscopy of molecular clouds provides a powerful tool for basic molecular physics. Of particular interest is the chemistry of the interstellar medium, which is readily studied at millimeter wavelengths where the spectroscopy of cosmic molecules rivals in richness the Fraunhoffer spectrum of the sun and stars. Observations of these lines play an important role in helping to understand how molecular clouds collapse to form stars, to identify the molecules primarily responsible for cloud cooling, and to determine the kinematic details of the process from the observed velocity fields. One very important result will be a great improvement in our understanding of star-forming regions in our own and other galaxies.

Millimeter astronomy was developed and pursued solely in this country until the early 1980's. Although no large millimeter wavelength instrument has ever been built by the United States, major facilities are now in operation in Europe and in Japan. The Millimeter Array will recapture the once dominant position of the United States in millimeter astronomy and will complement the major U.S. instruments that will be in operation by the end of the next decade in other wavelength bands.

The MMA will be especially well-suited to simultaneous multi-wavelength spectroscopy with high spectral resolution and will have a wide range of astrophysical applications, including solar system research, molecular spectroscopy, studies of protoplanetary systems, star formation, primordial galaxies, and the microwave background.

In the most distant parts of the Universe, the MMA will image thermal dust emission in galaxies out to redshifts of ten, yield images of dust emission in active galactic nuclei and quasars with a resolution about one hundred parsecs, detect carbon monoxide from galaxies out to large redshifts, and image the Sunyaev-Zeldovich effect from clusters of galaxies to provide an independent determination of the Hubble constant and the size and age of the Universe.

For nearby galaxies, the MMA will determine the masses and kinematics of optically obscured galactic nuclei with a resolution of a few parsecs, and image the distributions of the molecules containing carbon, oxygen, nitrogen, and sulfur and their isotopes.

Within the Galaxy, the MMA will observe stars of every spectral type and luminosity class, measure their photospheric emission and temperature gradients, and determine positions with astrometric accuracy. Observations with 0.1 arcsecond resolution will permit the identification of regions of star formation in dark clouds, resolve cloud fragments, protostars, and circumstellar accretion disks as small as 10 AU, image the density and velocity structure of protostellar and pre-planetary disks, and provide images of the chemical gradients in protostellar nebulae and circumstellar shells that reflect the chronology of stellar nucleosynthesis and envelope convection,

Inside the solar system, the MMA will probe the physics of particle acceleration in solar flares; image the atmospheric winds and the temperature profiles of Venus and Mars; resolve the phosphine emission in the Great Red Spot on Jupiter, hydrogen cyanide on Titan, and volcanic emission on Io; and obtain unobscured images of cometary nuclei, asteroids, and the Pluto-Charon system.

The Millimeter Array will be a national facility open to all qualified users, and will provide fast sensitive high fidelity sub-arcsecond imaging from 30 GHz to 350 GHz, wide-field imaging, sensitive simultaneous broadband operation, and a comprehensive "single-dish" capability. The proposed array will contain 40 transportable antennas, each 8-m in diameter, and will be reconfigurable to match the angular resolution to a wide range of astrophysical problems. The angular resolution will be 0.07 λ_{mm} arcsecond in the largest 3-km configuration. In its compact configuration, the MMA will have a resolution comparable to that of a 70-m antenna with a collecting area equal to that of a single 50-m diameter antenna. The rms sensitivity for point source continuum observations will be better than 1 mJy/(min)$^{1/2}$, and for spectral line observations at 230 GHz, 1.2 K/(min)$^{1/2}$ for a 1" beam and 1 km/sec velocity resolution. Design and prototyping work for the MMA is planned for the period 1991-1994, and construction from 1994-1998.

Medium Scale New Instruments:

Large Millimeter Wave Telescope: The Radio Astronomy Panel recognizes the need for a modern 50-meter class filled aperture radio telescope capable of operation to at least 230 GHz, located at a good site and available to scientists independent of their institutional affiliation. The Panel is impressed by the progress being made in the use of active optics to build a large millimeter radio telescope at relatively low cost. Such an instrument equipped with focal plane heterodyne and bolometer arrays will offer a huge increase in speed and sensitivity over currently available instruments and will provide an extremely powerful tool for the study of interstellar matter and star formation. The Large Millimeter Telescope will allow fast spectroscopic and continuum surveys of large regions of the sky, and may have application to planetary radar. The LMT will also provide a substantial enhancement to millimeter VLBI. It is expected that the LMT will cost about $35 M to construct, with about half of this to be paid by private or state funds.

VLA Expansion: The gap between the VLA and the VLBA may be bridged with a combination of tape recorder and fiber optic links between the two arrays and by adding new antenna elements. This will increase the resolution of the VLA at all frequencies; improve the dynamic range, field of view, and extended source sensitivity of the VLBA; and give a "scaled array" capability over a much wider range of frequencies than is now available. Of particular importance will be the ability to determine how the Stokes parameters of the radiation vary with frequency over a wide range of frequency at a fixed angular resolution. The Radio Panel recommends a phased plan which includes: (a) placing up to four VLBA tape recorders at the VLA ($1 M); (b) constructing up to four new antennas in New Mexico and Arizona ($21 M); (c) providing fiber optic links from the VLA to the four new antennas and to the one at Los Alamos, and expanding the VLA correlator from 27 to 33 stations ($11 M). These improvements will provide a greatly enhanced resolution and imaging capability over a wide range of frequency, and brightness sensitivity with many applications to radio observations of the Sun and planets, radio emission from stars, novae, protoplanetary nebulae and stellar winds, as well as from star-forming regions, and for the study of active galactic nuclei and quasars.

Space VLBI: The VLBA will give the highest resolution images of any astronomical instrument, and further improvement can be obtained only by going into space. Recent experiments using the TDRSS satellite have demonstrated the feasibility and power of space VLBI. Space VLBI was among the 1982 Astronomy Survey

Committee recommendations for "moderate new programs," but NASA has been slow to participate in any of the space VLBI missions planned for the 1990's.

The International VLBI Satellite (IVS) is being discussed by European, Soviet, Japanese, and U.S. radio astronomers for a possible launch near the end of the decade. IVS will include a 25-meter class antenna working to wavelengths as short as 3 mm located in high Earth orbit at altitudes between 20,000 and 150,000 km. IVS will provide an order of magnitude or more improvement in sensitivity and image resolution over the Japanese and Soviet missions planned for the mid 1990's. It will also be capable of sensitive, single aperture, spectroscopic observations of molecular oxygen in the 60 GHz band from above the earth's atmosphere, which is opaque at this frequency. Knowledge of molecular oxygen abundance, which is very uncertain, is important to understanding the chemical and dynamical evolution of molecular clouds.

IVS is being planned as an international facility which will include the participation of ground-based radio telescopes throughout the world. The current baseline for the study contains an ESA antenna and a Soviet service module and Energia launch vehicle. Other variants are possible which include a substantial US involvement, for example, a U.S. service module and part of the experimental package. The Radio Astronomy Panel considers space VLBI to be the highest priority project for radio astronomy from space during the 1990's, and it is important that the US take an active role in the early planning and mission definition studies for an international VLBI satellite.

Small-Scale Projects

> *The Panel recognizes the need for a continuing level of support for small-scale programs that can react to the rapidly changing developments in radio astronomy. We describe below several important areas which we are able to identify at this time. We do not prioritize these small-scale initiatives because we fully expect that new meritorious ideas will arise on a time scale less than that of the next decade review, and that the selection of specific programs should depend on normal agency review and the nature of funding opportunities as they arise.*

Large Southern Radio Telescope: Many important research programs require the highest attainable instantaneous sensitivity, and thus the largest possible collecting area. All of the biggest radio telescopes in the world are located in the northern hemisphere. The construction of a large aperture radio telescope in the southern hemisphere will give a powerful new capability for research in the southern skies, including atmospheric studies, access to solar system objects invisible from the north, the galactic center region, the Magellanic Clouds and the southern extragalactic sky. The recent design and construction of Gregorian subreflector systems, conducted as part of the project for upgrading the Arecibo telescope, show the great potential of modern applications of spherical antenna technology for achieving very large collecting area.

The proposed LSRT will work at short centimeter wavelengths and will have a collecting area comparable to that of the upgraded Arecibo telescope. A novel feed arrangement will give a wide declination range covering most of the southern sky. It is expected that the LSRT will be built and operated in Brazil by an international consortium, at a total cost of about $100 M and a cost to the United States about $10 M.

Cosmic Background Imager: The cosmic background radiation is perhaps the most important tool of observational cosmology. On angular scales greater than a few degrees, the background radiation reflects directly conditions in the early Universe at an age of only one hundred thousand years. On smaller angular scales, it may be distorted, both spatially and spectrally, by various processes connected with gravitational collapse and the formation of galaxies and other large-scale structures. Upper limits on the anisotropy are now at a level of a few parts in a hundred thousand. If anisotropies are not found at a level of a few parts in a million, then our basic understanding of the early Universe may need to be fundamentally revised.

Recent technological advances in the design of reliable low-noise bolometers and heterodyne receivers suggest that it is possible to reach the required levels of sensitivity with radio telescope systems especially designed for this problem. One possible approach is to use an array of about 50 horn antennas to eliminate the confusing effects of ground and atmospheric radiation. Another option is to use an array of bolometers located in the focal plane of a millimeter radio telescope. Such instruments will also be powerful tools for accurately measuring the microwave decrement due to the Sunyaev-Zeldovich effect. When combined with accurate X-ray data, these measurements will yield an entirely independent determination of the Hubble constant and the size and age of the Universe.

RADIOASTRON and VSOP: The USSR and Japan are each planning to launch VLBI satellites in the mid 1990's. U.S. scientists have been involved from the start in defining these missions, and many Eastern and Western European countries, Australia, and Canada are participating in various ways in their implementation.

In order to exploit the unique opportunities made possible by these foreign space VLBI missions, as well as to develop the necessary skills and expertise needed to plan for a future U.S. space VLBI mission, the Radio Astronomy Panel recommends that the large U.S. ground-based radio telescopes be made available as elements of the Earth-Space interferometer, that VLBA compatible recorders be supplied at Soviet and U.S. ground stations used to receive data from the space element, that U.S. radio telescopes be equipped to provide local oscillator and data links to the foreign spacecraft, and that U.S. scientists participate fully in developing and carrying out the VSOP and RADIOASTRON scientific programs.

The overall level of support for space VLBI in the next decade is expected to be about half a billion dollars, primarily in Japan and the USSR. Although VLBI techniques were developed in this country, and the only successful space VLBI experiments so far have been done with an U.S. satellite, there is no planned U.S. space VLBI mission. Fortunately, U.S. astronomers can fully share in the scientific returns of the Japanese and Soviet missions with a relatively small financial investment. Since the U.S. will have no involvement in the design, construction, or launch of any space hardware or in mission management, the Panel suggests that U.S. participation be handled through conventional grants to the participating scientists rather than administered as a NASA project.

Search for Extraterrestrial Intelligence: The Search for Extraterrestrial Intelligence (SETI) will be an exciting intellectual and technical challenge for the next decade. A successful "contact" would be one of the greatest events in the history of mankind. We are the first generation that could realistically succeed, and there is great public interest in SETI. The major issue is the appropriate level of resources to devote to SETI and how these resources should be divided between a large centrally managed program and the more traditional university-based research efforts. SETI is not part of radio astronomy, but the tools of radio astronomy are used for SETI, and radio astronomers have pioneered the development of observational SETI programs.

NASA has initiated the *Microwave Observing Project* which will begin searching in 1992 and last until the end of the decade. The MOP will expand the volume of parameter space investigated (direction, frequency, polarization, sensitivity) by many orders of magnitude over what has been done in the past. The search consists of two complementary strategies: the Sky Survey which will use 34 meter DSN antennas to scan the entire sky between 1 and 10 GHz, and the Targeted Search using the world's largest radio telescopes to examine about 800 nearby solar-type stars between 1 and 3 GHz. The *Microwave Observing Project* will be the first truly systematic SETI exploration of the microwave region of the spectrum and will cost about $100 M over the rest of the decade.

The Radio Astronomy Panel recommends the establishment of a university-based SETI research program to develop new ideas and architectures for signal processing algorithms and processors, to develop search strategies, to implement innovative new hardware, and to establish search programs complementary to the NASA *Microwave Observing Project*. The proposed new program would provide a medium for the exchange of new techniques and hardware and the training of students in advanced signal processing techniques, as well as a means to pursue a viable observational program. Present levels of support to the scientific community are of the order of $100 K per year, which is inadequate to sustain a productive effort. The Panel recommends that university-based SETI research be supported at an annual level of about $500 K which would be adequate to support one or two independent research groups, each with four or five undergraduate, graduate, and post-doctoral researchers.

Solar Radio Astronomy: The VLA, Arecibo, and the proposed Millimeter Array are powerful instruments for solar research, but the existing instruments lack frequency agility. The Panel recommends that consideration be given to equipping the VLA and Arecibo Telescopes with frequency agile receivers and feeds. Opening up the full radio spectrum will provide powerful diagnostic information, not only for the sun, but for a wide variety of stellar, galactic, and extragalactic objects as well. The Panel also endorses the planned extension of the solar-dedicated, frequency-agile array at the Owens Valley Radio Observatory.

Fast All Sky Radio Telescope: FAST is a proposed array of twenty 3-meter antennas which will monitor most of the sky at centimeter wavelengths with an rms sensitivity of 10 mJy in one or two days. It will be used to study time variability in compact active galactic and extragalactic radio sources. FAST will be the only high-resolution, all-sky monitoring instrument available in any spectral band.

Continuing Activities and Projects Already Underway

Facility Operation and Maintenance, Upgrading of Telescopes and Instrumentation

NSF support for radio astronomy has been inadequate for the operation, maintenance, and upgrading of the national and university-operated radio observatories and for the modernization of instrumentation and computing resources at these facilities. Of particular concern is the deteriorating state of the VLA, the inadequate support for millimeter and sub-millimeter astronomy, and the need for modern computing facilities.

By far the most powerful and most productive radio telescope in the world is the VLA with its extraordinary speed, sensitivity, resolution, and image quality. Since there have not been adequate funds for even the most basic maintenance, the railroad track, power distribution system, antenna structures, and other aspects of the physical plant are deteriorating. Much of the instrumentation of the VLA uses 15-20 year old technology because there has not been the refurbishment and upgrading at the level appropriate to a scientific instrument of this size, sophistication, and productivity.

> *The operation and maintenance of the VLA needs to be brought to a level appropriate to its broad scientific impact and great capital investment, and the seriously out of date instrumentation needs to be replaced with modern low-noise radiometers, fiber optic transmission lines, and a modern broad band correlator. These upgrades will improve the sensitivity by up to an order of magnitude, improve the frequency coverage and spectral resolution, and increase the maximum allowable image size.*

As a result of the years of inadequate support, it will now cost about $40 M to incorporate these badly needed modernizations. This is comparable in cost to the moderate sized ground-based projects being considered by the Survey Committee. However, spread over eight years, it represents a level of investment corresponding to two to three percent per year of the replacement cost of the VLA/VLBA. Routine maintenance and modernization programs of this type and at this level would normally not be included in a discussion of major new facilities, but the situation has become critical and has risen to a high level of visibility because of the nearly complete absence of funds for this purpose since the completion of the VLA about a decade ago.

In many respects the VLA dramatically exceeds its performance at the beginning of the previous decade. The speed is faster by a factor of two, image size is larger by a factor of four, the maximum dynamic range has been improved by a factor of 50, the number of spectral line channels has been increased from 8 to 512, and mosaic images larger than the primary beam of the antenna are now being made. These scientifically important gains in performance have come as a result of powerful new algorithms but at the cost of greatly increased computing requirements. As a result, the computing situation for the VLA has been critical for some years, and many excellent scientific programs are not done because of inadequate computing facilities. The power of the VLA system, the complexity of modern data reduction algorithms, and the need to annually support more than six hundred users now overwhelms the available computing capacity at the NRAO and elsewhere. Full exploitation of the power of the VLA and VLBA will require new hardware and software that can be readily shared between the arrays and their user sites, the installation of small supercomputers and imaging workstations at the VLA/VLBA operations center and in university laboratories, and the establishment of effective interfaces to the large supercomputer centers for the most CPU-intensive data.

The Panel also recognizes the exciting opportunities available during the 1990's for research at millimeter and sub-millimeter wavelengths made possible by recent developments in technology in this newly opened region of the radio spectrum, the need to aggressively develop the technology necessary for future instrumentation at millimeter and sub-millimeter wavelengths, and the need to maintain and expand the pool of skilled millimeter wavelength scientists. The upgraded Haystack radio telescope and the new Green Bank Telescope will provide powerful new opportunities at longer millimeter wavelengths, while the Caltech and MPI-Arizona sub-millimeter telescopes will continue to have unique capabilities, even after the completion of the MMA.

> *The existing millimeter and sub-millimeter radio telescopes and especially the millimeter wavelength interferometers need to be extended and enhanced, instrumentation based on the most advanced technology needs to be developed for these facilities, adequate support given for their operation, and additional resources made available to make these instruments accessible to a broad group of scientists independent of their institutional affiliation.*

Projects Already Underway:

At the beginning of the 1990's, five major radio astronomy projects are already underway. The timely completion of these instruments and provision for adequate funds for their operation will give tremendous improvements in the angular resolution, sensitivity, and frequency coverage over now existing radio telescopes, and will ensure the vitality of U.S. radio astronomy during the decade of the 1990's.

Very Long Baseline Array: The VLBA is a major new aperture-synthesis user facility in the form of a 10-element transcontinental array capable of imaging at the sub-milliarcsecond level. Construction of the VLBA received the highest priority for major new ground-based instruments in the 1982 Astronomy Survey Committee, and the project received initial funding in 1984. Annual funding at a much lower level than originally planned has extended the construction time from four years to nearly a decade. The first several antennas in the array are complete. Additional elements will be completed at a rate of two to three per year, but may not be fully utilized due to limited operating funds. The expected completion date for the full array is now 1992, five years later than originally planned.

Arecibo Upgrade: The Arecibo 1000-foot radio telescope has by far the largest collecting area of any centimeter wavelength facility in the world. Major improvements now in progress will provide significant enhancement of nearly all capabilities of the telescope for radio and radar astronomy and for atmospheric research. At present the spherical aberration of the telescope's primary mirror is corrected by using line feeds with inherently narrow bandwidths. Ingenious ideas behind a broad-band Gregorian feed system for Arecibo were first discussed a decade ago. Many details of a practical design were worked out over the next few years, and the concept was proven with detailed physical-optics computer modelling. A scaled-down "mini-Gregorian" feed system was put into operation in 1989. Tests have shown that this system efficiently illuminates a 350-foot portion of the Arecibo reflector and behaves in every way as expected. Funding for the full upgrading project, which will improve the sensitivity by a factor of 3 to 40 and will be cost-shared between NSF and NASA. Completion is expected in 1993. The enhanced telescope will have continuous frequency coverage between 0.3 and 8 GHz, with unprecedented instantaneous sensitivity.

Green Bank Telescope: The construction of a large, fully steerable, filled aperture radio telescope has been endorsed by essentially every review of the needs of U.S. radio astronomy. The 1982 Astronomy Survey Committee Radio Panel recommended an instrument in the 100-meter class which would work to wavelengths at least as short as one centimeter as an important priority for the 1980's. Following the collapse of the Green Bank 300-foot antenna in late 1988, NRAO accelerated its design study for a fully steerable, filled aperture instrument. The GBT is being designed with a novel "clear aperture" feed support system to reduce the effect of unwanted signals and active optics to permit use at wavelengths at least as short as 7 mm and possibly to even shorter wavelengths. A special congressional appropriation made funds available to NSF for the construction of the GBT, which is expected to be in operation by 1995.

The-Submillimeter Telescope (SMT): The University of Arizona Steward Observatory, and the Max Planck Institut fur Radioastronomie, FRG, are constructing a 10 meter diameter precision radio telescope to be located at an altitude of 3180 meters (10,425 ft) on Mt. Graham 75 miles northeast of Tucson. The SMT will use carbon-fiber-reinforced-plastic to achieve an overall surface accuracy of 15 microns. When completed in 1992, it will be the largest telescope with good performance at wavelengths as short as 350 microns. Funds for the construction and operation of the SMT are being provided primarily from the MPIfR and the University of Arizona.

Sub-Millimeter Wavelength Array: The Smithsonian Astrophysical Observatory is building the world's first sub-millimeter array, which will consist of at least six antennas, each six meters in diameter, operating primarily at wavelengths between 0.3 and 1.4 mm. The array will image the continuum and spectral-line emission from protostars, galactic nuclei, and solar system objects with an angular resolution in the range 0.1 to 10 arcseconds. The rms sensitivity at 0.3 mm wavelength (the shortest and most difficult operating band) for an integration time of 8 hours, velocity resolution of 1 km/sec, and angular resolution of 1 arcsecond is expected to be about 4 K. The sensitivity to continuum emission for the same integration time will be about 80 mJy.

Sub-millimeter observations are particularly sensitive to thermal emission from gas and dust, with temperatures in the range 10 K to 100 K. The powerful combination of high angular and spectral resolution spectroscopy offers the prospect of detecting and studying the gravitational motions in the gas around forming stars, the structure and motions of protostellar disks, and molecular outflows. The array will also be unique in its high-resolution imaging of neutral carbon lines at 0.37 and 0.61 mm wavelength. These lines, which have no millimeter wavelength components, can reveal physical conditions extremely close to a hot star where molecular lines are absent due to photodissociation. Emission from carbon and excited carbon monoxide lines will provide new tools to probe the spiral structure of galaxies. Sub-millimeter measurements of quasars and AGN's will help to distinguish radiation mechanisms of radio-quiet and radio-loud objects. Molecular line observations of planets and satellites will give new understanding of planetary chemistry and weather.

Design and development of the Sub-Millimeter Array has been in progress since 1987. Sites on Mauna Kea and Mount Graham are being evaluated. The current schedule calls for completion of the array by 1996. Funds for the construction and operation are coming from the Smithsonian Institution. The array will be available to all qualified scientists based on peer-reviewed proposals.

Long Range Programs and Technology Developments

Radio astronomy has historically advanced as technology was developed for measurements at shorter wavelengths, with more resolution, or with higher sensitivity. Our present frontier for new wavelengths is in the sub-millimeter, moving toward the far infrared. Quantum noise will stop this progression at about ten microns. Although radio astronomy is well developed at wavelengths longer than one millimeter, we expect significant advances at all wavelengths in the next decade.

At centimeter and longer wavelengths, the best receivers are either approaching the quantum noise limit or are so good that other sources of noise in the system will make the increase in sensitivity from further gains small. Cooled HEMT amplifiers will probably be the dominant type of receiver, and they allow much larger bandwidths than are commonly used now. Higher performance electronics for backends will make it possible to use these increased bandwidths to improve the sensitivity for continuum observations.

Even though improvements in individual receivers will not be large, the potential improvement in speed from multiple receivers in focal plane arrays increases directly with the number of receivers. Inexpensive, small HEMT amplifiers, possibly integrated in arrays, combined with less expensive backends will make multifeed systems practical. The first focal plane arrays have used multiple conventional feed horns, and therefore have a spacing of several beamwidths between beams. This approach is applicable to large area images, but overlapping beams which fully sample the focal plane also seem possible.

Although receiver technology at millimeter wavelengths approaches the quantum noise limit, there is much room for improvement. SIS receivers come within a factor of five of the quantum limit, but further development will be needed to achieve this level of performance over the whole millimeter and sub-millimeter band. Niobium junctions work well at wavelengths down to at least 1 mm, but higher temperature superconducting material such as niobium nitride will be better at shorter wavelengths. Better refrigerators are needed to make these receivers reliable and inexpensive to operate. Bolometers will be the radiometers of choice for single aperture continuum observations.

The most significant recent advances in antennas are seen in the design of the GBT which will have an unblocked aperture made possible with the use of modern structural analysis. The unblocked aperture will result in low ground pickup which will significantly reduce the system temperatures at lower frequencies and low side lobes which will allow more accurate measurements of the distribution of galactic neutral hydrogen. The GBT will also have surface panels accurate enough for millimeter wavelengths and remotely controllable adjustments for the panels. If suitable metrology can be developed for active surface control, operation at 2.6 mm wavelength will be possible in favorable weather.

Composite materials made of carbon-fiber reinforced plastic have revolutionized the design of sub-millimeter antennas. Inexpensive ways of fabricating accurate aluminum panels have also been developed, so that at millimeter wavelengths a cost-performance tradeoff must be made. The use of active surface control may allow building a very large antenna for millimeter wave operation. For much of the sub-millimeter band, operation above the earth's atmosphere is necessary—in an airplane, on a balloon, or in space.

Advanced Computing Facilities: The capability of computers continues to rapidly grow and their cost continues to fall. Evolving standards reduce the difficulty of sharing software, and the growth of networking will allow rapid access to images, easier exchanges with collaborators, and much more effective remote observing. New types of software will make searching for images and access to information from remote sites, easier and observatory operation more automatic. In the 1990's, radio astronomers will depend more than ever on high-performance computers to realize the full capabilities of their telescopes. This is especially true in planetary radar studies, pulsar research, and synthesis imaging where computers provide the "adaptive optics" needed to form correct images in the presence of corrupting effects of the earth's atmosphere. During the past decade, the computing needs of radio astronomical imaging have grown due to the large data volumes from array telescopes and from single-dish array feeds when used for spectroscopy. Moreover, the data processing required to extract all of the information from arriving signals in the presence of atmospheric and instrumental fluctuations are highly CPU-intensive, and fast response is essential for many applications.

Progress in radio astronomy will depend as critically on fast, high-data-volume interactive computing as much as it will on low-noise electronics or advanced antenna design. In order to exploit the full scientific potential of the country's substantial investment in radio telescopes, and to remain competitive with the modern computing systems being found increasingly in other countries, it is important to make a wide range of advanced computing systems available at our national observatories and in our university laboratories.

The computing power that is best suited to analyzing different types of interferometer observations spans an enormous range. The simplest VLA and VLBA continuum projects can be processed using inexpensive workstations, but most VLA projects are best handled by machines with the interactivity and performance provided by "small supercomputers" or high performance workstations. Some spectroscopy and wide-field imaging projects still require the largest supercomputers available. No one location, or even type of location (observatory, university department, or computer center), is appropriate for all VLA, VLBA, and millimeter interferometer data reduction. The wide bandwidths expected on the national computer network in the 1990's will make it easier for resources such as software, data bases and computing cycles to be shared between the arrays and their user community. This will exploit the unique merits of workstations, small supercomputers, and large supercomputers to use each efficiently for different types of VLA, VLBA, and millimeter interferometer data processing.

Meter Wavelength Astronomy: Although radio astronomy began at meter wavelengths, the scientific potential of the long wavelength bands has barely been tapped, largely because of the difficulty in obtaining adequate resolution and the distortions introduced by the earth's ionosphere. Improved resolution is needed at meter and decameter wavelengths to study the galactic non-thermal radio emission, the distribution of diffuse ionized gas in the Galaxy, the galactic halo, the interstellar plasma via the scattering and refraction of extragalactic sources, pulsar emission, "fossil" radio sources due to long lived synchrotron processes in galactic supernovae, radio galaxies and quasars, and non-thermal emission from the Sun and the planets.

In spite of these exciting scientific opportunities, for the past two or three decades meter wavelength radio astronomy has been outside the mainstream of astronomical research. Much more effort has focused on centimeter and millimeter wavelength research where it is easier to exploit technological advances to obtain good sensitivity and resolution. For a number of reasons, the situation is now changing. We now have a reasonable understanding of how to correct for propagation irregularities in the ionosphere, digital VLSI and modern computers make large array mapping techniques feasible, interference rejection techniques are beginning to be effective, and transmission of data over many kilometers is now cheaper and more reliable than in the past.

A 4-m wavelength receiving system is being developed for the VLA, and is expected to be in operation in time for the next sunspot minimum in the mid 1990's. But it is important to also begin now to develop techniques and prototypes for an array operating at even longer wavelengths, with a collecting area greater than 10^5 square meters and a resolution better than 10 arcseconds for both continuum and spectral line work. This can be done with a combination of university and national observatory collaboration, which will foster university development of techniques and the training of the next generation of telescope builders at the graduate and postdoctoral level.

In order to better image the wide range of phenomena that are observed at long radio wavelengths, resolutions need to be improved so that they are comparable with that of the VLA at centimeter wavelengths. This will mean establishing a program of space radio astrophysics during the next decade leading to the establishment of a Low Frequency Space Array, a free-flying hectometer wavelength synthesis array for high resolution imaging operating below the ionospheric cutoff frequency.

Lunar Opportunities: The Panel recognizes that over the next few decades national goals may lead to extensive exploration and colonization of the lunar surface. This may create exciting opportunities to build radio telescopes of very large dimensions. From the lunar surface it will be possible to observe at very low frequencies where the terrestrial ionosphere introduces increasing distortions as well as at very high frequencies where the earth's atmosphere becomes opaque. The far side of the moon, which is protected from man-made interference, is a particularly attractive site for low frequency radio astronomy. A particularly important use of the lunar far side will be as a base for interference-free SETI observations. We note with distress, however, that lunar orbiters and human activities, including radio astronomy research, on the far side of the moon, could generate their own RFI from telecommunication and computing devices. International agreements must be adopted in this decade to protect the far side of the moon for scientific research.

Lunar-based radio astronomy will be very expensive by normal standards, and probably cannot be justified during the next few decades in comparison with ground-based requirements. But if the country is committed to a major presence on the moon for other reasons, then there will be exciting opportunities to do radio astronomy, first on the near side and later from the far side, that would otherwise not be possible.

Sub-Millimeter Astronomy from Space: The Large Deployable Reflector, a future space telescope for sub-millimeter and far infrared wavelengths, is being discussed by NASA which is supporting the development of reflector and detector technology. In order to successfully complete this innovative and difficult project, NASA will require access to the most advanced technology in the world. Many of the necessary developments are in progress in university and other non-NASA national laboratories around the country. The Radio Astronomy Panel urges NASA to support work in these laboratories both for the benefit of the LDR program and for the spin-off benefit to ground-based millimeter and sub-millimeter radio astronomy. The Panel also recognizes the need for the Sub-Millimeter Moderate Mission (SMMM) for an initial high resolution spectroscopic exploration of the full submillimeter band of star-formation regions and distant galaxies as a precursor to LDR.

Radio Astronomy in Antarctica: Due to the high altitude, extreme cold weather, and low water vapor content of the atmosphere, the Antarctic Plateau may be the best site on Earth for astronomical observations at infrared, sub-millimeter, and millimeter wavelengths. The Antarctic Sub-millimeter Telescope and Remote Observatory, (ASTRO), a consortium of AT&T Bell Laboratories, Boston University, and the University of Illinois, is scheduled to start operating a 1.7 m telescope at the South Pole toward the end of 1992. A larger consortium has proposed to establish a major Center for Astrophysical Research in Antarctica (CARA) which would ultimately include ASTRO, infrared telescopes; an experiment to measure the Cosmic Background Radiation anisotropy; and an advanced telescope project to develop detailed plans for a permanent observatory having several instruments, possibly including a 10-30 m sub-millimeter telescope. Scientific programs will include key problems in cosmology, star formation, and the physics and chemistry of the interstellar medium.

Signal Processing: Pulsar signals are highly dispersed, rapidly time variable, and strongly modulated in frequency. In some instances, these effects are a nuisance to be removed, or averaged; in others, they are the object of investigation. In all cases sophisticated signal processing must be done either in hardware or in software, or both. Special signal processors based on filter banks and autocorrelators have been developed over the years to carry out these investigations. There is a continuing need for such devices that operate with faster sampling rates, with more frequency resolution, and wider bandwidths. Such a processor can also serve the needs of spectroscopy, particularly dynamic spectroscopy of radio stars, and radar. Interface to a high-speed, high-volume recording medium is critical for pulsar searching.

At radio frequencies below about 1 GHz, dispersion is best removed by coherent techniques. Pulsar dispersions often exceed the chirp rates used in radar, so the commercial devices for de-chirping are not adequate for the pulsar task. Special construction efforts using, for example, VLSI techniques may provide a solution to this need. Pulsar searching can also be done by real-time signal processing since the data acquisition and analysis tasks are easily divided between microprocessors in a parallel or pipelined architecture.

VLBI Recording and Data Transfer: VLBI observations are limited in sensitivity by the capacity of the recording medium. Since the early 1970's commercial television tape recorders have been used for VLBI, and during the past decade inexpensive and reliable consumer type video-cassette-recorders (VCR's) have come into widespread use. The VCR-based system is cheap, flexible, and easily available, but the bandwidth is restricted to a few megahertz, or about two orders of magnitude less than that of the VLA at present. A broadband VLBI recording system for geodetic studies based on a commercial instrumentation recorder has been developed by the Haystack Observatory with NASA support and will be used with modifications for the VLBA. But the VLBA recording system is expensive and the bandwidth still limited to about 100 MHz. A competitive system has been developed by the Sony Corporation for use with the Japanese space VLBI mission, but has the same limitations as the VLBA system.

The bandwidth of the VLBA recording system can be improved by adding additional headstacks to the existing tape transports, but this will further increase the costs of construction and operation. Ultimately, fiber optics or satellites will be used to provide real time links, thus obviating the need for cumbersome transport and handling of tapes, but this must await the commercial installation of national and international broad-band data links. For the foreseeable future, VLBI will depend on physically transporting the recorded data to the correlator, and it will be important to develop new recording techniques to allow high density, broad bandwidth recordings that are both reliable and cost effective.

Radio Frequency Interference: Celestial radio signals are extraordinarily weak, often less than one hundredth of one percent of the internal receiver noise. As a result of the rapid growth in use of the radio spectrum, particularly from space and airborne transmitters as well as the dramatic increase in receiver sensitivity over the past decade, radio astronomy observations are increasingly affected by interference. The protection of radio astronomy from man-made interference requires thoughtful spectrum management, careful observatory site selection, continued efforts toward site protection from internal as well as external sources of interference, and the development of techniques for reducing or eliminating the effects of interference from the received signals.

Optimum sharing of the radio spectrum with other services will require the participation of active radio astronomers in the regulatory and coordinating bodies such as the National Academy of Science Committee on Radio Frequencies. Because of the extreme sensitivity of radio astronomy receivers, it is often difficult to document specific sources of radio frequency interference. Special RFI search and monitoring stations, including at least one mobile station, should be established for this purpose. It is also important that radio astronomers use the most advanced technology available in order to best coexist with other users of the radio spectrum.

Social, Political, and Organizational Considerations

International Opportunities

Modern observational astronomy has become so complex that no country can expect to have state-of-the art instruments covering all parts of the spectrum and satisfying the needs of all types of observational programs. International collaborations present a wider variety of opportunities to individual scientists, permit the achievement of scientific objectives which may require a specific geographic location, provide an important forum for the interchange of ideas among people of different backgrounds and cultures on subjects that transcend scientific considerations, and may contribute to scientific and educational growth in developing countries.

The Radio Astronomy Panel recognizes the potential opportunities resulting from international collaborations to develop major new radio telescopes that would not otherwise be feasible. The Panel encourages, where possible, that observing time on major facilities throughout the world be available on the basis of competitive proposals without regard to institutional or national affiliation.

Perhaps the most straightforward form of international collaboration, and one involving a minimum of bureaucratic overhead, is the use of telescopes by visiting scientists from other countries. This not only provides observing opportunities that are otherwise not possible, but stimulates the exchange of scientific and technical ideas from which everyone learns and profits. Many research programs involve extensive and repeated observing sessions, as well as continued contact with colleagues having special technical or scientific expertise. These programs may be difficult to carry out by means of short observing trips, but will require extended periods of collaboration.

The VLA and VLBA, the upgraded Arecibo telescope, and the GBT will provide powerful observing opportunities for American radio astronomers at centimeter wavelengths. However, until the completion of the MMA, the Japanese and French-German IRAM millimeter wavelength facilities are likely to remain unmatched in this country. It has been the practice in the United States that observing time on radio telescopes at our national observatories and other major facilities be awarded without regard to nationality or country of residence, and the Panel recognizes the important role that this policy has played in maintaining the vitality of U.S. radio astronomy. It is hoped that the managements of foreign radio observatories will make the same opportunities available to U.S. radio astronomers, and that adequate funds be provided particularly to young scientists, to exploit these opportunities.

With the decreasing levels of worldwide tensions, the opportunity for international collaborations in the construction and operation of unique radio telescopes will become increasingly important. Careful attention will be needed to balance the opportunities for intellectual interaction and the savings in costs to individual countries with the bureaucratic and fiscal overhead that is not uncommon to large international projects. In developing plans for international cooperative projects, it will be important to minimize administrative constraints such as formulas for the distribution of observing time, for financial expenditures, for siting, or for the allocation of staff and management positions which are based on national affiliations rather than merit.

Very Long Baseline Interferometry: Radio astronomy has had a long and fruitful record of international collaborations such as the worldwide program in VLBI. The international cooperation in VLBI works because the science requires it. Moreover, since each country spends its money largely in its own country, complex spending formulae have not been a constraint. A number of major new VLBI facilities have been and are being built throughout the world as *national* efforts, but as part of the growing *international* VLBI network.

As a result of informal arrangements made by the scientists and observatories involved, any individual anywhere in the world can have simultaneous access to as many as twenty of the world's major radio telescopes located in more than a dozen countries, including the USSR and China. This is done by submitting a single, simple proposal to one of several VLBI consortia, which will arrange for the observing time, the shipping of magnetic tapes across international boundaries and, often, even for the correlation of the tapes in one of several processing facilities operated in the United States and Europe. Image construction and analysis is done using common software which has been developed by a dedicated group of scientists working at a variety of institutions around the world, who have frequently migrated among the active VLBI observatories and who have freely exchanged the results of their labors. Logistical and technical coordination is handled primarily by the scientists involved with a minimum of administration from the managements of the observatories, and none from government administrators. The system works well, and the scientific results have been spectacular. But, during the next decade, major international collaborations involving national and commercial agreements will be established for space VLBI facilities such as Radioastron, VSOP, and IVS.

Most radio telescopes being used for VLBI are located in the northern hemisphere and give poor image quality for sources at low and southern declinations. A VLBI element located in South America is needed to complement the northern hemisphere VLBI networks and the VLBA. The optimum location is close to the equator in the western part of the continent, and several South American radio astronomy groups have expressed interest in developing a southern hemisphere VLBI facility. It should be possible for them to construct a suitable antenna from their own resources, but supporting instrumentation will be needed from the United States.

The Large Southern Hemisphere Radio Telescope: Due to the interdisciplinary and worldwide interest in the construction of a large aperture facility that will meet the requirements of radio and radar astronomy as well as atmospheric science, the funding and operation of a large southern hemisphere radio telescope is being discussed by an international group of potential partners. Political and economic factors suggest a novel funding scheme to tap resources, not normally available to the scientific community, to convert the existing foreign debt of the host country into development, construction, and operating funds. The advantages to the consortium partners include a unique scientific instrument available to the worldwide community, together with technology transfer and debt reduction to the host country. Because of the interest of its scientific community, the geographical location, and the economic and industrial capacity, Brazil is considered the most appropriate host country for the LSRT.

Millimeter and Sub-Millimeter Astronomy: The joint Max Planck Institut - University of Arizona German built sub-millimeter radio telescope to be located on a high mountain site in southern Arizona is expected to produce one of the most powerful sub-millimeter facilities in the world. A new collaboration between Caltech and the University of Toronto has been established to develop the expansion of the OVRO millimeter interferometer. Mauna Kea in Hawaii has two sub-millimeter telescopes: the James Clerk Maxwell Telescope, (JCMT) operated by the UK, Canada, and the Netherlands, and the Caltech Sub-millimeter Observatory located 150 meters away. These two telescopes as well as the Smithsonian Sub-Millimeter Array may be used together for high resolution interferometric observations of star-forming regions and nuclear regions of galaxies as well as other compact sub-millimeter emission regions.

Balance Between the National Observatories and University Facilities:

Radio astronomy is an experimental science. Traditionally many of the most important discoveries were made directly as a result of new instrumentation built by skilled and devoted experimentalists. Today, the wide variety of astrophysical problems being studied by radio techniques requires both the national facilities that can support a broad spectrum of users as well as smaller innovative research facilities in which technologies of the future can be developed.

> *The Panel is concerned about the inadequate opportunities at university operated facilities for the development of new instrumentation, increasing pressures on the remaining facilities to operate more and more in the mode of the large national user facilities, and the decreasing opportunities for the training of scientists skilled in instrumentation.*

The availability of powerful national research facilities to individual scientists has resulted in the proliferation of small (one to three people) but viable radio astronomy groups in many university astronomy or physics departments that do not have their own observing facility. At the national observatories and large university-operated facilities, the instrumentation is often complex and at the state-of-the-art. The development of new instrumentation for these telescopes is generally the responsibility of professional engineers with little understanding of, or even interest in, the end use for astrophysical research. Nevertheless, the interested and capable scientist can exploit the tremendous technical resources at the national observatories and individually contribute significantly to the development of advanced instrumentation and techniques.

At the university observatories, the instruments are often, but not always, smaller, less expensive, generally more specialized, and have a smaller user base than those at the national centers. In principle, the university facilities provide a qualitatively different research environment which encourages a more experimental approach by talented researchers, including graduate students and postdoctoral fellows, who are highly motivated and have few other duties to divert their attention and effort from research. With less pressure on the efficient use of observing time, it is possible for an observer to modify a program or equipment in response to early results or to try something new or different. In the event of failure, the experiment can be repeated as needed. The environment is also ideal for long-term projects requiring repeatability, specialized equipment or techniques, or unusual scheduling.

A consequence of the construction of major new research facilities at the national centers has been the closing of smaller university instruments. Moreover, many of the traditional differences between the national centers and the larger university-operated observatories are becoming less distinct. As operating funds have become more restricted, the university groups are driven into alliances with other groups by forming small consortia, they are encouraged to support visiting observers as a condition to funding, their scheduling procedures become more formalized, and their operation assumes much of the flavor of large national observatories.

There is also increasing pressure for national radio observatories to expand their traditional role of operating only expensive, unique instruments and to begin operating smaller observing facilities as well. The operation of user friendly observing facilities is expensive, and the Panel believes it is not cost effective to provide extensive user support to small or modest sized facilities. The long-term health of radio astronomy requires proper balance between large unique facilities at the national centers and the smaller, but often also unique, facilities where many important discoveries are being made, where innovative new techniques are being developed, and where the next generation of observers and telescope builders is being trained.

Student Training: Most of the developers of new instruments and telescopes have received their training in the universities and institutes which operate their own facilities. But, there are now only a few remaining facilities involved in student training. The Panel is concerned about where the next generation of instrumentalists will come from—the men and women who will design and develop new facilities needed to maintain the country's preeminent position in radio astronomy.

The lack of suitable facilities for training technically oriented radio astronomers is exacerbated by a lack of well-defined career paths for people more interested in building instruments than using them. Moreover, low salary levels for needed professional technical support at our university and national center facilities have contributed to the loss of key personnel that will be increasingly felt in future years. As a result, many of the key positions in radio astronomy in this country are being filled by scientists trained in other countries. It is important for the future of radio astronomy in this country that universities with plans and ideas for the development of new telescopes, new techniques, or other new instrumentation be adequately supported, and that the university observatories and national observatories recognize the contributions of instrumentalists as well as users in their professional staffing and promotion policies.

Agency Funding and Management Policies:

The rapid growth of radio astronomy in the United States began in the late 1950's with encouragement and financial support primarily from the Office of Naval Research, the Air Force, and later from the National Science Foundation. The field enjoyed substantial growth throughout the 1960's and early 1970's, but starting with the adoption of the Mansfield Amendment in 1968, DoD funding for radio astronomy has greatly diminished and the NSF has had to assume essentially all of the support for radio astronomy. The growth of radio astronomy has continued in the 1980's, as measured by construction of new radio telescopes, increased numbers of students and active radio astronomers, the development of new and more powerful ancillary equipment and techniques, and research activity in general. But NSF support for radio astronomy has not continued to grow, and in fact has not even kept up with inflation. In terms of purchasing power, NSF funding of astronomy today is almost identical to that in 1977. But the demands on those funds are very much greater today than they were in 1977. Every institution and every individual that depends on the NSF for research support is affected by this. It is a major problem for all of astronomy.

Today, there are more people doing astronomy than ever before. New instruments have been built and put into operation, university research groups have increased in numbers and in size, and in numbers of students. This growth is driven by many factors, including the general population growth, but especially the challenges and excitement of the field; and it is appropriate that it should grow. How has this increased activity been sustained at the same time NSF funding has been constant or declining? More than a dozen radio telescopes have been closed over the past two decades to make way for new ones; NASA has increased its astronomy grants; and there has been some increase in state and private support of astronomy. All of these have helped a little, but basically the constant NSF budget is being spread more thinly over an ever increasing number of people and activities.

Radio astronomy is particularly affected by the inadequate level of NSF support because it is essentially a ground-based science and thus receives much less NASA support than other fields of astronomy. Moreover, the mission-oriented nature of NASA support is not a substitute for the NSF supported activities which are based on peer review without regard for the need to satisfy specific programs, missions, or national goals.

It is of course reasonable to close obsolete or ineffective facilities in favor of new ones. This has been and will continue to be done, but it is not enough. If ground-based astronomy is to survive and flourish it needs both a reasonable influx of capital investment in new telescopes and equipment, and an expansion of operating and grant support at least consistent with the natural growth of the field. There is no reason why *ground-based* astronomical research in the United States should shrink as a fraction of the total intellectual and cultural base, nor should it shrink relative to world science and astronomy in general; but both of these are now happening.

Good science is a highly individualistic effort, and the administrative system should impose a minimum of management or control beyond that necessary to assure reasonable accountability. The NSF has historically supported individual scientists based on peer review of proposals and without regard to the need to satisfy specific programs or missions. The NSF also supports major national facilities such as NRAO, NOAO, and NAIC, but provides no scientific control or management, leaving this to the Observatory Directors, the managing consortia, and their advisory committees.

NASA, by contrast, is heavily mission oriented with a tradition of strong program management. This may be appropriate for major space missions where design, construction, and operation phases may last as long as twenty years, and where a highly coordinated central management is needed. But, we are concerned that policies which have been formulated to insure the success of expensive complex space missions are also applied by NASA to the administration of scientists and science programs which are not directly related to the preparation and operation of actual space missions.

The current disparity between the NASA and NSF budgets allows NASA to make important contributions to ground-based astronomy that normally would be in the domain of the NSF. But we note that fundamentally important areas of ground-based radio astronomy are being funded out of "small change" from other, essentially unrelated large NASA space missions. This has led to striking anomalies in what gets done and what doesn't get done and to a possible distortion of priorities.

Due to the large difference between the scale of NASA and NSF funding, the dollars are driving the science rather than the other way around. Individual scientists are increasingly submitting proposals to use space facilities or proposals to access space data banks, because that is where the money is. Good science is done, but this may create the illusion of greater demand on these facilities, and in turn helps to generate even more financial support for these activities. This may appear to be attractive in the short term, but may have long-term, adverse implications for the progress of astronomy. It is driving scientists to pursue work related to NASA missions and NASA money, or related to broader national or agency policies such as the establishment of a space station, or a return to the moon, rather than where their scientific curiosity directs them. As a result, some astronomers are being driven away from ground-based astronomy to space astronomy, and this may have particularly serious consequences for radio astronomy.

The Panel is concerned about the deteriorating level of support for individual scientists and recommends a restoration to earlier levels of the traditional NSF research grant support. Particular attention must be paid to young scientists who are finding it increasingly difficult to break into the funding system.

Acknowledgements

The Panel benefited from written comments received from about 50 individuals and from presentations by J. Bally (Bell Labs), L. Bautz (NSF), R. Brown (NRAO), L. Caroff (NASA), R. Giovanelli (Cornell), P. Goldsmith (University of Massachusetts), J. Greenstein (Caltech), M. Hollis (NASA), P. Horowitz (Harvard), F. Jordon (JPL), M. Klein (JPL), T. Kuiper (JPL), T. Landeker (NRC, Canada), R. Martin (Arizona), C. Pellerin (NASA), R. Perley (NRAO), A. Readhead (Caltech), J. Salah (Haystack Observatory), A. Sargent (Caltech), P. Schloerb (University of Massachusetts), P. Vanden Bout (NRAO), P. Wannier (JPL), and H. Zirin (Caltech). Helpful written comments and suggestions were also received from a number of foreign scientists including J. Baldwin (Cambridge), N. Kardashev (USSR), M. Morimoto (Japan), Y. Parijiskii (USSR), V. Radakrishnan (India), R. Schilizzi (Netherlands), G. Setti (Italy and ESO), Wang Shouguan (China), and V. Slysh (USSR).

INFRARED ASTRONOMY PANEL

Report of the Infrared Panel

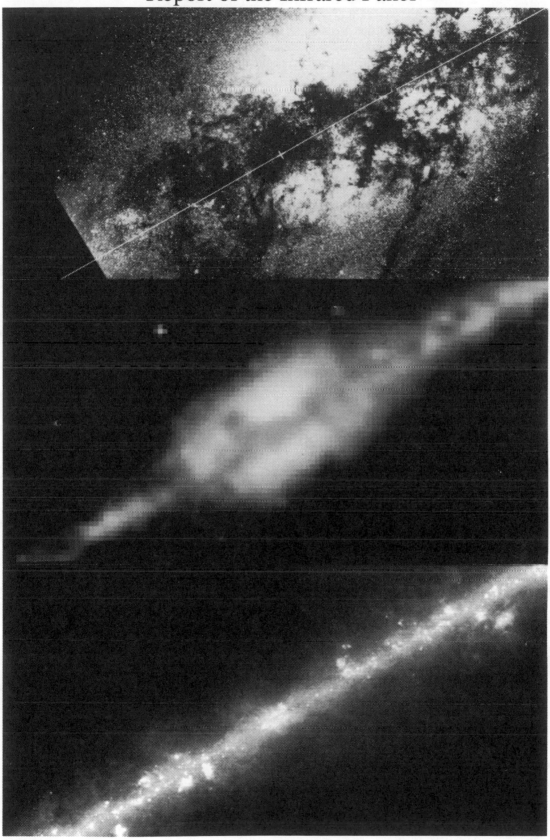

I. EXECUTIVE SUMMARY

Infrared astronomy is on the threshold of a revolution. The decade of the 1990's presents an unparalleled opportunity to address fundamental astrophysics issues through observations at infrared wavelengths (1 μm to 1000 μm) made possible by enormous technological and scientific advances during the last decade.   The formation of galaxies, stars and planets, the origin of quasars, and the nature of active galactic nuclei are among the key scientific issues to be addressed by these observations.

The major elements of the recommended program are:

THE SPACE INFRARED TELESCOPE FACILITY (SIRTF)

SIRTF is the single most important new astronomy project for the 1990's. A cryogenically cooled space observatory of 5 years lifetime, SIRTF will complete NASA's Great Observatories program by enabling observations across the 3 μm to 700 μm region with a sensitivity gain of 2 to 4 orders of magnitude over all current capabilities. SIRTF will redefine the state-of-the-art in infrared exploration of the universe.

THE STRATOSPHERIC OBSERVATORY FOR INFRARED ASTRONOMY (SOFIA) & THE IR OPTIMIZED 8-M TELESCOPE (IRO)

SOFIA, a 2.5-m telescope system in a modified Boeing 747 aircraft, utilizes the airborne environment to observe free from most of the absorption by telluric water vapor.

IRO, a national 8-m telescope fully optimized for infrared observations, exploits the remarkably dry and stable atmospheric conditions of the summit of Mauna Kea.

Together, SOFIA and IRO span the IR at high angular resolution with more than an order of magnitude increase in sensitivity compared to current ambient temperature telescopes. SOFIA and IRO will complement SIRTF by their ability to achieve higher spectral and spatial resolution, by their long lifetime, and by their ability to support evolving instrumentation.

Other highlights of this report include:

A DETECTOR AND INSTRUMENTATION PROGRAM

A vigorous, broadly based IR array detector development, evaluation and implementation program to exploit the revolution in IR array technology and recent advances in adaptive optics.

THE SUBMILLIMETER MISSION (SMMM)

An Explorer-class space experiment to obtain an unbiased spectral line survey of the rich 100 μm to 700 μm region.

THE 2 μM ALL SKY SURVEY (2MASS)

An all-sky three-color survey in the 1 μm to 2.5 μm region, reaching 50,000 times fainter than the existing two micron sky survey. Key issues in the structure of the local universe and the large scale structure of the Galaxy will be addressed.

A SOUND INFRASTRUCTURE

A strong base of support for laboratory and theoretical astrophysics, for the development of well trained instrumentalists through airborne and ground-based observational programs and for archival data analysis programs.

TECHNOLOGY DEVELOPMENT PROGRAMS

An active technology program directed towards the further exploitation of the space environment for far-IR/submm astronomy, and a ground-based imaging interferometry program leading to an optical/infrared VLA in the next decade.

II. PERSPECTIVE

How do stars and solar systems form? How does material synthesized in stellar cores enrich the interstellar medium and alter the subsequent evolution of stars and planets? Why do new stars form at such a dramatic rate in some galaxies? What produces the exotic behavior in the nucleus of our Galaxy, and is this activity related to the far more powerful QSO phenomenon? What are the sources of cosmological background radiations? How and when did the first stars and galaxies in the Universe form? From the sensitive probing of the first seeds of structure in the Universe, to exploring the nature of the processes that shape star and planetary system formation, observations in the IR address questions that form the basis for modern astrophysical inquiry.

Four physical principles underscore the critical role played by infrared observations in addressing these questions.

IR observations uniquely reveal cool states of matter.

The most common stars are cooler than the Sun, and emit much of their energy in the infrared. The Earth and other planets, including those around other stars, emit most of their radiant energy in the IR. Star-forming regions and massive interstellar clouds are cooler still, radiating essentially all their energy in the IR.

IR observations explore the hidden Universe.

Ubiquitous cosmic dust, an efficient absorber of optical and ultraviolet radiation, becomes increasingly transparent in the IR, where it re-emits the bulk of its absorbed energy. This is dramatically illustrated in the frontispiece. Our galaxy is transparent at mid-IR wavelengths, while optical radiation from the Galactic Center is attenuated by a factor of about a thousand billion.

IR observations access a wealth of spectral features.

Spectral features of atoms, ions, and virtually all molecules and solids are located within the IR. These features probe the conditions in celestial regions as diverse as the shocks in supernova remnants, obscured stars and the nuclei of galaxies, the atmospheres of planets, the cold interiors of dark clouds, and the circumstellar disks which contain the raw material of planets. The emitting material in these sources ranges from condensed forms of matter like ices and silicates to highly ionized gases and from that most abundant of simple molecules, molecular hydrogen, to highly complex hydrocarbons.

IR observations reach back to the early life of the cosmos.

The expansion of the universe inexorably shifts energy to longer wavelengths. The primeval fireball of high energy gamma rays produced in the Big Bang now appears as 2.74K blackbody emission that peaks near 1000 μm. Most of the energy emitted from stars, galaxies, and quasars since the beginning of time now lies in the infrared. How and when the first objects in the universe formed will ultimately be determined by infrared observations.

These four principles make infrared observations crucial to the solution of the most pressing problems of modern astrophysics. Technological advances during the 1980's revolutionized our ability to exploit the potential of the infrared, while the scientific advances of the decade profoundly changed our view of the sky. Major highlights included:

The successful deployment of crogenically-cooled telescopes in space.

The Infrared Astronomical Satellite (IRAS) and Cosmic Background Explorer (COBE) missions, developed as part of NASA's Explorer program, achieved enormous gains in sensitivity through a million-fold reduction in the level of background radiation using liquid-helium (LHe) cooled telescopes and instruments which demonstrated the fundamental technology required for exploiting the environment outside the earth's atmosphere.

The 1983 IRAS sky survey, a joint U.S./Netherlands/United Kingdom project, revealed for the first time the richness and variety of the IR sky. It laid the foundation for major scientific advances, impacted all fields of astrophysical research, and made IR observations accessible to the whole astronomical community; about 70% of the IRAS data users are other than infrared astronomers.

IRAS discovered disks of particles orbiting nearby mature stars. These disks, first discovered by studying the bright star Vega, appear closely related to the evolution of planetary systems, and may be debris of the planet accumulation process. IRAS also discovered dust bands in our solar system that appear to be debris of asteroid collisions, a likely source of the bulk of the zodiacal dust grains. The IRAS observations of hidden clusters of young stars and protostars in dark clouds throughout the Milky Way provided the first census of the luminous stellar content in these stellar nurseries. IRAS discovered structured emission from interstellar grains (termed "IR cirrus") that may be produced in part by large complex hydrocarbon molecules, a major new component of the interstellar medium.

IRAS found IR-bright galaxies, galaxies emitting more than 99% of their luminosity in the IR, radiating 1000 times or more the energy output of the Milky Way. These galaxies are more numerous than QSO's of the same luminosity, and may represent an early stage of QSO evolution. Because of their relatively high space density, their high luminosity, and their apparent link to systems undergoing collisions or mergers, ultra-luminous IR galaxies are potentially powerful tracers of luminous matter out to the edge of the Universe.

COBE, launched in November 1989, is in the process of transforming our understanding of the early Universe by collecting data on the infrared, submillimeter and microwave diffuse emission in space. Early in its life, COBE demonstrated precise agreeement between the observed Cosmic Background Radiation and a 2.74K blackbody. This result severely constrains models of the early Universe, and rules out a uniform hot intergalactic medium as the source of the X-ray background. The COBE instruments will be remarkably sensitive probes of the important processes in the early universe, and may reveal the first seeds of structure in the Universe and evidence of the epoch of galaxy formation.

A million-fold improvement in the performance of IR detectors.

The recent development in the US of large-format high-performance infrared sensitive arrays promises to make both cryogenic and ambient-temperature IR telescopes millions of times more capable than their predecessors of a few years ago. Each pixel of a modern IR array is some 10 to 100 times more sensitive than previous single detectors. Combined with the increased format, the gain in "speed" can be 10^7 to 10^9, making possible qualitatively new approaches to instrumentation design and use, thereby enabling entirely new classes of scientific investigations. In their infancy these arrays have already been utilized to image a wide range of environments, such as probing for stars forming in the nearest clouds (Figure 1) and searching the extragalactic sky for very distant galaxies (c.f. Figure 6). A few years ago, the 2.2 um image in Figure 1 would have required mapping with a single detector, taking many nights to complete; only a few minutes were required with a camera based on modern IR array technology.

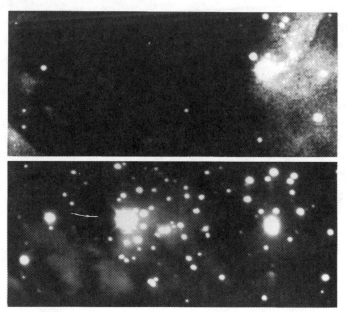

Fig. 1. Images of the star formation region NGC2024; 0.7 μm (top) and 2.2 μm (bottom), revealing young embedded stars in the IR. Courtesy of R. Probst, KPNO

Exploration of new phenomena through the flexibility of the airborne astronomy program.

IR observations from an altitude of 41,000 feet and above liberate us from many of the limitations imposed by atmospheric absorption. For the past 15 years, NASA's Kuiper Airborne Observatory (KAO) has been a showcase for the wealth of phenomena that can be observed from a mobile stratospheric platform. These include the discovery of the torus of gas and dust around the Galactic Center, the first observation of the water molecule in comets, and the first direct estimates of the masses of Fe, Co, and Ni in Supernova 1987A in the Large Magellanic Cloud.

The last decade has resulted in an explosion in the breadth and depth of our investigations of the infrared sky, driven by the initial exploitation of infrared observations from the space environment and by advances in the technical maturity of infrared detectors and associated instrumentation. Observational infrared astronomy is now poised to revolutionize our understanding of the most fundamental questions of modern astronomy.

III. SCIENCE OPPORTUNITIES

A. THE ORIGIN OF GALAXIES

How did homogeneously distributed matter in the early universe condense into galaxies? To address this question empirically we must observe the galaxy formation epoch, then trace the development of large scale structure in the distribution of galaxies as a function of redshift. We must understand how individual galaxies relate to the processes that explain these structures, and on a more local scale, we must fit the details of our own Galaxy into this overview.

How do cosmic backgrounds relate to the large scale structure of the universe?

Results from the COBE mission will provide an initial opportunity to look for evidence of primordial galaxy formation. The COBE all sky maps from 1 μm to 1 cm will be the first step toward a determination of the true cosmological background radiation. The best windows on the extragalactic universe are at 3 μm to 4 μm, between the scattered and re-radiated components of the zodiacal dust, and at 200 μm to 500 μm, between the thermal emission from Galactic dust and the cosmic microwave background. These windows are the most promising in the entire electromagnetic spectrum for the detection of radiation from primordial galaxy formation. Such radiation can provide cosmological information even in the absence of observations of individual primordial galaxies. Extragalactic infrared backgrounds can set important limits on the epoch and nature of galaxy formation as long as known galaxies at lower redshifts are accurately subtracted. This requires accurate knowledge of all populations contributing to the backgrounds. These populations will be accounted for by the infrared surveys of the 1990's. Folding in the deep source counts of near-IR and mid-IR observations, combined with estimates of their contributions in the far-IR should allow us to tell unambiguously if infrared background radiations arise from the glow of the universe at the time galaxies first formed.

Infrared sky surveys in the 1990's will be essential to probe the distribution of matter in the nearby universe. The near infrared is the optimum region to obtain a mass census of galaxies because infrared radiation is both insensitive to the extinction within galaxies (including our own) and sensitive to the stellar component which dominates the luminous mass. The "great attractor", for example, is in a direction greatly affected by obscuration from the Milky Way Galaxy. Any mass concentration there may be detectable only in the infrared. Infrared observations will be the only way to detect cool, solid objects in galaxy halos. Systematic observations of galaxies with old stellar populations and the deepest possible searches for faint matter are essential to understand the mass distribution in large scale structures.

When did galaxies form?

Recent observations in the optical have established that the epoch of initial nucleosynthesis, star formation, and galaxy formation is at a sufficiently high redshift that the initial formation epoch can only be observed in the infrared, because the redshift must be greater than 5. Pushing back the limits on this formation epoch, or finding it, is a major imperative for observational extragalactic astronomy.

A quasar is now known at a redshift of nearly 5. If a significant population of quasars exists much beyond this, such a population cannot be measured without observations in the near infrared. Although quasars currently represent the highest redshift objects observed, their relation to primordial galaxy formation is unknown because their ultimate luminosity sources seem unrelated to stars. Direct evidence, or limits on the existence of galaxies made of stars, is also an observational requirement before the formation of stars and galaxies in the early universe can be understood. Determination of a redshift based on the bluest line, Lyman alpha, requires infrared spectroscopy for redshifts greater than 7, while observations of the old stellar population require measurements at wavelengths of about $(1+z)$ μm.

Such observations of old stellar populations in distant galaxies will set limits on the epoch of star formation in the universe, but direct observation of the first generation of star formation is the ultimate goal. Depending on the redshift of the formation epoch, near to mid-infrared observations will be required to detect the intrinsic ultraviolet luminosity from hot, young stars. IR observations have already demonstrated that extensive star formation in local galaxies is invariably accompanied by dust absorption at UV and optical wavelengths and re-emission in the IR. Is this the case also at high redshift epochs? Will HST observations of young galaxies at great distances be affected by dust attenuating the ultraviolet continuum?

What energy sources power galaxies?

One of the most significant discoveries of the 1980's was that of ultra-luminous IR galaxies, systems in which some process - perhaps starbursts or accretion by massive black holes - produces enormous infrared luminosities on a scale previously identified only with quasars. IRAS survey results suggest that this energetic activity correlates with galaxy interactions or mergers. Such luminous targets can be traced all the way back to the formation epoch with observations in the 1990's. Comparison of infrared luminosity distributions with those measured from other surveys, particularly in X-ray or ultraviolet radiation, is essential to account for luminous galaxies and quasars otherwise overlooked because of obscuration.

What is the distribution of matter in the Milky Way and nearby galaxies?

The evolution of galaxies is marked by a continuing cycle of birth and death of stars. This leads to the evolution of the elements from nearly pure hydrogen and helium to material with sufficient heavy elements to form earth-like planets. IR studies out to the distance of the Virgo cluster and beyond will measure elemental and chemical abundances of many of the heavy elements, both in the reservoir of the interstellar medium and as newly-formed material ejected from supernovae, novae and red giant stars. The age, composition, and structure for our Galaxy are crucial benchmarks for understanding other galaxies. Our vantage point is immersed in the obscuring dust of the Galactic disk, but infrared observations allow us to penetrate the dust in order to study stars and interstellar matter throughout the Galaxy. IRAS far-IR and COBE near-IR images of the sky toward the Galactic center, together with a visual view, are displayed as the frontispiece of this report. The COBE and IRAS images show the distribution of stars and luminous dust clouds, respectively, in this region of sky - extraordinary demonstrations of the power of infrared observations to reveal the grand design of the Milky Way Galaxy.

B. THE ORIGIN OF PLANETS, PLANETARY SYSTEMS AND STARS

The essential questions concerning star and planet formation, processes which are central to our concept of the universe in which we live, remain unanswered. Most of the visible matter in the Universe is in the form of stars, and star formation is central to the formation and evolution of galaxies. Closer to home, the formation of planets and planetary systems is a prerequisite for the formation of life as we know it. Both of these birth processes occur deep within dense clouds of dust and gas opaque at visible wavelengths but transparent in the infrared.

How do stars form, and what conditions lead to protostellar collapse?

Star and planet formation begins with a dense molecular cloud core which collapses to form a protostar embedded in a circumstellar protoplanetary disk. The protostar grows by direct infall of material onto the star and by accretion from the inner boundary of the disk. The gas and dust remaining in the disk is the raw material from which planets form.

The rate at which stars form and the resultant distribution of stellar masses must depend on the physical properties of the molecular cloud; composition, gas density, temperature, velocity field, chemical and ionization state, and magnetic field. In the 1990's we will image molecular clouds with sufficient sensitivity, spatial resolution and spectral resolving power to measure the conditions throughout star-forming clouds, to detect the emission from individual embedded stars, to determine the luminosity function into the substellar range - far below the hydrogen burning limit of about $0.08 M_0$ - and to correlate star formation rates and stellar masses with the cloud properties.

The spatial and spectral resolution available at far-infrared and submillimeter wavelengths will enable the detailed study of numerous infalling cores in nearby molecular clouds. It is believed that the infall halts abruptly at an accretion shock, which marks the boundary of the protostar or the protostellar disk. By detecting the IR spectral lines from these dust embedded accretion shocks, and measuring their profiles, the observations will probe the non-spherically symmetric infall onto the protostellar system and reveal how planet-forming disks are assembled.

The infall phase ceases when an outflow from the protostar impacts the infalling material, reversing its direction and sweeping it outward. The outflow is frequently collimated, by an as yet unidentified process, into a bipolar or jet-like flow. In the 1990's, IR imaging and proper motion observations will make it possible to see the jets and outflows as close as 10 AU from the protostar, thereby probing the agent of collimation. By observing the IR emission from the shocks produced when the outflow encounters ambient gas or protostellar disks, we will discover how the outflow evolves, how it inhibits infall, how it affects the disk, and how it interacts with the ambient molecular clouds.

How do protostellar and debris disks evolve?

Circumstellar disks are common features of stars of all ages. Theoretical models of the evolution of disks around protostars envision the growth of the disk by accretion, and the possible development of disk instabilities which cause material to spiral into the protostar or form a binary system. As the accretion phase ends, the dust in the (now) protoplanetary disk settles to the midplane, coagulates and forms planetesimals which ultimately may accumulate into planets. The forming planets sweep up the disk material near their radius and gravitationally interact with more distant material, producing "gaps" in the disk similar to those seen in the rings of Saturn. Finally, only planets, moons, asteroids and comets are left, and the occasional collisions of these larger bodies produce the leftover planetary debris disks observed around older stars, just as similar processes in our own solar system sustain the zodiacal dust cloud.

The infrared capabilities of the 1990's will allow us to image protostellar and protoplanetary disks around stars as distant as the Taurus cloud, and to detect gaps which may signal planetary system formation. The orientation of the disks can be correlated with the cloud core orientation and the magnetic field direction to improve our understanding of the dynamics of the collapse. Rotation curves can be measured and a composition profile derived for disks at all evolutionary stages, using resonantly scattered near-IR line emission for the inner disk and longer wavelength emission for the outer disk. Evidence for dust coagulation as a function of distance from the protostar will follow from spatially-resolved continuum spectroscopy at mid-ir wavelengths.

The final stage in the evolution of the protoplanetary disk is the planetary debris stage. IRAS found that a large fraction, certainly more than 20%, of mature main sequence stars near the Sun possess orbiting solid material. Prime examples are the disks orbiting the stars Vega, Epsilon Eridanus and Beta Pictoris. These disks, which are composed of particles much larger than typical interstellar grains, remain the best evidence for the occurrence of planetary systems around other stars. In the next decade, we should be able to detect disks similar to that around Vega around tens of thousands of stars out to distances of several kpc, to resolve a zodiacal cloud similar to the Sun's for the nearest stars, and to obtain detailed images of disks - searching for interior voids which may signal the presence of planets. Determination of the composition of the disk material and the size and frequency of voids as a function of the age and other characteristics of the central star will provide strong constraints on the processes, mechanisms and time scales for planet formation.

The study of the origin, evolution, and prevalence of extra-solar planetary systems will be complemented by the continued investigation of our own Solar System, particularly the region beyond the orbit of Jupiter. The spectral coverage and sensitivity of the infrared instrumentation of the 1990's will permit the study of all classes of objects in the outer solar system - not only the outer planets and their large satellites but also comets and the minor satellites, which may be better samples of the primordial material of the outer solar system. Infrared spectroscopy of the atmospheric gases and the surface ices of these primitive objects will provide critical information concerning the composition and physical conditions in our own protostellar nebula.

How prevalent are Low Mass Objects and Brown Dwarfs?

Infrared observations will not only detect the gaps in disks which signal the presence of planets, but also can provide direct measurements of young giant planets and of brown dwarfs of all ages and masses. These substellar objects range in mass downward from 0.1 M_O to planetary masses of order 0.001 M_O (about the mass of Jupiter). The number and distribution of such objects with mass smaller than the smallest star is unknown, but their abundance and properties may answer important questions about the formation of stars and planets, about the behavior of matter at high pressure, and even about the missing mass in astrophysical systems. Infrared observations are uniquely capable of detecting thermal emission from not only substellar companions but also isolated substellar objects, which glow faintly in the infrared as their residual heat of formation diffuses away. Isolated brown dwarfs can be detected in nearby space, in molecular clouds and stellar clusters, and, possibly, in the halos of other galaxies.

IV. TECHNICAL OVERVIEW

The IR spans three orders of magnitude in wavelength from 1 μm to 1000 μm. Over this very large span, the experimental techniques, the properties of the atmosphere, the telescopes, and the detectors all change dramatically.

In the **near-IR (1 μm to 2.3 μm)**, observations can be made from the ground through three "windows" in the earth's atmosphere. Telescope and instrument techniques are quite similar to those developed and deployed at optical wavelengths, except that detectors are hybrid arrays using exotic infrared sensitive detector material like InSb or HgCdTe bonded to silicon readout devices. The recent revolution in this IR array technology enables a wealth of new scientific opportunities. At near-IR wavelengths, two near-term opportunties are highlighted in this report; the revitalization of existing telescopes with state-of-the-art instrumentation, including adaptive optics, and the 2 μm All-Sky Survey, 2MASS.

Near-IR observations from the ground are limited by airglow background, atmospheric absorption and atmospheric "seeing". The second generation near-IR instrument under development for the Hubble Space Telescope (HST), NICMOS, will exploit the absence of airglow emission and atmospheric absorption and seeing to make high sensitivity diffraction limited imaging and spectroscopic observations in the 1 μm to 2 μm range.

In the **mid-IR (2.3 μm to 30 μm)** ground-based observations are possible through four windows, but thermal emission from the atmosphere and from an ambient temperature telescope creates an enormous photon background against which observations must be made. A cryogenically cooled telescope operating outside the earth's atmosphere, like IRAS, COBE and SIRTF, is free from the limitations of atmospheric and telescope emission and atmospheric absorption. The natural mid-IR background in space, which originates from interplanetary and interstellar dust grains, is at least a million times fainter than that at any groundbased observatory; SIRTF's sensitivity is limited only by the statistical fluctuations in this natural background. Because SIRTF uses modern IR array technology, this facility has the power to transform our understanding of the basic questions of astrophysics. SIRTF is the cornerstone of IR astrophysics for the 1990's and beyond.

There are, however, some significant strengths of ground-based and airborne observations in the mid-IR; these include ready access to instrumentation, and large collecting area - useful both for light-gathering power for spectroscopy and for higher (diffraction limited) spatial resolution.

Modern mid-IR astronomy began in the 1960's with the introduction of the LHe cooled Ge bolometer on ground-based telescopes. Ground-based observations in the 2 μm to 30 μm atmospheric windows have evolved using adaptations to optical telescopes, progressing to the point that now most major optical telescopes incorporate some level of IR capability, although use of such facilities is generally compromised by inadequate IR adaptation or an inferior site for IR observations or, most frequently, a combination of both. In order to minimize the thermal emission, the telescope configuration and optics must be "optimized" for thermal IR observations, and the site selected to minimize the atmospheric absorption and emission. The recommended 8-m IRO is a unique IR optimized telescope on the best IR site known, the summit of Mauna Kea.

There is only a single large U.S. telescope dedicated to IR observations; the 3-m Infrared Telescope Facility (IRTF) on Mauna Kea, operated as a national facility by NASA's Planetary Exploration division. The upgrade of this facility to exploit the IR potential of Mauna Kea would create major new scientific opportunities for a large community and should be pursued aggressively and rapidly.

In the **far-IR (30 μm to 300 μm)** the earth's atmosphere is essentially opaque. As in the mid-IR, the residual atmosphere and telescope are strong sources of background emission. SIRTF is an extremely powerful observatory throughout this wavelength range.

Observations in much of this spectral regime and at many of the other infrared wavelengths obscured from the ground can be made from stratospheric platforms. NASA's airborne astronomy program began in the late 1960's, highlighted by deployment of a 0.3-m telescope in a Lear jet and the development of the KAO, a 0.9-m telescope in a C-141, in 1974. The KAO has been very successful for the past 15 years, providing critical observations ranging from stellar occulations at optical wavelengths to the study of collapsing clouds at 300 μm, but is now becoming scientifically outmoded because it can study only a small fraction of the IRAS sources. The recommended airborne observatory, SOFIA, which will replace the KAO, is key to the development of airborne infrared astronomy, providing a powerful and flexible facility for scientific observations in this field, capable of detailed study of all the IRAS Point Source catalog objects, and permitting development and deployment of state-of-the-art instrumentation.

Extrinsic germanium photoconductors provide the most advanced detector technology for far-IR wavelengths, used either as individual detectors or in small arrays. Low temperature bolometer arrays are also under development, particularly for use beyond 120 μm, and high frequency heterodyne receivers are becoming available at the longest wavelengths.

In the **submillimeter** (**300 μm to 1000 μm**), observations are again possible from the ground through four windows whose transparency is a strong function of water vapor content; the longer wavelength windows have higher transparency. Telescope and receiver techniques are often extrapolated from radio wavelengths. Novel heterodyne components and more precise surface figures for telescopes than are commonly achieved at radio wavelengths are required. Thermal emission from the atmosphere and telescope, while still by far the dominant sources of background, are less important for high resolution spectroscopy, and telescope diameter is important for photon gathering and improved diffraction-limited angular resolution.

Submm observations with the 10m Caltech Submm Observatory (CSO) on Mauna Kea have demonstrated the viability of astronomical observations in the atmospheric windows at 600 μm and 800 μm using extrapolation of radio techniques. Already the findings of submillimeter water vapor masers and high velocity molecular outflows in evolved stars have demonstrated the importance of such observatories. Additional submm facilities currently under development, including the 10-m submm telescope for Mt. Graham, the Submm Array Interferometer under development by SAO, and the South Pole Submm telescope are expected to establish the basis for future ground-based developments in the submm wavelength range down to 350 μm.

Freedom from atmospheric absorption is a strong driver toward airborne or space-based platforms. SOFIA will provide an excellent platform for exploratory submm observations and instrument development. However, critical molecules like H_2O and O_2 cannot be observed from within the earth's atmosphere, even at airborne altitudes, because of telluric absorption. A small submm explorer, the Submillimeter Wave Astronomy Satellite (SWAS), is currently under development and expected to be launched in 1994. This mission, the first dedicated to high spectral resolution submillimeter observations from space, will study the Galaxy in spectral lines of H_2O and O_2. The recommended Submm mission (SMMM) exploits the complete freedom from telluric absorption of the space enviroment. to provide our first unbiased spectroscopic view of the submm regime, while SIRTF will provide the most sensitive submm continuum measurements out to 700 μm.

Antarctica offers an intriguing possibility for a ground-based astronomical site with conditions of atmospheric water vapor content and ambient temperature that are much more favorable to IR observations even than those of Mauna Kea. A small submm facility, ASTRO, is currently under development to evaluate the submm potential of the South Pole and a similar scale experiment, SPIREX, is proposed to investigate the potential of the South Pole for observations around 2.4 μm. These small scale projects are ideal for addressing such issues as seeing, operability, atmospheric conditions and IR background levels, which must be well understood in order to evaluate the potential of this site for a major IR observatory.

V. PROJECT RECOMMENDATIONS

Table 1 summarizes the projects recommended in this report. The project recommendations and descriptions following this table are presented in the same order as they appear in the table. Current estimates of the project development costs are included; in the cases of international participation, the U.S. costs are shown.

TABLE 1. PROJECT SUMMARY

CATEGORY	PROJECT	CATEGORY PRIORITY	START TIMES DEVELOPMENT	OPERATION	COST (FY89$)
SPACE PROJECTS					
MAJOR					
	SIRTF	1	1993	1999	1.3B
MODERATE					
	SOFIA	1	1992	1997	230M
	SMMM	2	1995	1999	125M
SMALL					
LAB/ASTR AUGMENTATION		1	1991		8M
GROUND-BASED PROJECTS					
MAJOR					
	8M IRO	1	1993	1998	80M
MODERATE					
DETECTOR/INST DEVELOPMENT		1	1992	1993	50M
SMALL					
	2MASS	1	1992	1993	5M

A. SPACE PROJECTS

We very strongly recommend that the highest priority for astronomy in the 1990's be the development and operation of the Space Infrared Telescope Facility (SIRTF), the culmination of NASA's Great Observatories program.

The SIRTF observatory consists of a 1-m class cryogenically cooled telescope mounted on a free-flying spacecraft in high earth orbit. SIRTF will be operated as a national facility, with more than 85% of the observing time during its five year life time available to the community. SIRTF will achieve a one hundred to ten-thousand-fold improvement in sensitivity over current infrared capabilities. This gain, coupled with the imaging and spectroscopic power inherent in its large-format infrared detector arrays, will make SIRTF unique for the solution of key astrophysical problems ranging from the doorstep of the solar system to beyond the horizon of our current understanding.

SIRTF will be equipped with instruments utilizing modern two dimensional infrared arrays providing wide-field and diffraction limited imaging and spectroscopic capability over most of the IR spectral regime. SIRTF's cameras will provide imaging capability from 2 µm to 200 µm and photometry from 2 µm to at least 700 µm. SIRTF's spectrographs will cover the wavelength region 2.5 µm to 120 µm at low spectral resolution (R=100), and the 4 µm to 200 µm region at higher spectral resolution (R=2000). The sensitivity will be limited over most of this wavelength range only by the natural backgrounds in space, allowing SIRTF to achieve sensitivity gains of 100 to 10,000 over present capabilities. SIRTF will be launched by a Titan IV-Centaur into a circular orbit at an altitude of 100,000 km. The five-year lifetime will permit follow on studies with SIRTF itself of the many new scientific questions which will be posed by SIRTF's own discoveries, while an archival research program will make the SIRTF data accessible to the scientific community long after its data acquisition phase ends.

SIRTF has been under active study by NASA for more than a decade. The key technologies have been demonstrated; all that awaits is to build and launch this premier mission of the 1990s. With a development start in FY 1993, SIRTF can be launched by the year 2000, allowing a significant period of time for coordinated observations with NASA's other Great Observatories, HST and AXAF.

SIRTF's sensitivity increase, combined with the power of the detector arrays, achieves extraordinary gains in capability (Figure 2), one-million fold or more over the current state of the art in the infrared, and one-thousand fold or more over the performance anticipated for the ISO mission (planned for launch by the European Space Agency in 1993).

SIRTF SCIENCE HIGHLIGHTS

SIRTF will make fundamental contributions to virtually all contemporary forefront astrophysical problems. As illustrated in Figures 3 and 4, SIRTF will permit detection and identification of objects as disparate as galaxies at redshifts z>5 and brown dwarfs in the solar neighborhood; and detailed study of problems as different as the relationship between high luminosity infrared galaxies and quasars, and the nature of planetary system debris around nearby stars. The investigations will reshape our understanding of processes ranging from galaxy formation in the early universe to the formation of our solar system.

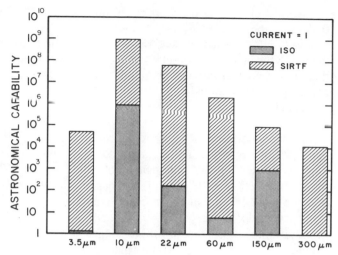

Fig. 2. Astronomical capability is defined as lifetime * efficiency * # pixels / (sensitivity)^2. Current capabilities are defined by IRTF, KAO and IRAS performance. Predictions for ISO and SIRTF use published data.

The identification of the epoch of first star formation in galaxies is critical to our understanding of the process of galaxy formation. Once a very young galaxy reaches an age of tens of millions of years after its first stars are formed, much of its luminosity is produced by cool red giant stars emitting a broad spectrum peaked at 1.6 μm. SIRTF can identify this near-infrared peak redshifted to longer wavelengths; for example, SIRTF can detect a young galaxy of average mass at redshift z=5, seen when the universe was less than 10% of its current age. An object of high redshift which shows no evidence for the peak is a candidate for identification as a galaxy still forming its first generation of stars.

In some ultraluminous infrared galaxies, bursts of star formation account for nearly all of the observed luminosity, but others appear to harbor dust-enshrouded quasars. SIRTF's imaging surveys will trace the evolution of quasars and ultraluminous infrared objects to redshifts well in excess of 5. Low resolution infrared spectra can identify characteristic features due to emission from dust and gas and determine the redshift, and hence the luminosity, of the infrared-bright objects. These studies will determine how both the absolute and the relative number of starburst galaxies and dust-enshrouded quasars varies with epoch and determine how the cosmic evolution of this population compares with that of the optically and radio selected quasars to which they may be related.

Brown dwarfs more massive than Jupiter and less massive than the 0.08M_O required for

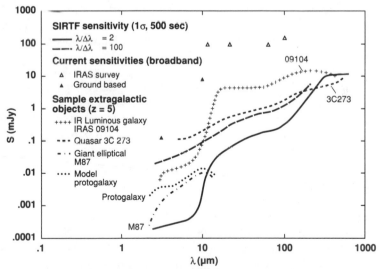

Fig. 3. The enormous sensitivity of SIRTF is shown by comparison with the expected brightness of a variety of classes of extragalactic sources at z=5.

a star to sustain nuclear hydrogen burning are expected to be visible in the infrared as they radiate the heat generated by their gravitational contraction. SIRTF's sensitivity will be such that objects with 10-30 Jupiter masses and ages less than 10^7 years can be detected out to the Taurus cloud. SIRTF can also discover older, less luminous brown dwarfs through deep imaging of nearby star clusters, unbiased surveys, or through targeted searches for companions of nearby stars.

Zodiacal clouds like the Sun's can be imaged by SIRTF around the nearest solar-type stars, while planetary debris disks like those found by IRAS can be studied around stars more distant than 1kpc. For the more prominent systems, SIRTF's images will show the orientation, structural features, and detailed morphology of the disks, including the inner dust-depleted regions suggestive of planets orbiting within the disk. Low resolution spectra of the debris - in both reflection and emission - will provide critical diagnostics for studies of composition, replenishment, and origin of the material.

Just as SIRTF can study material in the outer regions of other solar systems, it will extend the studies of the outer regions of our own Solar System. The ices and gases that condensed in the collapsing solar nebula further than 5 AU from the Sun, and that are now locked in the comets and planetary satellites, carry the chemical and physical history of the primitive solar nebula in the zones where the outer planets formed. SIRTF can obtain detailed spectra of the gaseous and solid materials in this distant zone; tracing the primitive solar nebula in this fashion will permit us to combine knowledge of the early solar system with the observed properties of stars in formation to understand more fully the formation of solar systems.

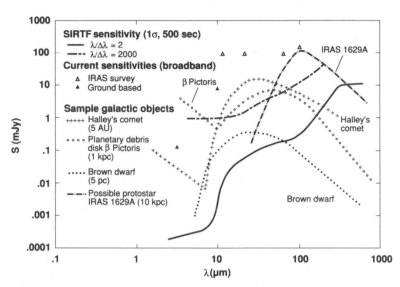

Fig. 4. SIRTF sensitivity to Galactic and Solar System sources.

SIRTF will also conduct surveys: Some of these will be targeted atspecific scientific problems - searches for candidate protogalaxies, for distant Kuiper belt comets in the ecliptic plane, or for embedded protostars in dark clouds in the Galaxy. Others will be totally unbiased, deep surveys aimed at searching for the as yet unnamed and unimagined phenomena which will lie within SIRTF's vast new horizon. SIRTF's spectrograph will be used extensively not only for followup observations of objects discovered in the imaging surveys, but also for complete infrared spectroscopic surveys of known classes of objects. These surveys and the archive of SIRTF's targeted observations will represent a legacy for astronomical study long after the end of the SIRTF mission.

We recommend the immediate development of SOFIA, a joint NASA-Federal Republic of Germany airborne observatory for infrared astronomy.

SOFIA is a 2.5-m telescope system mounted in a modified Boeing 747 aircraft. Flying over one hundred 8-hour missions per year at altitudes of 41,000 feet, above 99% of the water vapor in the earth's atmosphere, SOFIA will provide the astronomical community routine observations at most infrared wavelengths inaccessible from the ground. SOFIA's tenfold increase in collecting area over the KAO will enable study of any of the sources in the IRAS Point Source Catalog. A great strength of SOFIA is its flexibility: frequent access to the near space environment, realtime hands-on access and rapid interchange of focal plane instruments, and round-the-world deployment capability. SOFIA provides an excellent platform for the development of advanced instrumentation and for the education and training of the next generation of experimentalists.

SOFIA is a joint project, with the Federal Republic of Germany supplying the telescope system, supporting the operations at roughly the 20% level, and participating in the flight program at a similar level. NASA and the German Science Ministry (BMFT) have successfully completed preliminary design studies for SOFIA. A development start for SOFIA in FY1992 would allow observations to begin in 1997.

SOFIA SCIENCE HIGHLIGHTS

SOFIA's capability for diffraction-limited imaging beyond 30 microns and for high resolution spectroscopy over the entire 1 μm to 1 mm infrared band will allow studies of the composition, structure, and dynamics of planetary atmospheres, comets, and interstellar gas and dust; the initial luminosity function of stars embedded in nearby molecular clouds; the infall and outflow from protostars; and the nature of the luminosity sources in nearby starburst and AGN galaxies.

Far infrared observations with the spatial resolution of SOFIA will probe the distribution and nature of the embedded luminosity source(s) in the nuclei of nearby galaxies. Our own galactic nucleus may provide valuable clues to the phenomena occuring in galactic nuclei. Far infrared polarimetry and spectroscopy will map the magnetic field distribution and gas dynamics in the 2-10 pc ring of gas and dust surrounding the center of the Galaxy, possibly a magnetic accretion disk from which material spirals into the galactic center.

The bulk of the luminosity from protostars generally emerges in the 30 µm to 300 µm band (illustrated in figure 5), so only far infrared observations can measure the bolometric luminosity. SOFIA has the sensitivity to detect 0.1 L_O embedded stars at distances as large as 500pc. SOFIA's far IR spatial resolution enables studies of the luminosity function characterizing solar or subsolar mass stars in nearby star forming regions like Taurus. Resolved maps of the infrared continuum and line emission from individual protostellar sources determine the dust and gas density distribution around the protostar and the nature of the protostellar infall and outflow. Far-IR high spatial and spectral resolution observations of protostars in Taurus will measure the accretion shock spectrum and provide the definite detection of infall and accretion in low mass protostars.

The high spectral resolution capability of SOFIA is also necessary for the determination of the composition and dynamics of interstellar and solar system gas. The ability to detect characteristic molecular lines, such as the high rotational lines of CO and rotational transitions in H_2, is often limited by a small line to continuum ratio, and so high resolving power dramatically improves the sensitivity of the line measurement.

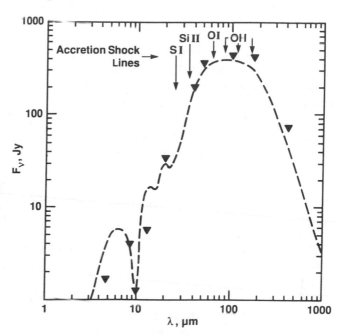

Fig. 5. The energy distribution of the possible protostar L1551 in the Taurus molecular cloud. The wavelengths of lines relevant to the study of accretion shocks are indicated.

Compositional studies range from mapping the elemental abundance variations in ionized and atomic gas in nearby galaxies, to determining the abundances of the major reservoirs of C, N, and O and the state of oxidation of the primitive solar nebula during the epoch of star formation through observations of comets. The unique capability of an airborne observatory to track occultations will allow absorption measurements of, for example, the structure of the atmosphere of Pluto, one of the few large planetesimals remaining in the solar system. An improved understanding of the primitive chemical processes of the outer solar system will translate directly into an improved perception of the processes at work in other solar systems and an understanding of how representative is our own solar system.

We recommend the Submillimeter Mission (SMMM), an innovative new mission for the Explorer program.

The Submillimeter Explorer utilizes a 2.5-m ambient temperature telescope with a liquid-helium cooled complement of instruments designed to obtain complete submillimeter spectra from 100 µm to 700 µm for a large number of galactic and extragalactic sources. SMMM will provide our first complete, unbiased spectroscopic view of the submillimeter portion of the electromagnetic spectrum.

Key elements of the SMMM technology program are well underway, and must be continued in the near term to be ready for a timely development phase in the mid-1990's. The areas of technology development that require support now are extending heterodyne receivers to wavelengths as short as 300 µm, and development of lightweight panels that can meet the high surface accuracies required for diffraction limited performance at 100 µm. The SMMM represents a substantial opportunity for a joint project with the French Space Agency, CNES, and also possibly with ESA. The SMMM could be pursued either as an Explorer class mission or an expanded moderate mission, depending on the results of the current Phase A studies, and the extent of the international collaborations that are finally negotiated by the interested parties.

SMMM SCIENCE HIGHLIGHTS

SMMM will probe the origins of stars and the chemistry of the interstellar medium, and will obtain a complete spectral atlas for molecules, atoms and ions in the 100 μm to 700 μm range in a wide variety of galactic and extragalactic sources. Complete submillimeter spectra of clouds of gas - ranging from quiescent atomic clouds, to dense cold clouds, to collapsing clouds, to those clumps containing protostars - will probe the critical chemistry, dynamics, heating, and cooling processes that occur in the gas and dust before and during gravitational collapse. Submillimeter spectra of nearby galaxies will provide the basis for global comparative studies of star formation, interstellar chemistry, and cooling processes.

These spectra will enable studies of chemical and isotopic abundances and hydride molecules for metals to atomic number of 30 or more, including the cosmologically significant species such as HD and LiH, and be able to examine their variation with position within the galaxy. SMMM will observe the carbon reservoir species CO and CI and the oxygen reservoir molecules H_2O and O_2 in a wide variety of gas phase environments, and identify dominant large molecules and small dust grains by observations of vibration-rotation spectra of large linear-chain and polyacetylene molecules and of vibrational modes of polycyclic-aromatic-hydrocarbon dust grains, thus examing the link between the lighter molecules observed at millimeter wavelengths and the small dust grains discovered at shorter infrared wavelengths.

In star formation regions, SMMM will enable thermal balance studies in dense protostellar enviroments and molecular shock regions where H_2O and hydride molecules dominante the cooling process. SMMM will enable studies of protostellar infall by observations of line absorption against continuum emission from dense cloud cores.

We recommend that the laboratory astrophysics program be substantially augmented in the 1990's.

Understanding the IR observations of the 1990's will require, as input, fundamental data describing the properties of atoms, ions, molecules, and dust grains. It is frequently the case that the largest uncertainties in theoretical models are due to uncertainties in these data, rather than to approximations inherent in the models. These essential data are provided in some cases by theoretical atomic, molecular, or condensed-matter physics, but predominantly by laboratory astrophysics. Laboratory astrophysics must be recognized as a special interdisciplinary area which requires significantly higher levels of support for the 1990s, with funding made available for laboratory start-up programs and for graduate students and postdoctoral fellows.

Atomic, ionic, and molecular spectral lines in the IR will offer a wealth of information concerning chemical and isotopic abundances and the physical conditions in the gas, but interpretation of this spectral data will require substantial progress in atomic and molecular astrophysics. Even for the most simple and fundamental molecule, H_2, there are great uncertainties regarding the cross sections for collisional excitation of the vibration-rotation levels from which we observe emission. Collisional cross section data for many isotopic species of stable molecules, as well as molecular spectroscopic data for highly reactive molecules and ions, are needed. There are currently scant laboratory data on highly vibrationally-excited states of molecules, radicals, and ions or on the larger organic molecules and ions which bridge the gap between the molecules observed in the radio and the polycyclic aromatic hydrocarbon (PAH) macro-molecules and small grains seen in the near-IR.

Most of the mid- to far-infrared radiation in the universe originates from interstellar dust grains. Both laboratory and theoretical work is needed on the physical properties of candidate grain materials, in order to interpret absorption, emission, and polarization measurements. The study of small grains, whose sizes (5 to 30 Å) put them in the transition region between large molecules and bulk grains, is especially important. For example, the optical properties of PAH clusters, and of hydrogenated amorphous carbon (HAC), require careful laboratory study in order to permit interpertation of the infrared emission features seen from interstellar dust clouds. Laboratory spectroscopy of mixed molecular ices of astrophysical interest is another area of high importance to the study of the origin and evolution of solid matter in interstellar clouds and in the solar system.

B. GROUND-BASED PROJECTS

We recommend the immediate construction of a national infrared-optimized 8-meter diameter telescope on Mauna Kea.

From this remarkably dry and stable site, by the use of modest adaptive optics techniques, IRO will exceed the sensitivity of a conventional 8-m telescope by more than an order of magnitude and will achieve diffraction limited performance of 0.07 arcsec at 2.2 μm. Within all atmospheric windows from 1 μm to 30 μm the telescope emission will add only minimally to that from the atmosphere itself, producing the most sensitive measurements from the ground at these wavelengths where sharpest imaging is naturally achieved.

The technical requirements for the IRO are well understood, and can be met by extension of the design for the NOAO 8-meter telescopes. The telescope must deliver a final image of less than 0.1 arcsec diameter. To minimize radiation from the telescope itself, the mirrors must be coated with a material of lower emissivity than aluminum, probably silver. Some adaptive optics capability will be routinely required, but even simple wavefront tilt compensation will allow imaging at a resolution of 0.1 arcsec. The choice of site is crucial: Mauna Kea is recognized as the best IR site in the world.

We note also that, as recommended earlier, the NASA IRTF, the dedicated 3-meter infrared telescope located on Mauna Kea, is the logical location at which to take the earliest advantage of infrared optimization techniques.

IRO SCIENCE HIGHLIGHTS

IRO will provide a level of clarity in imaging never before achieved from the ground, with angular resolution in the near IR an order of magnitude sharper than typically obtained at optical wavelengths. Infrared images of planets, satellites, comets and asteroids will reveal the composition and structure of surface and atmosphere, and can be used to monitor temperature variations or other changes, for example the volcanic activity on Io.

The IRO will enable detailed imaging and spectroscopic observations of forming stars in many nearby star formation regions, such as Taurus, Ophiuchus, Orion and NGC2264. Studies of the structure, energetics and composition of protoplanetary disks around young stars, with a spatial resolution of 10 AU in the nearest star-forming regions, will allow detailed characterizations of the disks and may show condensations or voids where planets are in the process of forming.

IRO will measure three dimensional motions for hundreds of stars within the central parsec of our Galaxy through a combination of proper motion and spectroscopic studies to an accuracy of 10 km/sec, and so provide a critical test for the presence of a black hole. IRO may even be able to measure directly the velocity of the gas rotating around such a massive object or to detect emission from a black hole accretion disk.

Perhaps IRO's greatest contributions will arise from its abilities to study distant galaxies as they first form stars in the early Universe. Models indicate that young galaxies at redshifts exceeding 5 should have 2.2 μm magnitudes in the range 22 to 24, and so be readily accessible to IRO; the deepest 2.2 μm image currently available (with a 5-sigma threshold of 21.5 mag) already contain galaxies not detected at optical wavelengths (figure 6). IRO can obtain detailed images of such galaxies, and measure spectra of them and other extremely distant examples discovered by SIRTF, in order to determine their morphology, redshift, composition, and ionization state.

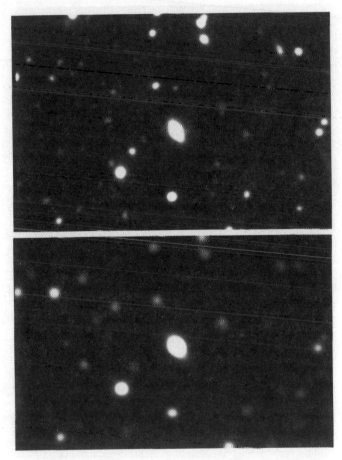

Fig. 6. A deep 2.2 μm image (lower panel), with a 5 sigma threshold 21.5 mag, reveals faint red galaxies not detected in the deep I-band image (upper panel). Courtesy of L. Cowie, Univ. of Hawaii.

We recommend a strong program to develop and evaluate IR array detectors and to deploy the best of these arrays in state-of-the-art instruments for existing telescopes.

Infrared array technology in the wavelength range from 1 μm to 30 μm is evolving very rapidly, and substantial development and evaluation work is needed to exploit this technological revolution. The technology involved has been largely developed as a result of the high interest in 1 μm to 30 μm IR arrays by the Department of Defense. NASA has carefully nurtured detector development activities both for SIRTF and for NICMOS, the infrared instrument on HST. It is extremely important to continue the testing and understanding of the properties of these arrays in order to maximize the scientific return from HST and SIRTF.

Currently, high quantum efficiency, low read noise, low dark current arrays are available in formats up to 256x256 in the wavelength range 1 μm to 5 μm. Of critical concern in this wavelength range is to extend the format to 512x512 pixels; an array of this size is at the edge of current technology evolution, and is feasible within 5 years. Arrays of comparable quality for the thermal infrared from 5 μm to 30 μm in formats of 128x128 or even 256x256 pixels are also within reach. Of particular concern for ground-based and airborne observations is the need for high speed, low-noise readouts for large format arrays working beyond 3 μm, where the thermal photon background from the atmosphere and ambient temperature telescopes becomes very high.

The heart of any observatory is its complement of instruments. Those that utilize the finest state-of-the-art detectors maximize the scientific return. Ground-based telescopes can be reborn, increasing observing efficiencies by orders of magnitude, with the introduction of new array based cameras and spectrometers. **We urge the NSF to assist observatories in procuring and deploying the best of the infrared arrays** for new generations of focal plane instruments. A particularly exciting opportunity enabled by modern IR arrays is the ability to systematically survey substantial areas of the sky with high sensitivity and efficiency. One particular survey, an all-sky broadband near IR survey, is highlighted in this report. There are other, more specific, surveys that are also of high scientific interest; for example deep images of large or unusual galaxies, galactic surveys in the H_2 lines around 2 μm or in the Brackett gamma line of HI at 2.16 μm, and very deep multicolor IR surveys for the faintest extragalactic objects.

Another compelling opportunity for the 1990's is the application of adaptive optics to ground based telescopes to correct continuously for the effects of atmospheric turbulence. Existing telescopes equipped with adaptive optics can address new questions while new large telescopes can be utilized to obtain images in the 1 to 5um range with unprecedented clarity. Adaptive optics has been substantially developed within the defense community and NSF has taken the lead in providing the astronomical community access to this technology. The complexity and cost of adaptive optics is much reduced at IR wavelengths because of reduced bandwidth requirements, relaxed reference star requirements and increased correlations sizes at the longer wavelengths. Even a modest wavefront tilt correction can provide substantial image size improvements. The IRTF is particularly well suited for early implementation and utilization of this technology.

We recommend the immediate initiation of a 2 μm all sky survey (2MASS) to a level of sensitivity 50,000 times greater than that achieved by the 1969 Two Micron Sky Survey.

2MASS is a prime example of a modest project with a very large long term payoff. A pair of dedicated one meter class ground-based telescopes, one for each hemisphere, equipped with modern near-IR array detectors, can completely survey the sky at three wavelengths between 1 μm and 2.2 μm in less than two years, detecting an estimated 100 million sources. The 2MASS survey will explore the large-scale structure of the local Universe by mapping the distribution of galaxies to a distance of 100 Mpc over the whole sky. This survey will be relatively unaffected by dust obscuration in our own and other galaxies and is uniquely sensitive to those classes of stars which dominate the mass.

The survey will explore large scale stellar structure of the Milky Way Galaxy and address basic stellar evolution questions by measuring luminous evolved stars throughout the Milky Way, in the Magellanic Clouds, and other galaxies in the Local Group. 2MASS will dramatically expand our current census of the coolest stars and probe the young stellar population within dozens of dense molecular clouds. It will provide basic support to a variety of NASA missions of the 1990's by highlighting new questions and identifying new targets for the major IR missions, and supporting X-ray surveys through identification of x-ray sources with reddened stars, cool dwarfs, and AGN's.

Studies to define the survey hardware and strategy are currently underway. Data processing and survey product generation utilize NASA's Infrared Data Analysis Center (IPAC). Project development could begin as soon as 1991.

C. PERFORMANCE COMPARISON

Together SIRTF, SOFIA and IRO provide the astronomical community with extremely powerful capabilities over the entire infrared spectral regime. In this section, we attempt to quantify and compare the performance expected for these three facilities. Sensitivity estimates are 1 sigma in 500 sec.

Imaging and Photometry (Figure 7)

Over the entire 3 μm to 700 μm range SIRTF provides uniquely powerful and sensitive imaging and photometry capability, orders of magnitude more sensitive than any other facility, existing or proposed. SIRTF, together with the NICMOS instrument on HST provides superb imaging sensitivity from 1 μm to 700 μm. The 8-m IRO offers more than an order of magnitude improvement in point source sensitivity over the current IRTF capability. SOFIA offers nearly an order of magnitude improvement over the KAO, and access to any of the IRAS PSC objects for detailed study. For studies of emission extended on a scale large compared to the angular resolution, SIRTF is some three orders of magnitude more sensitive than any warm telescope.

Angular Resolution (Figure 8)

SIRTF provides diffraction limited angular resolution beyond 3 μm. Because of their larger diameter telescopes, SOFIA and the 8-m IRO offer better diffraction limited capability. IRO provides the highest resolution imaging capability in the near IR windows of any ground-based optical or IR telescope, while SOFIA offers the highest available resolution in the IR bands unaccessible from the ground.

Spectral Line Sensitivity (Figure 9)

The performance is illustrated for a low spectral resolution of 100 and high resolution of 10^5. In the 3 μm to 200 μm region, SIRTF provides the highest sensitivity to broad spectral lines, for example those expected for extra-galactic sources. In the 1 μm to 30 μm windows, the 8-m IRO offers unsurpassed spectroscopic sensitivity for narrow lines ($\Delta v \sim$ 1km/sec) and very powerful capability for observing broad spectral lines in the near-IR. SOFIA offers flexible spectroscopic capability throughout the IR, particularly important at the longer IR wavelengths. SOFIA and the IRO can support the development of state-of-the-art spectrometers throughout their long lifetime. Again, it is assumed that the line source is spatially unresolved. For extended sources SIRTF offers very substantial additional advantages.

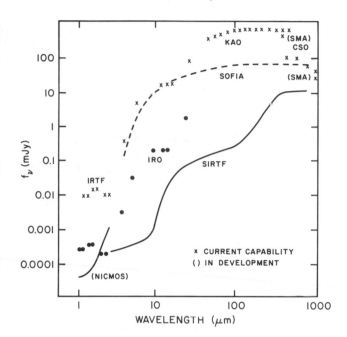

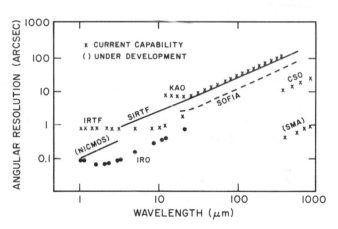

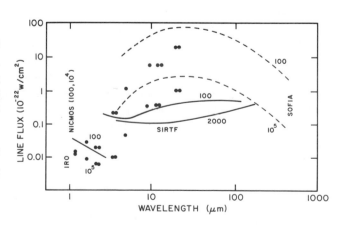

VI. FUTURE DIRECTIONS

This entire report is a testament to the huge rewards that follow directly from the deployment of optimized telescopes, from their equipment with state-of-the-art instruments, and from their location at sites with superior environmental properties. The way to the future must pursue those same strategies to ever higher levels of implementation.

Some long-term implications of this tripartite prescription are not difficult to deduce. On the ground, steps through modest pilot projects aimed at the eventual construction of IR imaging interferometers consisting of multiple telescopes; continued development and deployment of larger format array detectors; investigation of the possible use of Antarctica as an observing site. In space, larger single dishes and Lunar interferometers; longer lifetime refrigeration of larger format detectors; an orderly progression from low earth orbit, to high earth orbit, the Moon, and eventually to observatories located beyond the zodiacal cloud -- as distant from the Sun as Jupiter.

TABLE 2. FUTURE TECHNOLOGY RECOMMENDATIONS

PROJECT	PRIORITY	COST (FY89$)
SPACE TECHNOLOGY		
FAR-IR/SUBMM PROGRAM	1	75M
GROUND-BASED TECHNOLOGY		
O/IR INTERFEROMETRY PROGRAM	1	20M

We recommend a moderate scale program of ground based IR interferometry development and demonstration for the 1990's.

This program would support current technology development activities, build at least one interferometric imaging array of 3x 2 meter telescopes adequate to undertake imaging interferometry in the near infrared and to evaluate the practical limits in the thermal infrared out to 20 μm, and establish the technical basis for planning a Very Large Optical/Infrared Array for the first decade of the 2000's

The next major development in 1 μm to 30 μm, ground-based astronomy is very likely to be imaging interferometry. IR imaging interferometry will provide orders of magnitude improved spatial resolution, and qualitatively new kinds of information from very high angular resolution imaging of forming stars and planetary systems, starburst galaxies, and active galaxy nuclei. Interferometric imaging will also allow progress on study of circumstellar environments (YSO's, evolved stars, novae, nebulae) with unprecedented resolution.

The technical basis for optical/IR interferometry is well developed, with such components as delay lines, correlators, fringe tracking servos, and active optics for rapid tilt and piston correction in use at the Mt. Wilson Mk III optical astrometric interferometer and the Berkeley Infrared Spatial Interferometer together with several smaller technology development projects underway in the U.S. The technical objectives of the decade program include construction, metrology, and control of small and medium aperture telescopes to required tolerances and optimization of correlation techniques as well as site testing for correlation time, isoplanatic angle, and interferometer specific parameters. The experience of radio astronomers in aperture synthesis carries directly over to optical/IR, with the same software in use for both wavelength regimes.

The recommended technology demonstration projects of the 1990's will offer milliarcsec or better imaging capability with limiting magnitudes for 2-m apertures and existing detector arrays of about 15 mag at 2.2 μm and 5.5 mag at 10 μm. The major ground based facility for the first decade of the 2000's, the Very Large Optical/Infrared Array, will offer full aperture synthesis imagery of bright and faint sources.

This program should be carried out largely in the university community, where intensive participation by students will be possible. At least some of the projects should involve several groups with a range of research and technical interests. The large facility of the 2000's will be a unique national or international resource, and should offer the access and support expected of a national facility.

All of the projects proposed for the decade in this panel report hold promise for exceptional scientific returns. To reach some dreams, however, such as imaging the accretion disk around a black hole, or viewing an earth-like planet within a few AU of another star, we must obtain a spatial resolution significantly better than a milliarcsecond for very faint sources. This cannot be done with ground-based interferometry in the optical and infrared. The only location with adequate thermal, mechanical and image stability for such an interferometer is the lunar surface. The individual telescopes that are part of such an array need not be large, and so the construction of an optical/infrared interferometer is feasible as an early project when humans return to the Moon. Demonstration of the crucial technologies for interferometry at these wavelengths is the essential first step which must be undertaken on the ground during the coming decade if astronomy is to be poised to act quickly when the opportunity arises to construct a lunar-based interferometer.

We recommend that the NASA program for phased exploitation of the far-ir/submillimeter wavelength range be vigorously pursued.

The Decade review report of the 1980s included a major 30 μm to 1000 μm project, the Large Deployable Reflector (LDR), designed to complement HST, large ground-based telescopes, and SIRTF. LDR was envisaged as a 10-m class, ambient temperature telescope to be erected in space from the Space Shuttle and operated in low earth orbit. It is now generally agreed that low earth orbit is undesirable for such a facility, that the technology is not ready for this mission, and that another approach is required. The mature perspective which led SIRTF to high earth orbit guides us to suggest that the correct path for submm astronomy in space will evolve via the 0.5m diameter Small Explorer mission SWAS, already in development, through a 2.5m class Explorer mission (the submillimeter mission SMMM), to a major project such as a 10-m to 20-m diameter telescope operated in high earth orbit or a Lunar far IR/submm interferometer.

In order to accomplish this plan, support is needed in the form of a technology development program aimed initially at bolometers and photoconductor arrays and heterodyne receivers in the far infrared, and at telescope panels required for the early missions. Large photoconductor detector arrays for the wavelength range 30 μm to 200 μm are necessary for both imaging and spectroscopic applications. Sensitive, low temperature arrays of bolometers should be developed for imaging continuum sources at wavelengths beyond 200 μm. Submm heterodyne receivers are several orders of magnitude away from theoretical performance limits. High frequency mixers, local oscillators and ultra-high bandwidth IF frequency components all require further development. Once theoretical noise limits are reached, receivers should be combined to form focal plane arrays. Large, lightweight panel structures capable of maintaining diffraction limited performance at wavelengths shorter than 100 μm are needed.

It is essential that the university community be heavily involved in this technology program, and that the developments have applications in near term missions. Continued research and development that capitalizes on developments in industry and university laboratories is critical to making the most of the space environment. In this context it is important that laboratory devices be incorporated into instruments that can be used on ground-based and airborne telescopes. Such instruments return valuable science, fill gaps between major missions, and uncover problems at a phase of development when the cost of fixing problems is modest.

Subsequently the technology program should be directed toward developing increasingly long lived cryogenic coolers for space, and assembly and control of larger or multiple structures in high earth orbit or on the Moon.

The fundamental performance limitation for any submillimeter space observatory is the size of the diffraction limited beam. Achieving spatial resolutions comparable to those obtainable at near infrared wavelengths with IRO would require a large filled aperture or multi-element interferometer with size or spacing exceeding 100 meters. The only site with the necessary stability for such a telescope is the lunar surface, on which there is no obvious limitation to the obtainable size. Cost limitations are set primarily by the mass of materials to be transported. The ultimate potential of submillimeter astronomy from the Moon is so great that technology development in the coming decade for detectors, relevant cryogenics, and lightweight telescope components is a sound investment to prepare the technology base for a lunar observatory.

VII. INFRASTRUCTURE

The 1990's will witness a tremendous leap forward in our knowledge of the infrared sky, a sky rich with information. We will be able to convert these observations into understanding if, and only if, a strong infrastructure is in place and well supported.

Theoretical Astrophysics

To a considerable extent, interpretation of astronomical obserations involves comparison between the observations and predictions of theoretical models -- it is in this way that we infer the conditions present in remote regions of space and refine our physical understanding. The forthcoming explosion in IR imaging and spectroscopy will demand greater sophistication in the theoretical models. Theoretical progress is required on a number of overlapping fronts:

Improvements in our understanding of the local "microphysics" involved in heating, cooling, ionization, and chemistry in regions ranging from cool stellar atmospheres to diffuse interstellar plasmas. It is becoming increasingly important to develop the capability to compute realistic emission/absorption spectra, including detailed non-LTE excitation of many-leveled species such as H_2 and H_2O, and more detailed, but still approximate, treatments of larger species such as PAHs.

Accelerated work on the nature and formation/destruction of interstellar dust, and the important effects which dust grains exert on the interstellar medium, as well as improved understanding of the interaction of gas and radiation including the structure of photodissociation fronts, ionization fronts and X-ray heated regions.

Fluid dynamic modeling of the flows which occur in, for example: MHD shock waves; "turbulence" in molecular clouds; gravitational collapse of rotating, magnetized clumps in molecular clouds; accretion disks around protostars; outflows associated with star formation; and ionization-shock fronts formed when neutral clouds are exposed to ionizing radiation from newly-formed OB stars.

Global models that study the interplay between stars, gas, photons, and gravity which determines the structure of the interstellar medium in a galaxy, the rate of star formation, the initial mass function, conditions in starburst, interacting and merging gas-rich galaxies, AGNs, and QSOs.

It is imperative that adequate funding be provided to support a vigorous theoretical program.

Training and nurture of experimentalists

There is a grave risk that a program devoid of opportunities for young scientists at various stages of their careers will ultimately lack the senior people required to carry out major programs as they arise. Challenging opportunities providing "hands-on" experience are vital. Within infrared astronomy, both ground-based and airborne/spaceborne programs are required; the former can be provided through the support of a broad based instrument program and the latter through support of SOFIA and the balloon program. SOFIA will support approximately 40 investigator teams per year, including an on-going instrument development program. Small university groups will be able to carry out complete investigations on a short time scale at low cost. Balloon-borne telescopes will be an essential component of the next generation of CMB anisotropy experiments at near-mm wavelengths. In principle, maps of the CMB anisotropy with $\Delta T/T \sim 10^{-6}$ can be made in a single balloon flight. SOFIA, the ground-based instrument development program, and the balloon program offer excellent test-beds for new detectors and receivers, and an excellent training ground for young scientists.

Data Analysis and Archiving

The creation of an entire branch devoted to Mission Operations and Data Analysis demonstrates a commitment within NASA to a data reduction and analysis effort commensurate in scope with that of the project which generates the data. This level of support should become the standard for all of astronomy.

A strong archival data analysis program allows a large, broadly distributed community to participate in forefront astronomical research. An excellent model is provided by the Infrared Processing and Analysis Center (IPAC), which was established to process the data from the IRAS mission, generate and verify the scientific products and to support the community in the utilization and analysis of the data products. IPAC has been very successful in supporting broad community participation in the analysis of the rich IRAS data base. The IRAS database will continue to be a vital resource for the astronomical community. In the future, the COBE, ISO and SIRTF data should be available for archival interdisciplinary study. Large, extrremely well characterized data bases from selected ground-based observations, e.g. survey data, should also be considered for inclusion in the archival program.

OPTICAL/IR FROM GROUND PANEL

STEPHEN STROM, University of Massachusetts, Amherst, *Chair*
WALLACE L.W. SARGENT, California Institute of Technology, *Vice-Chair*
SIDNEY WOLFF, National Optical Astronomy Observatories, *Vice-Chair*

MICHAEL F. A'HEARN, University of Maryland
J. ROGER ANGEL, University of Arizona
STEVEN V.W. BECKWITH, Cornell University
BRUCE W. CARNEY, University of North Carolina, Chapel Hill
PETER S. CONTI, University of Colorado, Boulder
SUZAN EDWARDS, Smith College
GARY GRASDALEN, University of Wyoming
JAMES E. GUNN, Princeton University
JOHN P. HUCHRA, Harvard-Smithsonian Center for Astrophysics
ROBERTA M. HUMPHREYS, University of Minnesota, Minneapolis
DAVID L. LAMBERT, University of Texas, Austin
BRUCE W. LITES, National Center for Atmospheric Research
FRANK J. LOW, University of Arizona
DAVID G. MONET, U.S. Naval Observatory
JEREMY R. MOULD, California Institute of Technology
S. ERIC PERSSON, Mt. Wilson and Las Campanas Observatories
PETER ALBERT STRITTMATTER, University of Arizona
ALAN T. TOKUNAGA, University of Hawaii
DONALD C. WELLS, National Radio Astronomy Observatory
MICHAEL W. WERNER, Jet Propulsion Laboratory

JOHN McGRAW, University of Arizona, *Consultant*

Optical/IR From Ground

EXECUTIVE SUMMARY

The 1980s witnessed major advances in detector and computer technology, fabrication and polishing of large, lightweight optical elements, and telescope design. In combination, these advances will enable construction during the 1990s of a new generation of large (8-m and 4-m diameter) optical/infrared telescopes. These telescopes will provide an order of magnitude increase in angular resolution, and up to a two order of magnitude gain in sensitivity when compared with current generation telescopes, which are limited by the quality of their optics and by atmospheric turbulence or "seeing". By using actively controlled optical elements to compensate for seeing, and by improving the figure on all mirror surfaces, it will be possible to concentrate the light from unresolved astronomical sources into a diffraction-limited core, and to provide significant gains in image quality for resolved sources. These new generation telescopes must be seen as the harbingers of optical/infrared facilities which by the latter half of the 1990s promise to provide yet another order of magnitude increase in angular resolution.

Thus, for the first time, optical/infrared facilities will have the combination of sensitivity and angular resolution required to observe and analyze
- large samples of galaxies and clusters of galaxies at redshifts $z \geq 1$;
- galaxies at the epoch of their formation;
- protoplanetary disks surrounding young, solar-type stars;

and to provide thereby the observational basis for understanding the *origin* of large-scale structure in the early universe, of galaxies, and of planetary systems similar to our own.

Because these are problems which capture the imagination of both scientists and citizens who share an innate curiosity about our origins and place in the universe, it is hardly surprising that astronomers throughout the world are striving to take advantage of these technological possibilities: our colleagues in Europe and Japan have already been funded to build large O/IR facilities of modern design. For the first time, US leadership in ground-based astronomy–a constant of international science since the beginning of the 20th century–is being challenged. If US astronomy is to maintain its vitality and leadership, it is essential that a new generation of optical/infrared facilities be built during the next decade. We recommend the following program to maintain US competitiveness in O/IR astronomy during the 1990s, and to position the US to establish leadership in high angular resolution O/IR astronomy at the beginning of the next decade.

Large Scale Programs

Priority 1: *A coordinated program to combine federal funds (*$185M) with state and private funds to build and instrument large (8-m class) US ground-based telescopes. The key federal contributions are:
- support for construction of *a pair* of nationally accessible 8-m telescopes, one each in the northern and southern hemisphere. The northern hemisphere telescope will be a *uniquely powerful instrument*, optimized for performance at infrared wavelengths by providing diffraction-limited images (of angular size $\sim 0.05''$ at $\lambda \geq 1.6\mu m$ and minimizing the background thermal emission from the telescope. This unique facility must be built on the best site in the world – Mauna Kea, on the island of Hawaii. Its twin will provide the US astronomical community with the access to the southern celestial hemisphere essential in the era of the NASA Great Observatories, and can be optimized for performance at optical wavelengths.
- support for developing and building the advanced auxiliary instruments required for the new generation of large telescopes – including the recommended pair of national 8-m telescopes, and other US telescopes of comparable aperture which are under construction or in advanced planning stages.
- support to develop and deploy wavefront sensors and adaptive mirrors capable of providing diffraction-limited imaging at near-infrared wavelengths for the new generation of large telescopes.

Medium Scale Programs

Priority 1: *A coordinated program to develop high angular resolution O/IR astronomy which includes federal investment (*$50M) in order to:
- extend the wavelength range (from infrared to optical wavelengths) for full adaptive correction of atmospheric distortion above large telescopes;
- support engineering efforts to design an ultra-large ($D \geq 30$-m), adaptively-corrected single-aperture telescope;
- support university-based efforts to build and operate pilot interferometers;
- support a university-based effort to develop a sensitive O/IR interferometer array by mid-decade;
- support engineering studies leading to the design of a large, national O/IR array capable of imaging celestial objects with at least ten times the angular resolution of 8-m class telescopes and at comparable or greater sensitivity.

Priority 2: *A program to construct at least four, new-generation 4-m class telescopes.*
These telescopes will provide the basic tools necessary to carry out a wide variety of large-scale surveys, innovative observational programs, and basic research aimed at deeper understanding of known astrophysical phenomena. It is highly desirable that these telescopes be constructed by individual universities and university consortia. The most immediate community need is for two new generation 4-m class telescopes (one in each hemisphere) to support and complement the Great Observatories. Construction of four new-generation telescopes will require a combination of federal ($30M), state and private funds.

Small Scale Programs

Priority 1: *A program to carry out near-infrared and optical all-sky surveys with digital arrays (*$11M)
These surveys will produce complete, unbiased infrared and optical maps of both celestial hemispheres, and will thus provide an essential database for planning space-based missions and major observing programs on large, ground-based telescopes.

Priority 2: *A program to develop a National Astrometric Facility (*$5M)
The NAF will provide the ability to obtain ultra-precise (better than $0.001''$) positions of celestial objects both within and outside the solar system.

Infrastructure Support

Priority 1: *A program to fund the development, purchase and distribution of large format optical and infrared detectors (*$40M)

The ready availability of advanced panoramic detectors is absolutely essential to the development of sensitive instruments for the new generation of large telescopes, and thus to achieving the full power and potential of these major facilities.

Priority 2: *A program to fund the fabrication and polishing of large mirrors, specialized optics, and the development of mirror coating techniques* ($25M).

The availability of large ($\geq$ 4 m diameter) accurately polished lightweight mirrors, correcting optics, and mirror coatings underlies our ability to build new generation telescopes, and the auxiliary instruments required to enable their most efficient use.

OPTICAL/INFRARED ASTRONOMY IN THE 1990S

Research Environment

The advances in detector and computer technology of the 1970s and 1980s, combined with our ability to observe the universe throughout the electromagnetic spectrum by using powerful space- and ground-based telescopes, have revealed new phenomena (e.g. evidence of gravitational lenses; x-ray emission from clusters of galaxies and globular clusters; gamma-ray bursters; infrared luminous starburst galaxies; energetic winds emanating from obscured protostars), and heretofore unknown structures (e.g. the tapestry-like structure of the universe as revealed by 3-dimensional maps of galaxies; concentrations of unseen matter which locally alter the expansion of the universe; collimated jets of plasma driven by young stars; infrared luminous protoplanetary and perhaps post-planet building disks) and have forced astronomers to view the cosmos from new perspectives. The drive to understand these phenomena, and to place them in the context of current paradigms if possible (or to create new ones if necessary), motivates the development of ever more powerful experimental techniques.

The 1990s promise to be one of the most exciting decades in the history of astronomy. During the next ten years, NASA will launch the Gamma-Ray Observatory (GRO), the Advanced X-Ray Astrophysics Facility (AXAF), and the Space Infrared Telescope Facility (SIRTF). Together with the Hubble Space Telescope (HST) launched in April, 1990 these "Great Observatories" will provide revolutionary tools for exploring the universe at wavelengths throughout the electromagnetic spectrum.

As astronomy enters this new era, *ground-based optical/infrared facilities will play a central role.* Large ground-based telescopes are required to provide the sensitivity needed for spectroscopic analysis of phenomena discovered with the Great Observatories. Extensive observations with moderate-size and smaller survey telescopes are essential to (1) selecting targets for the Great Observatories; and (2) providing the context for understanding objects and phenomena discovered at other wavelengths (identification of gamma- and x-ray sources, for example).

Ground-based optical/infrared telescopes also promise to be powerful engines of discovery in their own right. Dramatic advances in detector technology, telescope design, and computer controlled optical elements during the 1980s, provide the technical basis (see below) for constructing a *new generation* of large ground-based optical/infrared telescopes and auxiliary instruments capable of providing 10 times the angular resolution and up to 100 times the sensitivity of currently available ground-based telescopes.

These same technological advances, combined with the results of innovative experiments carried out in the US and in Europe, promise that by the beginning of the next century, it will be possible to build interferometric arrays – the optical/infrared analogs to the cm-wavelength Very Large Array (VLA) and the proposed millimeter array (MMA). Such optical/infrared arrays will provide images with an angular resolution 500 times better than the finest images taken on the ground in 1980 and 10 times the quality of the images expected from HST.

Science Opportunities

The anticipated gains in angular resolution and sensitivity from new generation large telescopes and interferometers provide immense opportunities for new scientific discoveries and for achieving deep understanding of phenomena which lack ready explanation within current paradigms. Outlining the "science opportunities" provided by devices which will provide order of magnitude improvements in *two* parameters of "discovery space" –angular resolution and sensitivity– is a challenging exercise, depending as it must on extrapolating from the frontiers of knowledge reached only a few years ago. We have chosen to meet that challenge by *illustrating* the potential of these new tools for enabling major advances in understanding three areas – the origin of structure in the universe, the origin and evolution of galaxies, and the origin of stars and planetary systems.

The Origin of Structure in the Universe

Prior to 1980, observational cosmology could be defined as the search for two numbers: H_o and q_o, the expansion rate of the universe and its rate of change. Even though exact values for these two numbers have remained elusive, cosmology in the 1980s came to include the formation and evolution of large scale structures in the universe, and the origin and distribution of massive aggregates of galaxies and matter which (locally) distort the uniform expansion of the universe (the Hubble flow).

Advanced detectors enabled the determination of redshifts (recession velocities expressed in units of the speed of light) for large samples of relatively nearby galaxies and, in combination with extant imaging surveys provided the first *three-dimensional* (distance from earth, in addition to the projected location on the celestial sphere) maps of the distribution of galaxies in local universe. The result was both startling and unexpected: galaxies are not distributed randomly in space, but appear to lie along thin, web-like structures separated by great voids. Over what scales do these web-like structures persist? At present we don't know – because despite ten years of concerted effort we have been able to provide a 3-dimensional map covering only 0.01 percent of the potentially observable universe.

The 1980s also witnessed the development of new techniques to measure the relative distances to galaxies – for example the relations between velocity dispersion or rotation velocity and a galaxy's integrated luminosity or size. Plots of galaxy recession velocity (obtained from spectroscopic measurement of doppler velocity) against the distances derived from the these new indicators, revealed local changes in the expansion rate of nearby galaxies which astronomers attribute to a massive aggregate known as the "Great Attractor." Whether the "Great Attractor" is unique, or representative of a large number of such aggregates is unknown, because at present we have accurate relative distances to fewer than 1000 galaxies. Accurate measurements of distortions in the Hubble expansion rate are needed to determine the gravity field and match it to the galaxy distribution – to test for the existence of "dark" (unseen) matter in the universe, and if found, to measure its quantity and location.

During the 1990s, the construction of new generation 4-m and 8-m class telescopes will enable great strides in understanding the structure and distribution of matter within the universe. Three fundamental goals for observational cosmology in the 1990s are:

- to map the 3-dimensional distribution of "nearby" galaxies to redshifts z ~0.1. This program will require measurements of redshifts for 1 million galaxies selected from deep, uniform, all sky surveys. Because these systems will be 5 to 10 times fainter than those surveyed to date, it will be necessary to carry out the redshift measurements with new generation 4-m class telescopes equipped with multi-object spectrographs capable of simultaneous observations of hundreds of galaxies over a $2°$ field. Even with such powerful facilities, it will require ~6 years per celestial hemisphere to observe the ~ 1 million galaxies necessary to complete this redshift survey.

- to map the gravity field – deduced from distortions in the uniform Hubble expansion – out to a redshift z ~0.03. To carry out this program requires accurate relative distances to approximately 10,000 galaxies. Distances can be obtained with accurate photometry, made available from deep all-sky surveys, and measurements of galaxy velocity dispersions or rotational velocities. To make these latter measurements requires approximately 1 hour per galaxy on a new generation 4-m class telescope. Because these nearby galaxies are relatively isolated, simultaneous measurements with multi-object spectrographs cannot provide major increases in observing efficiency. The basic program will thus require more than 5 years to complete.

- to study the evolution of large-scale structure in the universe by mapping the distribution of galaxies and clusters of galaxies at $z \geq 1$, and comparing structures in the young, distant universe with "old", nearby structures. This requires that we complement our map of the nearby ($z \leq 0.1$) galaxy distribution, with redshift determinations for galaxies at redshifts near and beyond unity. Appropriate samples of distant galaxies must be assembled from small area, deep imaging surveys with 4-meter class telescopes, and x-ray (ROSAT and AXAF) surveys for galaxy *clusters*. To sample a region comparable to the 1 million galaxy "local" survey will require redshifts for $\sim$100,000 distant ($z \geq 1$) galaxies to a magnitude limit of B=24 in a 100 square degree area. Obtaining redshifts of this large a sample of faint galaxies will be *enabled* by new generation 8-m class telescopes. Even so, this project will require approximately *5 years* on an 8-m class telescope equipped with a multi-object spectrograph capable of observing 100 galaxies simultaneously. Such a survey would, however, be of profound importance: it would allow detailed comparison of the topology of structures in the universe now, and at an epoch when the universe was less than 30 percent of its current age.

The Origin and Evolution of Galaxies

During the 1980s, astronomers carried out challenging spectroscopic observations which led to the discovery of the most distant systems known in the universe: quasars with redshifts $z \geq 4$, and radio galaxies with $z \geq 3$. These observations provide evidence that some, perhaps the majority of galaxies must be assembled and must begin forming stars at $z \geq 5$, and thus provide strong motivation to extend current searches to higher redshifts (earlier epochs in the life of the universe) in order to locate galaxies just taking form from protogalactic clouds. Identifying the epoch of galaxy formation, defining the structure of young galaxies, and characterizing star formation and chemical element production early in the lifetime of these systems, represent fundamental steps toward understanding the origin and evolution of galaxies.

At $z \geq 5$, galaxies will be faint (K $\geq$ 22 mag), their angular size will be small (3″ to 10″), and light from the visible region of the spectrum will be shifted to the 2 μm window. Comparison of the structure and chemical composition of these distant young systems with that of more evolved systems at lower redshift – and thus charting their early evolution – will require sensitive, high angular resolution imaging and spectroscopy in the infrared. It is the infrared which carries fundamental information regarding the underlying structure of the galaxy through its sensitivity to light emanating from intermediate and low mass stars—the likely dominant constituents of these systems, optical measurements will likely be overwhelmed by (restframe) ultraviolet light emerging from complexes of newly-born massive stars. Infrared spectroscopy of red-shifted emission features such as [O III] and Hβ will provide the basis for determining chemical composition and estimating the vigor of star-forming events.

New generation 8-m telescopes will *enable* the study of forming galaxies:

- by locating candidate forming galaxies by means of deep optical and infrared images which measure galaxy color, emission line strength, and morphology (based on high angular resolution images). With current generation 4-m class telescopes, deep imaging of galaxies with K $\sim$ 21 mag requires 12 hours of integration. Surveys to the even fainter limiting fluxes required to locate galaxies at $z \geq 5$ must be carried out with 8-m telescopes. Because these faint galaxies must be viewed against the background emission from the telescope, it is also essential that one or more of these telescopes be designed for low emissivity operation in the infrared. Furthermore, the likelihood that forming galaxies will exhibit small-scale structure (nuclear starburst regions; giant H II complexes) places great premium on high angular resolution in order to achieve maximum sensitivity in imaging forming galaxies at high redshifts, and in characterizing their morphology.

- through their ability to measure redshifts for these extraordinarily faint systems. Redshift determinations for systems at $z \geq 3$ at optical wavelengths require hours of time on current generation 4-m class telescopes, and are not yet possible in the near infrared even on the largest available telescopes. The new generation of 8-m telescopes will provide the increase in sensitivity necessary to *enable* redshift measurements and spectroscopic analyses of systems beyond $z = 3$, and the power necessary to observe the large samples of forming galaxies over the redshift range $z \geq 5$ to $z \sim 2$ required to understand the galaxy birth process and the early evolution of these systems.

The Origin of Stars and Planetary Systems

Before 1980, it was known that stars form in cold, dark aggregates of matter known as "molecular clouds." Studies of the stellar populations of young stars just emerging from stellar wombs had provided astronomers with a rudimentary picture of how these objects evolve prior to igniting hydrogen in their cores and becoming stable, main sequence stars. During the 1980s, sensitive mm-wave, optical and infrared measurements from the ground, and the launching of the IRAS satellite provided a number of profound surprises and revolutionized our understanding of the star formation process:

- mm-line observations of molecular clouds revealed that stellar birth is a violent process, accompanied by energetic outflows, sometimes mapped by highly collimated jets of hot plasma;

- IRAS observations at infrared wavelengths revealed emission from extraordinarily young stars, still being assembled from material contained within dense, opaque protostellar cores which obscure the birth process at optical wavelengths;

- IRAS and ground-based infrared images of molecular clouds revealed that some clouds seem to produce new stars quickly and with high efficiency, while others form stars more slowly and convert only a small fraction of their store of molecular material into stars;

- Ground-based, IRAS, and mm-continuum photometric measurements, along with high resolution spectroscopic observations made it plain that a large fraction of solar-type stars are surrounded by disks of solar-system dimension and of mass comparable to or greater than the mass of material out of which our solar system is thought to have formed;

During the 1990s, a variety of new and powerful instruments – HST, ISO, SIRTF, SOFIA , single dish mm- and sub-mm wave telescopes, and mm interferometers – will enable qualitatively new kinds of measurements which should effect even more dramatic transformations in our understanding of star and planet formation.

Ground-based O/IR telescopes of diameter $\sim$8-m will play a central role in our quest to understand stellar and planetary birth by virtue of their potential to provide both diffraction-limited images from 1.6 μm to 20 μm , and light gathering power sufficient to permit ultra-high resolution spectroscopy of young stars and their circumstellar environs.

The new generation 8-m class telescopes will permit us for the first time:

- to obtain infrared images and spectra of extraordinarily young stars, associations, and clusters deeply embedded within their natal cores, and to understand the relationship between star formation efficiency, the stellar initial mass function and the physical properties of parent cores and molecular clouds. Equipped with adaptive optics, such telescopes will be sufficiently powerful to allow us to image forming stars and clusters even in cores obscured by up to 50 magnitudes of visual extinction and to separate young stars in newly-formed stellar clusters with densities exceeding 2×10^4 stars pc^{-3}, and to obtain the photometric and spectroscopic measurements necessary to place these objects in the HR diagram – not only in relatively nearby regions of low mass star formation, but also in more distant molecular clouds where much rarer high mass PMS and young MS stars must be studied. What kinds of clouds/cores give rise to high mass stars and how do they differ from regions where low mass star formation is the rule? Is star formation in the Milky Way and other galaxies bimodal? Under what conditions do bound clusters, multiple and binary stars form? How does the early evolution of high mass stars differ from their low mass counterparts?

- to obtain images of solar system-size circumstellar disks surrounding nearby solar-type PMS stars with unprecedented clarity, and to study the kinematic properties of such disks. By use of adaptive optics, we can trace the distribution of solid material within the disk at effective spatial resolutions of $\sim$ 5 to 10 AU, by observing light scattered by circumstellar dust. By virtue of their large collecting area, 8-m class telescopes will allow us to obtain disk rotation curves from analysis of R $\sim$100,000 infrared spectra of photospheric light scattered by circumstellar dust, or of resonantly-scattered molecular line radiation.

- to study the gas content and gas/dust ratio for circumstellar disks throughout the disk, and as a function of time. The photon gathering power and high spatial resolution of 8-m class telescopes will allow us to trace emission from CO, NH_3, SiO, and from small (50 atom) grains as a function of position within disks surrounding stars of differing age, and to compare the observed distribution of gas with that of the dust as inferred from infrared imaging and photometric measurements. Can we observe gas rich disks – the structures expected following the assembly of distributed disk dust into planetesimals? Over what range of ages can we detect disk gas and how severe a constraint does the gas survival time place on the timescale for assembling giant, gas-rich planets similar to Jupiter? What role do T-Tauri winds play in dispersing the disk gas?

- to obtain multiple high spatial resolution images of collimated stellar outflows ("jets") emanating from embedded young stars, in order to learn how these structures become collimated, and how they evolve. With adaptive optics, 8-m class telescopes will provide a spatial resolution of $\sim$ 0.05$''$ at 1.6 μm, which will allow us to obtain images and spectra of jets to within 5 to 10 AU of the stellar surface, and to define their relation to disk and protostellar core structures. By observing embedded and emerging

PMS stars of differing ages, we can learn whether outflows remain highly collimated throughout the phases of PMS star evolution when disks are present. If they do, how are gas and dust cleared from post-planet building disks?

- to obtain images and spectra of substellar mass, close companions of PMS stars. 8-m class telescopes will provide the photon-collecting power to derive accurate effective temperatures from moderate resolution spectra, and luminosities from broad-band photometry, for sub-stellar objects of luminosities as small as 10^{-5} $L_\odot$. With adaptive optics providing diffraction-limited images at 1.6 μm, such objects can be imaged at separations as small as 10 AU from a companion PMS star in the nearest star-forming complexes. By locating substellar mass objects associated with PMS stars, we will 1) be observing them at their earliest evolutionary stages when they are expected to be most luminous, 2) be able to determine their approximate ages (from the ages of their PMS companions), and thereby to confront theories describing the evolution of such objects, and 3) to determine whether low mass companions are likely formed within the circumstellar disk of a parent PMS star, or via a separate fragmentation and collapse process within a common molecular core. Do large planets and ultra low mass stars have a common origin or is one class of objects assembled within disks and the other form via fragmentation?

To observe the terrestrial planet-forming regions of pre-planetary disks will require an order of magnitude gain in angular resolution, and thus the development of ground-based optical/infrared interferometers. Such interferometers can provide the angular resolution and sensitivity necessary to probe the structure of planet-forming disks around several hundred nearby ($100 \leq d \leq 200$ pc) young solar-type of ages ranging from 1 Myr to 20 Myr, and will for the first time *enable* astronomers to image planetary systems in differing stages of development.

Technical Developments of the 1980s and Opportunities for the 1990s

Our ability to design the facilities and instruments which enable these fundamental investigations results from advances in telescope and detector technology during the 1980s.

Advances in Telescope Technology

In its review of astronomy at the beginning of the 1980s, the Field Committee strongly recommended investments aimed at developing the technology to enable the construction of a new generation of large optical telescopes. Vigorous efforts involving a combination of private, state and federal resources have led to a veritable revolution in our thinking about large telescopes: as we enter the 1990s, almost every aspect of telescope design is viewed from a dramatically different perspective:

- Two approaches for constructing large primaries were developed:
    - (1) active primary mirror surfaces comprised of multiple glass mirror segments
    - (2) monoliths of ultra-lightweight honeycomb borosilicate glass
- New polishing techniques were developed which enable reduction of telescope focal ratios, from f/3 to ~f/1, with a corresponding ~threefold decrease in overall telescope size relative to mirror diameter;
- Finite element analysis enables thorough and accurate analysis and optimization of mechanical and optical support structures;
- Advances in control systems (computers and electronics) enable the design of active control systems for monitoring and correcting mirror figures and maintaining the performance of the entire optical system;
- Advances in understanding the influence of wavefront distortion introduced by the local telescope environment provide the potential for dramatic improvements in image quality through reduction of "dome seeing";

These advances have enabled the planning of a new generation of large (8-m and 4-m diameter) optical/infrared telescopes which will provide higher image quality and superior overall performance than any telescopes built to date, and can be built at far lower cost per square meter of collecting area.

In order to remain at the forefront of astronomical research worldwide, it is necessary to take advantage of these technological developments to construct a new generation of optical/infrared telescopes in the United States.

Advances in Detector Technology

While no major ground-based optical/infrared telescopes were constructed in the United States during the 1980s, ground-based O/IR astronomy nevertheless witnessed a revolution in observing power, driven in large measure by the introduction of sensitive, large format optical (charge-coupled devices, or CCDs) and infrared arrays. US astronomers have incorporated these array detectors in a variety of instruments which have multiplied the sensitivity of extant telescopes by factors of tens to hundreds. These advances are a direct result of United States leadership in the development of sensor technology.

- *At optical wavelengths* CCDs (now of dimension 2048x2048 pixels) provide an order of magnitude improvement in sensitivity over previous detectors, and offer major improvements in geometric and photometric stability. Introduction of large format CCD detectors has enabled high precision photometric studies of stars and galaxies, monochromatic imaging, sensitive multi-object spectroscopy of stars and galaxies, and high (S/N $\gg$ 100) signal/noise echelle spectroscopy.

- *At infrared wavelengths,* array technology has progressed dramatically during the past 5 years. In the wavelength regime $\lambda \leq 5\mu m$, devices as large as 256x256 pixels have rapidly replaced the single detector systems which, until the mid-1980s, were the standard. It is now possible *for the first time* to image astronomical objects at infrared wavelengths, to obtain spatially resolved spectra of extended sources, and to build high spectral resolution cryogenic echelle spectrographs.

During the next decade, it will be necessary to build larger format optical and infrared detectors characterized by lower read noise, faster read times, and broader wavelength response. It is imperative that astronomy take full advantage of US strengths in advanced optical and infrared sensor technology to develop array detectors matched to the new generation of O/IR telescopes. Detector performance *and availability* are *sine qua non* for competitive instrument performance on all telescopes.

It will thus be necessary to make a strong commitment to continued development of advanced sensors, and to evolve strategies for bulk purchases and distribution of detectors for use by the US astronomical community.

Auxiliary Instruments

The availability of sensitive array detectors and the rapid evolution of sophisticated image analysis techniques enabled by advanced computer technology, has led to development of instruments far more complex and powerful than the photometers and spectrographs built prior to 1980. Among the most dramatic advances have been:

- the introduction of spectrographs capable of obtaining spectra of large numbers of stars or galaxies simultaneously. Multi-object spectrographs can reduce observing time for many survey programs by $\sim$100-fold.

- the development of cryogenically-cooled infrared spectrometers which enable two-dimensional spectroscopy of infrared sources at moderate spectral resolution, and promise within the next year $\sim$1000-fold gains in sensitivity for carrying out high spectral resolution studies.

The availability of more advanced array detectors will enable more powerful versions of these and other instruments to be constructed during the 1990s, and will extend their capabilities to broader wavelength ranges. In order to take full advantage of the potential gains offered by telescopes of advanced design, it will be necessary to develop instruments matched specifically to these telescopes.

The new generation of large telescopes will require a major investment in instrumentation which will differ in scale and design from the instruments of the 1980s.

Pioneering a New Frontier: High Angular Resolution O/IR Astronomy

The advances of the past decade enable the design and construction of a new generation of ground-based optical/infrared facilities which can provide order of magnitude gains in angular resolution and sensitivity Indeed, the 1990s promise to be the decade in which astronomers throughout the world exploit these advances to pioneer a new frontier: high angular resolution infrared and optical astronomy.

A deep, long exposure of an astronomical object, even one taken with a large optical/infrared telescope located at the best site in the world, at present produces images of point sources which are blurred to a diameter of $\geq 0.5''$, and will thus not reveal details on resolved sources on angular scales smaller than $\sim 0.5''$. This image blurring, or "seeing" results from distortions in the incoming wavefront produced by the combined effects of multiple, rapidly moving turbulent elements at all levels in the earth's atmosphere above the telescope. If the atmosphere were removed, that same telescope would *concentrate* the light from a star into a diffraction-limited core (of dimension $0.02''$ at 0.55 μm, and $0.05''$ at 1.6 μm for an 8-m telescope), and would *resolve* features comparable in size to the diffraction limit in extended sources.

The strong desire to overcome the limitations imposed by the atmosphere in order (1) to improve our ability to detect faint point sources; (2) to improve the image clarity for resolved, often distant sources and (3) to *enable* for the first time the resolution of structures hidden from view by the "blur" introduced by the atmosphere, were among the primary factors motivating the development of the Hubble Space Telescope. Even though the HST mirror is of only modest (2.3 m) size by current ground-based standards, its location above the earth's atmosphere can potentially provide a *diffraction-limited* image which results in a $\sim$ 100-fold increase in sensitivity for point sources and a 10 times increase in angular resolution at ultraviolet and optical wavelengths.

Even the resolving power of HST will be unable to probe the centers of active galaxies, to image forming galaxies, or to resolve the terrestrial planet-forming regions around solar-type stars. These and other exciting problems have motivated astronomers worldwide to design and build telescopes and instruments capable of providing even greater sensitivity and image clarity. During the 1980s, instrumentalists have developed a variety of novel techniques for improving the image quality achieved with ground-based O/IR telescopes. The results of their efforts have borne fruit, and during the 1990s will enable:

- sensitive imaging with diffraction-limited resolution ($0.05''$ at 1.6 μm and $0.02''$ at 0.55 μm) using the full aperture of 8-m class telescopes by means of a technique known as *adaptive optics* ; by comparison, HST will provide images of size $\sim 0.07''$ at 0.55 μm. At this angular resolution, it will be possible to study stellar populations in the nuclear bulges and disks of nearby galaxies, probe the planet-forming regions of primordial solar nebulae, and image solar system bodies to resolutions of $\sim$75 km at the distance of Jupiter.

- imaging with potential angular resolutions of $\sim$0.002'' at 1.6 μm and $\sim$0.0005'' at 0.55 μm with *interferometric arrays* of moderate-size telescopes separated by $\sim$200 meters. At this angular resolution it will be possible to image the narrow emission line regions of active galactic nuclei, accretion disks in close binaries, and expanding envelopes surrounding late-type stars.

Adaptive Optics

Adaptive optics is a technique which makes use of sensitive array detectors and high performance computers (1) to detect and model the amount by which an incoming plane wavefront from a celestial source is altered by the atmosphere (using a a device known as a "wavefront sensor"); and (2) to use that information to command a fast servo system to alter the figure of a flexible optical element (an "adaptive mirror") by an amount necessary to compensate for constantly varying atmospheric distortions.

There are *three* parameters that are basic to understanding the vocabulary of and concepts underlying adaptive optics. The *first* is r_0, the *atmospheric correlation length*. An incoming plane wavefront from a celestial source is distorted randomly by moving turbulent elements in the atmosphere above the telescope. One can think of the primary mirror as comprised of a large number of patches, each of dimension r_0, over each of which the wavefront is approximately (and instantaneously) flat, *but tilted* relative to its neighboring r_0 patch; the patch-to-patch *tilt* differences correspond to *phase differences in the incoming wavefront. The parameter r_0 is the characteristic spatial scale over which the rms phase differences (or wavefront tilts) are less than one radian. At 0.55 μm, $r_0 \sim$ 20 cm under conditions of of excellent (0.5'') seeing. The "wavefront sensor" uses a bright reference object (either the star itself, a nearby star or an artificial star) to model the distortions in the incoming wavefront wrought by the earth's atmosphere in terms of an ensemble of wavefront tilts over the $(D/r_0)^2$ patches covering the primary mirror (of diameter D).

The *second* parameter is τ_0, the *atmospheric correlation time*. If one imagines the turbulent elements responsible for atmospherically-induced wavefront disturbances as being swept rapidly past the telescope by the wind at some speed, v, then the characteristic time over which the wavefront tilt over a given r_0 patch changes by one radian is $\tau_0 \equiv r_0/V$. For a typical windspeed of 10m/s, τ_0 is typically 20 msec at 0.55 μm.

The *third* parameter is i, the *isoplanatic patch angle*. If one imagines a pair of stars, A and B, separated by an angle i, then i is the separation within which the *relative* tilts of the wavefronts emanating from A and B do not exceed one radian; i is thus the angle subtended by a patch of dimension r_0 viewed from the height of the atmospheric layer where the wavefront tilts originate. The characteristic height of this layer above the telescope is or order 10 km, so that at 0.55 μm, $i \sim 4''$.

The challenge of adaptive optics is to sense *wavefront tilts* in a time short compared with the *atmospheric correlation time* by using the imaged celestial source itself or an adjacent source which lies within the *isoplanatic angle,* and to signal corrections to a flexible adaptive mirror equipped with $(D/r_0)^2$ actuators. In Table 1, we summarize characteristic values for r_0, τ_0, and i, assuming 0.5'' seeing, an 8-m telescope and

**Table 1. Characteristic Parameter Values for 0.5″ seeing,
8-m telescope 10 m/sec windspeed**

λ (μm)	r_0 (cm)	$(D/r_0)^2$	τ_0 (msec)	(λ/D) (arcsec)	iso-angle (arcsec)
0.5	20	1600	20	0.013	4
0.9	40	400	40	0.026	8
1.6	84	90	84	0.042	17
2.2	118	46	118	0.057	24
4.8	300	7	300	0.124	62
11.0	816	1	816	0.284	142

a (typical) wind speed of 10m/s. This table illustrates the *dramatic differences* in the complexity of adaptive correction systems as a function of wavelength: it is *far easier* to effect adaptive corrections in the *infrared* because (1) fewer actuators are required; (2) the wavefront tilts can be measured more easily using fainter reference stars because the correlation time is longer and the effective "collecting area" that can be used is proportional to r_0^2; and (3) more reference stars are available for effecting adaptive corrections for faint sources because $r_0, \tau_0,$ and i are larger.

Early applications of adaptive optics by European astronomers working with the 3.5m NTT in Chile have already produced images at 2.2 μm in which a large fraction of the power is contained *within a core whose size is that of the diffraction limit of the telescope.* Experiments underway at the University of Hawaii suggest that dramatic gains in image quality at near-infrared wavelengths may be achieved with very simple adaptive mirrors having relatively few actuators, $N \ll (D/r_0)^2$. These results make us confident that *full adaptive corrections will be possible at $\lambda \geq 1.6\mu m$ within a few years.*

In contrast to the infrared, adaptive corrections at optical wavelengths require thousands of actuators to deform the adaptive mirror. Moreover, the paucity of bright reference sources expected within the smaller isoplanatic angle make it necessary to consider developing an artificial reference star for use at optical wavelengths. In this scheme, a bright laser beam is used either to excite sodium atoms in a layer at the top of the atmosphere, or to backscatter off air molecules. By suitably focusing the beam of the pulsed "laser star" onto the relevant disturbing layer in the atmosphere and timing the return laser pulses, one can sense and correct the wavefront tilts once every τ_0. Clearly, investment in a major technology development program will required in order to provide full adaptive corrections at optical wavelengths.

Ground-Based Optical/Infrared Interferometry

During the 1980s, astronomers made great strides in another very promising direction for achieving ultra-high angular resolution: *optical/infrared interferometry.* The technique of combining and interfering beams from widely separated telescopes has been used for more than four decades by radio astronomers to provide high angular resolution imaging at cm- and more recently at mm- wavelengths. The challenge of constructing an interferometric array is far more daunting at O/IR wavelengths primarily because the earth's atmosphere is far less benign at these wavelengths. Atmospheric distortion of incoming wavefronts from celestial sources vastly complicates efforts to track interference fringes, and to make use of the full aperture of each component of the array.

Despite these problems, great progress has been enabled by advances in both computer and detector technology, and by the ingenuity of pioneering experimentalists:

- fringe tracking for significant time periods was first achieved in the mid-1980s by a team of French astronomers operating a two-element optical interferometer at baselines ~100 meters. US groups have recently enjoyed major successes as well, and appear on the verge of developing imaging interferometers capable of observing bright optical and infrared sources.

- astronomers in the US made use of the separate 1.8m mirrors comprising the Multiple Mirror Telescope to establish the feasibility of interferometric imaging over fixed baselines of ~10m; This success with the MMT led European and US astronomers to design the Very Large Telescope, the Keck telescope, and the Columbus Project to provide interferometric capability with fixed baselines.

- a number of pioneering groups in Europe and the US have begun to build interferometric arrays of small telescopes operating both at infrared and optical wavelengths, and over variable baselines ranging up to several hundred meters.

- parallel advances in adaptive optics technology offer the promise of making use of the *full area* of large array elements, and thus of building imaging systems capable of *high sensitivity* observations at infrared, and later, optical wavelengths.

Interferometry, first with fixed baseline arrays of ~20-50m, and later with variable baselines extending to several hundred meters or longer, promises to provide infrared and optical images with angular resolutions exceeding 0.002".

The ability to study the optical/infrared sky at high sensitivity and at angular resolutions 10 times and later 100 times current capabilities will enable new of classes of astronomical research. Interferometers will allow us for the first time to *image* the surfaces of stars, to observe planets outside the solar system, and to image the regions surrounding the engines which power active galactic nuclei.

It is imperative to invest in a major effort aimed at developing high angular resolution astronomy at optical/infrared wavelengths (1) by developing and applying adaptive optics technology to enable sensitive diffraction-limited imaging by large ground-based telescopes, and (2) by combining adaptive optics and interferometry to produce sensitive imaging at resolutions 0.001" and greater. Support of this pioneering effort will be essential to maintaining US leadership at the frontiers of astronomy at the beginning of the next century. Such investments would represent a continuation of a strong US commitment to developing technically advanced radio interferometers which provide the highest feasible angular resolution– beginning with the Very Large Array (operating at cm wavelengths) in the 1970s, continuing with the Very Long Baseline Array (cm wavelengths) in the 1980s and culminating with construction of the Millimeter Array in the mid- to late- 1990s. The technology base developed in the 1980s combined with a vigorous development program in the 1990s, will provide the basis for designing an optical/infrared analog of the VLA during the the 1990s, and building the array during first decade of the next century.

Ground-Based Optical/Infrared Astronomy Outside the US

Our colleagues in Europe and Japan have recognized that leadership in astronomical research requires investment in a new generation of optical/infrared facilities. The Japanese government is committed to building an advanced technology 8-m class telescope on Mauna Kea in Hawaii. The European Southern Observatory (operated by a consortium of European countries) has just completed a 3.5-meter diameter New Technology Telescope. Equipped with an active system to control the mirror figure, and with adaptive optics to compensate for the blurring effects of the earth's atmosphere, the NTT has recorded the sharpest images ever made from the ground. ESO has also committed more than $200M toward the construction of a powerful, technically advanced optical telescope – the Very Large Telescope. The VLT, with an equivalent collecting area of 16-m, provides more than twice the light gathering power of the largest planned US facility. The VLT promises gains not only in light gathering power, but in angular resolution as well. When fully operational, the VLT can be operated as an interferometric array, capable of providing optical images with angular size 0.005". These facilities will provide astronomers in Europe and Japan with the tools needed to carry out frontier research in the era of the Great Observatories.

As scientists, we rejoice in the success of our colleagues in other countries and look forward to the discoveries which will inevitably ensue as the power of these new facilities is unleashed. As citizens, we are concerned that US leadership in astronomy (as in other sciences) is presently based largely on returns from investments and plans made in the 1960s and 1970s. *Without a strong commitment in the 1990s to join the competition for astronomical leadership worldwide, the relative quality of US astronomy will inevitably decline.*

Central to US competitiveness in astronomy is the development of world class ground-based optical/infrared facilities. In the following sections, we recommend an investment strategy – based on the technical and scientific opportunities we perceive for the next decade – aimed at preserving US competitiveness in ground-based O/IR astronomy during the 1990s and positioning the US to establish leadership in high angular resolution O/IR astronomy at the beginning of the next century.

RECOMMENDATIONS OF THE PANEL: LARGE SCALE PROGRAMS

Priority 1: A Coordinated Program For Large O/IR Telescopes

Background

Ground-based optical and infrared observations of faint celestial sources are currently limited in sensitivity by the amount of light telescopes can collect and measure, by the brightness of atmospheric and telescope emission, and by the blurring of images introduced by the earth's atmosphere. In the past, improvements in detector efficiency and instrument design could provide large increases in sensitivity at relatively modest cost. However, detector performance is now pressed close to theoretical limits, owing in great measure to US leadership in developing advanced optical and infrared array sensors with quantum efficiencies approaching unity. Hence, major improvements in *telescope* performance are necessary in order to provide the sensitivity to address frontier problems in astronomy.

Following a decade of successful technology development, optical and infrared astronomers are poised to build a new generation of powerful ground-based telescopes. Major advances in telescope collecting area, the optical quality of large mirrors and overall telescope systems, and in the image quality produced by earth-bound telescopes through the use of adaptive optics, promise a 10 fold increase in angular resolution and up to 100 fold the sensitivity of the largest extant facilities. These advances will *enable* qualitatively new astronomical observations. *We therefore recommend a vigorous program for the development of 8-m class telescopes in the United States. This program includes as an integral component, development of the adaptive optics systems and the auxiliary instrumentation required to realize the full potential of these frontier research tools.*

New Science Enabled by Greater Collecting Area

The first gains to be realized by large telescopes are in sensitivity–achieved simply as a result of their larger collecting area. We list below *examples* of key programs which would require years or decades of observing time on existing 4-m telescopes. Further progress on these programs thus awaits the increase in collecting area provided by 8-m diameter telescopes:

- determining the large-scale structure of the universe at early epochs through measurement of redshifts for large ($\sim$100,000) samples of galaxies at high redshift ($z \geq 1$);
- determining the distribution of quasars in space and time, especially at early epochs ($z \geq 3$) where quasars illuminated the universe for the first time;
- probing the chemical evolution of galaxies as a function of lookback time from high resolution spectroscopic observations of metal lines produced in multiple intervening galactic halos and seen against the background light of distant quasars;
- comparing the age of the Universe as derived from the Hubble expansion parameter with accurate age determinations for the oldest known observable systems, globular clusters; these age determinations will be enabled by sensitive spectroscopic observations which can provide accurate distances and chemical abundances for stars in Milky Way globulars;
- carrying out study of stellar oscillations through high signal/noise, high resolution spectroscopy, and determining the internal structure of stars other than the sun.

Most of these problems require not only increased collecting area, but the observation of large samples of objects. Some are feasible only if the development of 8-m class telescope is accompanied by the construction of advanced multi-object spectrographs capable of simultaneous observation of several hundred objects within fields of view of dimension 20′ to 30′.

New Science Enabled by Diffraction-Limited Imaging

The development of adaptive optics for use with 8-m optical/infrared telescopes will provide diffraction-limited images at near- and mid- infrared wavelengths, $\lambda \geq 1.6$ μm by mid-decade, thereby improving image quality by nearly tenfold when compared with typical atmospheric seeing.

By reducing the *size* of a stellar image from the seeing-limit to the diffraction-limit of the telescope, faint point sources will stand out far more prominently against the background produced by night sky emission and by the thermal radiation emanating from the telescope itself. Not only will sensitivity to faint sources be improved, but the diffraction-limited images provided by adaptive optics will also 1) resolve objects in crowded, source-confused regions; and 2) reveal and resolve structures heretofore hidden from view within the seeing disk. *Examples* of the power of adaptive optics used in combination with large ground-based telescopes to *enable* attacks on new classes of problems include:

- determining the morphology, star-forming activity, and chemical composition of galaxies at their formation epoch, by means of high resolution imaging and spectroscopic studies in the red and near infrared spectral regions;
- determining the stellar content and mass distribution of the Galactic Center through imaging and spectroscopic studies of this optically-obscured, densely-populated region at 2.2 μm;
- determining the efficiency of star formation and constraining the initial mass function from infrared imaging and spectroscopic observations of obscured regions of highly-efficient star formation – analogs to the Orion cluster, where the stellar density may exceed 20,000 stars/pc^{-3}, and the mean separation between faint stars is $\leq 1''$.
- determining the structure, mineralogy, and gas content of solar-system size disks predicted to surround young, solar-type stars, thus providing astronomical constraints on the evolution of primordial solar nebulae. Pre-planetary disks are expected to have diameters $\sim$100 AU or 0.7'' around the nearest young stars, and are thus hidden within the seeing disk of the bright parent object.
- obtaining images and spectra of massive planets and sub-stellar mass companions located within 1'' of parent stars with ages 10 Myr and less, when such sub-stellar mass objects are most luminous;
- obtaining spectroscopic and imaging observations of planets within the solar system at angular resolutions $\sim$0.05'', thereby enabling remote studies of surface mineralogy (Mars) and of atmospheric structure and composition (Jovian planets);

The gains offered by adaptive optics are potentially so great that we recommend that during the decade, all 8-m telescopes in the US be equipped with systems capable of *full adaptive corrections in the near-infrared*. Full corrections at *optical* wavelengths will involve more complex wavefront sensors and adaptive mirrors, and will likely be achieved only toward the end of the decade, following investment in a significant research and development program.

Expected Performance Gains

The quantitative gains expected from 8-m class telescopes depend upon the nature of the problem and the telescope performance. In Table 2, we summarize these gains for observations in which the signal/noise ratio of a given observation or the required observing time is limited a) by the flux of incoming photons from a source (for example, most high spectral resolution observations at optical wavelengths); and b) by the background emission from the night sky or the telescope (e.g. low resolution spectroscopy of galaxies; detection of excess mid-infrared emission from circumstellar disks surrounding young stars). We consider two cases

- that the observations are limited by atmospheric seeing as is the case for telescopes with no adaptive corrections for incoming wavefront distortions;
- that diffraction limited images are achieved; 8-m telescopes equipped with modest adaptive optics systems will deliver diffraction-limited imaging at wavelengths as short as 1.6 μm by mid-decade or before.

For photon-limited observations, the time required to reach a fixed S/N will be decreased by a factor of 4 for an 8-m compared to a 4-m diameter telescope. This gain will enable high resolution, high S/N spectroscopic studies of large samples of objects currently beyond the practical limits of 4-m class telescopes (e.g. observations of unadulterated atmospheric chemical compositions for globular cluster main sequence stars). For background-limited observations, the time required to reach a fixed S/N will again be decreased fourfold for telescopes lacking adaptive optics. This gain will, for example, enable the assembly of redshifts for large samples of galaxies at the limit of the observable universe ($z \geq 3$). With adaptive optics, the time required to reach a fixed S/N will be decreased by *16 times* for faint point sources observed against sky or

Table 2. GAINS WITH INCREASING TELESCOPE DIAMETER

	Seeing-Limited		Diffraction-Limited	
	Unresolved Source	Resolved Source	Unresolved Source	Resolved Source$(pix)^{-1}$
Photon-Limited Observations				
S/N	D^1	D^1	D^1	D^0
1/time (fix S/N)	D^2	D^2	D^2	D^0
Background-Limited Observations				
S/N	D^1	D^1	D^2	D^0
1/time (fix S/N)	D^2	D^2	D^4	D^0

telescope background. The gains for infrared photometric and spectroscopic studies (e.g. of faint stars or unresolved obscured galactic nuclei) will thus be spectacular.

Recommended Program for the 1990s

The scientific programs enabled by 8-m class telescopes equipped with adaptive optics make it certain that such facilities will be among the leading tools of research and discovery during the next decade. As noted earlier, US leadership in ground-based astronomy–a constant of international science since the beginning of the 20th century–is being challenged as never before. *If US astronomy is to maintain its vitality and leadership, it is essential that a new generation of large telescopes be built during this next decade and made available to members of the US astronomical community.*

The response to the challenge of maintaining leadership in O/IR astronomy has thus far come from private institutions and state universities who have raised funds for, and in some cases started to construct 8-m class telescopes. These independent large telescope efforts include (1) the nearly-completed Keck 10-m telescope (California Institute of Technology and the University of California), (2) the Columbus Project (University of Arizona, Ohio State University, and the government of Italy) to construct a pair of 8-m telescopes, (3) the Magellan Project (Carnegie Observatories, Johns Hopkins University, the University of Arizona) to construct an 8-m telescope in the southern hemisphere, (4) the Smithsonian Astrophysical Observatory and the University of Arizona project to replace the six 1.8m mirrors comprising the Multiple Mirror Telescope with a 6.5m monolith, and (5) the University of Texas, Pennsylvania State University program to construct an 8-m telescope specialized for spectroscopic studies. It is vital that these independently-funded telescopes be equipped initially and continue to be equipped with the most sophisticated instruments and detectors. This will require *federal* contributions over the next decade to complement the private and state funds already committed toward construction of these facilities. It is also vital to the health of US astronomy that the nation's astronomers have competitive access to uniquely capable *national facilities* of this size.

The O/IR panel therefore recommends a *coordinated program* to combine federal funds with state and private funds to *build* and *instrument* large (8-m class) US ground-based telescopes. This program should encourage innovative new developments (e.g. adaptive optics; advanced instrumentation) and the sharing of technologies and facilities to optimize the total national effort–independent consortium telescopes and national observatories–in optical and infrared astronomy. The key federal contributions to this program are:

- support ($120M) for the construction of *a pair of 8-m telescopes*, one each in the northern and southern hemisphere. The northern hemisphere telescope should be located on Mauna Kea, and be designed and operated to achieve optimized performance at infrared wavelengths. *An infrared-optimized telescope located on Mauna Kea will provide US astronomers with a unique and powerful facility.* Building and operating this telescope will require (1) that all telescope mirrors be silver-coated; (2) that procedures be developed to keep telescope mirrors dust-free in order to minimize thermal emission from optical surfaces; and (3) that the mirror be a monolith, with a polished surface of unsurpassed quality–sufficient to take full advantage of adaptive optics technology which promises diffraction-limited images at wavelengths 1.6 μm and longward by mid-decade. By demanding the lowest possible telescope emissivity and highest mirror quality, the national IR-optimized telescope will take full advantage of the superb infrared transmission and image stability of the Mauna Kea site. The combination of diffraction-limited imaging and high sensitivity will make the IRO uniquely capable of carrying out the most demanding observations of forming galaxies, stars and planetary systems.

The southern hemisphere 8-m national optical/infrared telescope *should be a twin of the northern hemisphere telescope* in order to achieve the cost savings which derive from commonalty of design. It will provide US astronomers with the ability to observe the many objects (e.g. the Magellanic Clouds, the center of the Milky Way) best studied from the southern hemisphere. The need for access to the southern skies is particularly pressing in the era of the Great Observatories, which can observe astronomical objects over the complete celestial sphere; without national access to a southern hemisphere 8-m class telescope, US astronomers will be at an enormous disadvantage in characterizing new phenomena discovered by these space-based facilities. The O/IR panel recommends that *initially* the southern hemisphere 8-m telescope mirrors be coated with aluminum in order to optimize performance at optical and near-ultraviolet wavelengths. The southern hemisphere 8-m telescope will thus complement the infrared-optimized northern hemisphere facility, and will enable astronomers to carry out a number of important scientific problems (e.g. studies of quasi-stellar object (QSO) absorption spectra) for with the IRO is unsuitable. At some point, it may be desirable to coat the optics with silver in order that the southern hemisphere 8-m be able to carry out sensitive infrared observations of objects accessible solely from the southern hemisphere.

- support ($10M - $15M) for immediate development and deployment (on all new generation US telescopes of diameter 4-m and larger) of wavefront sensors and adaptive mirrors capable of providing full adaptive corrections at near-infrared wavelengths. This program is critical, not only for achieving gains in angular resolution and sensitivity, but to *enable* or greatly simplify development of advanced auxiliary instrumentation for large telescopes through reduction of the image size (for unresolved sources) from the seeing-limit to the diffraction-limit.

- support ($40M-$50M) for developing and building advanced auxiliary instruments required both for the new generation of large O/IR telescopes. Instruments for these new generation telescopes will be large in scale, technically sophisticated, and in many cases far more expensive than analogous auxiliary devices on extant 4-m class telescopes. The larger instruments for these telescopes may cost in excess of $5M. Both the pair of national 8-m telescopes and the independent 8-m projects include in their budgets funds for an initial complement of basic instruments (e.g. optical and infrared cameras; faint object spectrographs). Our recommendation is directed at providing the funds necessary for the development and construction of *advanced instrumentation,* both for the pair of national 8-m telescopes *and for the 8-m telescopes under construction by private/state consortia.* Examples of such instruments include

    - multi-object spectrographs capable of obtaining spectra of several hundred galaxies or stars simultaneously;

    - cryogenic echelle spectrographs (which will take advantage of advances in IR array technology in order to provide 10,000 fold improvements in sensitivity over currently availably high spectral resolution devices).

The requirement over the decade for advanced instrumentation for these new facilities will necessitate federal funding at a rate twice that of the current annual NSF expenditures for O/IR instrumentation.

The recommended decadal investment in O/IR instruments for large telescopes *presupposes* a vigorous program to develop and distribute advanced optical and infrared array detectors (see below). Such a program is a vital prerequisite for this recommended instrumentation program and for effective use of large telescopes throughout the community.

The availability of funding to construct advanced instruments is a key element of our recommended coordinated program aimed at combining federal funding with state and private resources to fully develop the facilities and instruments required for a competitive program in ground-based optical/infrared astronomy during the 1990s. Our program recognizes the essential role played by the private telescopes in providing the US with the complement of large telescopes required to retain its competitive position in world astronomy: private and state resources exceeding $150M have already been invested in or will be committed to the construction of 8-m class telescopes. Funding for advanced instrumentation at the recommended level is required in order that these facilities, as well as the national 8-m telescopes be able to carry out competitive frontier observations throughout the decade.

Our recommended funding level assumes that very expensive or specialized instruments will not in general be replicated. In many cases, several teams of scientists from throughout the community may propose competitive instrumental approaches, only one of which may be funded. If the winning team proposes to locate a unique instrument at an "independent" observatory, national access should be made available either through mutually agreeable financial arrangements or through exchange of telescope time.

RECOMMENDATIONS OF THE PANEL: MEDIUM SCALE PROGRAMS

Priority 1: A Coordinated Program For High Angular Resolution

Background

During the 1980s, astronomers worldwide devoted considerable effort to developing ways to increase the angular resolution of ground-based telescopes:

- improving the local seeing introduced in the immediate telescope environment; through careful control of the thermal environment of the dome, the size of the seeing disk has been reduced from $\sim 1''$ to $\sim 0.5''$ at superior sites;
- controlling primary mirror figure and telescope focus through active corrections applied on timescales of minutes to hours; the European Southern Observatory (ESO) NTT has achieved $0.3''$ images at optical wavelengths using this technique;
- correcting for instantaneous ($\tau \leq 0.1$ sec) distortions in the incoming wavefront by use of wavefront sensors and adaptive mirrors; use of adaptive optics at the ESO NTT has produced near-infrared ($2.2 \ \mu$m) images in which a significant fraction of the light is concentrated in a diffraction-limited core of diameter $\sim 0.14''$.

Our recommended program for the development of "new generation" large O/IR telescopes assumes building, telescope, and instrument designs that incorporate all these techniques. As a result, we confidently anticipate that this new generation of large telescopes will produce diffraction-limited images at $\lambda \geq 1.6 \ \mu$m, and will thus routinely provide angular resolutions of $0.05''$ at $1.6 \ \mu$m and $0.2''$ at $10 \ \mu$m.

In order to improve angular resolution beyond this limit requires (1) extending diffraction-limited observations to wavelengths $\lambda \leq 1.6 \ \mu$m, (2) constructing larger telescopes, and (3) the development of interferometric arrays.

Experimenters are now working on techniques to extend the capability for full adaptive corrections to optical wavelengths. This effort will require development of sophisticated wavefront sensors, complex adaptive mirrors and laser reference stars. When available late in the decade and used in conjunction with 8-m telescopes, these systems will provide diffraction-limited images of diameter $0.02''$ at $0.55 \ \mu$m, 3 times the corresponding angular resolution at $1.6 \ \mu$m.

The design of single optical/infrared telescopes with apertures much larger than 8-m (perhaps 30 m to 50 m) now seems feasible given the advances in mirror technology and O/IR telescope design made during the 1980s. Such telescopes would provide diffraction-limited images of size $\sim 0.01''$ at $1.6 \ \mu$m and $0.003''$ at $0.55 \ \mu$m with consequent gains in both angular resolution *and* sensitivity to faint unresolved sources.

The 1980s also witnessed major advances in overcoming the difficulties inherent in ground-based interferometric imaging at optical/infrared wavelengths. Following the French astronomers at CERGA, and the experience gained in the United States by the astronomers supported by the United States Navy, experimentalists are now confident that an optical/infrared interferometric array comprised of ~ 5 elements distributed along baselines ranging up to least ~ 200m can be built during this decade. Such an array could provide images of size $\sim 0.002''$ at $1.6 \ \mu$m and $0.0007''$ at $0.55 \ \mu$m and sensitivity proportional to the aggregate collecting area of the telescopes comprising the array.

The prospect of increasing the angular resolution of optical/infrared observations by 2 orders of magnitude over the span of a decade would be nothing short of revolutionary. We therefore strongly urge a coordinated program for development of high angular resolution astronomy during the next decade. This program involves

(1) continued development of adaptive optics to extend the reduction in image size gained from application of this technique from infrared to optical wavelengths; (2) engineering studies leading to the design of ultra-large, adaptively-corrected single aperture telescopes; (3) vigorous support of ground-based O/IR interferometry; and (4) engineering studies leading to the design of an optical/infrared analog of the Very Large Array.

New Science Enabled by High Angular Resolution Observations

By extending full adaptive corrections from the infrared to optical wavelengths, 8-m telescopes will deliver diffraction-limited images of size ~0.02″ at 0.55 μm. Thus, in some applications, these telescopes will provide 4 times the angular resolution and 16 times the sensitivity of HST at optical wavelengths. These gains in angular resolution and sensitivity will be particularly critical to:

- (1) study of the stellar populations and source distribution in the nuclear bulge and star-forming regions in nearby galaxies;
- (2) detection of Cepheid variables and M supergiants and other standard candles against the arm and disk light of spiral galaxies;
- (3) detection and spectroscopy of globular clusters located in the halos of distant elliptical and spiral galaxies;
- (4) probing the planet-forming regions of primordial solar nebulae associated with young stars to distances as close as 3 AU from the surface of the parent solar-type;
- (5) imaging solar system bodies with a resolution of 0.02″ (75 km at the distance of Jupiter).

Problems (1) to (3) are representative of a class of measurements which require gains in both sensitivity and angular resolution in order to detect and analyze faint sources in crowded regions viewed against the sky background. Problems (4) and (5) illustrate the sheer power of imaging at 0.002″ resolution.

The next threshold in angular resolution that promises a dramatic increases in the range of accessible astronomical phenomena occurs at 0.001″ (1 mas) or 20 times the resolution anticipated for 8-m class telescopes at optical wavelengths once full adaptive corrections are available. With instruments capable of delivering 1 mas resolution at O/IR wavelengths, astronomers will for the first time be able to:

- obtain detailed images at a resolution of 1 pc of the narrow-line emission regions of active galactic nuclei and crude structural information about the broad emission-line region of AGNs;
- image accretion disks in close binaries at resolutions of 0.1 AU
- image expanding envelopes around late-type stars with resolutions of 0.1 to 1 AU;
- image pulsating stars and monitor changes in their angular diameters;
- image planet-forming disks, jets and winds associated with young solar-type stars with resolutions ~0.1 AU;
- image the photospheres of nearby main sequence stars and more distant giant and supergiant stars;

Recommended Program for the 1990s

At present, efforts to develop high angular resolution techniques at optical and infrared wavelengths in the United States lag behind comparable efforts in Europe. The scientific opportunities that derive from sensitive high angular resolution O/IR imaging are so compelling that we urge strong support of a program aimed at positioning the US astronomical community to assume leadership in this field by the beginning of the next decade. The major elements of this program include:

- vigorous support ($10-$15M) of efforts to extend the wavelength range for full adaptive corrections from the near-infrared to optical wavelengths. This funding level *assumes* the prior commitment of $10M-$15M to build adaptive optics systems for full wavefront correction in the near infrared—already included as an integral part of our recommended program to construct and instrument the new generation of large O/IR telescopes. Full wavefront correction at optical wavelengths requires much more complex wavefront sensors, adaptive mirrors involving thousands of actuators, and an artificial "laser star" to provide the reference for measuring wavefront correction. reference source (a "laser star") must be developed in order to measure wavefront distortions. Full adaptive corrections extending to optical wavelengths also makes it attractive to design large, single-aperture telescopes capable of enormous gains in sensitivity for background-limited problems. Investment in a program to develop adaptive optics is critical *not only* to achieving diffraction-limited imaging at infrared and optical wavelengths with large single-aperture telescopes, but to enabling *high sensitivity* interferometric imaging, particularly at optical wavelengths. Without adaptive optics, the largest telescope apertures that can be used effectively in interferometric arrays are comparable to the coherence length (r_0) for an incoming wavefront: 1m to 2 m at near-infrared wavelengths, but only 0.1 m to 0.2 m at optical wavelengths (see Table 1). With adaptive optics, the full aperture of each interferometer element can be employed.

- support ($2M) for engineering efforts aimed at designing ultra-large (D $\geq$ 30 m), adaptively-corrected telescopes capable of providing dramatic gains in sensitivity and threefold increase in angular resolution.
- immediate support ($10-15M) for a few, well-focused university-based efforts to build and operate pilot interferometers involving at least 3 small telescopes (diameter $\leq$ 1 m) located along variable baselines extending to between 50 m and 100 m. The primary goal of these interferometers is to develop efficient techniques to enable successful optical/IR interferometry on a routine basis (e.g. fringe tracking, beam combination, phase closure). Investment in these efforts will not only hasten development of these techniques, but will build the necessary infrastructure of graduate students and faculty to drive further development of interferometry. Despite their modest size and pioneering nature, we can anticipate important scientific returns in several areas, including:

  - measurement of fundamental astrophysical quantities such as stellar masses, luminosities and diameters;

  - imaging of the surfaces of giant and supergiant stars, and of dust envelopes surrounding evolved stars;

- support ($5M) of efforts to develop infrared and optical interferometry with 8-m class telescopes located on fixed baselines; these efforts will provide the experience necessary to carry out interferometry with large adaptively-corrected array elements;
- support ($15M) by mid-decade of a university-based effort to develop a sensitive O/IR interferometer array. This array should be comprised of at least 3 and perhaps as many as 5 telescopes of diameter ~2m located along continuously adjustable baselines of dimension $\geq$ 200 m. We can anticipate that such an interferometer will first take advantage of the more benign character of the atmosphere at longer wavelengths (greater correlation lengths, permitting the full use of apertures of size ~2m, and longer correlation times, permitting fainter sources to be used to determine atmospheric phase) to provide *high sensitivity near-infrared images* at an angular resolution θ ~0.002''–comparable to images achievable with VLBI techniques at cm-wavelengths. Later in the decade, when full adaptive corrections are possible at optical wavelengths, the recommended interferometer will be capable of making sensitive images in this wavelength regime as well;
- support ($10M) before the end of the decade to begin planning a national O/IR array. The program advocated above will provide the basis for planning an advanced optical/IR array to be constructed as a national facility between 2000 and 2010. The array will probably comprise a considerable number ($\gg$ 10) of apertures with a collecting area of $\geq$ 100 m^2.

This approach is modeled on the highly successful strategy which led to the development of millimeter-wave interferometry by university groups during the 1980s, and which promises to culminate in the construction of a national mm-wave array during the 1990s. Our recommended program in O/IR interferometry would continue the wise federal strategy which has gained world leadership for the US in high angular resolution astronomy – starting with the VLA, and VLBA and continuing through the development of mm-interferometry –*by providing the basis for beginning construction of a national O/IR Very Large Array at the beginning of the next century.*

Priority 2: A New Generation Of 4-m Class Telescopes

Background

Ground-based optical and infrared observations are central to developing a deep understanding of astrophysical systems and placing exotic phenomena discovered at other wavelengths in the context of the "known." While *discovery* most often follows the opening of new frontiers in wavelength, sensitivity or angular resolution, *understanding* usually requires a great deal of detective work by large numbers of scientists with access to appropriate investigative tools. As we enter the 1990s, the basic "tool" of *understanding* is an optical/infrared telescope of diameter ~4 meters. As we enter the 1990s, the number of observing hours available on 4-m class telescopes is woefully insufficient when compared with the time required to carry out the basic observational studies which lead to understanding. As a result, only a small fraction of the investigations thought to be most pressing can be granted telescope time, and often in quantity so minute as to preclude exploratory investigations or the assembly of databases adequate to ensure proper interpretation. The pressure will only increase as discoveries made with the Great Observatories place even greater demands on extant facilities for supporting and follow up observations.

Examples of the range of important scientific programs requiring extensive time on 4-m class telescopes include:

- characterizing the physical properties of sources discovered at other wavelengths (e.g. by Einstein, IRAS, ROSAT, GRO, AXAF, SIRTF) through imaging and spectroscopy;
- monitoring surface and chromospheric activity on solar-like stars;
- determining the interior structures of other stars from long-term spectroscopic monitoring programs designed to characterize stellar oscillation modes;
- searching for other planetary systems and sub-solar mass objects by means of long-term radial velocity studies of large samples of stars;
- carrying out synoptic spectroscopic and photometric studies of supernovae in external galaxies and active galactic nuclei;
- mapping the large scale structure of the universe out to $z \sim 0.1$ by determining the redshifts of 1 million galaxies;
- mapping the large scale structure of the universe out to much larger distances by determining redshifts of galaxies in carefully selected "pencil beams";

Recommended Program for the 1990s

The technological advances which enable the construction of 8-m class O/IR telescopes have greatly reduced the size, weight, and cost of 4-m telescopes, while enhancing their image quality and operational efficiency; the superb image quality obtained with the ESO New Technology Telescope attests to the potential of these new generation facilities. As a result of these advances, powerful facilities of this class can be built by individual universities or small consortia of institutions.

The O/IR panel recommends a program to build at least 4 new generation telescopes of aperture ~ 4 meters during the 1990s. Federal funds ($30 M) would be used in combination with state and private monies in order to construct these telescopes. The most urgent community need is to begin immediate construction of two general purpose 4-m class telescopes–one located in each hemisphere–to support and complement the NASA Great Observatories. University involvement in operation and management of these facilities will have the added benefit of providing opportunities for deep student involvement in both carrying out long term and/or exploratory programs and in developing novel instrumentation.

Our recommended program to construct four new generation 4-m class telescopes during the next decade is the *minimum* required to meet the most pressing needs of the community. The O/IR panel wishes to encourage in the strongest terms imaginative arrangements to build *additional* new generation 4-m class telescopes by combining state and private funds with modest levels of federal support.

A variety of models for siting and operating telescopes of this size have been suggested and many are viable. However, the economics of building and operating university telescopes of this scale requires that they be located at *a few* already developed excellent sites, thus saving development costs through sharing of roads, dormitories and support personnel.

The panel is particularly impressed by the advantages of arrangements whereby the facility is constructed with private and/or state funds on an NOAO site, while operation costs are assumed by increments to the NOAO operating budget from federal sources. This arrangement provides telescope time for (1) astronomers on the faculties of the institution(s) responsible for raising capital costs; (2) time for astronomers throughout the community who compete for access to the facility on the basis of peer-reviewed scientific proposals; and 3) takes full advantage of the nation's investment in the NOAO support infrastructure, which can provide the resources for economical, efficient and "user-friendly" operation. Such arrangements also provide the basis for continued economical operation of the existing 4-m and smaller telescopes at KPNO and CTIO – which the O/IR panel regards as critical to the scientific vitality of the US astronomical community.

RECOMMENDATIONS OF THE PANEL: SMALL SCALE PROGRAMS

Priority 1: Near-IR And Optical All-Sky Surveys

All sky surveys to fixed flux levels (e.g. the National Geographic/Palomar Observatory photographic sky survey and the recently completed IRAS infrared survey) are critical to efforts aimed at understanding the distribution of objects comprising the observable universe. Such surveys are essential because they provide *complete* pictures of the sky, unbiased by selection effects. Well-designed surveys provide the basis for vital scientific research long after their completion. For example, the galaxy redshift programs of the past decade which revealed the web-like distribution of galaxies in the universe were designed from catalogs of galaxies constructed from the first Palomar Observatory sky survey (POSS)–completed nearly 40 years ago. Advances in optical and infrared detector technology have made it possible to carry out deep all-sky digital surveys with high photometric accuracy. These surveys promise to provide fundamental databases for cataloging the distribution and brightnesses of galactic and extragalactic objects – the ultimate basis for designing a wide range of scientific programs for ground-based and space-based telescopes. *The O/IR panel recommends the completion of digital all sky surveys at (1) near-infrared, and (2) optical wavelengths before the end of the decade.*

A Near-Infrared All Sky Survey

An all sky survey at near-IR wavelengths
- can provide a complete picture of the structure of the Galaxy and the universe unhampered by the opaque dust which obscures optical views through the plane of the Milky Way.
- can probe the interior regions of optically-opaque star-forming regions,
- can locate radiation emitted by the coldest stars–objects which dominate the observable mass of our own and other galaxies.

A new, near-infrared survey can be carried out in a scanning mode (analogous to the IRAS survey)over the next five years on small telescopes (diameter ∼1 meter) using moderate-size infrared arrays. Such a survey would have a limiting magnitude of 14 at K (2.2 μm) – nearly 50,000 times fainter than the faintest source catalogued in the original Two Micron Sky Survey. The proposed infrared survey is urgently needed (1) for developing target lists and planning initial observing programs for the two major US infrared space missions of the 1990s: SIRTF, and the NICMOS imaging camera/multi-object spectrograph on HST; and (2) to act as the near-IR analog of the POSS for guiding development of observing programs for the 8-m IR-optimized telescope. A survey of this sensitivity will also enable a wide range of investigations in its own right. These include
- carrying out a census of galaxies *uniformly around the sky* , thus sampling the nearby universe and its gravitational potential field to redshifts z∼0.05. We note that the "Great Attractor" which has been invoked to account for asymmetries in the redshifts of galaxies out to distances of ∼100 Mpc appears to be located along a direction which is obscured from view at optical wavelengths by dust in the plane of the Milky Way;
- using the observed infrared brightnesses of spiral galaxies derived from the survey, along with neutral hydrogen line profiles, to determine accurate distances to large numbers of galaxies;
- probing the structure of the Milky Way galaxy by using K and M giants and infrared-bright asymptotic giant branch stars;
- providing deep maps of star-forming regions in Gould's Belt and in the Perseus, Sagittarius, and Carina arms of the Milky Way;
- providing the most sensitive means yet proposed (in terms of volume searched) to locate elusive sub-stellar mass objects ("brown dwarfs").

The survey must be carried out with two telescopes, one located in the Northern hemisphere, the other in the South. The estimated cost to build the required telescopes and cameras for the all-sky near-IR survey is $6M.

We recommend that the survey data be archived and made widely available to the international community of astronomers by using procedures analogous to those already developed at NASA's Image

Processing and Analysis Center (IPAC) which has provided exemplary service to the community in archiving and disseminating the IRAS database.

All-Sky Optical Survey

Since the invention of photography, astronomers have used deep sky images as their primary tool to learn about the morphology and distribution of objects which comprise the observable universe. The most important of these surveys has been the Palomar Observatory Sky Survey (POSS), completed in the early 1950s. This photographic survey and its successor (the deep Palomar II survey), and others like them in the southern hemisphere have been the mainstays of much astronomical research – providing the basis for identifying galaxies and clusters of galaxies, and cosmic sources of gamma-rays, x-rays, IR and radio radiation.

With the advent of large-format, sensitive digital detectors (e.g. 2048x2048 CCD arrays) it is now possible to replace photographic surveys by surveys that are both digital and well calibrated photometrically. We propose a survey aimed at obtaining simultaneous images of the sky at B (0.44 μm) and R (0.65 μm) which provide S/N = 10 for objects of B=23 and R=22 magnitude at a spatial resolution of 1″/pixel. It is desirable to complete a survey of both celestial hemispheres within 5 years.

Such a digital survey will provide:

- a uniform sample of galaxies to 20th magnitude (15 million galaxies at galactic latitudes, b $\geq$ 30°); this catalog will provide the basis for designing deep redshift surveys aimed at mapping the three-dimensional structure of the universe out to large distances;
- select clusters of galaxies to redshifts near unity (at z = 1, the 10th brightest galaxy in a cluster is B ~21 mag);
- obtain accurate photometry for brighter galaxies and stars, allowing the construction of photometrically uniform samples and color-selected samples;
- locate faint optical counterparts for radio, x-ray, gamma-ray, and infrared sources.

The survey requires a dedicated 1-m class telescope of conventional design. After completion of the initial all-sky survey, it will be possible to use the same equipment to carry out 1) deeper surveys in selected regions; 2) surveys with higher angular resolution in order to achieve high photometric accuracy for stellar objects in our own galaxy.

We recommend that the optical all-sky survey results be archived and made available to the international community of astronomers using the procedures established for the dissemination of IRAS and HST data at IPAC and STScI.

The estimated cost of the all-sky optical survey is between $3M and $5M depending upon whether extant facilities can be used or new telescopes must be constructed.

The panel wishes to note that the recommended infrared and optical all sky surveys are deemed the most pressing and important representatives of a much larger class of more specialized surveys (e.g. narrow-band imaging surveys at Hα and H$_2$ enabled by modern optical and infrared detector technology). It is thus essential that these two recommended surveys be designed with a clear goal of developing a community-accessible infrastructure for the reduction, analysis, archiving and distribution of databases assembled by more specialized surveys. It is these functions which are most costly over the long term, and yet most vital for ensuring the production of a uniform, well-understood and well-used database. It seems wise to make use of the current community investment in the facilities and scientific expertise assembled at IPAC and STScI to support such specialized surveys.

Priority 2: A National Astrometric Facility

Ground-based astrometry has made dramatic gains over the past decade, as a result of applying modern CCD detectors, image analysis techniques and innovative new instrumentation to astrometric measurements. Several groups have demonstrated accuracies in position determinations of 1 milliarcsecond (1 mas) or greater – a gain of 5 or more when compared to previous measurements.

Although this accuracy offers the potential to determine and calibrate fundamental stellar parameters, the number of stars for which such measurements are needed far exceeds the capabilities of existing astrometric programs. Moreover, ground-based and space-based programs are identifying many objects of interest with undetermined physical properties. Measurement of the parallax and proper motion of these objects is often a necessary prerequisite to further study.

Furthermore, the internal positional accuracy of images obtained with new space-based facilities (and by the mid-1990s, large ground-based telescopes equipped with adaptive optics) exceeds the accuracy of the optical reference frame (~50 mas) and the accuracy of the intercomparison of the optical and radio reference frames (~300 mas). Transferring coordinates between reference frames, particularly at radio and optical wavelengths, is critical to enabling the precise registration of images required for detailed multi-wavelength studies of spatially-resolved systems.

Finally, the search for planets, brown dwarfs and other underluminous objects must be undertaken over several decades and require both instrumental stability and long-term commitment of institutional resources to ensure their success.

The importance of these three classes of programs, the specialized nature of the instrumentation, the requirement for long-term stability and the need to supply the astronomical community with fundamental data lead the O/IR panel to *recommend the construction of a special purpose, dedicated 1.5m telescope and ancillary instrumentation. We urge that this facility be operated as a national observatory.*

The recommended *National Astrometric Facility* should be funded and operated in a manner which permits both rapid response to targets of opportunity (e.g. orbit determinations for solar system objects), and long-term commitment to programs requiring *decades* of observation (e.g. searches for planetary systems). It should be scheduled in response to proposals arising from the astronomical community. The National Astrometric Facility should also provide a test bed for development of innovative astrometric instrumentation and techniques.

The panel estimates that the construction and initial instrumentation cost of such a facility to be $5M. The first such facility should be built in the northern hemisphere at a first class site. In the long-term, a twin facility should be developed for the southern hemisphere.

RECOMMENDATIONS OF THE PANEL: INFRASTRUCTURE ISSUES

Develop, Purchase and Distribute Optical CCDs and Infrared Arrays

The development of optical charge coupled devices (CCDs) and near infrared arrays has revolutionized optical/infrared astronomy. At optical wavelengths, CCDs provide an order of magnitude improvement in sensitivity for most applications over the previous generation of detectors, and provide major improvements as well in geometric and photometric stability. With current noise levels approaching a few electrons rms, and quantum efficiencies near 80 percent, they are very nearly ideal detectors. Infrared arrays are less mature but no less revolutionary in their impact on the field. Near-infrared arrays (HgCdTe and InSb) have *enabled* infrared sensitive imaging and two-dimensional spectroscopy for the first time.

Our recommended program for the 1990s assumes as a prerequisite, adequate funding for the development and distribution of advanced panoramic detectors. Without this investment, it will be impossible to achieve the full potential of new generation, large telescopes. We outline below the elements of a national program to develop, purchase and distribute large format CCDs and infrared arrays.

Optical CCDs

The past decade has seen very rapid development, but still *unsatisfactory general availability* of large CCDs suitable for astronomy. Thanks to a fairly major NSF/NASA effort, the TI 800x800 chips developed for the Wide Field/Planetary Camera on HST have been distributed widely to the community. However, these devices have been far surpassed in performance by newer chips. Moreover, they are not well suited in overall dimension or pixel size to the new generation of large, 8-m class telescopes. *The O/IR Panel strongly recommends a program developing CCD detectors optimized for performance on large telescopes, purchasing them in sufficient quantity, and distributing them competitively to the astronomical community.*

It is essential that all large telescopes be equipped with CCDs if they are to achieve their full potential. To achieve this goal, the Panel has identified the following actions that need to be taken as soon as possible:

- A coordinated purchase of a particular CCD design in quantity and volume sufficient to interest commercial vendors in providing the fabrication and cosmetic quality required for astronomical detectors.

- A coordinated development program to understand the advanced CCD technologies needed for the next generation of astronomical devices and to develop those technologies so that they can be included in commercially available devices.

A wide variety of instrumentation needs on telescopes of all aperture sizes can be addressed by a single CCD design with the following characteristics.

- Pixel size determined by existing optical systems: As has been demonstrated by the NSF/NASA TI 800x800 CCD distribution, most existing optical systems are well matched by a 15 micron pixel size. Instrumentation on 8-m telescopes requires pixel sizes nearer 30 microns. Through the technique of on-chip binning, the same CCD can satisfy both requirements in most applications.
- Large Format: Wide field imaging and spectroscopy require minimum array sizes of 4096x4096 15-micron pixels, and many advanced applications require mosaiced-arrays of such devices.
- Charge Transfer Efficiency At Low Light Levels: The ideal device would require 99 percent collection efficiency at the far corner of the device, or better than 0.999992 efficiency per transfer assuming a three-phase, quadrant readout design.
- Low Noise: Amplifiers with rms noise levels of $\sim$4 electrons at rates of 50,000 pixels per second are now common. Requiring non-destructive readout with a few electrons rms seems feasible.
- Wide Wavelength Coverage: CCD detectors, by virtue of their native silicon photoconductor response, typically have quantum efficiencies of 60-80 percent in the green and red spectral regions. The response in the blue and ultraviolet is another matter. All approaches to maximizing blue response require that the device be *thinned* – that is, made thin enough so that light can strike the back side, away from the gate circuitry which is opaque to blue light. The alternative, covering the front surface with a photon energy down-converter, does not offer the tremendous gains offered by thinning.

While this CCD design is good, and its general availability requirement is urgent, it is not perfect. Research is needed in many areas, including flashgate technologies to improve the blue quantum efficiency of thin devices, anti-reflection coatings to optimize response in particular wavelength regions, mosaic technologies for combining devices, amplifier design for the minimization of noise, and gate structures to allow operation at warmer temperatures. A coordinated program to design, acquire, and test CCDs is needed to bring these advances from the laboratory to the telescope.

A major obstacle to concentrating effort on producing large format CCDs optimized for astronomical use has been the piecemeal purchasing patterns which are a result of the *lack of a concerted funding effort in this very critical area.* It is clear that expenditures small compared to the capital outlay for a large telescope would have an enormous impact on the availability of these crucial detectors. The Panel estimates that expenditure of $5M for the chip design discussed above would provide the few hundred devices required by the community. The cost of the research efforts and a second coordinated purchase of CCDs later in the decade would require an additional $5M. It is important to emphasize that the total $10M *is not* to be spent at $1M/year; it is critical that the first installment of $5M be provided immediately so that the large format chips be purchased and available for timely installation on the new generation 4-m and 8-m telescopes.

Infrared Arrays: The Future for the 1-5 μm Region

The NICMOS (Near Infrared Camera, Multi-Object Spectrograph) program has provided powerful 256x256 pixel arrays. Imaging and spectroscopy would benefit enormously from the production of larger arrays. InSb and HgCdTe arrays of dimension 512x512 seem technically feasible, but at present there is no support to develop larger successors to the NICMOS arrays. *We recommend a program to develop, purchase and distribute larger near-infrared arrays.*

Prior to a large-scale purchase, it is essential to carry out additional research and development aimed at providing better uniformity and broader wavelength response for these detector materials. It would also be extremely valuable to invest in the development of array designs that enable *moisaicing* of arrays.

Infrared Arrays: The Future for Mid-Infrared Arrays

The very high sky and telescope background at $\lambda \sim$10 μm produce enormous background fluxes ($\geq 10^9$ photons/s) which have made array implementation very difficult. Two strategies have been attempted:

- read out the detector as fast as possible (before the wells fill), or
- increase the well-depth by increasing the detector bias

Neither provides per pixel performance on par with the best achievable with single detectors. Because the problems are daunting, fewer groups are working at these wavelengths, and moreover, the spinoff from

SIRTF-driven research is minimal because the background levels for this cooled telescope in space are very much lower than for ground-based applications.

The development of these detectors is *essential* if we are to take advantage of the dramatic gains in mid-IR sensitivity and angular resolution enabled by diffraction-limited 8-m IR/optimized telescopes; these telescopes will provide ~10 times the angular resolution of SIRTF in the 5-20 μm regime. *We recommend a program to develop sensitive, low-noise mid-IR array detectors for use on ground-based telescopes.* The panel notes that:

- good detector materials for mid-IR arrays exist: Si BIB detectors provide good quantum efficiency from 5 μm to 28 μm;
- the need for large well-depth (to accommodate high background fluxes) necessitates care in the design of the readout multiplexer; current devices may have unit cells which are too small to provide the needed large capacitances;
- It seems plausible to produce small mid-IR arrays (32x32 or 64x64 pixels) in the near term. To produce larger arrays would appear to require fundamental design work.

It is essential to fund several groups to work with industrial firms to develop and test detectors of astronomical use. However, to induce development of advanced technology arrays will require significant investments.

The O/IR panel recommends a total investment of ~ $20M over the next decade in order to enable development, and later purchase, of the near- and mid-IR arrays that are so critical to achieving efficient returns from the new generation of large telescopes.

A Program to Support Large Optics Technology

Fabrication and Polishing of Large Mirrors

The coming decade is one of great promise for ground-based O/IR astronomy. A new generation of large telescopes is planned which, if built, will quadruple presently available light gathering power. Adaptive Optics hold the promise of diffraction-limited imaging by these telescopes, and offer more than an order of magnitude gain in angular resolution. *The key to these advances is the successful production of a number of primary mirrors 8-m in diameter–larger than any in existence.*

The fabrication of optics for these telescopes presents an enormous challenge. Methods have now been demonstrated on an intermediate scale for casting rigid glass honeycomb blanks. New techniques have been developed for polishing aspheric surfaces to the extremely high accuracies now required to match image quality at the best sites. All planned telescopes other than the Keck telescope will use these large monolithic mirrors.

As we enter the 1990s, construction of the first facility in the world for casting and polishing 8-m mirrors is nearing completion at the University of Arizona Mirror Lab. There still remains the task of proving the casting and polishing technology at the full size of the 8-m mirrors, and the production of six such mirrors during the decade. *The O/IR panel recommends continued support of the UAML at a rate of* ~ $2M/yr throughout the decade.

Fabrication and Polishing of Specialized Optics; Coatings

The full potential of new telescopes will be realized only if they are equipped with excellent instruments. The best designs call for optical elements beyond present optical fabrication capability. *The O/IR panel wishes to encourage the funding (*$500K/yr) of a coordinated effort involving university and private groups along with the NOAO to develop the technology for polishing difficult aspheric surfaces. It will also be necessary to invest in efforts aimed at developing efficient anti-reflection coatings and depositing them uniformly on large optical elements, along with efficient and durable reflective coatings aimed at minimizing telescope emissivity.

Toward a New Generation of Large Filled Aperture Telescopes

While much activity during the next decade will be devoted to the construction of 8-m class instruments and to the development of ground-based interferometry involving multiple small (~2-m class) telescopes, it would be wise as well to invest effort in developing the technology and encouraging imaginative approaches for construction of the next generation of large, single aperture telescopes.

One possible design for a 32-m class O/IR telescope is based on the mechanical structure of a radio telescope. In this concept, the dish is tiled with small polished glass segments, whose position would be rapidly adjusted to maintain diffraction-limited imaging, compensating not only for gravitational flexure as the telescope is moved, but also for wavefront aberrations caused by the atmosphere. To keep down fabrication cost, and to simplify alignment, the primary surface would be spherical (similar to the Arecibo telescope). To achieve good images, large and aspheric secondary and tertiary mirrors are required. New techniques for polishing large 8-m primaries for the telescopes of the 1990s appear extendable to these requirements.

Fundamental to the utility of large single apertures is the development of adaptive optics: the advantage of diffraction-limited imaging increases very rapidly with aperture. It follows that a very big telescope tailored for adaptive optics would be a very powerful tool. We can be fairly certain than a 32-m dish could be controlled to image much of the sky to the diffraction limit at 2.2 μm with the light from field stars. If the technology for sensing atmospheric wavefront errors from artificial (laser) stars is perfected, then diffraction-limited imaging over the whole sky will be possible at optical wavelengths. The corresponding resolution of 0.004″ with all the light from a 32-m aperture focused into an image this small would constitute an extraordinary advance in astronomy.

We recommend funding at the rate of $\sim$ \$1M/yr over the next decade in order to support the design of large aperture telescopes, and to develop mirror fabrication and polishing technology required to enable the construction of such telescopes.

A Program to Archive and Disseminate Astronomical Databases

Modern astronomical images and spectra are now obtained almost exclusively in digital form. The O/IR panel believes that it is important to develop archives of coherent data sets *so the community of astronomers can derive maximum benefit from these data now and in the future.*

The value of archived data is amply illustrated by the following examples:

- the Palomar Observatory Sky Survey has proven an immensely valuable database both in its original photographic form, *and in its duplicate copies which have been disseminated widely in the astronomical community.* The survey has provided catalog and finding lists of a wide variety of celestial objects from galaxies, to H II regions, to Herbig-Haro objects. It has proven invaluable for selecting guide stars for the Hubble space telescope. In brief, it provides a paradigm for the wide-spread utility and scientific longevity of *astronomical surveys.*
- The databases from the HEAO and IRAS satellites have proven to be vital sources of astronomical research well after these satellites ceased taking data. It is also noteworthy that although the original science instrument teams for HEAO and IRAS are responsible for many of the exciting initial discoveries made with each satellite, *members of the scientific community working with these databases have been responsible for the bulk of the scientific return from these projects.* Moreover, wide involvement of the scientific community has led to better calibration and understanding of these databases as well as to richer scientific return. These databases provide a paradigm for the great value of community involvement in working with databases originally developed by teams of scientists working on a variety of key projects.

The O/IR panel recommends that funding be provided in order to ensure:

- that databases created by *surveys* (e.g. all sky digital surveys, catalogs of redshifts), and by *key projects* (long-term programs carried out by individual observers or teams of astronomers) be archived and following a 1 to 2 year period of exclusive use, be made available to the scientific community.
- that every observation taken on a large telescope in the US is catalogued and characterized and that the catalog of observations be made available to the scientific community.

The panel urges that the data be archived locally first, and then transmitted to more central archives such as those at IPAC and the STScI, and eventually to the NSSDC. The O/IR panel urges that *as a matter of policy* that the format of all archived data will be FITS (flexible image transport system).

The panel considered the possibility of archiving *all* data taken at large telescopes, but rejected the notion of universal archiving until procedures developed for surveys and key projects are developed, understood, optimized and costed.

A Program for Training New Instrumentalists

There is a major demand for talented young astronomers to design and build instrumentation for the next generation telescopes. The current shortage of trained instrumentalists will become even more acute unless the community can stimulate its students to take an active interest in state-of-the-art technology for astronomical observations. Training in the techniques of instrument creation has traditionally involved apprenticeship to an individual or group building and using new equipment. We recommend a three-level program aimed at providing such apprenticeships:

- *(1) Involve graduate students in instrumentation development at NOAO.*

The NOAO should initiate a program aimed at active involvement of advanced graduate students in the development of instrumentation at the national observatories. This program would provide support for students (and in some cases their advisors) in residence at the national observatory. During this time, such students would work closely with the team of scientists and engineers involved the design and construction of a major or innovative instrument. We also urge that NOAO seek frequent collaboration with university groups to develop instrumentation for the national observatories. Active collaboration with such groups would allow NOAO to tap and support the talent distributed throughout the university community, and to support students actively involved in these instrumentation efforts.

- *(2) Provide explicit support for graduate students at institutions developing instruments for moderate and large telescopes.*

At least 10 major university groups will be building and instrumenting moderate or large facilities during the 1990s. Students enrolled at these institutions have an unusual opportunity to become involved with the construction of frontier instrumentation.

However, all instrument large development projects are cost-constrained, and increasingly must meet demanding time schedules. These pressures act both to limit student involvement in large projects in favor of more experienced engineers and technicians, and to reduce their freedom to learn through making creative mistakes. Funding agencies must recognize these realities, and develop positive incentives to involve students in instrumentation development at these institutions. At the very least, the financial pressures which may prejudice graduate student support should be eliminated through initiation of an NSF program – analogous to the Research Experience for Undergraduates (REU) program – designed explicitly for support of graduate students involved in such projects. PIs for large instrument projects should be able to apply competitively for "IEG" (Instrumentation Experience for Graduate Students) grants. Furthermore, evidence of encouraging student involvement should be viewed as an important positive factor in the review process for instrumentation grants funded by NASA and the NSF.

- *(3) Provide support for graduate students at all institutions*

Universities with access to relatively small telescopes (d $\leq$ 1.5m) can also play a critical role in training future instrumentalists if they are able to provide (1) steady access to telescopes with equipment *designed wholly or in substantial part by students;* (2) the infrastructure to enable student-built instruments to produce useful scientific results, including the availability of modern detectors, and adequate supplies of parts, tools, and test facilities; and (3) guidance by senior instrument designers who bear major responsibility for the student's career.

We recommend a new program for supporting operating costs and instrumentation development for small, University-operated telescopes. The program would provide funding for developing modern instruments for these telescopes, and for upgrading telescope control and data acquisition systems. The instruments need not be *unique,* frontier instruments. Rather, the request for funding should be judged on the basis of (1) potential for scientific return; (2) evidence of strong student involvement in the design, construction and use of the instrument; (3) the potential and track record of the senior mentor. This program would fill a gap between the NSF *Instructional Laboratory Improvement* grants and the traditional grants under the NSF Astronomy instrument program. These funds would be specifically targeted at support of creative programs aimed at *graduate training* of future instrumentalists.

We believe that initiation of this program would have a number of benefits. First and foremost, it would provide students with the involvement in a project over which she/he has primary control over design, construction and use. Second, it would result in the revitalization of a variety of local facilities, providing opportunities for student and faculty research on programs requiring imagination and a willingness to trade time for aperture. We cannot overemphasize the importance of giving students control of their own projects early-on in their formative years. Students involved in this program will be drawn to the pleasure of running a significant project from start to finish – and some will become the instrumental leaders of tomorrow. Funding of $\sim$ $500K/yr over the next decade will be required.

UV-OPTICAL FROM SPACE PANEL

GARTH ILLINGWORTH, University of California, Santa Cruz, *Chair*
BLAIR SAVAGE, University of Wisconsin, *Vice-Chair*

J. ROGER ANGEL, University of Arizona
ROGER D. BLANDFORD, California Institute of Technology
ALBERT BOGGESS, NASA Goddard Space Flight Center
C. STUART BOWYER, University of California, Berkeley
GEORGE R. CARRUTHERS, Naval Research Laboratory
LENNOX L. COWIE, Institute for Astronomy, University of Hawaii
GEORGE A. DOSCHEK, Naval Research Laboratory
ANDREA K. DUPREE, Harvard-Smithsonian Center for Astrophysics
JOHN S. GALLAGHER, AURA
RICHARD F. GREEN, Kitt Peak National Observatory
EDWARD B. JENKINS, Princeton University
ROBERT P. KIRSHNER, Harvard-Smithsonian Center for Astrophysics
JEFFREY L. LINSKY, University of Colorado, Boulder
H. WARREN MOOS, Johns Hopkins University
JEREMY R. MOULD, California Institute of Technology
COLIN A. NORMAN, Johns Hopkins University
MICHAEL SHAO, Jet Propulsion Laboratory
HERVEY S. STOCKMAN, Space Telescope Science Institute
RODGER I. THOMPSON, University of Arizona
RAY J. WEYMANN, Mt. Wilson and Las Campanas Observatory
BRUCE E. WOODGATE, NASA Goddard Space Flight Center

UV-Optical from Space

EXECUTIVE SUMMARY

Observational capability from space in the UV, Visible and IR spectral regions will be crucial for substantial progress on fundamental astronomical problems that range from the study of planetary systems to understanding galaxy formation. The combination of uninterrupted wavelength coverage (ultraviolet through the visible to the mid-infrared), high dynamic range, diffraction-limited images and dramatically reduced near-IR backgrounds with passively-cooled systems (to $< 10^{-6}$ that from the ground and HST) would result in major advances in scientific understanding.

A balanced program of both Observatory-class instruments and smaller directed missions of Moderate and Explorer size is essential. The panel strongly recommends that both components be developed concurrently in a dynamic long-term program.

A long-term program of Observatory-class missions is recommended, starting with the full realization of the potential of HST through optimized operation and the rapid implementation of state-of-the-art image-correcting cameras and spectrographs. The critical role that Observatory-class instruments play requires that a successor to HST be flown within a few years of the end of HST's nominal life. The panel strongly recommends that such a successor be planned for launch in the first decade of the new century. This would be the 6 m LST, the Large Space Telescope. Operation beyond Low Earth Orbit would lead to large gains in the science return and substantial savings in construction and operational costs. Finally, the program has as its goal a telescope of astonishing power, the 16 m NGST (Next Generation Space Telescope).

Many high-priority science goals cannot be addressed with general-purpose large telescopes. Paralleling the Observatory program must be a vibrant program of Moderate and Small missions. Progress in the Explorer program is currently so slow that its primary goal of rapid access to space for innovative scientific programs has been lost. We recommend that the Delta-class Explorer program be enhanced to lead to more frequent missions and shorter turnaround. The missions should also be managed more directly by the PI team; substantial educational and training benefits would accrue from such a change. While the final selection of missions should be carried out through the normal peer-review process, examples of outstanding science goals include a UV imaging survey, a wide-field astrometric telescope, a multi-waveband mission and a very high spectral resolution instrument.

Interferometry promises to revolutionize the study of compact objects with small length scales such as AGNs, QSOs and interacting binaries. The panel recommends beginning a systematic program of technology development and ground-based experimentation that leads to a space-based imaging astrometric interferometer with baselines of tens of meters early in the next decade.

A more complete outline of the recommended program can be found in §II (Implementation of the Science Program). The recommended program is:

SIZE	PROJECT	Cost	Start	Finish
Large:	LST – 6 m HST Successor	$2000M	1998	2009
Moderate:	Explorer Enhancement	$300M	1993	2000
Moderate:	HST Third Generation Instruments	$150M	1994	2000
Moderate:	Imaging Astrometric Interferometer	$300M	1997	2004
Small:	SMEX UV Survey.	$30M	1995	1998
Small:	Space Optics Demonstration	$30M	1993	2000
Small:	Supporting Ground-based Capabilities	$25M	1993	2000
Technology:	Technologies for Space Telescopes	$30M	1993	2000

Large:

LST (Large Space Telescope): The LST is a 6 m Observatory-class telescope incorporating UV to IR imagers and spectrographs. Passive-cooling and high-performance optics result in large gains in scientific capability over HST. A high priority goal is location beyond Low Earth Orbit (e.g., HEO – High Earth Orbit). This telescope is an excellent candidate for strong international participation. For operation by 2009 a start date of 1998 is considered necessary. Advances in technology and HEO operation will break away from the HST cost curve and lead to an expected cost of $2000M.

Moderate:

Rapid Explorer Deployment: The Delta-class Explorer program should be enhanced by more frequent missions and shorter development schedules. A wide range of forefront science can be carried out with such missions: contemporary examples are EUVE and Lyman-FUSE (the Extreme UV Explorer and the Far UV Spectroscopic Explorer). The programs should be chosen through peer review. An essential element is education and training of space scientists and engineers. The base of technical and managerial experience in space science must be increased by direct oversight and management of these programs by the PI team. Cost savings will accrue from such an approach. The expected incremental cost of the program is $300M.

HST Third Generation Instruments: The full potential of HST can only be realized with state-of-the-art instruments that compensate for its optical problems. A further set of such instruments should rapidly follow the WF/PC II (Wide Field/Planetary Camera) and the Second Generation Instruments. A high throughput camera is an example. The expected cost of the instruments is $150M.

Imaging Astrometric Interferometer: The potential returns from interferometry are high. An instrument with baselines beyond 20-30 m that demonstrates the needed technologies and offers substantial scientific gains should be developed for a new start late in the decade. The expected cost is $300M.

Small:

Small Explorer missions: A UV low spatial resolution all-sky imaging survey would return excellent science in this class. The UV survey would be particularly valuable for mapping diffuse UV emission from the hot and cold ISM. The expected cost of a Small Explorer mission (SMEX) is $30M.

Optics Development and Demonstration: Lightweight optics with high performance surfaces and active control are central to the goals of future telescopes, large and small. Technology developments and demonstrations are needed. The expected cost is $30M.

Supporting Ground-based Capabilities: Ground-based telescopes will continue to play an essential complementary role in support of space observations. They will continue to be particularly important for spectroscopic follow-up and for development and demonstration of new detector and instrument technologies. All-sky coverage is essential. The ground-based program should be augmented by $25M for 4+ m class telescope projects that support space observations.

Prerequisites:

Lyman-FUSE: The high scientific returns expected from the EUV and UV spectroscopic capabilities of Lyman-FUSE requires that it be completed and launched by the middle of the decade.

HST Optimization: It is crucial to realize the potential of HST through implementation of instruments that correct for its optical problems. Thus the WF/PC II should be completed as soon as possible, and the state-of-the-art Second Generation Instruments should be put on an accelerated completion and deployment schedule. Optimized operation is now even more crucial given the reduced efficiency of HST.

Near-term Science Program: Support for data analysis, ground-based observations, and modeling and theory is badly needed to give the scientific returns consistent with the capabilities and the large commitment of national resources to space science missions. Stable research funding of the science community and students is essential if the scientific productivity is to be commensurate with the investment in facilities. The ongoing missions such as HST, EUVE, IUE and ASTRO have and will make substantial scientific contributions. They require adequate support.

Technologies for the Next Century:

The Next-Century Program: Explorers, interferometers and large space telescopes depend critically on technology developments. Development of those technologies now will lead to major savings in cost, schedule and management complexity for the next generation of instruments. Such technological developments will be crucial for observatories of the scale of the NGST, the 16 m Next Generation Space Telescope.

Structure of the Panel Report.

The report is organized as follows:

I. THE SCIENCE PROGRAM.

The UV-Optical in Space Panel has been considering the science goals for a broadly-based program of space missions that cover the wavelength range from the EUV at ~ 100 Å into the near-IR at $\sim 5\text{-}10$ μm. The moderate (Explorer-class) and large (Observatory-class) missions outlined here allow for fundamental advances in almost all areas of astrophysics. They range from the detection of planetary systems around nearby stars to cosmology. The science program developed by the panel is outlined below.

Planetary systems.

The detection of planets beyond the solar system was given great impetus in the last decade with the discovery by IRAS of radiation from cold disks of dusty material around nearby stars. Detection of the light from planets, and the subsequent spectroscopic observations, are important goals for the coming decades and can very likely be attained through the design of space-borne instrumentation directed to that end. The technological requirements are challenging. A low background telescope with apodizing, occulting or interferometric instruments will be required. The telescope will need to be cooled (passively) to $\sim 100°$ K or less, and to have extremely smooth, low scattering optics. These optics must be smoother than required for diffraction-limited performance. While the detection of planets like those in our own solar system is likely to be beyond the reach of HST, a 6 m telescope with higher performance optics and passive cooling would be able to detect gas giants like the outer planets (Jupiter and Saturn) in our own solar system out to beyond 10 parsecs, even against the huge background from the central star.

To take the next step and detect Earth-like planets requires a substantial step in capability, especially since detection alone is not enough. We need to take spectra that can confirm whether the atmospheres of the detected planets could support life. The spectroscopic measurements must establish whether their atmosphere is in a chemical disequilibrium state that might indicate the presence of life (e.g, by detecting the 9.5 μm Ozone band).

The technical challenges confronting the detection and measurement of Earth-like planets are substantial. Such planets would be found $\approx 0.25''$ from a star at a distance of a few parsecs. An optimal approach would be to detect such an object at $\sim 10\ \mu$m with a cooled, 16+ m telescope where the first dark diffraction ring corresponds to the planet's orbit. Apodization or interferometric instruments would be used to greatly enhance the contrast of the planet against the light from the star. Then the telescope's low resolution spectroscopic system (with $\lambda/\Delta\lambda \sim 100$) would be used to obtain a spectrum to search for the signature of ozone (O_3) at 9.5 μm. Other weaker features, e.g., methane, could also be the goal of such a spectroscopic analysis, albeit with data of higher S/N and spectral resolution. Even with low resolution spectra it appears likely that satisfactory discrimination could be obtained between different atmospheres. For example, the thermal emission spectrum from the inner planets in our own solar system (Mars, Earth and Venus) differs greatly.

It is probably no exaggeration to say that the impact on humanity of successful detection of a planet with an atmosphere like that of the earth would be similar to that of Galileo's detection of the moons of Jupiter, following the invention of the telescope.

Star formation and origins of planetary systems.

Stars form through a series of complex physical processes. While absorption limits the study of much of the activity in star-forming complexes to long wavelengths, substantial investigations can be carried out in the near-IR and shorter wavelengths. The resolutions available at such wavelengths are a better match to the characteristically small length scales in protostars. Of particular scientific interest at these wavelengths is the study of protostellar disks and outflow jets. The resolutions and sensitivity of space telescopes will allow the derivation of physical conditions (temperatures and densities) along with kinematical and structural maps. 1000 AU diameter disks could be mapped with ~ 15 independent resolution elements at 1 kpc with the 6 m LST.

While our own Galaxy provides a laboratory for the study of star forming complexes at a level of detail that is impossible in other galaxies, different environments in other galaxies make such regions of particular interest. We would like to know how the formation rate for massive stars and the form of the mass function at high masses are affected by physical conditions that are found only in other galaxies. These include normal galaxies like our neighbors such as M31, NGC 205 and M33, and the LMC and SMC, as well as the more massive systems such as M101. Moreover, galaxies undergoing violent kinematic interactions such as "starburst" galaxies are of particular interest since they may be representative of conditions that were common during the formation and early evolution of galaxies. Large space telescopes that combine high spatial resolution with low backgrounds are essential for tackling this problem.

In general, the spatial resolution that will become available with HST, with the 6 m and ultimately with the 16 m will be crucial for unravelling some of the complex structural and dynamical characteristics of protostellar structures. The 6 m LST would resolve 3 AU in the nearest star-forming complexes, or 8 AU at Orion at 0.5 μm in the visible. At 3 μm the resolution would be ~ 20 AU and 50 AU respectively. Furthermore, the low backgrounds and high dynamic range available with space telescopes, particularly passively-cooled telescopes, will broaden the range of problems that can be tackled. With cooling to $\sim 100°$K, the background out to $\sim 10\ \mu$m can be > 18 mag fainter *per resolution element* $(< 10^{-7})$ than that from the ground.

Structure and Evolution of the Interstellar medium.

Fundamental information on the composition of interstellar gases and evidence for spectacular differences in their physical states has resulted from observations of the rich assortment of UV absorption lines in the spectra of background stars. Unfortunately, few of the spectra have had the velocity resolution to clearly differentiate the different underlying structures within gaseous regions. These regions could be responding to a variety of undetermined physical influences, such as ionization, dissociation and recombination fronts, thermal- and photo-evaporation interfaces, subsonic wave phenomena, and ionization and cooling from shock fronts. These phenomena can trigger important chemical changes or shifts in excitation and ionization over very small scales in distance or velocity. The interplay of such processes is important in the exchanges of mass and energy between different phases of the medium, and can be influential in regulating the large scale properties of gases within the disk and halo of the Galaxy.

With the high resolution spectrograph on HST (particularly with STIS, the Second Generation spectrograph) faint objects can be measured with a wavelength resolving power of about 10^5. Evidence for complex, narrow absorption lines in the visible indicate that studies of physical processes in the interstellar medium (ISM) would be greatly rewarded by having results from UV spectrographs which could observe bright sources with a resolving power of a few million. The technologies for UV echelle gratings and 2-dimensional detectors are mature enough that it is now possible to achieve good sensitivity and resolution to study bright stars (to 5th magnitude) with a small instrument (effective collecting area ~ 10 cm^2).

Much remains to done through imaging studies of the ISM. Observations on a variety of spatial scales in the UV would also help in elucidating the composition, physical properties, and spatial distribution of dust, interstellar H_2, and supernova remnants. These observations of interstellar emissions and scattering complement the spectroscopic measurements. Instruments with wide field coverage and high sensitivity to diffuse emission are needed to complement the capabilities of ASTRO and large telescopes such as HST and LST. Explorers, both SMEX and Delta-class, could lead to major gains in our knowledge.

Just as hot stars are used to probe the interstellar medium of the Galaxy, QSOs and AGNs can be used to probe the interstellar medium of much more distant galaxies. Contained in the investigation of QSO absorption lines are fundamental astrophysical questions: What is the history of chemical enrichment in various galaxy environments? How are the physical conditions in the gaseous medium influenced by the host galaxy properties and by the extragalactic radiation field in which the host galaxy resides?

Observations with HST and Lyman-FUSE will allow direct comparison of physical conditions in low-redshift absorbers with those in high-redshift absorbers as measured from ground-based data. From observing lines-of-sight through the halo of the Galaxy and local group galaxies, the gas phase depletion pattern can be determined as a function of stellar metallicity. Because the low redshift absorbers can be associated directly with host galaxies, the problem of current gas-phase metallicity as it relates to the history of star formation can begin to be addressed.

The key to relating QSO absorption lines to galaxies lies in obtaining observations in the redshift range in which both can be well studied ($z < 1$). HST will only begin to pursue such studies at high spectroscopic resolution because of photon limitations and the low surface density of bright QSOs. Larger aperture space telescopes such as LST will result in very substantial increases in the number of sources that can be studied. The physical conditions in the absorbing interstellar medium of galaxies could be directly related to other observable properties of the host galaxy. Unreddened B stars could be used to examine the interstellar medium of Local Group galaxies for velocity structures in star-forming regions, for metallicity as a function of star formation rate and history, and for gas phase depletion as a function of local metallicity.

With their high throughput and their capability for high-dispersion UV spectroscopy, LST and even larger telescopes such as the 16 m NGST are well-suited to such studies. A strong scientific motivation for LST is to raise QSO absorption line studies to the level where the chemical and dynamical evolution of interstellar and halo gas can be determined as a function of cosmic time.

Stellar astrophysics.

One of the great triumphs of modern astrophysics was the development in the 1960s and 70s of a broad understanding of the evolution of single stars from the gaseous cradle to the dense stellar grave. This rush of progress followed from parallel developments in observation and theory. Technical advances made it possible to measure stellar luminosities and temperatures accurately and simply, and to model the evolution of stars in the Hertzsprung-Russell diagram with detailed numerical calculations.

Our understanding of the evolution of close binary stars, however, is still at the phenomenological level. We can recognize main sequence stars at the start of their joint lives with the primary filling its Roche lobe; we notice peculiar chemical signatures in some red giants that could be explained by binary mass transfer;

we see spectacular displays from the accretion disks of dwarf novae; we suspect that the origin of Type I supernovae is an endpoint of close binary evolution. A physical understanding of close binary evolution, however, still awaits the crucial observations, in this case high angular resolution images of stars interacting gravitationally.

Near-term observational progress in this field will come from recognition of hot and compact companions of stars in the UV with HST, and from resolution of binaries with milliarcsec separations with the first imaging interferometers. In the longer term we can look forward to imaging the nearest W UMa and U Gem systems and investigating the accretion disks of cataclysmic variables. These observations call for 10 to 100 μas resolution.

High angular resolution astronomy will also challenge our understanding of single stars by revealing surface structure and variability. These observations will call for more sophisticated models than the current one-dimensional approach.

Much more insight into the cooler phases of stellar evolution (most importantly, the evolution of young pre-main sequence stars) followed the extension of observational techniques into the infrared. The low background in space will allow substantial further gains. Near-IR observations could place valuable constraints on the existence of baryonic dark matter.

There are important areas of stellar evolution that similarly await improved capability in the ultraviolet. The evolution of very massive objects remains a subject of intense theoretical interest from the points of view of both chemical evolution and black hole formation. The UV excesses in elliptical galaxies clearly call for a real understanding of post-asymptotic-giant-branch evolution. A suitable laboratory for studying the evolution of massive stars and post-AGB evolution exists in the Magellanic Clouds. An appropriate survey is needed, as the hottest post-AGB stars currently go unrecognized because their optical energy distributions are insensitive to temperature. An excellent example of the significance of the Magellanic Clouds for stellar evolution was furnished by SN1987A, an event to which space astronomy was able to respond remarkably well, and which produced a major leap forward in our knowledge of the evolution of massive stars. High angular resolution observations are of special importance in the study of massive stars (such as R136a in the 30 Dor star formation region) which need to be separated from their coeval companions.

Stellar populations.

In the standard Big Bang cosmological model, baryonic matter forms as diffuse gas consisting almost entirely of hydrogen and helium. Due to remarkable effects associated with gravity and nuclear processes, baryons at the present epoch have become concentrated into stars within galaxies containing a full range of chemical elements. Since lower mass stars have nuclear burning lifetimes that exceed the age of the universe, information about the full evolution from a gaseous to a stellar universe is retained in the stellar populations of nearby galaxies. Conversely, conditions at the time of galaxy formation, when the universe was making the transition from gaseous to stellar states, can be ascertained from a solid understanding of stellar populations in combination with sensitive observations of faint, high redshift objects.

Both of these approaches require an extension of knowledge of fundamental properties of normal stellar populations. In the Galaxy the description of the stellar mass distribution function needs to be completed; e.g., by reaching stars below the nuclear burning mass limit through determinations of the space densities of brown dwarfs. A census of low luminosity, compact stellar remnants is also required, especially in very old systems such as globular star clusters. Ancient white dwarfs, neutron stars, and black holes are significant, both as semi-dark forms of baryonic matter and as indicators of the properties of stellar populations when the Galaxy was young. Explorations of the extremes of the Galactic fiducial stellar population will require multi-wavelength, high sensitivity observations, e.g., in the near infrared with HST and its successors to find brown dwarfs, and across the EUV-UV-Visible to find old white dwarfs. Explorer-class missions could also contribute to aspects of this problem. The search must reach $V = 29$ in the nearest globular clusters and a magnitude fainter in the halo far from the Galactic disk. *Claims that the halo of the Galaxy is non-baryonic cannot be taken seriously until a census is complete to at least this level.*

A direct complement and check on the study of galaxy evolution by "looking back" at galaxies of significant redshift is the determination of the star formation history of nearby galaxies. By measuring the main sequence luminosity function down to one $L_\odot$ it is possible to determine the time-history of the star formation rate in a stellar system over the last 5 billion years. To date the experiment has only been performed in the solar neighborhood and in the Large Magellanic Cloud, but our understanding of the star formation history of all spiral galaxies is generalized from these results. An important goal is to extend the experiment to the full sample of galaxies available in the Local Group. This requires imaging to V = 30 with spatial resolutions of $\mathcal{O}$(10-20) milliarcsec – these requirements are beyond what will be achievable with HST.

With such a limit, color-magnitude diagrams could be obtained in the nearest galaxies (M31, M32 and M33) to $M_V > +5$, below the main sequence turnoff for globular cluster-like populations. This would enable age determinations to be obtained in the oldest populations of these galaxies. These ages constrain the formation epoch of galaxies, and also constrain cosmologies. The initial mass function (IMF) could be determined through much of the Galaxy (to masses ~ 0.1 M$_\odot$ out to several kpc). IMFs could be derived from the luminosity function to $M_V \sim +18$ in the nearest globular clusters and to $M_V \sim +12$ in the LMC and the SMC. The LST would meet the goal of imaging to V=30 mag. With a high-throughput camera it would give S/N=10 measurements in 10^4 s in the visible (V) for large numbers of stars fainter than 30 mag over a field of several square arcmins.

Once stellar populations have been properly sampled in nearby galaxies and the results placed in a physical model, we will be in a better position to interpret hard won multi-waveband measurements of the light from distant galaxies. Empirical models of stellar populations can be obtained most directly by resolving individual stars throughout the UV-Visible-IR. Detailed physical studies of stellar populations are currently carried out within the Galaxy. High angular resolution, multi-wavelength observatories in space will extend these studies to Virgo supercluster galaxies, where a full range of morphological and evolutionary galaxy classes can be sampled.

In addition, our knowledge of the frequency of formation of massive stars is biased to those which are luminous at visible wavelengths. Really massive stars would not necessarily be those which are most luminous visually; most of their radiation is emitted in the ultraviolet region of the spectrum which is accessible only from space. Wide-field surveys in the UV would not only determine the distributions of hot white dwarf and subdwarf stars in the Galaxy, but would also map the massive star distribution in the Magellanic Clouds and other Local Group galaxies.

The galactic and extragalactic distance scale.

For the best part of this century astronomers have been ham-strung by the need to estimate the distances to stars outside the solar neighborhood by indirect methods. This has often led to circular arguments by assuming the astrophysical theory we were trying to test by observation! In the next decade we can begin to find our way out of this historic impasse. By means of optical interferometry it will be possible to measure the distance of every detectable star in the Galaxy geometrically. Ten μas precision in positions is a realistic goal for an astrometric Interferometric system. If we can determine the distances of a sample of Cepheids and RR Lyrae variable stars in the Galaxy, we will improve the precision of the extragalactic distance scale (which will be studied by HST in the coming decade), and refine our estimate of the expansion age of the Universe. Such astrometric precision would also permit measurement of the proper motions of Local Group galaxies (and far beyond – even to Virgo!) and yield an accurate dynamical age and mass of the Local Group.

For megaparsec distances the most reliable distance indicators are Cepheid variable stars. Using Cepheids, HST with the WF/PC II will measure distances to the Virgo cluster (redshift 1000 km s^{-1}). In the last decade significant peculiar velocities have been measured for galaxies at redshifts of 3000 km s^{-1}. An accurate determination of the Hubble Constant requires that the expansion field be corrected for large scale flow velocities. These are at present mapped with secondary distance indicators. The LST with its significantly higher angular resolution would permit detection of Cepheids out to Coma. The LST would permit our model of these flow velocities to be tested directly with Cepheid distances, an important check on HST's H$_0$.

A new technique that is based on measurement of the fluctuations in surface brightness that arise because of statistical clumping of stars appears to result in very precise measurement of distances – possibly to accuracies of 5%. This technique also has the potential for allowing valuable constraints to be placed on the age and metallicity distribution in the stellar population. While HST, in conjunction with observations from large ground-based telescopes, will probably allow precise measurements out to Coma, extension beyond that will require larger space telescopes. The possibility that galaxies are clumped and structured on scales larger than 10,000 km s^{-1} and so will require accurate distance determinations to beyond such velocities, is a further argument for LST, and its successor, the 16 m NGST

Nature of galaxy nuclei, AGNs, and QSOs.

It has been known for over a quarter of a century that many galaxies possess highly luminous nuclei. QSOs are the most dramatic examples. More recently, it has become clear that this is a widespread phenomenon and that the majority of galaxies, including our own, exhibit activity at some level. The most common explanation of these phenomena involves black holes with masses between a 10^6 and 10^9 M$_\odot$ orbited by extensive accretion disks. However, the black hole models remain unverified. A new generation of UV-Visible space telescopes would place our understanding of active galactic nuclei on a firm footing and should help us to develop theories of galaxy evolution that parallel existing theories of stellar evolution.

A 10^8 M$_\odot$ black hole has a size comparable with that of the Earth's orbit. So for the closest active galaxies, at distances of order 10 Mpc, this subtends an angle $\sim 10^{-7}$ arcsec, far too small to be resolved by any telescope. However, its presence can be detected indirectly, through its dynamical influence on the surrounding stars which move at $\sim 10^{-3}$ the speed of light and are therefore influenced out to a 10^6 black hole radii, or ~ 0.1 arcsec for nearby galaxies with a 10^8 M$_\odot$ black hole. Observation of this effect has been attempted, with promising results, on a few nearby galaxies both active and inactive, using ground-based observations where only a few resolution elements are possible. HST, with its tenfold increase in resolving power but a lower sensitivity than ground-based telescopes, should allow some improvement to be made in these measurements.

Observations with high dynamic range and angular resolution superior to ~ 0.1 arcsec will be necessary in order to weigh nuclear black holes and elucidate their fueling in a variety of galaxies (or to the contrary, demonstrate that black holes are not responsible for nuclear activity).

Most AGN radiation appears in the UV and probably originates from the inner parts of an accretion disk. UV emission is also responsible for exciting the broad emission lines that are the primary indication of nuclear activity. Without data in the UV it has not been possible to measure AGN powers to better than a factor of ten. In order to understand the compete spectrum of an AGN, it will be necessary to observe high-redshift QSOs in the UV with magnitudes fainter than ~ 19 mag, too dim to be seen with Lyman-FUSE. This would enable photon energies as large as 60 eV in the QSO rest frame (within a factor of two of the energies of photons observable in nearby AGN with X-ray telescopes like AXAF) to be detected. These measurements should also provide crucial tests of theories of accretion disks.

Most QSO and Seyfert infrared radiation is now believed to be re-radiated by dust in orbit around the active nucleus. For the closest active galaxies, the expected size of the 10 μm infrared source is ~ 10 pc or ~ 0.1 arcsec. Fainter infrared emission may be detectable from closer low-luminosity AGNs.

Radio-loud QSOs and radio galaxies produce non-thermal jets, some of which have been detected at visible and UV wavelengths. Lower power bipolar outflows have been observed from Seyfert galaxies and are known to be related to the narrow emission line regions. Continuum imaging of visible and UV jets could greatly clarify the particle acceleration mechanism. Line imaging may allow a determination of the outflow speeds, which cannot be measured using radio techniques.

The required advances in angular resolution needed for both the spectroscopic and the imaging studies must come from space-based interferometry. For example, a 50 m baseline deployable interferometer could yield images with resolutions of 2 milliarcsec at 400 nm (5×10^{17} cm at Virgo). Ultimately 10 km baseline interferometers on the moon may resolve accretion flows at optical wavelengths.

Formation and evolution of galaxies at high redshifts.

A high-priority scientific objective for large space telescopes is the study of galaxies at high redshift. The imaging and spectroscopy of galaxies at redshifts $z > 1$ is crucial for understanding how galaxies form and evolve. While the epoch of galaxy formation is very uncertain, existing data suggests that the redshift range of 2-4 is a particularly promising region for exploration. The recent results for radio galaxies, the deep galaxy counts and color surveys, the long standing results from QSO studies, and the significant evolution seen at redshifts $z < 1$ all support such a view, as do evolutionary N-body models.

In particular, these models and the data suggest that galaxy formation involves the interaction and assemblage of galaxies from clumps. Galaxies as we know them today would build up through interactions and merging of such clumpy subunits. While such a view should be adopted with caution, it does give guidelines as to the capabilities needed to detect, map and establish the physical and kinematic conditions in protogalaxies. What is clear is that both resolution and sensitivity will be crucial, for spectroscopy as well as for imaging. The characteristic scale of structure in galaxies is hundreds of parsecs. Resolutions that match these scales are of particular importance for understanding the evolution of galaxies, since much of the structure in galaxies occurs on such length scales of 100 pc to 1 kpc (e.g., star forming regions, spiral arms, merger "arms and tails").

Thus resolutions substantially better than 1 kpc are required, with sufficient collecting area to obtain images and spectra of objects fainter than 27 mag. IR observations will be important as the redshift moves much of the information out beyond 1 μm. Cooled, low background telescopes are therefore necessary.

HST will offer significant gains over ground-based observations when it is equipped with image-correcting instruments, but will typically be limited by its light-gathering power to objects at $z < 1$ for this extremely difficult problem. The 6 m LST and particularly the 16 m are highly suited to this task. They have the combination of resolution and sensitivity needed to detect and measure structure in $z > 1$ "galaxies". LST will provide an essential intermediate step in capability from HST to allow continuing progress in this fundamental area.

LST gives resolutions of 15 milliarcsec at 400 nm. A resolution of 15 milliarcsec corresponds to ~ 150 pc resolution in *galaxies at any redshift*, given currently accepted cosmological parameters. This is remarkable. This is within a factor 2 of the resolution with which we see Virgo galaxies from the ground (1 arcsec is ~ 75 pc at Virgo). The resolution on high z objects would still be ~ 1 kpc in the zodiacal background minimum at 3 μm (see below). The 16 m NGST would improve on this, and bring its greater collecting area to bear as well.

Simulations of the imaging power of the 16 m NGST are shown in Figure 1. These simulations were made by Jim Gunn for "The Next Generation Space Telescope" workshop held in Baltimore in 1989. Galaxies at a redshift $z \sim 1$ appear as though they were nearby objects being observed from the ground! While HST's abilities in this area are impressive compared to that available from the ground, the contrast with the 16 m is striking.

The simulations were made using images for a typical nearby ScI-II spiral galaxy. They were made assuming no luminosity evolution at a redshift $z = 1$ and for images taken in the rest-frame B-band (observed at 9300 Å with a 20% wide filter). They were assumed to be taken with a CCD camera with realistic throughput and noise characteristics. With the wide field of the 16 m and mosaics of CCDs similar to those now being obtained, hundreds of such galaxies could be imaged in each exposure. The galaxy frame (of NGC 2903 at 10 Mpc) used in the simulations was taken in 0.8 arcsec seeing conditions; any worse and the image would not have been adequate to simulate the capability of the 16 m at $z = 1$!

In addition to the resolving power and sensitivity of these telescopes, the low IR background will become increasingly important for the study of high redshift objects. It is in the near-IR at around 3-4 μm that the sky is darkest from space. This minimum occurs in the transition zone between zodiacal scattering and emission. The background at 3-4 μm with a passively-cooled space telescope will be $\sim 10^{-6}$ of that from the ground, a substantially greater gain than the 10^{-3} decrease expected in the 1-2 μm region accessible to HST.

The light from the stellar populations in very distant high-z galaxies will be redshifted into this region, as will many of the important emission lines. Observations at these wavelengths are then of particular importance since comparisons with nearby well-studied galaxies will require spectra and colors that correspond to visible light in the rest-frame. Already, but with great difficulty because of the high background, 2 μm observations from the ground are being used to try to determine the contribution of an "old" stellar population in high-z objects.

The Second Generation IR camera for HST (NICMOS) with its 1-2.5 μm imaging capability will allow very substantial gains to be made, but the background from the "room temperature" HST and the limited resolution ($\sim 0.25''$ at 2.5 μm) will not allow HST to fully exploit the potential of space observations in this region. The combination of low background, resolution and spectroscopic performance from a large collecting area will likely prove essential if we are to probe this redshift range in a direct way. In other words, telescopes such as LST and the NGST will probably be essential if we are to understand fully galaxy formation and evolution in the redshift range $z \sim 1\text{-}5$ (or beyond to $z > 5$?).

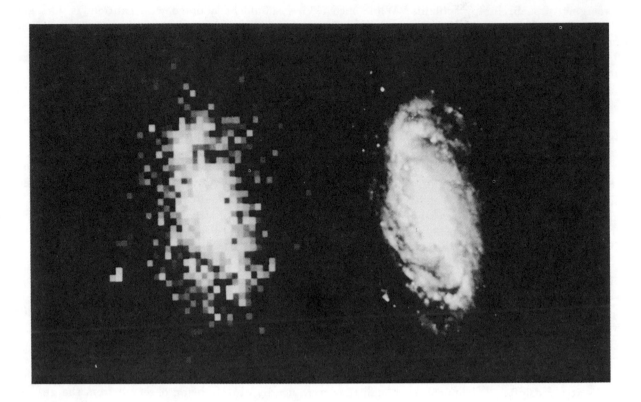

Figure 1. Simulations of the potential imaging performance of HST and of a 16 m diffraction-limited telescope, for a $z = 1$ Sc spiral. a) Left panel. A three orbit (approximate 2 hours) exposure with HST and the WFC of the replacement camera, the WF/PC-II. The (undersampled) resolution is about 1 kpc. b) Right panel. The 16 m with a 2 hour integration. The imaging performance is spectactular. This is the resolution with which we image *Virgo cluster galaxies from the ground*. Except for noise it is almost indistinguishable from the original CCD image of the 10 Mpc galaxy. The LST would have intermediate performance.

Cosmology.

The fundamental cosmological parameters remain uncertain. Accurate measurements of the geometric parameters (the Hubble constant H_0 and the "deceleration" parameter q_0), the age of the Universe t_0 and the density of the universe Ω_0 are needed. In addition, accurate abundances of the cosmologically important elements, deuterium, lithium, and helium are also needed. Several of these problems have been addressed earlier, particularly the determination of the Hubble constant and the limits on ages from age-dating the oldest stellar populations. A key issue is the comparison of the cosmological clock with the stellar nuclear clock.

The Deuterium abundance is a high priority for Lyman-FUSE, and one of its outstanding scientific goals. The abundance of deuterium is a sensitive diagnostic of conditions in the early universe and its determination is of particular interest for issues relating to the baryonic content of the Universe.

New approaches to establishing distances to distant galaxies that will utilize the spatial resolution of HST, and subsequently LST, would allow tighter limits to be placed on q_0. Examples are the use of distant supernovae and those techniques that depend upon combinations of structural parameters and kinematic data, such as the Fisher-Tully relation for spirals. A similar relation is now known for ellipticals. Comparisons of q_0 with Ω_0 (determined locally from large-scale motions), and also of t_0 with H_0 and Ω_0 are likely to have interesting cosmological implications (i.e., the value of Λ). Fundamental tests of the simplest Big Bang cosmological models will be made by combining data on the formation epoch of galaxies, inferred from observations of distance galaxies and from the ages of populations in nearby galaxies, with these determinations of H_0, q_0 and Ω_0.

The nature of the dark matter remains enigmatic. While many tests will be performed by ground-based surveys (e.g., for brown dwarfs and microlensing) the spatial resolution available from space telescopes will play a crucial role in defining the dark matter distribution in galaxies and clusters (and elsewhere) through gravitational lensing studies.

Another area of considerable interest for its cosmological implications is the question of structure in the universe. The large-scale clustering of galaxies poses particularly challenging theoretical constraints. These are becoming increasingly difficult to reconcile with current models. If significant structure really exists on scales greater than 10^4 km s^{-1} careful mapping of the distances of large samples of galaxies will be required. Developing accurate distance measurements will require application to samples at $z \sim 0.1$. Many of these techniques require high spatial resolution and so need space telescopes, such as HST, LST and interferometers.

Such telescopes will also have a particular role to play in investigations of the evolution of structure with time, both directly at high redshift and through measurement of QSO absorption lines. The characteristics of galaxies at higher redshift as a function of galaxy density will be of particular interest.

The determination of the shape, amplitude and origin of the fluctuation spectrum will be potentially the most significant link between astrophysical constraints and work of the particle physics community with the supercollider. The observational difficulty of answering these cosmological questions and deriving the fundamental parameters will require a concerted long-term ground and space-based program.

II. IMPLEMENTATION OF THE SCIENCE PROGRAM.

The UV, Visible and near-IR region lies at the center of contemporary astrophysics. This is due in part to the vast historical base of information and understanding that has been developed over the last century of observations with ground-based telescopes, and the last few decades of exploratory UV observations. It is also due to the physical nature of the universe, in that much of the baryonic matter in the universe can be studied and diagnosed through its radiation in this region. Space astronomy in the 1990s and into the next century offers opportunites in this regime for fundamental improvements in our ability to probe and understand the universe. The Great Observatories program, with HST at its heart, exemplifies the central role that UV, Visible and near-IR region plays. EUVE and Lyman-FUSE further demonstrate the potential of missions in this region.

With the launch of HST a new era has begun for space astronomy. The initial steps have proven to be more difficult than should have been the case. However, HST's optical problems, while serious, can be overcome in part, and will likely result in a telescope that can meet most of its original objectives. The importance of large space observatories should not be lost in the discussions that result from this failure. The problem did not result from the requirements outstripping the available technology, but was a failure of testing and oversight. The complexity of the management of large projects should not be underestimated. The HST experience indicates that the management, review and oversight process needs to be given as much visibility as the technical challenges that face a project of such scope. The success of the high energy physics community both here and in Europe in managing large, complex scientific projects at the forefront of technology shows that such projects can be successfully carried out. It is crucial that the lessons from these projects and the experience with HST be taken to heart by all the participants in future projects. In particular, the full involvement of the scientific and engineering resources of the science community is essential.

A coherent long-term program driven by scientific and technical opportunites needs to be developed and implemented to build upon this first Observatory mission. A central element of such a program would be a continuing series of long-lived missions that are directed at tackling and solving the fundamental issues of astrophysics. These include the formation of stars and planetary systems, the formation and evolution of galaxies, and the structure and geometry of the universe. To do so will require that we fully exploit the potential of space observatories. While new technologies and launch capability will be needed, substantial gains can be made through incremental utilization of developments in these areas.

While observatories like HST are central to the program, a crucial element is a parallel program of continuing missions that explore and exploit areas that are not possible with large observatories. Examples are spectroscopic and imaging systems in the far UV and the EUV; the potential of high spatial resolution offered by interferometers; the gains from precision astrometry of large samples of objects; concurrent observations across a broad wavelength region, particulary of time-dependent phenomena; and all-sky surveys for both point sources and for low-surface brightness structures. As COBE has recently shown and as EUVE and Lyman-FUSE will show, data of great importance to astrophysics can be obtained within the very cost-effective directed missions of the Explorer class.

Such missions also can play a critical role in both the technical and managerial training of scientists, instrumentalists and technologists. Such training and experience is necessary for the synergism needed between the science community, industry and government that is critical for the long term health of a space science program, and for future Observatory-class missions. A complementary, viable and dynamic program of Explorer-class missions managed and implemented by PI teams is an essential component of the overall program.

The Observatory-class mission program begins with HST. Its potential must be realized through the implementation of image-correcting instruments, starting with the WF/PC II. The Second Generation instruments, the near-infrared camera and spectrograph NICMOS, and particularly the imaging spectrograph STIS should be redesigned, developed and implemented as quickly as possible. A state-of-the-art Third Generation instrument set should follow the Second Generation instruments. An advanced camera would be an example of such an instrument. The importance of optimizing the operational effectiveness of HST should not be overlooked. It will pay handsome scientific dividends. HST must be supported until its successor is available, so that its crucial capabilities in the UV, Visible and the near-IR are not lost in the era of the Great Observatories.

The problems associated with HST emphasize the need for an early development of its successor, the Large Space Telescope, LST. The LST would be a 6 m-class passively-cooled UV-Visible-IR space telescope. Substantial gains in capability and science productivity are possible through implementation of developing technologies, and through its location in High Earth Orbit (HEO). It is recommended that this be a truly international partnership amongst the major space nations. With the global interdependence of nations, and the dramatic changes which are leading to common interests, such a scientific venture could act to cement these common goals in this last frontier.

Planning and development of its successor must begin soon if it is to be available within 5 years of the nominal end of HST's life. The experience gained with HST and the maturing of instrumental and optics technologies should allow us to build a substantially more powerful successor to HST on a shorter timescale and at reduced cost relative to HST. *We recommend that a compact, passively-cooled telescope be planned for deployment in 2009.* With concept development and planning beginning in 1993 and a new start in 1999, the launch of LST in 2009 is a realistic goal.

The compact design of LST in HEO and the use of optics and control technologies unavailable for HST should lead to an overall weight comparable to HST. Its instruments can be developments of the mature technologies that will be used for the Second and Third Generation HST instruments. By passive-cooling to $\sim 100°K$, LST will have a background at the zodical light minimum at 3 μm that is *six* orders of magnitude less than from both HST and the ground. Images of 10 milliarcsec at 0.25 μm could be expected from improved optical performance.

Finally, the program leads to the implementation of an observatory of astonishing power, the Next Generation Space Telescope (NGST), a 16 m UV-Visible-IR telescope. The technical and logistical challenges of such an observatory are substantial, but they are by no means insurmountable. We should also utilize some of the technological developments for the 16 m in the 6 m successor to HST. By so doing we will not only obtain a telescope of substantially improved capability, but will further refine the technologies in a working observatory. The modular nature of a 16 m telescope and the required long life make the Lunar Outpost a natural location for the NGST, particularly if it is combined with a very large km-scale interferometer. The km-scale interferometer complements NGST and allows for study of outstanding astrophysical phenomena on microarcsec to milliarcsec scales that cannot be reached by other means.

The goal of these wide ranging efforts is deeper understanding of the universe. Scientific understanding comes through careful data analysis, extensive modeling and concurrent theoretical developments. Only if such activities are adequately funded can the great investment in facilities be appropriately redeemed. Long-term funding for students, researchers and for the equipment necessary for their research programs is a crucial element of the program. We applaud the very substantial gains made in this area within NASA over the last few years. Such funds should continue to be routinely associated with new missions and observatories at a level that allows the scientific returns to be commensurate with the cost of these missions.

All-sky access to ground-based telescopes for a variety of observations, but notably spectroscopic follow-up, will continue to be needed. Spectroscopic observations are invariably time-consuming, but provide the physical information needed for understanding. The unique capabilities provided by high-performance spectrographs on space observatories, will need to be complemented by efficient spectrographs on large ground-based telescopes; they will be needed into the indefinite future. Continuing access to large 4+ m ground-based telescopes in both hemispheres, and the funding necessary to support those programs is an essential element of the astrophysics program.

High performance optics are a central component of all instruments in this wavelength range. The scientific goals demand more difficult, larger and smoother optics. Weight will continue to be a concern, particularly as the gains to be realized from HEO operation and beyond become a driver for missions. A further concern driving one beyond LEO (Low Earth Orbit) will be the increasingly high probability of damaging collisions with space debris. As optics and hence structures get larger the difficulty of pointing and tracking increases. Moving away from fine pointing by body pointing to fine pointing with an optical element is a very desirable goal. All these concerns lead to the need for lightweight optics with high performance surfaces with active elements. The need for a technology development and demonstration program is clear. It will return great value across missions of all scales in the UV-IR region.

III. THE OBSERVATORY-CLASS MISSIONS.

The science objectives call for an enhanced program of long-lived, large telescopes with imaging and spectroscopic capability from the UV to the mid-IR. HST is the first of this new class of Observatories. It would be followed by the 6 m LST and ultimately by the 16 m NGST. The crucial elements for this decade, beyond HST's image-compensating instruments and its optimization, are the third-generation instruments for HST and the start of LST, the 6 m successor to HST.

HST.

HST is the premier instrument of NASA's Great Observatory series and will be the cornerstone of the space UV, Optical and Near-infrared program for the 1990s. In performance and overall capability, HST has the potential to surpass all other space and ground facilities operating between 0.12-2 μm in this decade. Only in visible and near-IR spectroscopic capability will it be challenged by the new 8-10 m ground-based telescopes. It is crucial that that potential be realized by new instruments that will restore and extend its imaging and spectroscopic performance. HST's sensitivity and resolution will link future observations at other wavelengths from both ground and space. Its intended lifetime (15 years) will span the entire generation of Great Observatories (HST, GRO, AXAF, and SIRTF).

The scientific returns possible from HST require that we restore its optical performance with new instruments, and maintain and refurbish the Observatory and its supporting systems until its successor has been launched or is in the final development phase (late Phase C/D).

The replacement camera, the WF/PC II, should be completed and implemented as soon as possible. It is also crucial to accelerate the development of the two Second Generation scientific instruments, the NICMOS and the STIS. These should be completed rapidly and deployed as early as possible – by 1995-6. They not only bring very large gains in capability to HST, but will help to restore much of the lost capability.

The breadth of the science program of HST attests to the power of observatories in space. For example, HST is expected to play a key role in finally determining an accurate value for the Hubble constant. The use of QSOs to explore the intergalactic medium in the young universe as well as of the gaseous components of young galaxies will also be greatly enhanced by our ability to explore like environments at current epochs. We will be able to relate the absorption-line systems to objects that can be studied directly from HST and the ground, thereby helping us to understand those objects in the young universe that contribute to the high-redshift absorption-line systems.

The parallel mode surveys that can be made with the cameras on HST have the potential for making major discoveries. While the images will be of great value for elucidating the structure of distant galaxies and for establishing the distribution of stars throughout the Galaxy, they are likely to surprise us since the images will be taken serendipitously.

The operational support will be as challenging and as critical to its scientific productivity as the performance of the major spacecraft elements. NASA and the scientific community have made considerable investments in the Observatory and the science and mission operations systems which will sustain the 15 year scientific mission. It is critical that these efforts be continued, and, in addition, that the M&R (Maintenance and Refurbishment) program maintain and improve the Observatory so that its 15 year mission is productive and fully complementary to its sister Great Observatories.

The goal of these efforts is to maximize the scientific productivity of HST over its 15 year lifetime. This involves providing responsive and efficient spacecraft and science operations, including scientific capabilities and improvements to science efficiency that are not yet available. Near-term examples that are being implemented to meet the original goals of HST are improved support of planetary observations and parallel observational capability with the cameras. By optimizing science planning though improved planning software, a 66% increase in the amount of useful observing time is potentially available. Continued refurbishment of the spacecraft, orbit and ground systems will be essential if the high level of scientific productivity is to be maintained.

A strong data analysis program is required, as is the rapid diffusion of the scientific results. Important elements of this are the development of the HST data archive, and the adequate support of General Observers and Archival Researchers working on analysis of HST data. Every effort should be made to educate and involve the next generation of scientists in HST activities.

Third Generation Instruments for HST.

A third generation of instruments should be planned for an M&R mission later in the decade. These also will play a major role in ensuring the continuing high productivity of HST. While the selection should be made through peer review, examples of likely instruments could be:

An Advanced Camera. A more sensitive camera with better sampling of the corrected PSF (Point Spread Function) in both the UV and the visible would be a major improvement over that expected for the WF/PC II. Higher performance in the UV would require a large visible-blind array detector with good dynamic range or the development of visible-blocking filters. Higher performance in the visible requires CCDs with efficient coatings over broad wavelength ranges and lower readout and dark noise. Incorporation of a narrow-band imaging capability across a wide field could provide new insights in many areas, for example, AGNs, star forming regions and distant clusters. This might be accomplished with a Fabry-Perot system or an Acousto-Optical Filter.

A high performance Spectrograph. STIS is an excellent example of the spectroscopic gains that can be made with contemporary technology. Further gains are possible. One potentially very valuable area for improvement is multiobject capability with a programmable multislit system. This would overcome the current restriction to observing one target at a time. The multiobject system could utilize fibers, slits, or more generic technologies that would allow arbitrary-shaped entrance apertures to be defined (based, for example, on liquid-crystal-like technology). Substantial gains in effective throughput would result from such techniques. Spectropolarimetry is also an area that could be valuable to explore. Cosiderable scientific return could result from having linear and circular polarimetric capability, far UV capability, very high spectral resolutions (10^6), and from fully utilizing the high angular resolution of HST in the UV and visible.

LST – the 6 m Successor to HST.

The scientific case for enhanced Observatory-class capability in the UV-IR region is overwhelming. The panel strongly recommends that a 6 m-class telescope be launched in the first decade of the next century.

Substantial gains in capability can be realized at a cost comparable to HST. Building upon the advances in technology and instrumentation, the 6 m LST would offer a substantial performance boost over HST, while not being strongly dependent on as-yet to be developed technology. The improvements in optics technology, the maturing of detectors and instruments, and the technical and managerial experience gained in the nearly two decades since HST was designed, will make LST an affordable project, even though the primary optic is significantly larger. Tremendous gains in sensitivity in the near-IR would accrue from a passively-cooled optical system. The operational complexity and cost can also be substantially reduced by placing the telescope in HEO.

As noted throughout the science discussion in §I, LST can have a substantial impact in almost every area of endeavor, from cosmology to planetary systems. While the testing of the existence of truly Earth-like planets will await 16 m class telescopes and interferometric techniques, very substantial gains can be made with LST in the area of planetary system studies in the Solar neighborhood within ~ 10 pc. It is clear that major issues can be addressed in the areas of the ISM and its evolution with time, in stellar astrophysics, in star forming regions, in studies of activity in galaxies, and in stellar population studies. In this latter area thresholds are passed that allow for the study of populations in nearby galaxies that will establish how such galaxies form and evolve. The constraints placed on the ages of the oldest populations also help determine the age of the universe. Since the Hubble constant will be well in hand by the time LST is launched this also allows constraints to be placed on other fundamental cosmological parameters (e.g., Ω_0). Its image quality and infrared sensitivity combine to give it the power for studying galaxies in their youth, and even during formation, that far exceeds that of HST and any ground-based facilities.

With improvements in optical fabrication techniques, the wide-angle scattered light in the UV and visible from residual small scale surface structures could be substantially decreased below that expected for HST (after correction of the spherical aberration). This would have major implications for the science program. The UV gains would result in improvements in the imaging capability – 10 milliarcsec resolution at 0.25 μm is a reasonable goal for a 6 m. The dynamic range that would result from lower scattering would have immediate advantages for detecting large planets and protostellar disks, as well as for the study of QSOs and AGNs. It would be ideally suited to stellar studies with its sensitivity (10:1 S/N) of 30 mag in 10^4 s in the V band for unresolved sources. Its power for galaxy studies, particularly distant galaxies, can be estimated from Figure 1, where its capabilities would fall between HST and the 16 m NGST.

The use of active optical elements would greatly reduce the difficulty of developing a pointing and tracking system to satisfy the more stringent requirements of this telescope. The structural requirements would be relaxed also by the use of an active optical system. The varying thermal loads of LEO would be eliminated, as would the effects of aerodynamic drag. The complexity of the power system is eased in HEO. Furthermore, body-mounted solar arrays could be used.

Passive cooling could greatly enhance its IR performance. If the telescope could be cooled to the vicinity of 150°K, it would allow background-limited performance to beyond 3 μm where the zodical background is at its minimum (between scattering and emission). Cooling to ~ 100°K would result in outstanding gains in the mid-IR at 10 μm and beyond. Passive cooling to this level would be quite possible in HEO. Its IR sensitivity would be high. If it is cooled to 100°K its 3 μm sensitivity (S/N = 1; 500 s integration) would be ~ 15 nJy, and its 10 μm sensitivity would be 2 μJy, ~ 30X and 3X, respectively, the sensitivity of SIRTF in the near-IR. With its higher resolution ($> 6\times$ that of SIRTF) it will build upon the science programs and discoveries of SIRTF.

Difficulties will arise in fabricating and testing large optics for use at those temperature, but an even more challenging goal (< 100°K) is being set for the 16 m, and so technological development needs to occur in this area. The scientific gains are large. The background would be reduced by *six* orders-of-magnitude at 3 microns compared to that typical of ground-based telescopes.

With the developments in detector technology, particularly with CCDs and near-IR detectors, photon-limited wide field imagers and spectroscopic systems that approach theoretical limits appear to be quite feasible. Particular attention will need to be given to the operation of detectors in the higher radiation environment of high earth orbit. This is an open question, but it has been argued that the low read noise and the high charge transfer efficiency of contemporary detectors allow multiple short exposures that minimize the effect of high particle rates. The Second and Third Generation instruments for HST will lead to mature technologies for detectors and instruments that can readily be applied to a smaller complement of versatile instruments for the LST.

The complexity of the ground system for HST is driven largely by the many constraints of LEO operation (which in turn was dictated by the use of the Shuttle for deployment and on-orbit maintenance). A substantial fraction of HST's cost has occured as a result of these constraints and complexities. Operations costs and the costs of maintenance will eventually dominate the total program cost. Not only does operation in LEO incur large cost penalties, but the intrinsic efficiency loss also compromises the science program and the science productivity of the telescope. The cost of operating and maintaining HST will exceed $150M per year, not including the support for data analysis. A high priority goal for future space science missions

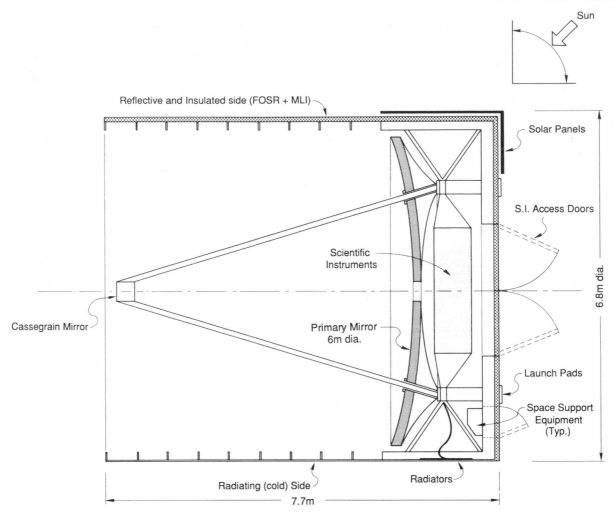

Figure 2.   A schematic of the 6 m HEO LST. Apparent are the compact dimensions that result from the use of both a fast primary and the short baffle that is practical in HEO. A preliminary analysis of the weight of such a telescope indicates that it would be *comparable to that of HST*.

should be access to HEO. Operations costs for a HEO LST would be reduced to a small fraction of that above.

The total cost of HST has been quoted as being in the vicinity of $1.5B to $2B. What is usually not realized is that a very substantial fraction of the cost of HST has been incurred by the ground and operational system software and hardware costs, and by the engineering analysis, spare parts inventory and additional management needed for the M&R program. Both these elements are driven primarily by HST's location in LEO. An additional factor was the lack of maturity of instrumental and spacecraft technology and the lack of experience with such a large, complex spacecraft. While it is not clear what the actual costs are, reasonable estimates place the ground and operational system costs at ~ $400M and comparable amounts for the M&R program. Thus the actual cost of the flight hardware system of HST is closer to $1B than $2B. This provides a valuable baseline number for discussions related to the cost of LST.

Why should the cost of LST not be proportionally larger by the usual scaling laws? There are several very good reasons why the LST as discussed here would lie on a very different cost curve from HST. An obvious one is that HST is the first of the UV-Visible Great Observatories, and that it is based on technology that is now nearing 20 years in age. We have learnt a lot since HST was conceived, and technologies in many areas have advanced significantly (e.g., optics, electronics, computers and control systems, and instruments). Such technological advances will make a very significant difference to the construction and operation of LST. Another useful guide to cost has been spacecraft weight. A preliminary analysis of the likely weight of an

Figure 3.   An artist's conception of the 6 m LST in HEO.

HEO LST, (of the design shown in figures 2 and 3) is that it will be *comparable to HST*. While not obvious at first sight, this result is quite plausible.

The technological developments that lead to substantial weight savings over HST are many. In addition, substantial efficiencies accrue from operation in HEO. Together they make a dramatic difference. First, new optics polishing and fabrication technologies (ion polishing; stressed lap polishing) will lead to lighter, higher-performance optics. Second, a simpler structural support for the secondary with active location to compensate for modest thermal and aging variations leads to a lighter and less demanding optical assembly. Third, the fast focal ratio leads to a short structure and a very short baffle because of the HEO location. Earth and Sun angles of 90° are realistic in HEO. Fourth, the instruments can be comparable to those in HST, and could well be modest developments from the HST Second and Third Generation systems. Fifth, the power requirements are lower and much less complex because there is no rapid charge-discharge battery cycling. Sixth, more durable and reliable body-mounted solar panels would be used. Seventh, HEO operation plus an active optical element for fine pointing combined with large area detectors for field acquisition and guiding will greatly simplify the PCS (Pointing and Control System). This has been one of the most demanding elements of HST. Finally, by taking the step to a non-man-rated, non-maintainable Observatory (except possibly for robotic replacement of cryogens, for example, if active cooling systems do not reach maturity) considerable further savings can accrue.

While we have discussed the LST as being a single all-purpose UV to mid-IR telescope, it has been suggested that it may well be cheaper to design and configure two spacecraft, one for the UV-Visible and the other for the Visible-IR. This is not obviously the case. Cost tradeoff studies need to be undertaken to establish the most cost-effective and timely route to fruition of the program. Meanwhile the LST will be discussed as being a single UV-Visible-IR telescope.

The surfacing of a major optical problem with HST has emphasized again that the successful completion of a project of this scale is more than just the application of sophisticated technology. The optics problem is but the most obvious of many such examples throughout the HST program. Problems and mistakes are to be expected in projects of such scale. The process must have mechanisms that allow for the early identification and rapid correction of such occurences. The complexity of such projects brings with it a challenge for the oversight, review and testing that is comparable in difficulty to the technology developments. This challenge must be faced squarely, and must be given the same level of attention as that required for the technological developments. The lesson to be learnt from HST (and from the success of even larger high energy physics projects and others) is that the management must be done by experienced people with deep technical backgrounds and a long-term commitment to the program. In addition, it must involve fully the end users of the mission in the process. That is, it must involve the scientific and engineering resources of the scientific community. The lessons from HST should not be neglected as we move ahead with large projects like LST.

The current state of launch vehicles is in flux. The envelope and weight requirements for such a spacecraft could easily be met by the larger members of the US ALS/HLV (Advanced Launch System/Heavy Lift Vehicle) ELV (Expendable Launch Vehicle) program. Shuttle-C also could be configured to accommmodate such an envelope. Whether it will is unknown. The Soviet Energia could easily launch such a payload to HEO. The options for an appropriate upper stage for orbit circularization need to be investigated further. Further orbit tradeoff studies along the lines of those done for SIRTF need to be carried out.

The essential and central role of an UV-Visible-IR observatory in astrophysics makes this a natural program for an international collaboration. It should receive very widespread scientific support. Furthermore, it has the capability, longevity, and "presence" to be attractive as a truly international scientific venture for the first decade of the new century.

Next-Generation 16 m Telescope.

Some of the most demanding scientific goals are beyond the power of even a 6 m space telescope. For a range of questions of fundamental significance it is clear that a telescope as large as 16 m is required. The NGST, Next Generation Space Telescope, would be a passively-cooled, diffraction-limited, telescope instrumented for the UV-Visible to the mid-IR. It would have unprecedented power for tackling a wide range of the most fundamental astrophysical problems, from the detection and spectroscopic observation of Earth-like planets to the structure of (forming?) galaxies at redshifts z approaching those of the highest redshift QSOs.

The investment in such a telescope would require it to be a long-lived facility. As such it would naturally form a part of the science program for the Lunar Outpost. In addition, very real advantages accrue for such a telescope from the use of the Moon as a base. The stability and isolation from noise sources afforded by the moon could be crucial if its ultimate performance is to be achieved. It could also be combined with the proposed very large interferometer. The combination of the NGST and the km scale interferometer is one that pays substantial scientific dividends. The NGST would enhance the sensitivity of the interferometer with its large collecting area. The ability to provide data for complex strucutres on scales of 5 milliarcsec to arcminutes also enhances the utility of the interferometer. It then provides a natural overlap in resolution with data from ground-based telescopes and from space telescopes in other wavebands.

While initial studies may focus on the development of such a telescope as part of the Lunar Outpost program, these studies should be generic in nature. If the Lunar Outpost program fails to develop an attached, large, long-term scientific capability, the developments can be utilized for space missions.

The telescope would have a lightweight, segmented primary with active wavefront sensing and control for diffraction-limited performance, into the UV if technically feasible. The optics are particularly challenging because of the objective of detecting and studying Earth-like planets. This will require low-scattering optics for high dynamic range. As for the 6 m, the structure and optics would be passively-cooled but now to less than 100°K to maximize the contrast for the planet detection and measurement program. Per resolution element the background would be 21 mags *less than that on the ground* to beyond 10 μm, i.e., $\sim 10^{-8}$ that for ground-based observations of unresolved sources.

With a collecting area that is 44 times that of HST, the sensitivity for unresolved sources is remarkable. The visible background is $\sim$ 32 mag per resolution element, i.e., $\sim 10^{-4}$ that for ground-based observations. In the V-band, unresolved objects could be measured with 10:1 S/N in 10^4 s to 32.5 mag! The sun could be detected with 10:1 S/N at 3.5 Mpc, and a 10-day cepheid at 200 Mpc. A $z = 2$, $M_V = -21.5$ *normal galaxy* containing an 18th mag QSO could be detected at 5:1 S/N, all for 10^4 s integrations. The simulations in Figure 1 graphically demonstrate the power of this telescope. Its 3 μm sensitivity (S/N = 1; 500 s integration) would be $\sim$ 3 nJy, while at 10 μm it would < 0.5 μJy.

With reasonable projections from current detector technology, mosaics of detectors could be used in wide-field cameras to give fields of several arcmin in size with diffraction-limited images. Resolutions would range from less than 3 milliarcsec in the UV to 50 milliarcsec in the zodical background "window" at 3-4 μm. High performance spectrographs would complement the imagers. As in HST, multiplexed operation of the instruments would allow for surveys of unprecedented sensitivity.

IV. MODERATE AND SMALL MISSIONS.

The Observatory-class missions cannot stand alone. Many scientific programs call for specific capabilities that cannot be accommodated in the large versatile telescopes. Examples such as EUVE and Lyman-FUSE abound. Frequent access to space for high scientific merit programs that can address a specific problem will remain central to the goals of the UV-Visible-IR space astronomy program. Specific examples follow.

Delta-class Explorers.

Missions of the Delta-class have played a crucial role in many areas of astrophysics. They will continue to do so. The panel strongly recommends that the Delta-class Explorer program be enhanced so as to allow more frequent missions with shorter development times.

The following §V, which deals with the ongoing program, discusses some excellent examples of current Explorer missions (EUVE and Lyman-FUSE).

Examples of possible future Delta-class Explorers clearly show that the scientific payback is very large from an enhanced Explorer program. While the panel believes that the following examples would all provide excellent and much needed scientific returns, they also feel that the most appropriate approach to the selection of Explorer missions, both Delta-class and smaller, is through peer review. The examples of forefront scientific missions (in no priority order) are an all-sky UV Survey that matches the resolution of the current ground-based Schmidt surveys, a wide-field astrometric system with sub-milliarcsecond accuracy on small and large scales, a multi-waveband system with the ability for simultaneous observations from the X-ray through the UV, and a very high spectral resolution instrument for ISM studies.

The lack of an all-sky survey in the UV is striking. While EUVE will map the region below 900 Å, the 1000 Å to 3000 Å range remains unmapped. Ideally the $\sim$ 1 arcsec resolution all-sky Schmidt surveys now available in the visible from the ground should be matched. A 1 m-class wide-field instrument would reach hot stars as faint as $m_V = 26$, and would provide a vast database of scientific value in its own right, as well as providing sources for further study with HST and LST. It would provide a quantitative database of UV point and diffuse objects that will enable statistical studies of hot stars, UV-bright extragalactic sources and diffuse sources throughout our own and other galaxies. Such a quantitative survey also needs to be undertaken from the ground in the visible with CCD detector mosaics. The resulting large color baseline (UV-V) would have particular value for stellar population studies. While the demands on detector and data handling technologies are large and may have limited the feasibility of such a program in the past, the rapid maturation of such technologies has made such a program much more realistic.

A similar survey should also be carried out at 2 μm. Future emphasis on the IR through SIRTF, LST and the NGST makes this an important goal. The first such survey should be carried out from the ground. However, the reduction in sky brightness in space by three orders of magnitude from that on the ground would result in very substantial gains in the depth of such a survey. Continuing development of IR array detectors will make a second space-based survey a practical proposition in the future.

Accurate reference frames, distances and space velocities are fundamental. Beyond a certain threshold the opportunities for very significant and even fundamental gains in knowledge are available. An astrometric system that can determine positions of stellar objects with submilliarcsec accuracy is an example. This would ideally be usable both for establishing a wide-field reference frame and for small scale mapping. It would have substantial impact on our confidence in the distance scale, age dating and evolutionary results drawn from stellar population studies and on the structure of the Galaxy. It would be a very valuable precursor to the more ambitious and more costly interferometric systems which can lead us into the microarcsecond range.

Many astronomical objects emit simultaneously across a wide range of wavelengths. For objects in which such activity occurs on timescales of seconds, minutes and even hours, concurrent observations are difficult to obtain because of the problems of scheduling multiple instruments. An instrument which allows simultaneous spectroscopic observations from the X-ray to the UV, and possibly into the Visible would play a particularly valuable role in understanding time-variable phenomena in a variety of astrophysical systems. The scientific objectives would include understanding the physical processes occuring in accretion disks, such as boundary layer phenomena, and the conversion of magnetic energy to relativistic particles and heat, the characteristics of the matter and radiation environment around highly condensed stars, the physics of the activity on stellar surfaces and within their coronae, and, on longer timescales, the dynamics and physics of the activity in active galactic nuclei. Simultaneous observations by co-aligned modest-sized instruments on the same satellite in HEO will permit uninterrupted long-duration observations of the intrinsic variability of plasma at 10^4 °K to 10^8 °K in these sources. This is not feasible with separate satellites in low Earth orbit. Furthermore, a quick response could be made for targets of opportunity such as novae, supernovae, and outbursts in cataclysmic variables and X-ray binaries.

Another area which could benefit from an Explorer-class mission is a spectroscopic study of the Interstellar Medium at very high resolving powers. At the velocity resolution needed, less than 1 km s^{-1}, the spectral resolutions required are of the order 10^6. Even though bright stars can be used for such observations, any substantial mapping requires many sources and considerable time to carry out the program. This area has been the focus of numerous rocket experiments, but an orbital mission would clearly result in substantial gains in the number of lines of sight and features that could be studied.

The advantages of orbits beyond LEO for Observatory-class misions also apply to these smaller missions. The returns from IUE attest to the value of HEO. The scientific return from the HEO missions is sufficiently high that as the plans for future launch capability are developed every effort should be made to include the means for ready access to HEO for Explorer-class vehicles. This is particulary prudent since the probability of damage in LEO from collisions with space debris is increasing significantly and is projected to approach unity for long-lived missions.

Another issue is the currently planned use of a single Explorer platform for three Explorers, namely EUVE, XTE and Lyman-FUSE. An expanded Delta-class Explorer program would provide the funds to allow the acquisition of a dedicated spacecraft for Lyman-FUSE. This is an attractive option given the potential scientific productivity of Lyman-FUSE and the risks associated with two manned Shuttle missions to service the platform (see also the Lyman-FUSE discussion in §V).

The primary goal of the moderate-sized program is to improve access to space for the many high-priority scientific objectives. Enhancement in this area would be directed towards:

(1) Improving the frequency of missions;

(2) Reducing the time to implement the mission;

(3) Giving the PI more direct oversight and responsibility for the mission, with the goal of greater training and reducing time and cost.

The scientific case for more frequent Delta-class Explorer missions is very strong. While augmentation of the program is primarily focused towards a higher mission frequency, two other elements, (2) and (3) above will play a crucial role in containing costs and improving the extremely high educational and technical training value of such missions.

Reducing the time an Explorer mission spends in the queue preparing for launch is very important. The timely response to new scientific initiatives through the exploitation of contemporary technology is a defining goal of the Explorer program. The rationale for the program is undercut when high-priority missions strech out well beyond a decade, as Lyman-FUSE has done. To lessen the lag before missions can be carried out will require both an increase in resources for the program as a whole and changes in program and budget planning. Ideally, the length of an Explorer Project should be determined only by the technical development schedule, but in practice the time to implement a project is determined by programmatic and budgetary considerations. As a consequence, the lifetime of a project is often much longer than it needs to be. The Explorer Program should be examined to find planning and management techniques which will reduce the average time in the queue. This will probably mean fewer missions in the queue at any moment, but more rapid motion through the queue. As part of this change, projects presently in the queue should be accelerated in order to reduce the lag time. Acceleration will require more overall resources. However, it will also require good planning and finding the best way to concentrate resources on a limited set of missions, although for a shorter time.

Acceleration of the Explorer Program has several advantages. Shorter implementation times lead to more rapid scientific progress. In addition, the cost of a project will be reduced for two reasons: first, there is always a core group of critical personnel which must be supported despite programmatic changes and delays; and second, long implementation times increase the probability of changes (e.g., in launch vehicle or mission operations) which increase the total cost of the mission. More rapid progress will also encourage the most talented and creative instrumentalists to participate. Indeed, their participation is necessary to produce the highest quality instrumentation for this program. Thus, a different approach to the programmatic and project management will not only speed the development of the science, it will produce better missions and reduce costs per mission.

The third element of the enhanced Explorer program is the much deeper involvement of the PI and his or her team. With such involvement, these programs can educate those involved and develop the experience base for the management and oversight of technically complex missions. The PI team should have greater autonomy and decision-making authority in Explorer-class programs. This also implies that the implementation plan will play a significant role in the selection process. High scientific merit and technological readiness will need to be supplemented by a thorough and convincing implementation plan. The deep involvement in these programs of the science community at universities, labs and centers is crucial for the long-term vitality of the field. Explorers provide a means of training and of developing the technical and management skills that are needed if the UV-Visible-IR space science community is to continue as a vital element of the space science program.

Imaging Astrometric Interferometer.

Long-baseline interferometry would have a revolutionary impact on the study of compact astronomical objects such as AGNs and interacting binaries. A first-generation orbiting interferometer would provide a major scientific step in imaging power and provide an essential demonstration of the demanding technology required for the further development of such systems. An early phase of technology development and demonstration is required that would naturally lead into a deployable 20-50 m baseline instrument late in the decade. In particular, this would entail demonstrations of the technologies for large stable structures.

Long baseline synthetic aperture UV-Visible-IR telescopes offer the potential of many orders of magnitude improvement in two areas, angular resolution and astrometric accuracy. In order to realize significant gains over existing and planned space-based observatories, the baseline of the interferometer should be a factor of 5-20 beyond the 2.4 m aperture of HST. A traditional problem of large optics is maintaining submicron tolerances for structures many meters in size. Current space interferometer proposals advocate the use of a laser optical truss as the primary tool, to control the dimensions of the optics. The accuracy of the optical components in the laser system is the key to the astrometric application of interferometry.

Technology for space interferometry is being developed and should be accelerated so that a moderate mission can be started in the latter half of this decade. Further technology development in very large structures (hundreds of meters), hyperprecision (sub-Å) metrology, large area optics, and low temperature optics offer additional orders-of-magnitude improvements in sensitivity, angular resolution, and astrometric precision. A moderate interferometer mission would be the first step in a sequence of space observatories that would provide order-of-magnitude increases in performance in one or more directions in parameter space.

An example of one concept that would have high scientific payoff is a moderate-sized UV-optical interferometer with three collinear baselines, each of order 10-20 m, with 30-50 cm apertures. Two of the baselines lock onto reference stars to stabilize the platform allowing long coherent integration times on the third baseline. Laser metrology would be used to monitor instrumental effects. Astrometric accuracy would be 10 μas for narrow fields and 100 μas for wide fields. Such accuracies would allow for planetary searches, determination of the cosmic distance scale from parallax measurements on Cepheids, and linkage of the optical and radio reference frames. In its high dynamic range imaging mode (10^3:1) it would allow study of such objects as AGNs, QSOs, extended atmospheres, interacting binaries, and gravitational lens objects fainter than 20 mag at resolutions less than 4 milliarcsec at 4000 Å.

The science goals include imaging of the narrow line region in AGNs and QSOs and measurements of the dust tori, searches for gravitationally lensed objects, and imaging of unique objects such as SS433. It would have great value for stellar astrophysics for the study of interacting binary systems, for the measurement of stellar diameters for most bright stars, and for the measurement of limb darkening in supergiants. Its resolution would also allow continuing studies of planetary and solar system objects (e.g., asteroids) at scientifically-interesting resolutions.

The astrometric potential is also very exciting. Astrometric searches for extrasolar planets and dark companions would benefit from the dramatically improved resolution. The distance scale and its many error-introducing steps could be placed on a much firmer footing by precise parallax measurements to Cepheids and other objects, and by allowing the distances to Local Group galaxies to be determined through a combination of kinematic and proper motion observations. Knowledge of the space motions of the constituents of the Galaxy will also have important implications for the formation and evolution of spiral galaxies.

Major scientific returns will result from the high spatial resolution that could be obtained only through interferometry. However, the development and operation of interferometric instruments will be a very challenging one. This is particularly true of the very large baseline, multi-element systems required for elucidating the very complex structures that form a major part of the scientific objectives of interferometry (e.g., AGNs). A realistic program of technology development and demonstration, initially on the ground and then in space is essential. While likely to be substantially larger in cost than traditional Explorers, an interferometric system such as that discussed here should be considered as the technological hurdles are overcome. *The panel strongly supports a program of technological development leading to a space-based imaging astrometric interferometer.*

Small Explorers.

Several disparate but important programs comprise the SMALL category of our recommended program. While small, they have an impact far beyond their size and so deserve strong support. They are: a UV all-sky survey; optics development and demonstration; and enhanced support for related ground-based observations.

SMEX UV Survey.

Most of the emphasis in this wavelength domain has been placed on the larger Delta-class Explorers. This reflects the relative maturity of observational capability in this waveband. However, there are a few areas where the smaller Scout-class Explorers (SMEX), or a somewhat more capable Pegasus-class Explorer, could yield very interesting scientific returns. One such program would be a UV low spatial resolution all-sky imaging survey. This appears to be scientifically very meritorious and is well-matched to Pegasus or Scout capabilities.

While there is a clear need for a 1″ resolution point-source survey that matches the multicolor Schmidt surveys available from the ground, a survey at low spatial resolution that is particularly sensitive to low surface brightness diffuse emission would pay substantial scientific dividends.

An Ultraviolet Background SMEX whose goal is a low spatial resolution survey primarily for low surface brightness diffuse emission would return substantial science at a very reasonable cost. This could be a small free-flying Scout-launched mission in an equatorial orbit, dedicated to an all-sky imaging survey at modest resolution (e.g., 15 arcsec). It would result in a large database of faint ultraviolet point sources, and would also survey the diffuse ultraviolet background over the entire sky. The inclusion of spectroscopic capability with modest resolving power ($\Delta\lambda \sim 5$ Å) would also be desirable.

Such a program would make important observations of hot interstellar gas, interstellar dust, fluorescence radiation from interstellar molecular hydrogen, intergalactic ultraviolet radiation, hot stellar populations in our own and external galaxies and QSOs; it would also provide a cornucopia of targets for the Great Observatories.

Optics Development and Demonstration.

Central to all instruments in this wavelength range are high performance optics. As the field progresses, the scientific goals will demand not only larger but also smoother and more complex optics. A technology development and demonstration program for lightweight optics with high-performance surfaces would benefit missions of all scales in the UV-IR region.

A quantitative, deterministic approach to optical figuring has been the goal of optical development programs for decades. Only recently have major developments taken place that indicate that such a goal is within reach. Two developments indicate that the ability to polish extremely smooth, strongly aspheric surfaces could be at hand. The techniques are the stressed-lap approach and ion polishing. Both appear to offer the possibility that surfaces such those needed for future generations of UV-Visible-IR telescopes could be obtained within a reasonable amount of time and for a reasonable budget.

While such deterministic polishing methods are important for the production of the required optical elements, many other steps are also needed. In particular, additional challenges are lightweighting of the elements, generation of surfaces, support of the surfaces, active compensation (in some cases), and fabrication and testing of optics to be used in passively-cooled systems. However, the polishing process and the attendent testing is a key element, and if the new approaches are proven they will greatly alleviate the overall difficulty of the task. For example, ion polishing offers a way to deal with the very difficult problem of "print-through" of the web-structure in lightweighted optics.

The nearly 20 years since the HST design was set in place sees us on the verge of major advances in optics technology. A program that focuses on the demonstration of these techniques with realistic optics will affect missions on all scales throughout this spectral region. *Optics development and demonstrations should be at the forefront of a technology program early in this decade.*

Supporting Ground-based Capabilities.

Ground-based observations have played a major role in bringing the science programs of space missions to fruition. The need to place such observations into the context of the development of nearly a century of astronomical understanding usually requires observations in the visible region. In particular, the need for spectroscopic observations and optical "identification" have driven great demand for ground-based facilities. As space observatories become more sophisticated and cover progressively larger ranges of the spectrum this latter demand will probably decrease. However, the need for spectroscopic follow-up will remain far into the future. Even with very large telescopes in space the cost-effectiveness of large ground-based facilities for time-consuming spectroscopic observations will require that such telescopes be built and supported. In general the rule that "anything that can be done from the ground should be done from the ground" is a wise and cost-effective one.

The results from an increased Explorer program, HST, and the other Great Observatories will place substantial pressure on ground-based facilities. This will be particularly the case for spectroscopic observations of faint sources. All-sky access to large ground-based telescopes for spectroscopic observations will be crucial if the desired scientific returns are to be made. Telescopes of the 4+ m-class will be under the most pressure. Additional support for the construction and instrumentation of such telescopes should be seriously considered as part of the overall space science program. In addition, support of the related use of such facilities should be available through funding of the analysis of space science data. The goal is clearly to maximise the scientific return and all the necessary steps need to be supported in a unified way.

Instrumental development and demonstration is another area where ground-based programs are valuable and cost-effective. Much can be learnt and many problems solved before the start of the very difficult, time-consuming and expensive process of fabricating instruments for space missions.

From a scientific viewpoint the need, effectiveness and value of a balanced and synergistic program of ground and space missions is obvious. Every effort should be made to ensure that such an approach is taken. The costs are minimal compared to the space missions and the scientific returns are large.

V. PREREQUISITES – THE CURRENT SCIENCE PROGRAM.

The ongoing program with IUE, ASTRO, EUVE, Lyman-FUSE, STIS, NICMOS, and HST operational improvements is a strong one. It must be well-supported to ensure that the scientific returns are made. The key elements are Lyman-FUSE, HST Optimization, and the near-term missions and their support.

Lyman-FUSE.

The Lyman Far Ultraviolet Spectroscopic Explorer mission will use high resolution spectroscopy below the 1200 Å HST limit to measure for the first time faint sources both throughout the Galaxy and at very large extragalactic distances. This important spectral window is virtually unexplored except for the Copernicus mission, which was limited to bright nearby stars.

The primary goal of the Lyman-FUSE mission is to obtain high resolution spectra ($\lambda/\Delta\lambda \sim 30000$) with unprecedented sensitivity (10^5 times that achieved by Copernicus almost 20 years ago) in the critical spectral window from 1200 Å down to 912 Å, the wavelength of the photoionization continuum of atomic hydrogen. Recent advances in optical and detector technology, which make the primary goal possible within a modest experimental package, also enable the spectral coverage to be extended down to 100 Å with a sensitivity much greater than that of EUVE. This can be done with only a minimal increase in complexity. The combined wavelength range thus bridges the gap between that covered by HST and AXAF at moderate cost.

The spectral window opened by this mission will provide a unique access to many critically important species for astrophysics. Interstellar and intergalactic deuterium, a crucial fossil nucleus formed in the early hot universe, can best be studied in the primary bandpass of this mission. In addition, Lyman-FUSE will obtain the direct measurements of molecular hydrogen, the primary constituent of cold interstellar clouds - the birthplace of stars and planets. For objects as diverse as planetary ionospheres, the interstellar medium, and QSOs, the unique diagnostics in the Lyman-FUSE spectral range will allow measurements of gas and plasma at temperatures over a wide range, from a few tens of degrees to greater than ten million degrees.

This mission will achieve major advances in a wide range of important scientific problems, because transitions in this spectral region include most of the important interstellar and intergalactic absorption lines. Furthermore, this region includes emissions from gas and plasma over an extremely wide range of temperatures. The most pressing scientific issues are concerned with the physical processes in three broad areas of astrophysics. They are (i) the physical processes in the early Universe, in essence the measurement of deuterium in a variety of environments, (ii) the physical processes that control the origin and evolution of galaxies, namely the examination of ISM and stellar processes that influence it, and (iii) the physical processes that control the origin and evolution of stars and solar systems, in particular study of the cold clouds in the interstellar medium, and of the dynamics of the formation process itself.

The utilization of a dedicated spacecraft for Lyman-FUSE instead of the Explorer Platform should be evaluated. EUVE will be launched on top of an Explorer Platform spacecraft by a Delta vehicle. The follow-on mission, XTE, will replace EUVE on the Explorer Platform using a Shuttle to exchange the two. The present plan for Lyman-FUSE calls for it to follow XTE as the final mission on the Explorer Platform. Because the Explorer Platform will be complete and the XTE will be in intensive development, this is a good time for reevaluation of the true costs and programmatic risks associated with this novel approach, including the use of the Shuttle. New technological approaches presently under development may make dedicated spacecraft, tailored to a particular experiment, more attractive. It is possible that the integration costs and costs associated with using the Shuttle may balance the cost of a spacecraft both for Lyman-FUSE and for future missions. In addition to reduced programmatic and budgetary risks, there is a potential for significant improvement in the scientific return by optimization of the spacecraft and orbit as well as by a longer-lived mission.

HST Optimization: Operations, WF/PC II, STIS and NICMOS.

Recovery of HST's Imaging Capability.

The highest priority must be given to recapturing HST's imaging performance. While the effect of the degraded images on the performance of the cameras is obvious, virtually all of HST's instrumental capabilities are affected by the poor images; for example, the spectroscopic throughput and the ability to distinguish between sources in crowded fields are compromised, as is the spectral resolution by the use of large apertures. The replacement camera, the WF/PC II, should be completed and deployed as soon as possible. The modified Second Generation instruments should also be funded at a high level to allow completion and deployment as soon as is practical. The spectroscopic performance of STIS is sorely needed.

HST Operations.

The complexity of operation in LEO has resulted in HST being launched with an operational system that does not yet reach the theoretical limits for on-target efficiency. Given the high cost of HST and its limited life every effort should be made to optimize the operational systems of HST. This includes, for example, bringing into operation as soon as is possible the means to observe moving targets and the means to operate and obtain data from two instruments simultaneously ("parallel" mode). These will result in substantial improvements in the science return from HST by allowing observations of planets, their companions and other bodies in the solar system, as well as parallel surveys with the camera systems. This latter capability is likely to result in major discoveries with HST, particularly with the new cameras. Continuing support for development and refurbishment of the ground system will similarly prove valuable as experience grows with the operational systems.

The combination of the implemention of the planetary and parallel modes, higher operational efficiency, the image-correcting camera WF/PC II, and the Second Generation instruments, NICMOS and STIS, will lead to very substantial gains in scientific productivity – exceeding orders-of-magnitude in some areas.

STIS.

The Space Telescope Imaging Spectrograph (STIS) is a high resolution, high sensitivity spectrograph that will incorporate several advances over first generation Space Telescope instruments. Its availability on HST will result in substantial gains in productivity. Use of large format two-dimensional detectors will allow large multiplex advantages over the current spectrographs on HST. Both photon-counting and CCD technologies will be used in the spectrometer, so the instrument will be sensitive in the ultraviolet, visible, and near-infrared wavelength regions, covering 1050-11000 Å in 4 bands. Spectral resolving powers between 80 and 140,000 will be available. Camera modes will also be available in all 4 bands. High angular resolution will be obtained by sampling the HST diffraction-limited image with 25 milliarcsec pixels, and by compensating for off-axis aberrations.

Spectrographs of the capability of STIS will play a major role in a very wide range of scientific programs. It is expected to have particular significance for studies of the stellar and gaseous kinematics of galactic nuclei, both active and "normal". Such data will play an important role in understanding the physical properties of such nuclei and in establishing the density distributions at scales unobtainable from the ground. Answering the long-standing question of the existence of black holes in such nuclei is at the forefront of the goals for this instrument.

The structure of the ISM and the nature of young stellar objects and their associated jets and disks are also problems which are well-matched by the long-slit and UV-Visible capabilities of this instrument. The interactions of jets and winds in forming stars, the characteristics of protoplanetary disks, the nature of flares in stars and mass loss from hot stars are all areas of stellar astrophysics that HST plus STIS will tackle.

NICMOS.

The Near Infrared Camera and Multi-Objective Spectrometer (NICMOS) will add a near infrared imaging and spectroscopic capability to HST. This Second Generation Instrument will extend the wavelength range accessible by HST into the near-infrared where the background is reduced by three orders of magnitude from that on the ground. NICMOS contains three cameras in the 1-2.5 μm spectral range and three spectrometers covering the 1-3.0 μm spectrum. The spectrographs have resolving powers that range from 10^2 to 10^4. Its cameras will exploit the diffraction-limited imaging performance of HST with pixel sizes of 43 and 64 milliarcsec, while also having the option for a larger field of 51 arcsec in its PSF matching mode (0.2 arcsec pixels). A wide range of filters and polarizers will be available. NICMOS has a 5.5 year lifetime that is set by its cryogenic capacity.

The scientific goals of the system range widely. It is a versatile powerful instrument that opens up a new waveband at a cost much lower than a dedicated spacecraft. Its key scientific objectives encompass deep surveys, distances to and beyond local supercluster galaxies, the study of dust enshrouded regions of galaxies, star forming regions, and studies of valuable spectroscopic diagnostics in our solar system. The huge reduction in background over ground-based telescopes makes it a powerful tool for studies of highly redshifted galaxies. Use of NICMOS in the parallel mode for surveys can be expected to lead to the unexpected. It can extend distance measures using the brightest red supergiants to as far as the Coma cluster, while also playing a valuable role for other distance indicators where interstellar absorption is a problem. It has great value for studying star forming regions, nuclei of active and starburst galaxies, and other regions where dust absorption severely limits the data that can be obtained in the visible.

The performance gains for HST are of such magnitude that NICMOS and STIS should be accelerated and implemented as soon after the WF/PC II as is feasible.

The Near-Term Science Program.

A wide range of very exciting and powerful new missions are in development or have been proposed in this report. However, these programs must build upon the base of previous scientific programs and ongoing missions. A healthy science program requires continuing access to space missions and support of the ongoing scientific activities. Thus the missions such as IUE, ASTRO, EUVE and a variety of small programs such as the rocket program provide crucial near-term scientific returns, as well as addressing scientific goals that are impractical or technically impossible with the larger and rarer missions.

EUVE.

The EUV region of the spectrum remains largely unexplored. The Extreme Ultraviolet Explorer (EUVE) is dedicated to an investigation of the EUV band from 80-900 Å. The complement of instruments include three sky mapping telescopes, one deep survey telescope and three spectrometers. The three sky mapping telescopes will make complete maps of the sky in four distinct spectral bands during the first 6 months of the mission. EUV sources will be catalogued and their positions determined to accuracies of a few arc minutes. The deep survey telescopes will map a strip along the ecliptic where the sky background is very low, thus allowing higher sensitivities to be achieved. The deep survey will be carried out in two spectral bandpasses spanning 80-400 Å. EUVE will obtain the first full sky maps covering the whole EUV band.

EUVE also includes three novel spectrometers which will allow spectroscopic observations of the brightest EUV sources from 80-700 Å. The resolution of the spectrometers will typically be $\lambda/\Delta\lambda \sim 300$. The EUVE spectrometers will be operated through a NASA Guest Observer program. These spectrometers will allow the detailed line spectrum of coronal emission sources to be obtained. Similar coronal studies to those carried out on the sun can be extended to nearby stars. Observations of continuum sources, such as hot white dwarfs, will allow investigation of the photospheres of these objects as well as of the intervening interstellar medium. The EUVE spectrometers will allow measurements of the HeI and HeII edges at 504 Å and 228 Å.

The EUVE mission will be a valuable precursor to Lyman-FUSE, and a valuable database for UV programs on HST and subsequent missions.

IUE.

The International Ultraviolet Explorer Satellite ranks among the most productive Explorer missions, and has introduced a broad community of astronomers to ultraviolet spectroscopy during its 13-year lifetime. IUE has a 45-cm diameter f/15 Cassegrain telescope, with two echelle spectrographs with spectral resolutions of $\lambda/\Delta\lambda \sim 10,000$ and $\lambda/\Delta\lambda \sim 200\text{-}500$, and SEC Vidicon Cameras with CsI and CsTe photocathodes covering the spectral region between 1150 and 3200 Å. Fundamental discoveries have been made in nearly every area of observational research, including planetary astronomy, stellar evolution, atmospheres and chromospheres, the physics of the interstellar medium, stellar populations and galaxy evolution, active galactic nuclei and the intergalactic medium.

Barring significant degradation of current capabilities, IUE will continue to provide a valuable complement to the scientific return from HST for a relatively modest annual operating cost. The unique advantages of IUE relate particularly to variable phenomena: broad simultaneous wavelength coverage in its high resolution mode as opposed to the very limited range of a single HST high resolution spectrograph observation; much longer periods of uninterrupted on-target time because of its HEO location; access to targets of opportunity within an hour of notification (critical for the early phases of supernova development); scheduling flexibility that allows for simultaneous observations with other spacecraft; and opportunity for long-term synoptic monitoring not readily available to HST proposers. As ground-based observatories maintain telescopes of several different apertures to support a range of qualitatively different scientific objectives, so it is appropriate to continue an active observational program with IUE during the HST era.

ASTRO.

Four instruments make up the Astro Observatory which is flown as a Shuttle-borne sortie mission. It is the first observatory that can simultaneously take ultraviolet pictures of objects, study their ultraviolet and X-ray spectra, and determine their brightness and structure through UV photometry and polarimetry. Using the X-ray and UV spectrographs together provides spectral coverage from 1-3200 Å simultaneously with millisecond timing, a capability that is unique to the Astro Observatory. It should thus demonstrate the scientific value of such a multi-wavelength mission.

The Hopkins Ultraviolet Telescope examines the ultraviolet spectrum from 400-1800 Å. This is the only spectroscopic coverage we will have of the astrophysically important 912-1200 Å region until Lyman-FUSE is launched in the mid-1990s. It is below the HST cutoff. The Ultraviolet Imaging Telescope takes detailed ultraviolet (1200-3200 Å) images with a circular 40 arcmin field of view and a resolution of 2″. Very little imaging information is available in this spectral region. The Wisconsin Ultraviolet Photo-Polarimeter Experiment will use the polarization of ultraviolet light to measure the strength of magnetic fields, the geometry of stars, and the nature of the material between stars. The Broad Band X-Ray Telescope will make the first high-quality, high-energy (0.3 - 12 keV) spectra of many X-ray sources and will be able to measure the important 6.7 keV iron lines.

The Astro mission is expected to make about 250 pointings, yielding more than a thousand measurements on astronomical sources. Astro was conceived during an era when space science was focused on Shuttle missions, and it was expected to have a multi-mission lifetime. The move away from man-rated limited-life missions is a welcome and long-overdue one. While Astro missions are costly, comparable capability is unlikely for many years, possibly beyond this decade, and so a second mission should be supported if justified by the scientific returns from the first mission.

Small and Sub-orbital missions.

In a well-balanced space astronomy program, there is a clear need to support the use of small, innovative and sometimes very specialized instruments. Inexpensive experiments operating on high-flying aircraft, balloons and sounding rockets are important for developing and testing new technologies and performing unique observations that may be unsuitable for the larger, more generalized facilities, such as the Delta-class Explorers and Observatories that serve a broad community of observers. A key operating advantage for suborbital research is that experiments and their support systems may be simple, easily modified, and can assume risks that are unacceptable for the more major missions. Moreover, the time scale from concept to flight is usually only a few years, which is ideal for training graduate students and post-doctoral fellows who will become the newest generation of investigators with a proficiency in designing, building and flying space hardware.

Sometimes, experiments developed for suborbital missions open a new field of observing and, when there is a potential for further development, they serve as an economical proof-of-concept for later, more mature facilities. Likewise, for an instrument which has operated successfully on a number of sounding rocket missions (usually lasting only about 5 minutes apiece), it may be appropriate to have it adapted for flight on an orbital mission of moderate duration, so that it can achieve a worthwhile incremental gain over its past accomplishments. While such a payload could function either as an attached payload on a Shuttle flight (as are some planned small UV payloads), or preferably, on a free flying, very simple spacecraft deployed and retrieved by the Shuttle in orbit, alternative approaches are desirable. Small experiment programs should not be constrained and burdened by the strict scheduling and costly safety requirements of the Shuttle. Inexpensive rockets which can achieve orbit, such as the Pegasus rocket, should be developed as suitable delivery systems.

From a general perspective, suborbital programs are ideal for establishing and supporting new space astronomy groups. These new research groups can play a valuable role in adding vitality and ingenuity to the national effort, as well as creating an environment for entrepreneurial activity that may not otherwise be possible. NASA should continue a vigorous flight and research program which uses small payloads and not assume that they must be supplanted by the larger, generalized Explorers and Observatories.

Data analysis, modeling and theory funding; Archives.

The added support for data analysis, modeling and theory through the enhanced data analysis program (the NASA MO&DA program) is an important change from the past. We applaud the designation of budgets for such activities that is commensurate with the very large investments that are made in the flight and ground systems. It is crucial that such support be continued, and that it be protected from overruns in ongoing development programs. The goal of the science program is to increase scientific understanding, to encourage the interest of the population at large in science and technology with the goal of enhancing our technical education level, and of sharing with the public the excitement that arises from our developing understanding of the universe. Without a timely scientific return from the science missions we run the risk of compromising the timely funding of future missions.

These goals cannot be met without strong, continuing support for the analysis of the data and its understanding through a strong theory program. These programs should be broadly-based to allow the access and input of data and results from the broad spectrum of astronomical inputs that are needed for solving a particular scientific problem. This may require supporting the use of data from several missions, from ground-based telescopes, from archives, and from sophisticated modeling. The goal is, of course, scientific understanding, and not just data analysis. Thus the funding approaches should not be too narrowly defined.

Archives are likely to play an even more important role in the future, particularly as the volume and complexity of the data from space missions increases. These archives need to be readily accessible and need to be supported by scientists with a broad understanding of the nature and limitations of the particular data. This may often be by the scientists closely associated with the initial mission. Fortunately, techological developments will allow for decentralised archives and easy access through wide-bandwidth communications. There will clearly be a need for standardisation in the interfaces and for a generic and cost-effective archive technology so that substantial volumes of data can be disseminated quickly and cheaply.

VI. TECHNOLOGIES FOR THE NEXT CENTURY.

New technologies are essential for the next generation of large telescopes, interferometers, and future smaller missions.

Substantial future progress in space telescopes and instrumentation will require technological advances. A wide variety of demanding technologies are needed for state-of-the-art telescopes and instruments in the UV-Visible-IR wavelength region. Advances in these areas can benefit essentially all missions, small and large. Any program should incorporate a development phase and a clear demonstration that the goals of the program have been met. Several areas where substantial gains in capability are needed and which appear to be feasible are given below.

High-performance optics.   Diffraction-limited optics are an integral part of space astronomy. A tough but not impractical goal is improving optical surfaces beyond the current state-of-the-art as represented by HST and the ESO 3.5 m NTT mirror. The optics need to have low-scattering surfaces for high dynamic range observations (e.g., for QSO "fuzz" observations and for planet detection). A challenging component of this is the fabrication of off-axis segments. Technologies such as stressed-lap polishing and ion-beam polishing have the potential for affordably manufacturing large as well as small optical surfaces with aspheric deviations far beyond what is currently practical. A program whose explicit objective is to demonstrate the polishing of a low-scattering, strongly-aspheric off-axis element should be instituted early in the decade.

Lightweight optics.   Lightweighting will remain crucial for space systems, especially as interest increases in HEO and Lagrangian point locations. The same is true of systems destined for the moon. Lightweighting techniques are intimately tied to the polishing technologies that are available. Since lightweighting can result in "print-through" to the surface, its application has been limited by the difficulty of removing such effects. The ability of ion polishing to potentially remove "print-through" allows for more aggressive lightweighting.

Active optics.   Active sensing and control of optical surfaces will be valuable in reducing the structural requirements and to compensate for thermal and dimensional changes, particularly in passively-cooled systems. Active elements can also play a valuable role in fine pointing and tracking, thereby lessening the demands placed upon the pointing and tracking control system, with attendant savings in complexity, weight and cost.

Lightweight and active structures.   Lightweight structures are essential for space missions. Considerable advances in performance can be obtained through use of new techniques for passive structures and through active structures. Such structures will play an important role for control of the location of critical elements in interferometers and in large optics systems (e.g., the secondary).

Tracking and pointing.   The tracking and pointing system for HST was a very demanding aspect of the spacecraft. New approaches involving large area detectors and sophisticated on-board processing have the potential for substantial savings in weight and cost. Both technologies were not mature at the time HST's design was frozen. The use of active optical elements for fine pointing would also have significant cost and technology advantages.

Passive and active cooling.   Passive cooling of the whole telescope to improve IR performance can return remarkable gains in sensitivity ($> 10^6$ beyond 3 μm). Active coolers will be needed to eliminate the need for expendable cryogens for IR detectors and instruments. These pose challenging vibration and lifetime problems and need further development. Passive cooling also sets a challenging problem for the manufacture and testing of large optics. This needs to be an integral part of the optics development program.

Detectors.   Detectors for the UV, the Visible and the IR are a crucial part of any system. We are making great strides in all three areas, but need to keep up the momentum. While consumer and defense requirements have been major drivers for technological developments from which we have benefited, the characteristics required for astronomical missions are usually different. Resources are needed for further development of the detectors to ensure that the required performance characteristics are met.

Instrument optics and coatings.   Increasing overall system efficiency will require substantial efforts to maximize the performance of instrument optics and coatings. New optical techniques are needed to allow the development of cameras with larger fields, better images and higher throughput – particularly with wide-band coatings. The wide-bandwidth of space instruments places particular demands upon coatings and filters. Further development is needed.

Computers and electronics.   The remarkable gains in computional capability over the last decade can provide sophisticated on-board processing leading to significant performance gains in the operation of the telescope and instruments. Simplification of optical-structural systems can also result with attendant cost savings. Concerns about the reliability of such systems could be alleviated by self-checking, redundant systems.

VII. LUNAR-BASED TELESCOPES AND INSTRUMENTS.

The Lunar Outpost program offers opportunities for a long-term astronomical program of remarkable scope and power. Of some concern to the astronomical community is the timescale for the implementation of such a program, and the potential for large gaps in time between orbital and subsequent generations of lunar-based facilities. It is imperative that a rational and realistic program be developed that allows for overlap between ongoing and new instruments. Such continuing capability allows for a well-planned transition and allows for contingency in case of delays. The lunar outpost program is ambitious, as it should be, but its impact on space science and the potential for damaging a vibrant scientific program which has considerable value for the scientific base of the nation should not be underestimated.

Currently the Lunar Outpost program in the UV-IR range incorporates three missions that allow phased implementation of capability. The initial project is a compact, simple instrument, the Lunar Transit telescope, which is followed by the phased implementation of an interferometer and a telescope, and culminates in a 16 m telescope of astonishing power, and a Visible-IR interferometer of some 5-10 km in size that is of comparable power.

The Lunar Transit telescope.   This telescope is a wide-field non-pointed imaging telescope with a series of CCD detectors optimized for operation from 0.1 to 2 μm. The focal-plane arrays record and transfer data at the apparent sideral rate and carry out a survey across a strip of the sky. The result is a wide-bandwidth, deep multicolor imaging survey. Repeated scans allow for the detection of variable objects. Limiting magnitudes are V $\sim$ 28 magnitude for a 2 m telescope with an image size of 0.1 arcsec at 0.5 μm. Its compact design and lack of moving parts makes it particulary attractive for the Emplacement phase of the development of the Lunar Outpost. Examples of the scientific returns range from the systematic detection and study of gravitationally-lensed objects, to deriving the structural and photometric properties of large numbers of galaxies ($\gg 10^5$), to the variability of QSOs and AGNs, and the detection of numerous supernovae.

The 16 m Lunar NGST.   The NGST is a passively-cooled, diffraction-limited telescope. Its imaging and spectroscopic capabilities for a variety of scientific programs across the UV to the IR would be quite remarkable. Its scientific goals would range from the detection and spectroscopic measurement of Earth-like planets to the study of galaxies at redshifts $z > 1$ where substantial evolution and possibly even the formation of galaxies could be observed directly. Its combination of sensitivity, wavelength coverage and resolution would result in quantum jumps in knowledge in areas as diverse as the evolution of the ISM to the formation of stellar systems to the evolution of galaxy clusters.

The telescope structure and optics would be passively-cooled to 100°K or less in the lunar night, lowering the background in the 3-4 μm zodiacal "window" to less than 10^{-6} of that from the ground. Its low background, resolution and low-scattering optics at 10 μm would allow it to both image and take spectra of planets at the separation expected for Earth-like planets around the nearest stars.

The telescope would have a lightweight, segmented primary with active wavefront sensing and control for use into the UV. The segmentation would allow for a modular approach to consruction of the telescope by robotic means or by astronauts. Designs in which instruments can be placed on or beneath the Lunar surface are also being considered. This has advantages for stability, maintainability, reduced particle flux, and for instrument upgrades. With such a major facility the ability to upgrade and change the instrumentation during its presumably very long life is an important factor. The stability of the lunar surface and the ability to isolate sources of noise would improve its pointing and tracking performance.

It would be sited near the large interferometer. By combining the 16 m with an interferometer high sensitivity can be achieved from the large area of 16 m, thereby increasing the number and type of sources that can be studied at interferometric resolutions. By providing imaging on scales of 5 milliarcsecond to arcminutes, it also provides a natural overlap between space interferometry and ground-based observations.

The Lunar Optical-IR Interferometer.   This km-scale interferometer exploits the stability of the lunar surface and the lack of an atmosphere to develop an interferometer of striking scale and hence resolving power. Techniques can be developed and tested on the earth and in orbit that will allow the appropriate choices of technology to be made as the Outpost develops.

The science potential ranges from submilliarcsec imaging of accretion disks around massive black holes in AGNs to images of extra-solar planets from Jupiter-sized to Earth-sized. The number of candidates could be very large within within 20 parsec. For stellar astrophysics the goals range from images of surface features of stars, both for normal stars as well as unusual objects such as X-ray binaries with neutron star companions. Elucidating the structure, the kinematics and the mass distributions close to the center of galaxies is another goal that is probably not achievable by other means.

Its astrometric performance is similarly striking with capability to sub-microarcsecond levels. This could give direct parallaxes to nearby galaxies, gravitational deflection of starlight to second-order, and determine the isotropy of the Hubble flow. Its astrometric capabilities would include the detection and determination of the masses of extra-solar planets.

The basic concept of the program is that of an optical VLA with a substantial number (> 10) of 1-2 meter-sized individual elements, with baselines to several km. This would allow a resolution for imaging of 10 μas and for astrometry of 0.1 μas. An initial 3 element system during the Emplacement phase would grow to many times that in the Utilization phase. Compared to Earth-based systems the gain is about a factor of 100 over instruments such as the VLT. It would be passively-cooled during the lunar night for good IR performance.

These telescopes have great scientific merit, and represent ambitious, but appropriate goals for the Lunar Outpost. Technology development towards these goals is needed and timely, provided emphasis is given in the early phases to demonstration projects of technologies that have broad utility for space science missions. The technologies highlighted in §VI meet these goals.

INTERFEROMETRY PANEL

STEPHEN RIDGWAY, National Optical Astronomy Observatories, *Chair*
ROBERT W. WILSON, AT&T Bell Laboratories, *Vice-Chair*

MITCHELL C. BEGELMAN, University of Colorado, Boulder
PETER BENDER, University of Colorado, Boulder
BERNARD F. BURKE, Massachusetts Institute of Technology
TIM CORNWELL, National Radio Astronomy Observatory
RONALD DREVER, California Institute of Technology
H. MELVIN DYCK, University of Wyoming
KENNETH J. JOHNSTON, Naval Research Laboratory
EDWARD KIBBLEWHITE, University of Chicago
SHRINIVAS R. KULKARNI, California Institute of Technology
HAROLD A. McALISTER, Georgia State University
DONALD W. McCARTHY, JR., University of Arizona
PETER NISENSON, Harvard-Smithsonian Center for Astrophysics
CARL B. PILCHER, NASA Headquarters
ROBERT REASENBERG, Harvard-Smithsonian Astrophysical Observatory
FRANCOIS RODDIER, University of Hawaii
ANNEILA I. SARGENT, California Institute of Technology
MICHAEL SHAO, Jet Propulsion Laboratory
ROBERT V. STACHNIK, NASA Headquarters
KIP THORNE, California Institute of Technology
CHARLES H. TOWNES, University of California, Berkeley
RAINER WEISS, Massachusetts Institute of Technology
RAY J. WEYMANN, Mt. Wilson and Las Campanas Observatory

Interferometry

KEY POINTS ON INTERFEROMETRY

- The extension of multi-telescope interferometry and aperture synthesis to infrared and optical wavelengths will enable in this decade significant progress in stellar and galactic physics, and will lead to great interferometric infrared/optical arrays of the future.
- The real-time correction of atmospheric turbulence with adaptive optics will allow the new generation of large telescopes to reach several magnitudes fainter, and to resolve spiral arms of galaxies anywhere in the universe.
- Laser interferometer gravitational wave detectors on earth and in space in this decade and the next can probe the dynamics of relativistic systems in the galaxy and the early universe.

Introduction

In this decade, the first infrared/optical interferometer arrays, adaptive optical systems, and laser interferometer gravitational wave detectors will be developed, implemented, and employed for fundamentally new types of observations. These new instruments will undertake systematic imaging in the infrared and visible of stellar surfaces and circumstellar material and of bright galactic nuclei, and will search for gravitational radiation from neutron stars and other condensed objects.

In the first decade of the next century, we foresee construction of great interferometric observatories on the earth and in space, performing astrometric measurements to microarcsecond precision and obtaining imagery of the faintest sources with angular resolution substantially better than 1 milliarcsecond. We also forecast the opportunity for an ultrasensitive, gravitational wave detector in space, designed for detailed observations of individual systems, and for detection of subtle gravitational wave tracers of the early epochs of cosmological history.

These instruments and observatories will be possible as a result of recent advances in optics, metrology and precision control, as well as improved understanding of gravitational wave sources and of atmospheric turbulence.

Overview of the Programs

Infrared and Optical Interferometry

Astronomers are increasingly aware of the opportunities which IR/Optical interferometry will offer. Following rapidly the path which radio astronomers traveled 20 years ago, optical interferometry is now

 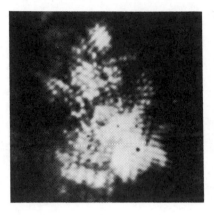 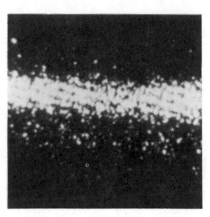

Figure 1. (Left) Optical interference fringes observed with an aperture mask (4.5 meter baseline) on a conventional telescope (Readhead et al 1988); (Middle) between subapertures on the Multiple Mirror Telescope (Hege et al, 1985); (Right) between independently mounted telescopes (15 meter baseline) at the CERGA observatory in France (Koechlin, 1984).

being carried out, albeit in a limited way. Aperture synthesis, which combines beams from multiple apertures to achieve the resolution of a much larger aperture, has been extended to the optical regime. Already an astrometric interferometer operates regularly on Mt. Wilson, actively controlled with a precision exceeding the optical tolerances of many large telescopes, while nearby an infrared interferometer monitors the formation of dust in the shells of stars. Interferometry is the most accurate technique for measurement of stellar diameters, with results from France, Australia, and most recently in the United States surpassing the accuracy of lunar occultation techniques.

Infrared/Optical interferometry will have a profound impact on astronomy. Current seeing and aperture limits to resolution will be surpassed by orders of magnitude. Telescope arrays planned for this decade may revolutionize stellar astronomy, yielding unprecedented detail about stellar surfaces, atmospheres, shells, companions, and winds. These future instruments will allow imagery with $m$sec resolution of oblateness of rotationally distorted stars, of chromospheric structures, of jets from young stellar objects, and of narrow line emission regions in Seyfert galaxies. Arrays of the next decade, on the ground and in space, will advance IR/optical interferometry to a sophistication comparable to that achieved by the radio astronomy community in the Very Large Array.

Interferometry also promises remarkable opportunities for astrometry. Already ground-based interferometry is approaching a precision of 1 $m$sec. Space missions of the 1990's and beyond should improve this performance initially by at least two orders of magnitude. It will be possible to determine an accurate parallax for any observable point source in the galaxy, and to measure proper motions of stars throughout the galaxy and the local galaxy group.

Numerous research groups, including several in the U.S., have initiated construction of arrays of two or three telescopes for imaging interferometry and astrometry in the visible and infrared. We recommend significant support, to assist rapid continued progress in this area. Specifically, we recommend support for a range of facilities operating in the visible and infrared with small and medium-aperture telescopes. Such breadth of activity is critical to the development of the field. By the end of the decade it will be essential to have in operation an array of five or more telescopes of medium aperture (1.5-2.5 meters). This array is required to achieve important infrared science objectives, to fully develop interferometry in the extreme multi-r_o condition, and to serve as a critical stepping stone to a very large optical array. This array of medium apertures will extend the reach of interferometric imaging to many YSO's and galactic nuclei, returning the science and technical experience needed for developments of the next decade.

As with radio interferometry, IR/optical interferometry will reach its full potential with large, well populated arrays of moderate to large aperture telescopes. We therefore recommend, for the latter part of the decade, the development of a plan for a Very Large Optical Array, to be built in the period 2000-2005. Composed of perhaps 20 medium-aperture telescopes, each equipped with adaptive optics, this array will achieve aperture synthesis imaging with sub-$m$sec resolution of active galactic nuclei, novae, stellar accretion disks and QSO's.

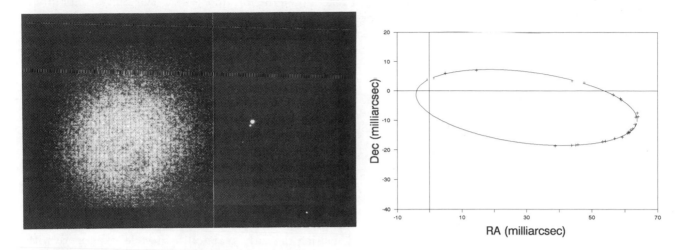

Figure 2a (left). Optical aperture synthesis image of the binary σ Her. (Left) A long time exposure, seeing disk diameter about 2". (Middle) The image reconstructed with closure phase relationships, printed at the same scale. The separation is 70 $m$sec, the dynamic range 30:1. (Nakajima et al, 1989)
Figure 2b (right) The orbit of the binary star β Ari, determined from interferometric measurements recorded with the Mark III interferometer. Maximum separation about 65 $m$sec. (Pan et al, 1990)

In the moderate cost category we recommend an astrometric interferometric space mission. This mission should offer a capability to achieve precision astrometry of the brightest stars in nearby galaxies. A number of concepts have been proposed, some based on interferometers that can also be used for imaging while others are dedicated astrometric devices. All support the expectation that interferometric techniques will yield astrometric precision on the order of 3-30 μsec over the entire sky.

The ultimate interferometric performance will be achievable above the atmosphere, beyond the limits imposed by atmospheric absorption, scattering, and turbulence. A modest array, operating in the ultraviolet, could image objects as faint as 20th magnitude with resolution of a few $m$sec, particularly useful for study of galactic nuclei. We recommend planning and preparation for both an intermediate sized imaging interferometer and an advanced array. The astrometric mission of the 1990's should demonstrate many of the metrology and control techniques required for future advanced arrays in orbit and on the lunar surface. Other aspects will be developed in the work on ground-based arrays. There remain issues that require further study, and we recommend an effort in development of this technology for the next decade.

The United States is considering a major commitment to the establishment of a permanent human presence on the moon. NASA, recognizing the outstanding merits of the lunar surface for interferometry, has recommended optical and sub-millimeter arrays for early implementation on the lunar surface. In the context of a broad lunar program, we endorse this plan. The moon offers a uniquely large and stable platform. A Lunar Optical Array should have a large number of apertures and long baselines. In the earliest phase, a small number of operational telescopes could be activated, perhaps optimized for a high priority capability such as high dynamic range imaging or sub-μsec astrometry, with such science goals as detection and study of planets in other systems, or measurement of the accretion disks in active galactic nuclei. In subsequent years, additional telescopes could be added to reach the final configuration, offering imaging of the faintest sources.

Adaptive Compensation for the Atmosphere

The terrestrial atmosphere has been and remains a primary limitation to observational astronomy. The possibility of removing the effects of atmospheric turbulence has long been considered, but only now are the necessary sensitive detectors and fast, reliable electronics becoming available at moderate cost. With the emergence of adaptive optics for solar and stellar imaging, adaptive techniques appear certain to play a major role in ground-based astronomy of the 1990's.

Adaptive optics (AO) will become a standard component of large telescope systems (≥ 4 meters) during the next decade. It will perform most effectively and generally in the infrared, where near-diffraction limited operation will be possible over a substantial angular field with relatively modest AO systems. Moving toward the visible, the corrected field will decrease, probably to a few arcseconds, and the cost of the optical system will increase. Although not applicable to all observational problems, AO will offer significantly improved resolution and increased sensitivity for many applications. It will provide a dramatic gain in angular resolution, and will increase the efficiency of high resolution spectrographs. These gains will be invaluable in many areas, for example the study of young stars and of galactic nuclei.

Adaptive optics is ripe for implementation in astronomical instruments. We recommend a program to apply to astronomy (often by simplification) adaptive optics technology which has already been developed under DOD support. The development of technology for laser reference techniques should continue. The DOD has developed a momentum in the field which astronomers will wish to preserve and steer to their benefit. We recommend the immediate implementation of adaptive correction technology to selected existing large telescopes. This will yield improved image quality at all wavelengths over the whole sky, but most dramatically and with assured significant scientific return in the near infrared.

Although the last decade has seen rapid progress in adaptive optics technology, many technical and cost tradeoffs critical to astronomy remain unstudied. During the early years of the decade, experience with first generation adaptive systems will lead to an adequate characterization of the atmospheric parameters critical for adaptive optics - seeing, time constant, and isoplanatic angle. This will provide the basis, later in the decade, for optimized systems functioning effectively in the visible range for many applications.

Gravitational Wave Observatories

The theoretical basis of gravitational radiation is well established, and the mathematical formalism and our astrophysical knowledge are now adequate to estimate the emission from many known and hypothesized astrophysical sources. Strong gravitational radiation is expected from individual binary systems composed of condensed components such as white dwarfs, neutron stars and black holes. Rotating neutron stars with an equatorial ellipticity may be detectable. A wide range of astrophysical information, accessible in no other way, may be revealed from gravitational wave observations, such as the evolution of asymmetries in the core of a supernova, and the rate at which condensed stars fall into supermassive black holes out to $z=3$.

The technology is now at hand to achieve the direct detection of gravitational waves, to measure the waveforms and determine the direction, frequency, and polarization, and to deduce the size and shape of the sources. A LIGO (Laser-Interferometer Gravitational-Wave Observatory) planned for this decade will provide a facility for the implementation of many generations of increasingly sensitive high frequency detectors. It seems likely that the LIGO will be a rich source of information about neutron stars and neutron star physics. LIGO should detect supernovae or collapse of a stellar mass to a black hole at 10 Mpc, if 0.01% of the energy of collapse is emitted as gravitational waves. It also should detect coalescence of two neutron stars out to 1000 Mpc and the merger of intermediate mass black holes out to the Hubble distance, as well as aid in determining the distance scale.

Massive objects and binaries will emit low frequency radiation which is difficult to detect from the earth, owing to terrestrial disturbances. A LAGOS (Laser Gravitational-Wave Observatory in Space) would detect individual binaries down to the confusion limited background noise of galactic and extragalactic binaries, and the merger of condensed stars with supermassive black holes at cosmological distances. LAGOS also might bring unique cosmological information with detection of relic gravitons from the early universe.

We expect the LIGO program to go ahead, with a substantial probability of direct detection of gravitational waves before or soon after the year 2000. Supernovae or coalescence of degenerate stars and/or stellar mass black holes are the most likely sources of detectable radiation. An antenna on the lunar surface would offer greatly improved discrimination of source direction. If ground-based searches for gravitational waves are successful at the expected levels, we foresee intense interest in extending the measurements to many other types of sources. The greatest return in astrophysical information about gravitational wave sources could be expected from a space observatory. With sensitivity extending to low frequencies, it might detect large numbers of compact binary systems, coalescence of condensed stars with massive black holes at cosmological distances, and a possible remnant background from the Big Bang. We recommend a planning and technology development effort to prepare the way for a LAGOS initiative in the early years of the next century.

Summary of Recommendations

The following table summarizes the recommended programs (with LIGO omitted because it has already been approved by the National Science Board).

Recommended Programs for the 1990's		
Ground-Based Interferometry		
Large	Optical/Infrared Interferometry with Multiple Telescopes	$50M
Medium	Adaptive Optics Development and Implementation	$35M
Small	Planning for Very Large Optical/Infrared Array	$3M
Space Interferometry		
Large	Interferometric Astrometric Mission	$200M
Medium	Technology Development: Advanced Space Interferometer	$20M
Small	Technology Development: Gravitational Wave Observatory in Space	$20M

Infrared and Optical Interferometry

No astronomer needs to be convinced of the fantastic capability of a space-based optical and IR interferometer operating with km-length baselines and with sufficient sensitivity to produce maps of faint sources down to 0.1 $msec$ resolution. Spatially compact phenomena of great astrophysical interest are far from simple in structure. For example, collimated jets of plasma are now known to be principal conduits of energy and mass outflow from protostars, mass-transferring binaries, and supermassive black holes. Accretion disks appear to be the main conduits of inflow in all of these objects, but we know very little of what these structures are really like. Supernova and nova explosions are highly anisotropic from their earliest stages, and the surface layers of "normal" stars seethe with starspots and flares. The development of radio VLBI has shown that the modeling of sources based on fringe amplitudes alone is far less valuable than the imaging which became possible when closure phase techniques were introduced. Our program in interferometry should aim to develop true imaging capability from the start.

Figure 3a shows the types of phenomena that can be resolved at different levels of angular resolution. The most dramatic increase in the range of accessible phenomena occurs at resolutions exceeding 1 $msec$, corresponding to the resolution now available with radio VLBI. With such resolution, one could not only resolve the disks of nearby main sequence stars, but could make detailed images of giant stars and of stellar winds. In the IR the structure of disks and incipient bipolar outflow in star-forming regions and the early evolution of novae and planetary nebulae could be studied. Milliarcsecond resolution will give crude structural information about the broad emission-line regions of active galactic nuclei, and detailed information about the narrow-line regions which are thought to represent the transitional zone between the active nucleus and the ISM of the host galaxy.

Resolution of 10-100 μsec would yield a new set of breakthroughs. Not only could the broad-emission line regions of AGNs and quasars be mapped, but it would become possible to investigate the inner regions of the accretion disks themselves (in the visible and UV). Accretion disks in close binaries could be imaged, along with the mass-transferring streams which feed them. Supernovae out to the Virgo cluster could be studied as early as three months after the explosion, and surface phenomena on main sequence stars and nearby white dwarfs could be mapped.

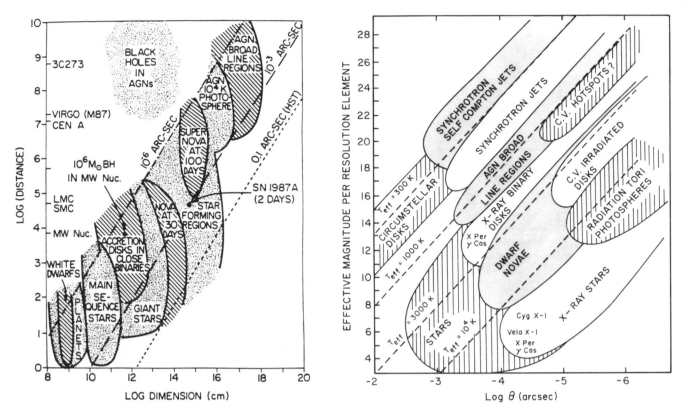

Figure 3a (left) Classes of objects and their angular sizes (diagonal lines are for constant angular size) for various ranges of size and distance.
Figure 3b (right) The required sensitivity (bolometric magnitude) for various types of objects, related to the size and characteristic temperature of an instrumental resolution element. Note that the regions begin at the lower left with the typical size of the largest members of each class.

Our justification for pursuing interferometry as a high priority must be based on these qualitative barriers which can be breached in no way other than imaging. Some goals, such as the imaging of faint AGN's at the highest resolution, will require sophisticated facilities of the future. Fortunately, simpler facilities recommended for this decade will make critical contributions toward many science goals, especially the astrophysics of young and evolved stars, bright galactic nuclei, novae, and the narrow emission-line regions of active galactic nuclei.

To be useful as an imager, a high resolution instrument must also have adequate sensitivity and dynamic range. Since phenomena associated with compact objects are often highly variable on timescales as short as minutes to hours, the time required to construct an image may also be a consideration. The bolometric brightness per resolution element may be expressed as a magnitude per pixel,

$$m \approx 43 - 10 \log T_{eff} - 5 \log \frac{\theta}{10^{-3}}$$

where T_{eff} is the effective temperature of the object being resolved and θ is the angular resolution in arcsec. The required sensitivity for various types of objects is plotted in Figure 3b. At the msec level, the required sensitivities are extremely modest: $\leq$ 7th magnitude per pixel for stars and AGN, 17th magnitude for protostellar nebulae. At resolutions of 10^{-5} arcsec, a sensitivity of 28th magnitude is adequate to image objects as cool as 300 K. Even interferometric imaging of extra-solar planetary systems may be possible, although a multi-pixel image of the planet would be extremely difficult.

Requirements on dynamic range are less certain. The most stringent constraints may come from objects such as close binaries, where there is often a mismatch between the surface brightness of the mass-transferring star and that of the accretion disk. A dynamic range as large as $10^2 - 10^4$ may be required in these cases, but a smaller dynamic range may suffice to map more homogeneous structures.

Although imaging will probably prove to be the most compelling long term motivation for developing interferometry, the most profound product of the early years of interferometry may well be ultra-precise astrometry made possible in space by freedom from terrestrial disturbances. Thus, we turn our attention to a space-based astrometric optical interferometer. Possible measurements should include proper motions, parallaxes, detection of dark companions, and relative motions in crowded fields (eg globular clusters). We expect that the instrument would have a measurement uncertainty of less than 30 μsec. It would measure the separation of stars that are well separated, yielding absolute parallax and interconnection of reference frames.

The RR Lyrae and Cepheid "standard candles" are critical to the determination of the Hubble constant. At present, the realistic uncertainty in the Cepheid scale is about 15 percent, and it is a prime objective of HST to make a modest improvement on this accuracy. Microarcsecond parallax measurements would reduce this uncertainty by at least an order of magnitude. Quasars are generally assumed to be at "cosmological distances" and therefore to show no proper motion of their centers of mass. An instrument with a few μsec precision could measure the relative motions of quasars, not only testing the cosmological-distance hypothesis, but also investigating the large-scale motions in the early universe.

In the area of Galactic structure, there are several applications of precision astrometry. We could measure the position, parallax, and proper motion of many of the massive young stars that mark the spiral arms. These data would map the arms without the distance uncertainty that now degrades such maps and would even show the motion associated with the density waves that are believed to be responsible for the existence of the arms. Measurements to stars within a few kpc of the Sun would yield a portion of the rotation curve for the Galaxy and thus a constraint on the mass distribution. Proper motions of the Large and Small Magellanic Clouds would make possible independent determinations of the total mass of the Galaxy and thus would tell us the amount of "dark matter" it contains.

Because of their brightness, we have an almost complete sample of the galactic globular clusters. These objects can serve as probes of the galactic potential to large galacto-centric distances. One only need measure a few stars in each cluster so as to average their typically 10 km/s motion with respect to the cluster center of mass. These measurements would also make possible the determination of the orbits of the clusters. The cluster orbital parameters could be used to investigate correlations between metal abundance, perigalactic distance, cluster radius, and orbital eccentricity, all having strong consequences for theories of the formation of the Galaxy. Finally, the membership of special stars such as Mira variables, AGB stars, helium-poor or helium-rich stars, or blue stragglers could be investigated geometrically, i.e., adding parallax to the angular position test.

Precision astrometry could be used to study stars, their formation and evolution, and their structure. Parallax measurements would substantially improve the number of accurate mass determinations for interferometrically resolved binaries, and orbits could be obtained for both components in an inertial frame, thus yielding the individual masses rather than the usual sum of the masses. Further, for nearby stars, the method of perspective acceleration could be used to determine the gravitational potential at the surface of the star. The resulting improved stellar mass and distance estimates, which would span the spectral types, would yield a sharper mass-color-luminosity relation. By concentrating on binary systems in clusters, we could empirically add age to the mass-color-luminosity relation and further test stellar-evolution models.

The study could be extended to special objects such as Eta Car, Cyg X-1, SS433, and the dozen Wolf-Rayet stars in double-lined systems that are near enough to study. The optical positions of the radio emission objects could be determined to higher accuracy than the radio positions are now determined; the spatial coincidence of these positions could thus be checked.

The astrometric determination of the wobble of a star around the star-planet center of mass would provide a sensitive means of searching for extra-solar planetary systems. Such a search, if conducted over an extended time period, would either find these systems or show them to be substantially less common than now expected, leading to a revision of our ideas about the formation of stars. Second or third generation imaging interferometers with large apertures may be able to image those planetary systems.

Astronomers are just beginning to obtain direct observational information about pre-planetary and remnant disks around young stars. Their nature and evolution can be established though high resolution near infrared studies of the material close to stars of widely different ages. These will provide a critical complement to millimeter interferometer measures of gas, which probe cooler circumstellar material at greater distances from the stellar surface. Of course, the question of the existence, distribution, and characteristics of other planetary systems, bears directly on the question of the existence of life remote from our solar system, and the possibility that there is other intelligent life in the universe.

The deflection of light by the solar mass is one of the standard tests of general relativity. If interferometric observations near the limb of the sun are possible, microarcsecond astrometry would permit the accuracy of this test to be increased by at least three orders. Such a deflection test would approach the sensitivity needed to measure the contribution to the deflection from the square of the solar potential. According to general relativity, this term is 11 μsec at the limb and falls as the inverse square of the impact parameter. When such a test is possible it would be the first "second-order" solar-system test of general relativity.

To summarize, the further development of IR/visible/UV interferometry should be undertaken with the goals of precision astrometry and high angular resolution imaging. Such capabilities will revolutionize many areas of astrophysics by providing the first morphological information on structurally complex systems. Attaining the goal of ultra-high resolution imagery will probably take some time, certainly longer than the nominal period to be covered in this decade survey. Fortunately, there will be important scientific returns from the earliest stages of the program.

A Ground-Based Program

Optical and Infrared interferometry is the subject of world-wide interest and activity, and the U.S. is fortunate to have several groups actively developing and using small interferometric facilities. The Berkeley Infrared Spatial Interferometer is currently operational at 10 micrometers (μm), and is now employed to measure diameters of late type stars and map their circumstellar shells. The Mt. Wilson Mark III interferometer, operating in the visible, is in use for wide field astrometry of stars, and also for several programs to measure stellar diameters and binary orbits. The Infrared Michelson Array, at the University of Wyoming, has recently seen first light and fringes. The Infrared Optical Telescope Array is under construction, with installation on Mt. Hopkins expected in 1991. It will operate in the visible and near infrared. While each of these projects is underway, none of them is as yet adequately funded to add the critical third telescope and associated instrumentation, required for imaging applications.

In addition, several groups working with aperture masks and large telescopes have demonstrated the reconstruction of optical images by aperture synthesis from U-V plane data.

Additional arrays are under construction or in an advanced planning stage. The Naval Observatory Astrometric Interferometer and the Big Optical Array are sponsored by the DOD. A design study at the Center for High Angular Resolution Astronomy at GSU has led to a plan for an interferometric array of small telescopes.

Owing in part to the terrestrial atmosphere, qualitatively different interferometric issues arise at visible and infrared wavelengths, with heterodyne and direct correlation, and for small and medium telescopes. Therefore, we give highest priority to supporting a multiplicity of projects; there is no other way to engage an adequate community of experimentalists and to train students for the development of the field. Similarly, progress on scientific issues over a wide front will require several operational arrays.

We recommend support for existing small interferometers and development of larger ones. These will further the understanding of technical issues, and will directly yield significant scientific benefits in the measurement of stellar diameters, binary star parameters, and many other areas of predominantly stellar astronomy. We recommend the construction of at least one array of approximately five telescopes of approximately 2 meter aperture, with baselines of at least 100 meters, for demonstration and use in infrared imaging. This will enable the systematic study of star formation, reconnaissance of the outer structure of AGN's, and many other topics involving moderately faint sources.

The proposed programs for the 1990's will lay the foundation for an advanced IR/optical array in the next decade. For such an array a considerable number of apertures, with a total collecting area comparable to a 16 meter filled aperture telescope, and an actual unfilled aperture size of many hundreds of meters, are envisaged. This project, a Very Large Optical Array, will extend ground-based optical interferometry to its faint limit, and will probably have a snapshot capability. We cannot at this time specify the VLOA in any detail. However, serious planning and design can and should begin later in this decade, with the objective of preparing a plan and proposal for the next decade survey ten years from now. This recommendation finds support in the long range planning efforts of NSF's ACAST (Long Range Planning Subcommittees, 1990).

The current DOD and private funding for optical interferometry is critically important, but requires significant augmentation, including non-DOD national funds. The experimental projects recommended for the 1990's should reasonably be concentrated in the university community, where cost-sharing is readily arranged, and where students will have maximum participation. The VLOA of the next decade will be a unique facility, and would appropriately have national sponsorship and community access.

A Space Program

The Interferometry Panel finds that there is great scientific potential in space-based optical interferometry. Much of the required technology is either in hand or rapidly emerging, but some specific technologies require development. This work should be started in the early 1990's so as to be ready to support a field that is ripe for productive scientific exploitation. We envision an orderly progression from small to large instruments, with the experience gained in the construction of ground-based imaging and astrometric interferometers and large aperture telescopes playing a significant role in the design of the large imaging instruments in space that we recommend for the 21st Century. The recommended progression has three steps: (1) an astrometric mission; (2) a multi-aperture imager (30-100 m baseline); and (3) a major imaging facility with a synthesized aperture of at least one km. These recommendations correspond to components of the recommended post-1995 astronomy and astrophysics program reported by the Space Science Board (1988). It is also consistent with the recommendations of NASA 's Planetary System Science Working Group (see Appendix).

(1) *During the 1990's we should design and build the first space-based optical interferometer.* This should be an astrometric instrument of the Explorer or intermediate class with a measurement accuracy goal of a few μsec, and certainly better than 30 μsec. Such an instrument would be a powerful new multi-disciplinary tool for astronomical research. It could open new areas of astrophysical investigation and change the nature of the questions being asked in some old areas. It would address several pressing scientific questions ranging from the existence (or absence), prevalence, and characteristics of extra-solar planetary systems, to galactic structure, mass, and mass distribution. It should directly determine the Cepheid distance scale to better than 2%, would be a potent tol for the detection of planetary systems out to at least 200 pc, and could perform a stringent test of general relativity. The importance of such an instrument has been recognized previously by the scientific community in various NASA and National Academy reports (Space Science Board, 1981; Physics Survey Committee, 1986; Space Astronomy MOWG, 1981).

Although much valuable work could be done with a narrow-angle instrument, there is a considerable advantage to an interferometer that can directly measure large target separations. Such capability opens the way for the determination and correction of instrument bias via 360 degree closure as well as for direct parallax determination without need for a zero-parallax reference object. This is important since, in the case of narrow angle astrometry, finding and certifying a zero parallax object for use with a chosen target in conjunction with μsec astrometric measurements poses problems that arise both because of the small (few square arcmin) field of view for reference objects and because in some directions (e.g., the galactic center) extragalactic objects are obscured. Similarly, measuring the proper motion of and within globular clusters is facilitated by being able to use reference objects outside of the cluster-obscured region.

For a wide-angle instrument and a given target star, the field available for the reference star is generally large and may approach a steradian. Thus a small set of well studied, bright stars (V $\simeq$ 10) can serve as reference objects for most observations. Sufficiently redundant observations within the reference set will yield a rigid frame in which the separation would be known a posteriori for any pair of stars, even those not simultaneously observed. Within the reference set, every star becomes a reference for every other star. Thus this frame could be tied to both the extra-galactic frame and the radio frame by a modest number of observations.

For a wide-angle instrument and a given astrometric target (star, asteroid, QSO etc) the field available for reference or comparison objects is large and may approach a steradian. Access is thus ensured to selected reference sources, including distant extra-galactic, zero parallax, zero proper motion objects, well studied stars, and other program sources, allowing all to be tied into a rigid frame.

(2) *During the 1990's we should develop the technology for an imaging interferometer that could be deployed in space early in the next decade.* This logical successor to HST should have a resolution about an order of magnitude greater than HST. Although considerable analysis is required before the architecture of such an instrument is established, we project that it will have a maximum baseline length of at least 30 and

possibly as much as 100 m, and total collecting area comparable to HST. The aperture might consist of 20 or more individual segments, arranged in a configuration determined by a tradeoff of resolution, sensitivity, time required to form an image, and other factors.

This intermediate-size instrument would yield a high level of scientific return by virtue both of its order-of-magnitude resolution advance and of its ability to image faint objects. It would benefit from the ongoing work on ground-based imaging interferometers; it would also benefit from the development of the astrometric interferometer.

(3) *During the early part of the next century, it should be possible to build a great imaging interferometer in space.* This km-scale facility with multiple large apertures (say two to six meters) would represent a major undertaking and could serve as the focus for an international scientific collaboration. The most desirable approach for building the great imaging interferometer is not yet clear. There may be considerable economic advantage to building this instrument on the Moon, depending on a clarification of the extent and time scale of the lunar scientific opportunities. A second option is to use separate free-flying spacecraft for each aperture, with the geometry of the array monitored and controlled by laser beams. The array might be located at one of the libration points of the Earth-Moon system or in geosynchronous orbit. This approach may be attractive, provided that it is possible to build such a "floating imager" at moderate cost. A third possibility is to connect the individual aperture modules by long beams. Studies of the different options will be necessary during the 1990's in order to identify the optimum approach.

Compensating for the Atmosphere with Adaptive Optics

The attainment of diffraction limited imaging is perhaps the most important challenge in ground-based astronomical instrumentation. Over the last 350 years, we have improved the sensitivity of our telescopes 10 million-fold, but at optical wavelengths atmospheric turbulence still prevents the resolution of even the largest telescopes from consistently exceeding that of a telescope with an aperture of only a few inches. We need high angular resolution in astronomy for the information available in improved imagery, and to reach fainter limiting magnitudes.

Many astronomical objects have structure much less than an arcsecond in scale. Such objects include all distant galaxies, the cores of normal and active galaxies, regions of dense star formation, most circumstellar disks, and all but the most gross features of planets. The Hubble Space Telescope should provide dramatic improvements in visible and UV spatial resolution, but only modest improvements in the infrared. However, an 8 meter ground-based telescope equipped with adaptive optics will provide angular resolution at 2 μm equal to the resolution of HST at 0.5 μm. Although adaptive optics will improve the performance of ground-based telescopes at all wavelengths, the gains in the infrared will probably be the most productive for astronomy in the 1990's, and the infrared science objectives are emphasized here.

Some of the most exciting applications of adaptive optics will be in the imaging and spectroscopy of galaxies. AGN's and the cores of "normal" galaxies are especially amenable to study with this technique. For instance, measurements of velocity dispersion near the center of normal galaxies such as M31 or M32 indicate that they may contain 10^7 $M_\odot$ black holes. Such data can be checked and extended to other nearby galaxies by using adaptive optics to carry out high spatial resolution spectroscopy. The central regions of galaxies are clearly special, and an 8 meter telescope will be able to explore these out to 3 Mpc with a resolution of 1 parsec. A 3.5 meter telescope should be able to resolve spiral arms anywhere in the universe. The surface brightness of an elliptical galaxy is very high, and it should be possible to carry out spatially resolved, velocity dispersion measurements for distant elliptical galaxies, enabling us for the first time to study the dynamics of galaxies in an earlier stage of evolution.

Supernova searches can be extended to high redshift galaxies by working in the infrared with adaptive optics. Even more exciting, especially for 8 meter telescopes, is the possibility of detecting very distant supernovae. Since SN probably occur before a galaxy forms a disk, we may see SN from primordial galaxies before we see the galaxies themselves. Also, it may be easier to detect a time-variable point source than a very low surface brightness extended source. The detection rate for SN at z=1 to 3 would be low for the local SN rate, but could be in the range 1/night of observation if the star formation rate in primordial galaxies were as large as 60 $M_\odot$ per year.

Large telescopes with adaptive optics will allow us to explore the morphology and kinematic structure of circumstellar disks, especially around late-type stars. These observations, essential for studies of stellar and

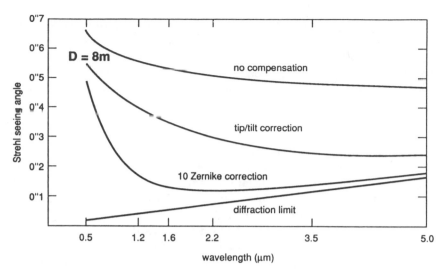

Figure 4 (left). A demonstration of diffraction limited imaging at 2.2 μm with adaptive optics and a 1.5 m telescope. The image is of the binary γ_2 Ari, separation 0.55" (Rousset et al, 1990).

Figure 5 (right). The calculated seeing width as a function of wavelength for an 8 m telescope with various degrees of adaptive compensation. The uncompensated Strehl width at 0.5 μm is assumed to be 0.66 arcsec. (F. Roddier, private communication)

planetary formation, can be achieved from ground-based telescopes. IR speckle observations have already shown the existence of infalling halos with solar system size inner diameters around stars such as HL Tau and R Mon. Adaptive optics allows us to explore such structures with a resolution of 10 AU for the nearest, with sufficient sensitivity to study a representative sample of objects with different ages. Studies of the circumstellar gas kinematics are necessary to determine how stars and planets form. Keplerian velocities are of the order of 10 km/sec, so we need spectral resolution of 100,000 and high spatial resolution to determine whether matter is falling into or leaving the star. Emission from the CO molecule at 4.8 μm should be well suited for this measurement.

Another important problem is the initial mass function of young stars as they condense from clouds of gas and dust. The IMF is crucial for many studies of star formation and galaxy evolution. In the Orion molecular cloud OMC1, the separation between stars at the present detection limit is a few arcseconds, and many of these are multiple. High spatial resolution is needed to obtain the HR diagram.

Brown dwarfs are objects too small to initiate hydrogen burning. Although there are strong theoretical grounds for their existence in large numbers, no confirmed detection has been yet made. Since the dwarf is close to the brighter primary, high angular resolution and dynamic range are required. With adaptive optics and an expected dynamic range of 9 magnitudes it should be possible to detect 40-50 Jupiter-mass objects out to a distance of at least 5 pc. This is a comparable sensitivity to the complementary technique of Doppler measurements (Latham et al, 1989).

For point objects improvements in seeing scale directly into improvements in limiting flux, owing to the reduction of background possible with improved concentration of light in a point source image. Higher angular resolution on large telescopes is in fact the most cost-effective way of observing fainter point objects. Thus a diffraction limited 3.5 meter telescope can in principle reach the same limiting flux as a 30 meter seeing limited telescope in the visible. Incorporation of diffraction limited imaging on an 8 meter telescope extends its limiting flux by more than an order of magnitude in the near infra-red.

The method of adaptive optics includes all orders of wavefront correction, beginning with the simplest, tilt correction (similar to fast guiding). The order of correction employed, and the resulting image improvement, will depend on the wavelength and telescope diameter. Thus in the infrared, low order correction alone may achieve near diffraction limited images, while at shorter wavelengths low order correction will provide improved image quality, while still far from the diffraction limit.

A major limitation of adaptive optics is the limited field of view over which the technique is applicable. Light from two sources in the sky travels through different turbulence and our ability to correct the wavefront over an appreciable field is limited by this effect. The region over which the optical properties of the atmosphere are correlated is called the isoplanatic patch. The size of the isoplanatic patch is a serious problem since often the target object itself is too faint to readily determine the instantaneous atmospheric wavefront. A nearby bright source must serve this function. We can then only observe bright objects or objects that happen to be near a bright star. The probability of finding a bright enough guide star within the isoplanatic patch varies with wavelength, so that in the infrared we can indeed use an off-axis guide star to make the complete corrections. The area of sky which can be covered depends strongly on the order of correction desired and the wavelength, and varies from near 100% at 10 μm (large isoplanatic angle, large time constant) to less than 1% at 0.5 μm (isoplanatic angle of only a few arcsec, short time constant). For low orders of wavefront distortion, such as would be compensated with partial adaptive correction, the correlation extends over much larger angles. Thus, for example, for tilt error, with an 8-meter telescope, the correlation coefficient should exceed 0.5 over angles of several hundred arcsec (this angle depends on the model of the atmosphere, seeing, etc).

To overcome the need for a nearby astronomical reference source Foy and Labeyrie proposed, in 1985, to generate an artificial star by focusing a single frequency laser in the Sodium layer of the upper atmosphere. The laser-generated artificial star could then be positioned near any object in the sky. The major technical problem are the laser and the correction of many elements in the aperture for work in the visible. An artificial reference source cannot provide a measurement of the lowest order errors, the piston (relevant only for interferometry) or the tilt, thus a celestial reference source is still required. However, with an artificial reference source to ensure a plane wavefront across the telescope aperture, a much fainter celestial reference may be used, and only low order information is required of it.

A Program for Development and Implementation of Adaptive Optics

A large and continuing DOD investment goes into adaptive optics. This effort represents an important potential resource for astronomy, although actual DOD systems may be too complex, expensive, or specialized for wide implementation in astronomy. Although several sources are available for the principal components of adaptive optics systems, turnkey or blackbox systems are not currently feasible or desirable. The astronomy community must develop the expertise to build and utilize these relatively complex systems.

Several groups in the U.S. are currently working with adaptive optics for astronomy. These include NOAO (solar and stellar programs), Johns Hopkins (adaptive optics with coronography), the University of Chicago and University of Illinois (laser reference techniques), University of Hawaii (wavefront sensing, adaptive mirrors), and Steward Observatory (neural net processing, adaptive systems). Such programs are important to development, adaptation, and implementation of technology and training of students. Support is required to continue and expand such activities.

We recommend support during the 1990's for the implementation of adaptive optics on a number of telescopes, particularly including large telescopes, where the potential gains are greatest. The early years of the program should emphasize the areas of infrared imaging and spectroscopy and solar astronomy. The scientific return in these areas is expected to be readily achieved, once adequate resources are made available. Later in the decade, more sophisticated systems will extend diffraction-limited visible performance to large telescopes. Combined with the multi-telescope interferometry programs described above, AO will yield a large gain in performance of distributed optical arrays.

Adaptive Optics is a revolutionary and cost effective emerging advance in ground-based astronomical instrumentation. At the moment it suffers from a high entrance fee in terms of intellectual investment and capital. We can target five areas which need support over the next decade.

1 *Rapid demonstration of partial correction.* Adaptive optics systems with 10-100 actuators may be demonstrated and used extensively for science during the next five years. These will be near state of the art and will be implemented initially for solar work at visible wavelengths and for 3-4 meter class telescopes in the infrared.

2 *Development of lower cost systems.* It should be possible to develop moderate cost adaptive systems oriented to astronomers requirements. Our objective should be to extend the most useful benefits of adaptive optics to many telescopes and interferometric arrays at a cost not much different from a major instrument. The key here is the development of new types of hardy, cost-effective adaptive mirrors and

wavefront sensors. Membrane mirrors, bimorph mirrors, and curvature sensing are techniques which appear promising. These systems will probably be built by astronomers at their home institutions.

3 *Laser reference systems.* Implementation of high order adaptive correction, or extension to faint sources of partial correction at visible wavelengths, will require laser reference systems. It will be a challenge to devise cost-effective implementations of these potentially expensive techniques for use in astronomy. Such work may use hardware developed for military systems, and may profit from collaboration with experienced DOD contractors.

4 *Algorithms.* The development and application of mathematical techniques is required both to control the system and to improve the images once they have been detected. Although much of the light can be contained in the outer halo of the point spread function, especially at shorter wavelengths with partial correction, simulations show that the halo is smooth and has low surface brightness after a long exposure. Deconvolution and contrast enhancement methods, such as the van Cittert or more powerful non-linear techniques similar to those used by radio astronomers, can dramatically improve the dynamic range of the processed images.

5 *The training of an adequate core of personnel.* Adaptive optics systems are complex. Small errors in understanding can and have invalidated some work. DOD systems often have a near permanent staff from the manufacturers to keep the sub-systems running. At observatories at least some staff on the mountain must be well informed and able to adjust, service, and improve the systems.

At present the only working adaptive optics system in night time astronomy (excepting simple tip/tilt correction) is in Europe. Funding and support of this technology is essential to maintaining the competitiveness of existing U.S. telescopes, even of the 8-m telescopes that are on the drawing board. The cost of a balanced program in the U.S. is estimated at $35 million over the next ten years, less than the cost of a single large facility.

Gravitational Waves

Observations of the inspiral rate for the binary pulsar PSR 1913+16 have given strong support to the predictions of general relativity for the strength of gravitational radiation from accelerating massive bodies. Valuable limits on the intensity of gravitational waves reaching the Earth have been set by a combination of terrestrial antennas of different kinds, including Doppler tracking of distant spacecraft and timing measurements on millisecond pulsars. We will describe here two proposed gravitational wave observatories using laser interferometers as broadband antennas, and the opportunities for obtaining entirely new types of astrophysical information with them.

One proposed observatory is LIGO (the Laser-Interferometer Gravitational-Wave Observatory), a high-frequency, earth-based project, which is now nearing the end of a decade-long planning, design, and research and development stage. This project would construct a vacuum facility to support many generations of gravitational-wave detectors, and would construct an initial detector system. Construction of LIGO has been approved by the National Science Board, and funding has been requested from the Congress. The other proposed observatory is LAGOS (the Laser Gravitational-Wave Observatory in Space), for which a preliminary feasibility study has been completed. It will be proposed to NASA for technology development during the 1990's and possible flight in the decade 2000-2010. These observatories would cover the frequency range from roughly 10^4 Hz to 1 Hz (LIGO) and the from 1 Hz to 10^{-5} Hz (LAGOS).

Gravitational waves are emitted by the coherent, bulk motions of large amounts of matter (e.g. collapsing stellar cores) and by coherent, nonlinear vibrations of space-time curvature (e.g. collisions of black holes). The strongest extragalactic waves bathing the earth are not likely to exceed h $\sim 10^{-20}$, where h is the strength of the perturbation in the metric. The strongest sources are likely to be black holes and neutron stars – e.g., the violent births of black holes and neutron stars in stellar implosions, and the inspiral and coalescence of binary neutron stars and black holes in distant galaxies. The characteristic frequencies of vibration and rotation for neutron stars are less than or of order a few kilohertz; and those for a black hole of mass M are

$$f \sim \frac{10kHz}{M/2M_\odot}$$

(where $2M_\odot$ is the smallest possible mass for a black hole that forms by stellar collapse). Thus, the strongest waves are likely to lie at frequencies of 10 kHz and below.

Because the strongest sources are compact concentrations of highly dynamical mass, they typically will lie in regions obscured by surrounding matter (e.g., in the core of a supernova explosion or at the center of a galaxy or in the big-bang origin of the universe). Fortunately, gravitational waves are highly penetrating. For example, primordial gravitational waves should have last scattered near the Planck time, $\sim 10^{-43}$ seconds, when the initial conditions of the universe were being set by the (little understood) laws of quantum gravity

High Frequency Gravitational Wave Sources (LIGO)

Coalescence of neutron-star and black-hole binaries. Of all sources in the LIGO's high-frequency band, the final inspiral of a binary neutron star is best understood: Because the binary orbit is nearly Keplerian, the details of the waves are known with confidence; and binary pulsar observations have provided enough information about birth rates of binary neutron stars to pin down the distance to which one must look in order to see several coalescences per year. That distance is 100 Mpc, give or take a factor of a few. Advanced detectors in the LIGO should be able to detect the inspiral waves out to a distance of about 1000 Mpc, where the event rate is presumably many per year (see Figure 6a).

From measurements of the waveforms emitted by the final inspiral and coalescence, one can extract the masses of both objects. The inspiral waveforms also reveal, directly, the distance to the source – without any assumptions. Thus, such inspirals serve as "standard candles".

Supernovae. Those supernovae (primarily type II) that are triggered by the collapse of a stellar core to form a neutron star should produce bursts of gravitational waves. Although the rate of such supernovae is well known from electromagnetic studies, the strengths of their waves depend crucially on the unknown degree of asymmetry of the collapse. The LIGO has hope of seeing highly asymmetric collapses ($\sim 10^{-3} M_\odot c^2$ radiated) out to the VIRGO cluster, but would have difficulty seeing nearly spherical collapses ($\sim 10^{-10} M_\odot c^2$ radiated) even in our own galaxy. Observational studies of the gravitational waves from supernovae could bring information not obtainable in any other way about the asymmetry in the collapse and about the physics of very young neutron stars.

Rotating Neutron Stars. Advanced detectors in the LIGO could detect the gravitational waves from a neutron star anywhere in our galaxy if its equatorial ellipticity exceeds $\epsilon_{min} \sim 10^{-9}$ ($P_{rot}/1$ msec$)^2$, where P_{rot} is the star's rotation period. If there are many such neutron stars, LIGO will be a rich source of information about neutron-star physics.

When a neutron star with magnetic field $B \simeq 10^7$ Gauss is being spun up by accretion from a companion, the spinup may be halted by an instability triggered by gravitational radiation reaction. Interesting constraints on the viscosity of neutron-star matter (which impedes the growth of the instability) and on the equation of state would follow from observing such objects.

Black – Hole Births. Black holes with masses of a few $M_\odot$ up to a few tens of $M_\odot$ are thought to be born in the collapses of massive ordinary stars; and more massive holes may be born in galactic nuclei, in collapses of supermassive stars, and/or in star clusters. The rate of such events is highly uncertain (it is not even firmly known that black holes exist); and the mass spectrum of the resulting black holes is highly uncertain. Advanced detectors in the LIGO should be able to detect highly nonspherical black-hole births throughout the universe, for hole masses between 50 $M_\odot$ and 1000 $M_\odot$.

Cosmological Studies. Gravitational-wave observations might bring us valuable cosmological information in several areas. One concerns fluctuations in the early universe. Gravitational waves emerging from the Planck era of the big bang – even waves so weak that they are nothing but vacuum fluctuations – can be amplified, during the earliest epochs of the universe's expansion, to make them detectable. The amplification is "parametric"; it results from coupling of the waves to the large-scale background curvature of the expanding space-time. Observations of such waves, or failure to observe them, will provide valuable information or constraints on theories of the very early universe, such as "inflation." The potential sensitivity of the LIGO is about 10^{-10} of the closure density at frequencies near 30 Hz. For comparison, pulsar timing is now nearing $\Omega_{GW} \sim 10^{-7}$ at $f \sim 10^{-8}$ Hz; anisotropy of the cosmic microwave background gives limits of $\Omega_{GW} \sim 10^{-12}$ in a very narrow window at $f \sim 10^{-16}$ Hz; and the time delay in the gravitational lens 0957+561 gives a limit $\Omega_{GW} \sim 10^{-8}$ at $f \sim 3 \times 10^{-18}$ Hz.

Low Frequency Gravitational Wave Sources (LAGOS)

Binary Stars. The total number of binaries in the galaxy with frequencies between 10^{-5} and 10^{-2} Hz is roughly 10^8. This includes detached binaries, contact binaries, cataclysmic variables, neutron star binaries, and close white dwarf binaries. The numbers expected are fairly well known, except for the close white dwarf binaries, which have not yet been observed. Specific predictions of the number density of white dwarf binaries are available but are highly uncertain, in part because such binaries must undergo two periods of common envelope evolution during their history, and could be disrupted. Evolutionary calculations suggest that substantial numbers of black hole-neutron star binaries may also be present.

The number density of binaries in our galaxy drops off rapidly at higher frequencies, and at some frequency near 10^{-3} Hz there will be less than one binary per cycle-per-year frequency bin. Somewhat above this frequency, most binaries will be clearly resolved, and their locations can be determined by analyzing the change in signal strength as they go through the sharp nulls in the antenna pattern. The highest frequency galactic binary signals are expected to be about 3×10^{-3} Hz for neutron star binaries, and about 2×10^{-2} Hz for close white dwarf binaries if they are present at 10% of the calculated number density. The strengths of the expected signals for these two types of binaries are shown in Figure 6b. Below 10^{-3} Hz, some binaries which happen to be unusually close will be observable above the unresolved galactic binaries.

Information concerning the unresolved galactic binaries can be obtained from the shape of the observed spectrum. However, these binaries also provide noise, which makes it more difficult to observe other kinds of signals such as possible extragalactic backgrounds and possible coalescence signals from active galactic nuclei. The unresolved galactic binary signal strength is shown in Figures 6a-c, both without close white dwarf binaries (solid curves) and with 10% of the calculated number density for them (dotted curves). If the close white dwarf binaries are present at nearly the calculated level, an isotropic stochastic background signal should be observable between 10^{-3} and 10^{-2} Hz due to the integrated signal from all the other galaxies. Each shell of a given thickness contributes roughly equally on the average, so the signal would come from sources nearly all the way out to the Hubble radius.

Coalescence signals from AGN's. Probably the most interesting burst source for LAGOS is inspiral signals from white dwarfs, neutron stars or $\sim 10 M_\odot$ black holes orbiting around central supermassive black holes (SMBH's) in AGN's or quasars. If perhaps 10% of the AGN or quasar core immediately after SMBH formation is in such compact objects, inspiral signals may be observable between 0.3 and 10 millihertz during the period while the SMBH is growing to roughly $10^7 M_\odot$. However, if close white dwarf binaries are present at nearly the calculated density, fluctuations in the integrated signal from many of them will limit the antenna sensitivity. In this case, the range of frequencies and distances over which signals could be observed would be limited, and the event rate would be reduced.

Cosmological backgrounds. The sensitivity of LAGOS in searching for cosmological backgrounds such as possible amplified relic gravitons from inflation would again depend on the strength of the noise background from close white dwarf binaries. For a density of such sources which is 10% of the calculated value, the sensitivity would be about 10^{-11} of the closure density. However, with few close white dwarf binaries, the sensitivity would be roughly 100 times better.

A Gravitational Wave Observatory Program

A laser interferometer gravitational-wave detector, in its simplest conceptual variant, consists of masses that hang by wires at the corner and ends of an L. A gravitational wave pushes the masses back and forth relative to each other, changing the difference in the length of the detector's two arms by an amount ΔL that is proportional to the arm length L and to the strength of the metric perturbation h. By laser interferometry one directly reads out $\Delta L/L$ and, from its time evolution, the waveform $h(t)$. The interferometry is typically done in one of two ways: each arm is operated as a delay line, with a light beam bouncing back and forth in it many times, or each arm is operated as a giant Fabry-Perot cavity with finesse as large as 10,000.

More sophisticated optical configurations, called "broad-band recycling," "dual recycling," and "resonant recycling" have the potential to improve the sensitivities markedly.

For most noise sources (e.g. seismic noise, gravity gradient noise, and thermal noise, which dominate at frequencies f $\leq$ 100 Hz), the displacement noise ΔL in the interferometer is independent of the arm length L; and, correspondingly, the gravitational-wave sensitivity h = $\Delta L/L$ improves with increasing arm length. Prototype interferometric detectors, with arm lengths from 1 meter to 40 meters, have been under development since 1970 and vigorously since about 1976. Feasibility studies, technology development, and planning for full-scale detectors have been carried out in the United States since 1981, with funding from NSF. In 1984 plans were initiated for the design, construction, and operation of a full-scale system called the LIGO; and in 1986 the Physics Survey Committee ("Brinkman Committee") strongly endorsed the LIGO.

The unequivocal ground-based detection of gravitational-wave bursts amidst instrumental and environmental noise requires cross correlation of two detectors at widely separated sites. For this reason, the LIGO will include two facilities, far apart in the continental United States, in which cross correlated detectors will operate. To determine the direction to a source and to separate its two polarizations (i.e. to extract the full details of the wave) will require cross correlating three, and preferably four detectors at widely separated sites. The American effort will have to rely on similar detectors in Europe, Japan, and/or Australia for the third and fourth detectors of a world-wide network. Numerical simulations of such a network predict, for gravitational-wave bursts, angular resolutions of a few tens of arc minutes for interesting sources.

Figures 6a-c show the expected sensitivities of (i) the first detectors planned for the LIGO, and (ii) more advanced detectors that might operate in the LIGO a few years after the first ones. The technology for the advanced detectors is expected to be well in hand within the next several years, but key aspects of their design cannot be tested adequately in the existing 40-meter prototype vacuum system, and must await the full scale LIGO.

The development and operation of a network of resonant acoustic bar detectors should continue to be supported. The detectors are approaching sensitivities which could detect galactic supernovae. The present prototype interferometric detectors, although beginning to rival the sensitivity of current bar detectors, will not be operated in observational mode during the development of the large baseline systems.

Laser Gravitational Wave Observatory in Space

Ground-based interferometric detectors cannot be used in the low- frequency band because displacement noise for their test masses is an insurmountable problem at low frequencies. To achieve high sensitivity in this band requires an interferometer in space with very large arm lengths. The proposed LAGOS antenna makes use of carefully shielded test masses freely floating inside three separate spacecraft, which are 10^7 km apart. The cluster of three spacecraft is located near the L-5 point of the Earth-Sun system. The lengths of the two interferometer arms will remain equal to 0.2% for proper choices of the initial orbit parameters.

Changes in the test mass separations are monitored by laser phase measurements. One watt of power from a cavity-stabilized Nd-YAG laser pumped by laser diodes is sent out from the central spacecraft to each end spacecraft through 0.3 m telescopes. Similar lasers in each end spacecraft are phase-locked to the received signal and transmit back to the central spacecraft. The phase of the returned signals is measured as a function of time with respect to the laser in the central spacecraft.

To the extent that the test masses can be protected from spurious accelerations at frequencies of 10^{-5} to 1 Hz, laser wavelength variations can be corrected for by measuring the sum of the apparent variations in length for the two interferometer arms. The corrected laser wavelength is then used to determine changes in the difference in arm lengths due to gravitational waves and instrumental noise. The effect of the residual laser wavelength variations is reduced because they are nearly common mode for the two arms. This correction procedure works at all frequencies in the range of interest, except for narrow bands around harmonics of the 0.017 Hz round-trip frequency for one arm.

The expected antenna performance is shown in Figures 6a-c. The best performance is achieved for frequencies between 10^{-3} and 10^{-2} Hz. The performance gets worse above 10^{-2} Hz because of the gravitational wavelength getting shorter than the arm length, and worse below 10^{-3} Hz because of spurious accelerations of the test masses and thermal fluctuations in the telescope. The performance degrades even more rapidly below about 2×10^{-5} Hz because the thermal isolation becomes inadequate.

A single antenna in space should be sufficient for measuring most kinds of expected low-frequency

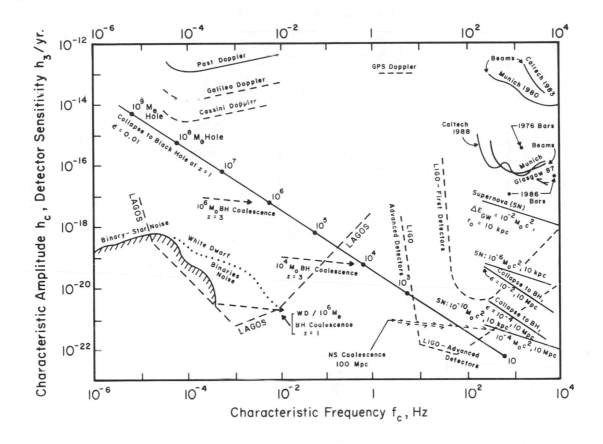

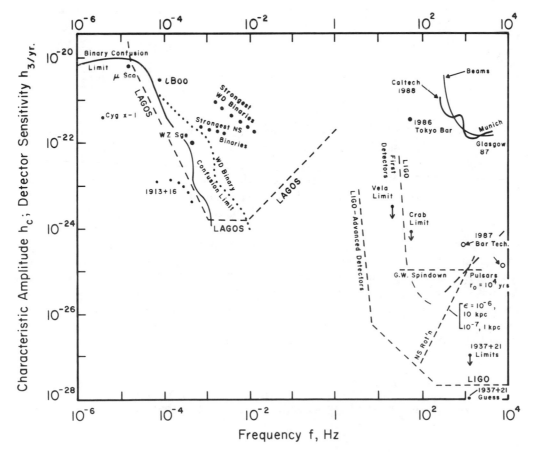

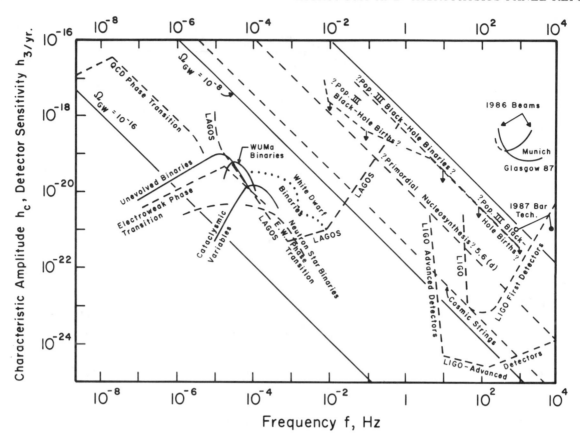

Figure 6a (opposite, top) Gravitational-wave bursts: estimated source strengths and detector sensitivities. Plotted vertically are the characteristic amplitudes h_c of gravitational waves from several postulated burst sources and the sensitivities of several existing and planned detectors. Abbreviations are BH, NS, and SN are used for black hole, neutron star, and supernova. The sensitivities are for 90% confidence of detecting very rare bursts.

Figure 6b (opposite, bottom) Periodic gravitational waves, and detector sensitivities for 90% confidence of detection in 10^7 seconds of integration time.

Figure 6c (above) Stochastic gravitational waves and detector sensitivities. (These figures adapted from Thorne, 1987, where additional discussion will be found.)

gravitational wave signals. Periodic signals from roughly 10^3 binary stars in our galaxy, near 10^{-3} Hz and possibly up to 10^{-2} Hz, should be easily observed. Below about 10^{-3} Hz, there will be a random superposition of signals from roughly 10^8 binary stars in our galaxy, which cannot be resolved in frequency even with a few years of data. However, the passage of the galactic center through the nulls in the antenna pattern four times a year will verify that the detected spectral amplitude curve is not due to unexpected instrumental effects. It is only for isotropic extragalactic background that the limitation of having a single antenna is likely to be substantial.

Until the LAGOS antenna is developed, the principal technique for searching for low frequency gravitational waves will continue to be Doppler tracking of distant spacecraft using microwave transponders. The sensitivity will be increased substantially by tracking in X-band on both uplink and downlink, a step to be taken for the first time with the Galileo mission; and can be increased still further by tracking at Ka-band, as is being considered for the Cassini mission. Since searching for gravitational waves is a small additional cost for missions to the outer planets, we recommend continuing to carry out searches with future missions, especially those that will have Doppler capabilities at X-band and/or at higher frequencies.

Technology Development During the 1990's

Conceptual design studies for LAGOS were carried out during 1985-1988 under the NASA Innovative

Research Program. Preliminary technology development studies were started in 1989. The present mass estimates are 200 kg on-orbit mass each for the two end spacecraft and 400 kg for the central one.

There are several areas in which major technology development is required. The first is the development of a very low thrust station-keeping and altitude-control propulsion system. The thrust levels required are far below those for any previous mission, but are needed continuously to counteract solar radiation pressure on the spacecraft, so high fuel efficiency is desired in order to permit a 10 yr mission lifetime. The primary requirements on the system are that it keep the spacecraft centered on the test mass, and that it maintain the spacecraft orientation accurately with respect to the laser beam transmitted by one of the other spacecraft.

A Disturbance Reduction System is needed to protect the test mass inside each spacecraft from spurious accelerations which could mask the effects of gravitational waves. The requirement is to keep the fluctuations of the spurious test mass accelerations about 6 orders of magnitude below the acceleration an uncorrected spacecraft would have due to variations in the solar radiation intensity and in the solar wind.

LAGOS requires construction and space-qualification of 1 Watt stable lasers with long lifetime and high efficiency. The development of diode-pumped Nd-YAG lasers has been proceeding rapidly, and a five year lifetime for each laser with about 15% efficiency and high reliability appears feasible. The lasers would be locked to rugged Fabry-Perot interferometers with low expansion spacers and optically contacted mirrors to provide good short-term stability.

Two other areas are the development of a high-stability laser beam steering system and the investigation of improved thermal insulation methods.

A gravitational wave antenna to go on the Moon soon after the year 2000 also has been proposed. By working with LIGO and with other terrestrial antennas, it would improve the angular sensitivity for determining the location of burst sources. Such a lunar antenna also would require technology development during the 1990s. The frequency range would be from perhaps 0.3 Hz to a few kHz.

Prospects for International Collaboration

International collaboration is likely to play a role in the larger projects recommended here. We note that there is considerable interest in the European space community in astrometry from space (viz. Hipparcos) and an imaging interferometer features prominently in the ESA Horizon 2000 program.

The search for gravitational radiation and especially the full development of gravitational wave astronomy would benefit from a worldwide network of interferometers to gain source position and wave polarization information. On-going efforts in Europe and proposed programs in Australia and Japan are to be encouraged.

Related Issues

(1) Adaptive optics and optical interferometry require imaging detectors of the highest quantum efficiency and lowest noise. This requirement is strongly driven by the need to detect position and phase reference information within an atmospheric time constant, usually much less than a second.

(2) The large telescopes planned for this decade could constitute a unique resource for interferometric experiments or instruments of the future, provided some consideration is given to their siting and configuration in the planning and construction phase.

(3) The transfer of technology from DOD to the astronomy community is important to the timely and cost-effective development of adaptive optics for astronomy.

(4) The Explorer program has a nominal cap for total mission cost to launch of $100M. There is a large gap between this level and the level of a typical "new start" taken to congress. This inhibits scientifically important missions which naturally fall in this cost range, as an Astrometric Mission probably does.

References

Hege, E.K., Beckers, J.M., Strittmatter, P.A., and McCarthy, D.W. 1985, *Appl. Opt.* **24**, 2565.
Koechlin, L. 1984, in *Kilometric Optical Arrays in Space*, ESA SP-226, ed. N. Longdon and O. Melita (ESTEC, Noordwijk), 99.

Nakajima, T., Kulkarni, S.R., Gorham, P.W., Ghez, A.M., Neugebauer, G., Oke, J.B., Prince, T.A., and Readhead, A.C.S. 1989, *A.J.* **97**, 1510.

Latham, D.W., Mazeh, T., Stefanik, R.P., Mayor, M., and Burki, G. 1989, *Nature* **339**, 37.

Long Range Planning Subcommittees, 1990, *Reports*, National Science Foundation.

Physics Survey Committee, 1986, *Gravitation, Cosmology, and Cosmic-Ray Physics*, National Academy Press.

Readhead, A.C.S., Nakajima, T.S., Pearson, T.J., Neugebauer, G., Oke, J.B., and Sargent, W.L.W. 1988, *A.J.* **95**, 1278.

Rousset, G., Fontanella, J.C., Kern, P., Gigan, P., Rigaut, F., Léna, P., Boyer, G., Jagourel, P., Gaffard, J., and Merkle, F. 1990, *Astron. Astrophys* **230**, L29.

Space Astronomy MOWG, 1981, *Space Astronomy Program Plan for the 1980's and 1990's*, NASA.

Space Science Board, 1981, *Strategy for Space Research in Gravitational Physics in the 1980's*, National Academy Press.

Space Science Board, 1988, *Space Science in the Twenty-first Century: Imperatives for the Decades 1995 to 2015*, National Research Council.

Thorne, K.S. 1987, in *300 Years of Gravitation*, (Cambridge U. Press, Cambridge).

Appendix - PSSWG Statement for the Interferometry Panel

The Planetary System Science Working Group (PSSWG) was formed by G. Briggs (NASA SL) in early 1988 to investigate means of detecting and characterizing extra-solar planetary systems. The PSSWG, under the chairmanship of B. Burke (MIT), expanded its role to include studies of the nature of planetary systems and the relation between their formation and formation of stars. This scope includes the physics of pre-planetary nebulae and circumstellar envelopes, and is closely related to the domain of the newly formed "origins" program.

In addressing scientific goals, PSSWG has concluded that the determination of the prevalence and natures of other planetary systems, in a broadly based study, is essential to an understanding of how such systems form and evolve. Further, such a study would eventually show whether our own Earth is a rare exemplar or one of myriad planets in our Galaxy that could support life. The study will require more than one generation of interferometric instrumentation, with successive generations permitting deeper searches. For studies of nearby stars, the stages might include (in order of challenge): Jupiter-sized planets three to ten AU from their stars; Uranus-sized planets in the water-ice condensation zone; and Earth-sized planets at one to two AU from their stars. At each stage, it is essential that the search "drain the lake," not just provide a "fishing expedition."

PSSWG has considered both short-term (next ten years) and long-term (following twenty years) strategies. In the short term, the prime objective would be the discovery of extra-solar planetary systems. An essential characteristic of any system chosen for this task is that a negative result be scientifically valid. As this new phase of planetary science develops, the work would aim at the preliminary characterization of the systems discovered and the statistical study of the kinds of stars that support planetary systems. The PSSWG has found that, considering the instruments that could be built in the short term, the principal approach for reliable detection of extra-solar planetary systems is by means of space-based astrometry. In particular, the goal of detecting a Uranus-sized planet having a radial separation of five AU appears to be beyond the range of ground-based techniques. However, with space-based instrumentation, it would be possible to achieve orders of magnitude improvement over the present ground-based astrometric accuracy of a few mas. Thus, this critical test is within reach and should be performed. The groundwork should also be prepared for developing more advanced interferometric systems that can carry out the in-depth planetary systems studies that will develop in the following decades.

FIGURE CREDITS

Figure 1 (Left) reprinted from Readhead, A.C.S., Nakajima, T.S., Pearson, T.J., Neugebauer, G., Oke, J.B., and Sargent, W.L.W., 1988, *Astron. J.* **95**, 1278, by permission of the authors and the *Astronomical Journal*, published for the American Astronomical Society by the American Institute of Physics; (Middle) reprinted from Hege, E.K., Beckers, J.M., Strittmatter, P.A., and McCarthy, D.W., 1985, *Appl. Opt.* **24**, 2565, by permission of the authors and the Optical Society of America; (Right) reprinted from Koechlin, L., 1984, in *Kilometric Optical Arrays in Space*, ESA SP-226, ed. N. Longdon and O. Melita (ESTEC, Noordwijk), 99, by permission of the author and courtesy of the European Space Agency.

Figure 2a reprinted from Nakajima, T., Kulkarni, S.R., Gorham, P.W., Ghez, A.M., Neugebauer, G., Oke, J.B., Prince, T.A., and Readhead, A.C.S., 1989, *Astron. J.* **97**, 1510, by permission of the authors and the *Astronomical Journal*, published for the American Astronomical Society by the American Institute of Physics.

Figure 2b reprinted from Pan, X.P., Shao, M., Colavita, M.M., Mozurkewich, D., Simon, R.S., and Johnston, K.J., 1990, *Astrophys. J.* **356**, 641, by permission of the authors and the *Astrophysical Journal*.

Figure 4 reprinted from Rousset, G., Fontanella, J.C., Kern, P., Gigan, P., Rigaut, F., Léna, P., Boyer, G., Jagourel, P., Gaffard, J., and Merkle, F., 1990, *Astron. Astrophys.* **230**, L29, by permission of the authors and Springer-Verlag.

HIGH ENERGY FROM SPACE PANEL

BRUCE MARGON, University of Washington, *Chair*
CLAUDE CANIZARES, Massachusetts Institute of Technology, *Vice-Chair*

RICHARD C. CATURA, Lockheed Palo Alto Research Laboratory
GEORGE W. CLARK, Massachusetts Institute of Technology
CARL E. FICHTEL, NASA Goddard Space Flight Center
HERBERT FRIEDMAN, Naval Research Laboratory
RICCARDO GIACCONI, Space Telescope Science Institute
JONATHAN E. GRINDLAY, Harvard-Smithsonian Center for Astrophysics
DAVID J. HELFAND, Columbia University
STEPHEN S. HOLT, NASA Goddard Space Flight Center
HUGH S. HUDSON, University of California, La Jolla
STEVEN M. KAHN, University of California, Berkeley
FREDERICK K. LAMB, University of Illinois at Urbana-Champaign
MARVIN LEVENTHAL, AT&T Bell Laboratories
ROBERT NOVICK, Columbia University
THOMAS A. PRINCE, California Institute of Technology
REUVEN RAMATY, NASA Goddard Space Flight Center
HARVEY D. TANANBAUM, Harvard-Smithsonian Center for Astrophysics
MARTIN C. WEISSKOPF, NASA Marshall Space Flight Center
STANFORD E. WOOSLEY, University of California, Santa Cruz

High Energy from Space

Introduction and Executive Summary

Barely 25 years have elapsed since the first space-borne observations of high energy phenomena in the Universe. In that time interval, an astonishing amount of progress has been made. At the beginning of the era, we had only brief, indirect glimpses of sites of violent astrophysical activity, *e.g.*, cosmic rays, or optical observations of supernovae and radio galaxies, and virtually no understanding of the underlying physical processes. Now high energy astrophysics is a fully established discipline: both our catalogs of observations and the maturity of our theoretical interpretations of these data are fully comparable to those in most other areas of astrophysics.

Despite these encouraging beginnings, the discipline of high energy astrophysics has weathered a 10 year hiatus in flight opportunities, a hiatus that has drastically slowed the rapid progress of scientific discoveries of the 1970s, and that has had a chilling effect on the entry of young researchers into the field. During this time, the main centers of activity in high energy astrophysics have shifted to Europe and Japan. Nevertheless, the discipline has remained intact in the US, and workers are eager and ready to take full advantage of the remarkable opportunities of the 1990s to regain this lost ground.

It is a safe prediction that the decade of the 1990s will see an amount of forward progress equal to that of the entire past 25-year history of our discipline. Although some of this stimulus will come from areas common to many fields of astronomy, *e.g.*, the explosive increase in available computing power, a dominant source of progress will be NASA's Great Observatories. The *Gamma Ray Observatory* (GRO) and the *Advanced X-ray Astrophysics Facility* (AXAF) will each represent jumps of orders of magnitude in observational sensitivity over past experiments, and inject not only large amounts of new data on high energy phenomena, but also tremendous intellectual challenges in interpretation of these data. Correlative observations from the *Hubble Space Telescope* (HST) and eventually the *Space Infrared Telescope Facility* (SIRTF) can only make these problems more interesting and exciting.

It is impossible to forget, however, that both AXAF and GRO were recommended in the Field report: in effect, they are the long delayed program of the 1980s. There is no doubt that they will revolutionize the field, but they alone do not constitute a complete program. First, there are scientific areas of major importance that are not addressed by these missions. Second, like all disciplines, high energy astrophysics needs a mix of small, medium and large programs to provide vitality and scientific flexibility. Third, we must begin now to plan for the the first decade of the next millenium – if nothing else, surely the past ten years have taught us that.

In addition to AXAF and GRO, there are several less ambitious missions now approved or under study. The only approved moderate mission is the *X-ray Timing Explorer* (XTE). We have identified two further moderate missions that address science of the highest priority, both of which should be started and launched in this decade. Also approved are a variety of much smaller opportunities on foreign missions. These are of great importance to US investigators, and provide a particularly cost-effective way of doing science. Such collaborations must be encouraged and expanded during the 1990s. Several other American payloads now under study include exciting prototypes of a new class of small, inexpensive mission. We have identified a surprisingly large number of important problems in high energy astrophysics which can be addressed by such small missions in the coming decade.

A complete program also requires development of the technology and infrastructure to pave the way for the new discoveries of the 21st century. Therefore, we discuss a program of technology and instrumentation development to prepare for the next generation of X-ray and γ-ray astronomy missions. Lastly, we discuss certain programmatic issues and stress the need for continued support of the research base of the discipline through funding of theory, research, and analysis programs.

In summary, our vision of a vigorous program for high energy astrophysics in the next decade includes

- Launch and extended flight operations of AXAF and GRO, and wide community involvement in their observing programs
- At least two new Explorer-class missions in addition to XTE, to address particularly exciting opportunities in X- and γ-ray astronomy
- Exploitation of smaller, less expensive space missions for important specialized problems
- An ambitious program of technology development in optics, detectors, and related hardware
- Changes in selected programmatic approaches that affect the research base in our field

This program can in the coming decade challenge scientific issues as diverse as stellar chromospheres, relativistic stars, the intergalactic medium, dark matter, the energetics of active galactic nuclei, and the large scale structure of the Universe. Observations of high energy phenomena from space will be fundamental to the astrophysics of the 1990s.

Important Scientific Problems for High Energy Astrophysics

Stellar Activity

One of the most important discoveries by the *Einstein Observatory* was that normal stars of nearly all spectral types are unexpectedly strong X-ray emitters. The mechanisms that produce these X-rays are poorly understood and indicate fundamental problems in present theories of stellar evolution, the interiors and atmospheres of stars, and the mechanisms of coronal heating. The observations required to solve these problems are unique to high energy astrophysics. Future X-ray missions will provide valuable classification data, and spectroscopy will bring the power of plasma diagnostics that has been so fruitful in solar studies. Further observations that combine high sensitivity with spectroscopy and broad field imaging are necessary to provide sufficient X-ray data to correlate observables with stellar properties such as spectral type, rotation rates, optical luminosity, age and multiplicity, and to monitor variability and study activity cycles.

The Interstellar Medium In Our Own and Other Galaxies

As the repository of material from which stars form and to which they return nucleosynthetically-enriched material at the end of their lives, the interstellar medium (ISM) plays a dominant role in governing the evolution of galaxies. Over the past 15 years we have come to recognize the ISM as a violent, dynamic environment whose structure is determined by the mechanical energy ejected from stellar winds and supernovae and whose energy density is divided roughly equally among several components with a large scale magnetic field, high energy cosmic rays and turbulent motion of massive clouds each containing as much as energy as the light generated by the Galaxy's hundred billion stars.

We have discovered several distinct phases of interstellar matter with temperatures varying between 10 K in the cores of molecular clouds to $>10^6$ K in a coronal component which is dominant in the solar neighborhood, but we do not know the distribution of matter among these various phases and have no self-consistent global model for the medium's evolution. Likewise, we have only the most rudimentary understanding of the variations of interstellar environments in different galaxy types and of the interaction of interstellar gas with the intergalactic environment.

With our recognition over the past decade of the central role of hot coronal gas and our ability to map this component in the nearest external galaxies, along with the discovery of diffuse positron annihilation and ^{26}Al lines, it has become clear that high energy astrophysics has a central role to play in unravelling the structure and charting the evolution of the ISM. The coming decade will see dramatic progress. New soft X-ray spectroscopic experiments will determine the temperature structure and chemical composition of the local hot gas, while sensitive X-ray imaging will allow us for the first time to study this component in a range of galaxy types and orientations, to elucidate the role it plays in governing galaxy evolution. For the first time we will gather a complete census of the hot cavities created in the ISM by recent supernovae (both in our Galaxy and in other Local Group members). GRO and other γ-ray missions will open new windows on the ISM, mapping the cosmic ray distribution through their interaction with molecular clouds, and providing a global picture of recent element creation by charting the distribution of radioactive ^{26}Al.

A γ-ray spectroscopy mission could provide a completely new dynamical tracer of the Galaxy through observation of the positron annihilation line. Over 10^{43} positrons, resulting mostly from the decay of radionuclei produced in processes of nucleosynthesis, annihilate in the interstellar medium per second. The shape of the 511 keV annihilation line, observable with high resolution γ-ray spectrometers, can differentiate between annihilation in cold cloud cores and warmer interstellar gas. In the cold cores the line is broader because positrons form positronium in flight, which annihilates while still moving with relatively high velocity. Since positrons could be prevented by magnetic fields from penetrating into cold cloud cores, unique information on magnetic fields can be obtained by mapping the galaxy in annihilation radiation.

Supernovae and Endpoints of Stellar Evolution

The study of compact objects and supernovae is fundamental to astrophysics, being essential to the understanding of the life-cycle of massive stellar systems. Compact objects are the degenerate end-points of stellar evolution, and include white dwarfs, neutron stars, and stellar-mass black holes. Supernovae are the most visible and violent manifestation of the death of stars, and include both the gravitational collapse of massive stars and the nuclear detonation of white dwarf stars.

The bolometric luminosity of both Type I and Type II events is powered largely by the input of energy from positrons and γ-rays resulting from the decay of radioactive elements. Thus, X-ray and γ-ray observations can provide unique information on some of the most basic questions related to the understanding of supernovae: How much and what type of explosive nucleosynthesis occurs in Type I and Type II events? What is the mechanism for the explosion? What are the key characteristics of the explosion, for instance, total energetics and mass ejection? What was the nature of the progenitor system? That is, what was the mass, composition, and structure of the progenitor, and was it in a binary system?

SN 1987A has surely left behind a relativistic remnant. X- and γ-ray observations over the next decade will play a fundamental role in identifying and studying this unique object, the youngest collapsed star known.

Because production of high-energy particles (and therefore X-rays and γ-rays) is common in the vicinity of compact objects, either via accretion processes, or acceleration in strong magnetic fields, the study of X-ray and γ-ray emission addresses numerous fundamental questions, including: What is the inventory of compact objects in the Galaxy? That is, what is the number density and age distribution of white dwarf, neutron star, and black hole systems? What are the masses of compact objects? What magnetic fields are characteristic for compact objects and what is the time-evolution of these fields? How do systems involving compact objects evolve, in particular, interacting binary systems and globular cluster systems? What is the basic physics of accretion disks?

Nucleosynthesis

Second only in intrinsic interest to the origin and structure of the Universe itself, the origin and evolution of the chemical elements has long been a central theme in astronomy. It is now generally agreed that the elements other than hydrogen and helium, which come from the Big Bang, are byproducts of stellar evolution. The 1990s will witness the continued development of the theory and measurement of nucleosynthesis as a *quantitative* science. The key observational tool is emission-line spectroscopy. Gamma-ray lines from supernovae give direct evidence of newly synthesized elements, from which the yields of supernovae of various masses and types can be deduced. Examples of the problems amenable to study with the technology of the 1990s are γ-ray lines from ^{56}Co (the progenitor of iron) made in Type Ia supernovae; ^{26}Al, made, at least partly, in type II supernovae; and possibly ^{22}Na made in novae and ^{44}Ti in young supernova remnants, including perhaps SN 1987A. In addition to information on abundance, the profiles of γ-ray lines can reveal the mechanism, velocity distribution, mixing, and asymmetry of supernova explosion. High resolution X-ray spectroscopy of young supernova remnants can yield the abundances of the ejecta of all types of supernovae, providing information on the enrichment of all the material returned to the interstellar medium. Spatially resolved X-ray spectroscopy will show the degree of spatial segregation of the ejecta, for comparison with models of stellar evolution and of supernova mixing processes.

Relativistic Plasmas and Matter Under Extreme Conditions

A relativistic plasma is matter dominated by $e^+ - e^-$ pairs. Such a plasma is thought to be present around compact objects when the emission, consisting of radiation above the pair production threshold (511 keV), is large and the size of the emitting region is small. In this case the $\gamma\gamma$ opacity is large enough to lead to copious pair production. The hallmark of pair production is pair annihilation leading to 511 keV line emission. There is evidence for such emission from a compact object at or near the Galactic Center and possibly also from Cygnus X-1. Relativistic plasmas are also expected near the massive black holes that are thought to power AGNs. Our own Galactic Center may prove to be the Rosetta Stone of the physics of AGNs, with high resolution γ-ray observations playing a decisive role.

Degenerate dwarfs and neutron stars provide relatively nearby cosmic sites where the properties of matter under extreme conditions can be studied. For example, neutron stars commonly have surface magnetic field strengths $\sim 10^{12}$ G, 10^6 times stronger than the strongest magnetic fields that have been produced on Earth. They have central densities $\sim 10^{14} - 10^{15}$ gm cm^{-3}, equal to or exceeding the densities of atomic nuclei, and produce radiation fluxes $\sim 10^{28}$ erg cm^{-2}, $\sim 10^{17}$ times the radiation flux at the surface of the Sun. The proton fluid in neutron stars is thought to be the highest temperature superconductor in the Universe, with a critical temperature $\sim 10^9$ K.

Despite the impressive progress in understanding these objects that has been made during the past two decades, many fundamental questions remain unanswered. How are neutron stars formed? What is at the center of a neutron star? How hot are they, and how rapidly do they spin when they are formed? What are the most important processes that cool them? What are their life histories? What are the electrical, thermal, and magnetic properties of the incredibly dense neutron and proton fluids in their interiors? Where do neutron star magnetic fields come from? How do they survive?

Study of the relativistic pair plasmas that are thought to surround some black holes will require space-based instruments that can measure the radiation they emit over the energy range from about 10 keV to several MeV. One of the most intriguing aspects of neutron stars and black holes is that their properties can change within a microsecond, or perhaps even less. Thus, X- and γ-ray photometric and spectroscopic instruments with high time resolution will be required to advance our understanding of these objects. Space-based instruments that can observe targets uninterrupted for hours or days, and revisit targets frequently, will also be important.

Nature of γ-Bursts

The 1990s may well be the decade when one of the longest-standing mysteries in high energy astrophysics is solved — the nature of the mysterious cosmic γ-ray bursts. Flashes of γ-radiation that appear suddenly in unpredictable locations, γ-ray bursts may last from less than a tenth of a second to several minutes and then disappear, usually forever. Though over 400 bursts have been recorded, none has been unambiguously identified with any persistent source at any wavelength. Most theorists believe that these events are associated with magnetic neutron stars in our Galaxy (which would make them the only detectable signal from a very large number of neutron stars), but debate continues as to the roles of accretion, thermonuclear explosion, starquake, and other more exotic sources of energy. If in our galaxy, the luminosity of a typical γ-ray burst, coming from a region only 10 km in radius, is 10^5 times that of the Sun. Some theorists believe, however, that γ-ray bursts are at cosmic distances, more than 200 Mpc to display the observed isotropy, in which case the energy release is more like that of a supernova. The 1990s are a time when major progress is expected toward solving this problem, because GRO will determine the source distribution (to 1° for each burst observed) with sufficient accuracy to prove or disprove an association with our own Galaxy. The *High Energy Transient Experiment*, which will view γ-ray bursts in several wavebands (γ-ray, X-ray, and perhaps, ultraviolet), will pin down the burst sites even more accurately, hopefully restrictive enough that larger instruments can be brought to bear to isolate the quiescent counterpart. Interesting constraints on the source and physics of γ-ray bursts will also be given by studies of their spectra, time history, and frequency.

Identification of Black Holes

Stellar mass black holes are inferred to exist in a few X-ray binary systems in our Galaxy. However, no totally convincing proof of the existence of black holes, or method of unambiguously differentiating them from neutron stars, is available. Progress thus far has been made primarily through a combination of soft X-ray observations, and optical observations of the source counterparts. The hard X-ray signatures of accreting black hole binary systems may provide important future constraints on the mass and radius of the compact star. Hard X-ray continuum spectra can measure apparent temperatures and thus constrain M/R. Positron and iron line features, expected to be broadened if from near the hole, can measure velocities, and thus central mass. Timing measurements can constrain minimum hole size and thus mass.

A key to unlocking this fundamental problem is a detection capability over a very broad hard X-ray band, with very high sensitivity, at least modest spectral resolution, and possibly polarimetric capability. With high angular resolution as well, studies of black holes can be extended from our Galaxy to nearby galaxies, and the unknown total number of black holes in galaxies then investigated for the first time.

Active Nuclei, Including Our Own

The most energetic objects known in the Universe are the nuclei of active galaxies (AGNs). Their luminosities are observed to exceed by a large fraction the integrated stellar luminosity of all 10^{11} stars in a normal galaxy, yet observations strongly suggest that this energy is generated in a volume of radius $\lesssim 10^2$ AU. Chiefly by a process of elimination, accretion onto massive black holes is most often invoked as the energy source, with masses of $10^6 - 10^9$ $M_\odot$ normally assumed. Precisely how this process might work is unknown, and remains a central mystery in astrophysics, to be addressed in the next decade.

High energy observations in X- and γ-rays have a unique role to play in the understanding of the physics of AGN, because much of the energy radiated from the sources is at these wavelengths. Also, the great penetrating power of these photons allows us to study processes occurring close to the central object. The expectation of much of the community is that some unique fingerprint of massive black hole activity may emerge from high energy observations, allowing us to firmly establish their presence and ultimately understand the details of the accretion process.

The overall photon spectrum from a significant sample of these objects has yet to be measured, particularly at higher energies. High quality spectra may reveal the presence of $e^\pm$ annihilation features, or other gravitationally redshifted emission lines which will provide diagnostics of the potential well of the collapsed object; accurate continuum spectra may constrain the detailed physical processes at work. Many workers believe that more than one mechanism is involved, and that characteristic spectral breaks or kinks will become evident. Indeed, it is likely that different parts of the central region of AGNs are being studied at different wavelengths.

A study of the time variability and total luminosity of a large sample of AGNs is also crucial in constraining the size scale and energetics of the central object.

In recent years, a growing number of workers have come to believe that some small, quiescent version of an active nucleus might commonly reside at the center of ordinary galaxies, including our own, largely unnoticed because of its low luminosity. Tantalizing but not yet conclusive evidence that the center of our own Galaxy might harbor a black hole has been produced by infrared and radio observations, and the issue remains controversial. It is interesting to note that for 20 years, γ-ray astronomers have reported an intense and variable source of antiparticles somewhere near our own galactic center. A concerted effort to precisely position this source is now underway, and it may well be that our own galactic center will play a key role in our eventual understanding of AGN.

Accretion Physics

Within the past ten years, accretion onto compact objects has come to be recognized as one of the most important energy generation mechanisms in the Universe. The process is ubiquitous and operates on many different astronomical scales. Within our own Galaxy, accretion fuels the most powerful high energy stellar sources, the X-ray binary systems. As noted above, on a much larger scale, accretion onto massive black holes is believed to provide the energy source for nearly all varieties of active galactic nuclei, the most intrinsically luminous cosmic sources currently known.

In view of the importance of this process, understanding the physics of accretion has become one of the greater challenges for modern astrophysics. Despite over two decades of intense effort, however, a number of fundamental uncertainties still remain: (a) How are accretion disks formed? (b) What process is responsible for the dissipation of angular momentum in disks? (c) Is accretion unstable? If so, how and in what parameter regimes? (d) What accounts for the extreme phenomenological diversity of known accreting sources? (e) What happens at the disk boundary layer with the compact object? (f) How does the accreting flow interact with the intense magnetic fields of neutron stars and white dwarfs? (g) What happens when the mass accretion rate exceeds the Eddington limit? (h) Can disks form and collimate jets around compact objects? (i) What is the role of internal fields in accreting flows?

High energy observations will play a crucial role in answering many of these questions. Since most of the energy release for accreting sources is in the X- and γ-ray regions of the spectrum, observations at these wavelengths provide the most direct and unambiguous signature of the accretion process. For example:

- X-ray timing observations on all timescales ranging from μsecs to years provide detailed information on instabilities in the accreting flow. For the galactic sources, improved phase coverage for binary systems will determine the geometric structure of accretion disks.

- Moderate-to-high resolution X-ray spectroscopic observations provide extremely detailed information on the interaction of the intense radiation with the circumsource environment. This in turn yields direct constraints of the physical structure of the accreting matter, including the density, temperature, and ionization dependences on position with respect to the central source. Elemental abundances can also be determined in this way.

- Improved hard X-ray spectroscopy yields detections of cyclotron absorption and emission features originating in the magnetospheres of accreting neutron stars. The rotational phase dependence of these features and of the continuum leads to determinations of how the accreting material penetrates the magnetosphere.

- γ-ray spectroscopy should provide the first detections of electron-positron annihilation features for most accreting sources. Many recent models suggests that the formation of $e^+ e^-$ pairs may be a crucial "regulating" process in the most energetic regions of the accreting flow. New γ-ray observations can test and constrain such models.

- X-ray polarimetric measurements would provide direct and unambiguous constraints on the geometry and structure of neutron star magnetospheres. Polarization measurements can also constrain the orientations of accretion disks with respect to our line-of-sight. These can be compared with the orientations of "jets" associated with the source to provide constraints on jet-formation models.

Large-Scale Structure

Study of clusters of galaxies at large redshifts is potentially one of the most interesting contributions of X-ray astronomy to cosmological studies. The sensitivity of X-ray surveys can be such as to permit identification and study of clusters at much larger distance than is possible in visible light. The X-ray luminosity of a cluster is a function of its mass, the amplitude of the density fluctuations at the epoch of formation, and its chemical and dynamical evolution. Study of large samples at different redshifts is required to disentangle these effects.

Study of the X-ray spectrum directly yields the redshift, due to the prominent iron emission lines, and the temperature, from the continuum shape. The virial mass can be derived from temperature and surface brightness distribution. Correlation of cluster distributions over contiguous large area of the sky ($10° \times 10°$) can reveal the presence of very large structures at remote epochs, and elucidate the process of their formation and evolution. Deep surveys with wide field optics will be required to supplement the AXAF capabilities for this type of scientific investigation.

Intracluster Medium

The discovery and study of an intergalactic medium in galaxy clusters was a major triumph of X-ray astronomy in the 1970s and early 1980s. This diffuse, hot gas (typically 10^{-3} atoms cm^{-3} at 10^8 K) contains as much or more matter than all the stars in all the galaxies in the cluster, and yet it reveals direct evidence of its presence and its properties only in the X-ray band. Past investigations show that the morphology of the gas in several dozen clusters takes various forms and that the chemical enrichment of a handful of systems is roughly half the the solar value. In some galaxy clusters or groups the gas is seen to be cooling and presumably condensing into stars at rates as high as 1000 $M_\odot$ yr^{-1}. In the 1990s, the study of the spatially extended and emission line rich thermal X-rays of the intracluster medium (ICM) through both imaging and spectroscopy, will make major contributions to our understanding of the structure and evolution of galaxies and clusters. One example is the use of X-ray studies to map the gravitational potential in systems ranging from well relaxed older clusters to multiple and merging subclusters (see also *Dark Matter*, below). Another is the mapping of ICM distribution and composition with redshift to trace the evolution of cluster potentials over cosmological timescales and to record the history of global nucleosynthesis and the ejection of stellar material from the member galaxies. Studies of cooling flows show a process of galaxy formation similar to that which must have occurred in the early Universe. And information on the properties of the ICM can be combined with radio observations of its effect on cosmic microwave radiation to give an independent measure of the Hubble constant (via the Sunyaev-Zeldovich effect).

Nature of Dark Matter

The existence of dark matter in many and probably all galaxies and galaxy clusters is now well established. It is highly popular but more speculative to conclude that this unseen matter constitutes 90–99% of the stuff of the Universe. However, the distribution of dark matter is poorly known and its nature is a total mystery; elucidation of these properties is a central problem of contemporary astronomy and astrophysics.

Pioneering studies with the *Einstein Observatory* demonstrate that X-ray observations can make a unique and important contribution to this question. The reason is that observations of hot gas in many galaxies and in most or all clusters and groups of galaxies traces the gravitational potential of the dark matter with a precision that far exceeds that of any other method. What are needed are high quality X-ray images and spatially resolved spectra from which one can deduce the distribution of gas pressure and hence, the gravitational pull that opposes it. In clusters the possible role of gas turbulence can also be discerned by X-ray spectra of sufficient resolution. In the 1990s X-ray studies can therefore provide for the first time detailed mapping of the dark matter in many parts of the Universe. Furthermore, these maps could help reveal the nature of the matter as well, because its distribution gives clues about the masses of its constituent particles and how strongly the particles interact with one another (dissipational versus dissipationless).

The X- and γ-ray Background

Almost 30 years ago, X-ray detectors on board a spinning rocket discovered both the brightest non-solar X-ray source, and an apparently diffuse X-ray background (CXRB). The source was ultimately demonstrated to be associated with a low-mass stellar binary system containing a neutron star; to date, the CXRB has defied unambiguous identification. Its spectrum mimics that of a 40 keV optically thin plasma, which is problematic for the determination of its origin in at least three respects:

1. A truly diffuse 40 keV origin is theoretically awkward;
2. No candidate populations of sources (including AGN, some type of which is still the odds-on favorite) seem to exhibit spectra that are as hard;
3. The only sensitive searches for CXRB point-source candidates have bee n made at energies near 1 keV, well below the energy at which the CXRB has its maximum energy density (this energy density is second only to that of the 3 K microwave background, at a level about two orders of magnitude lower).

Clearly, any experimental effort to identify the origin of the CXRB must be able to measure similar spectra from candidate sources near the energies at which their outputs peak. Another "bump" in the isotropic cosmic radiation spectrum may exist near 1 MeV, presenting an analogous identification problem in the γ-ray region that may be confirmed, but not solved, with the data available from GRO.

There is also a diffuse γ-radiation whose origin is unknown. Theoretical models for this emission included: redshifted π^0 decay γ-rays from matter-antimatter annihilation in the early Universe, Compton scattering of relativistic intergalactic electrons by the 3 K background radiation, and the superposition of emission from many unresolved active galactic nuclei. Some contribution from the latter is certainly present, but their contribution cannot be reliably determined until a larger number of active galactic nuclei have been studied in the γ-ray regime. A careful study of the uniformity and energy spectrum should also provide a major step towards determining which theoretical model accounts for most of the diffuse radiation.

The Existing Experimental Program

Advanced X-Ray Astrophysics Facility (AXAF)

Our panel unanimously endorses the paramount scientific importance and necessity of AXAF, which must be completed and launched in this decade.

The capabilities of the *Advanced X-Ray Astrophysics Facility* (AXAF) will address the major outstanding astronomical and astrophysical problems discussed above to an astonishing degree. The AXAF is a remarkable and unique scientific endeavor and is the centerpiece of the X-ray astronomy program presented here. The ultimate fate of the Universe could be inferred from accurate AXAF measurements of the rate of expansion (the Hubble constant) and the change in rate (the deceleration parameter). The existence of exotic particles predicted by supersymmetric theories might be confirmed (or ruled out!) by observing their effects on the hot, X-ray emitting gas found in both galaxies and clusters of galaxies. The equation of state of bulk matter at nuclear densities, totally inaccessible to experiments on Earth, can be tested by studying the X-radiation from neutron stars.

These are but a few examples of what may be accomplished with AXAF, one of the Great Observatories which also include the *Hubble Space Telescope*, the *Gamma Ray Observatory* and the *Space Infrared Telescope Facility*. Operating in concert, and complementing each other in their respective wavelength capabilities, the Great Observatories should place mankind in a unique and historical position to understand the Universe.

Plans are now advancing for an AXAF Science Center (ASC) to act as the scientific interface between the AXAF project and the community. Experience with institutional arrangements for previous large programs has shown that establishing such a center at the earliest possible opportunity in the lifetime of a project not only ensures that the scientific objectives are met and maximizes the scientific return of the program, but may also provide significant opportunities for minimizing the run-out costs of the entire undertaking. We therefore urge that the ASC be placed in operation at the earliest possible opportunity.

Gamma Ray Observatory (GRO)

The *Gamma Ray Observatory* (GRO), scheduled for launch in late 1990, is a major new mission for γ-ray astronomy, covering six decades of energy, from 30 keV to 30 GeV, with a suite of four instruments. The range of astrophysics objectives targeted by GRO is extremely diverse, and includes high-energy phenomena in the vicinity of neutron stars and stellar-mass black holes, such as accretion processes, pair-plasmas, magnetic fields, and particle acceleration; determination of the origin of γ-ray bursts; the energetics and emission mechanisms of AGN; nucleosynthesis in massive stars, novae, and supernovae; energetic particle interactions in molecular clouds and the ISM; γ-ray line and continuum emission from solar flares; and the diffuse cosmic γ-ray background.

It is essential for the scientific success of GRO and for the future health γ-ray astronomy that this mission, as part of the Great Observatory program, enjoy widespread participation from the astronomical community. To this end, a particularly strong GRO Guest Investigator Program must be encouraged, supported by the

GRO Science Support Center and the four Principal Investigator instrument teams. NASA must ensure that the requisite resources are in place to support broad use of GRO by the astronomical community from the start of the mission through its completion. Access to both high-level and low-level data products on a reasonable time scale will be essential for Guest Investigators, as will be the availability of suitable documentation and calibration data sets. Support for correlative and theoretical studies related to the GRO mission science is also necessary to obtain the maximum scientific benefit from the mission. GRO may well be capable of an extended mission significantly beyond the baseline duration; continued support for such operations is essential.

XTE

The *X-Ray Timing Explorer* (XTE) is an important mission, having unique capabilities for moderate resolution spectroscopy and photometry over the entire energy range from 1 keV to 100 keV, with microsecond time resolution. It will substantially advance our understanding of the physics of accretion flows, the properties of relativistic plasmas and matter under extreme conditions, and the central engines of active galactic nuclei. The XTE program has been started, and should be completed in a timely fashion.

There are good scientific reasons to expect that a full understanding of processes occurring near the event horizons of black holes and the surfaces of neutron stars will require instruments with even better sensitivity and higher time resolution than XTE. Data collected with XTE will help to guide the planning of missions carrying such instruments.

HETE

The *High Energy Transient Experiment* (HETE) is a mission currently scheduled for launch in 1994 as a "Gas Can" by shuttle. It is composed of three instruments offering continuous, nearly full hemisphere coverage in the ultraviolet (4 eV to 7 eV), X-ray (2–25 kev), and γ-ray (6 kev – several MeV, provided by the French). The total weight of the experiment and free-flying spacecraft is $\sim$100 kg, and the estimated cost is \$13 M. The chief scientific goal is the panchromatic detection and monitoring of γ-ray bursts and bright X-ray bursts, with an eye towards obtaining accurate source locations for the former. The angular position accuracy is 6′ in X-rays and 3″ in UV, sufficient to offer good prospects for optical and/or radio identifications.

American Participation in Foreign Missions

There are several currently approved opportunities for US participation in foreign missions. In many ways, these *are* the low-cost, "small" Explorers of the 1990s, and this is a definite benefit of this type of collaboration. On the other hand, such activities cannot take the place of US-lead missions for maintenance of our technology base and personnel expertise, and data return to American investigators is often very limited.

NASA participation in the recently-launched German ROSAT mission will provide unique opportunities for imaging X-ray sources that have not been available to US astronomers since the end of the *Einstein Observatory* in 1981. The five-fold oversubscription to the first proposal solicitation for ROSAT, and the wide scope of scientific questions addressed in the first round of selected proposals, has demonstrated the value this mission will have in complementing the US program. The all-sky, high spatial resolution survey to be made by ROSAT should provide a unique resource for a variety of future investigations.

ASTRO-D is a cooperative program between Japan and the USA, scheduled for launch in February 1993 on board a Japanese-furnished satellite launched by a Japanese-furnished launch vehicle. It contains four BBXRT-type foil telescopes with a total geometric area that exceeds that of AXAF (and is some four times greater at 7 keV). Two of these telescopes have GSPCs furnished by the Japanese at their foci, while the other two have X-ray sensitive CCDs furnished by the USA. The mission has great generality, devoted to pointed observations of X-ray sources with all four imaging spectrometers coaligned; the angular resolution of the telescopes is $\sim 2'$ FWHM, which is constant over the entire 30' FOV, with energy resolution ~ 100–200 eV over the range 0.1–12 keV. After an initial proprietary season of about nine months for the PI teams, US guest investigators will be entitled to 15% exclusive use of the facility, and another 25% in collaboration with Japanese investigators.

Spectrum X-Gamma is a Soviet mission with extensive European, and some American, participation, due for launch in 1993. As presently configured, it will carry several large X-ray telescopes and an array of complementary instrumentation. US groups are supplying hardware in support of an X-ray polarimetry experiment and an X-ray all sky monitor. The mission offers American investigators a chance to conduct some significant high sensitivity X-ray observations well in advance of AXAF. In addition, the US hardware participation on Spectrum X-Gamma provides an important precedent for increased cooperation in this field between the world's two major space powers.

The *X-ray Multi-Mirror Mission* (XMM) is a "facility-class" X-ray observatory under development by the European Space Agency for launch near the end of the 1990s. XMM is a "high-throughput" X-ray spectroscopy mission which provides a very significant complementary capability to that provided by AXAF. The scientific instruments selected for XMM involve substantial US hardware participation. This will ensure access to this facility for US Guest Observers, and, therefore, provides an extremely cost-effective way for Americans to conduct important X-ray astronomical observations.

We strongly endorse the US participation in these projects, and encourage the inclusion of an International Programs line item in the NASA budget to regularize the funding of such efforts.

Attached Shuttle and Space Station Freedom *Payloads*

The *Diffuse X-ray Spectrometer* (DXS) is a relatively simple, low-cost attached-Shuttle experiment which will directly address an important and poorly explored question: the physics of the low-energy component of the diffuse X-ray background. An understanding of the physical conditions in the hot component of the ISM responsible for this emission will impact many areas of astrophysics.

The Shuttle mission ASTRO with its X-ray spectroscopy component *Broad Band X-ray Telescope* (BBXRT) offers an important and uniquely new capability to high energy astrophysics. Reflight of this payload involving guest observer participation would be highly desirable in the early years of the decade, but should probably not be pursued beyond mid-1993 when the free-flying ASTRO-D will begin to provide an enhanced capability for moderate resolution spectroscopy and imaging in the 1–10 keV band.

The *X-ray Background Survey Spectrometer* (XBSS) is an attached Space Station *Freedom* payload which will greatly complement and expand DXS results by determining the spectrum of the soft diffuse X-ray background over 100% of the celestial sphere (as well as with increased resolution and superior wavelength coverage than DXS). This experiment is particularly well-matched to the Space Station, and should provide results fundamental to our understanding of the ISM.

A second Station payload is the *Large Area Modular Array of Reflectors* (LAMAR). The high throughput imaging and spectroscopic observations to be performed by LAMAR are of the highest scientific priority, and the successful execution of this science requires the full effective area of the current LAMAR concept. In addition, the possibility of using the foci of LAMAR optics modules as readily accessible test sites for novel detectors is attractive. However, the Panel is concerned that complexities of the Space Station, and the interface of LAMAR to the Station, may greatly delay the project and increase the cost very substantially. Thus, we urge that alternative missions to accomplish the important LAMAR scientific goals remain under consideration, and that particularly close attention be paid to LAMAR cost and schedule, to retain the

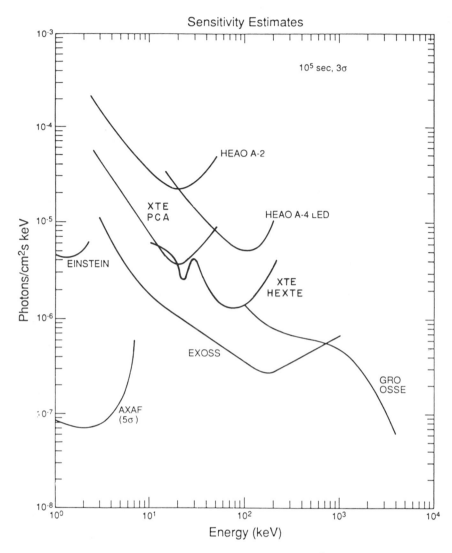

FIGURE 1   Sensitivity of selected X- and γ-ray astronomy missions for typical 10^5 s observations. EXOSS is an example of a concept for a future ~1 m^2 collecting area hard X-ray imaging experiment.

possibility of switching to a high throughput concept unconnected to the Station, should this become necessary to accomplish the science in a timely and economic manner.

Major Missions for the 1990s

The highest priority major programs in high energy astrophysics from space are the rapid completion and flight of the *Advanced X-ray Astrophysics Facility*, together with integration of the AXAF Science Center into the worldwide astronomical community as a smoothly functioning entity; and the successful flight and reduction of data from the *Gamma Ray Observatory*. We cite these highest priority items succinctly simply because the case is so clear-cut. Figure 1 compares the sensitivity of selected instruments on these two missions with past experiments. These two components of NASA's Great Observatory program will without question transform our field, and shape the course of research in this area for one or more future decades.

We can already be virtually certain that technology development will raise attractive possibilities for second generation AXAF instruments, probably on the timescale of the years 2000–2005. These developments may permit use of AXAF in certain exciting scientific areas not currently addressed by the payload, *e.g.*, polarimetry, and also greatly expand the capabilities of the Observatory in imaging and spectroscopy.

It is also already clear that beyond the year 2000, there will be a requirement for at least one major new mission in both the X- and γ-ray regimes. We can make this statement with confidence even before the flights of AXAF and GRO, because examples of the sources to be studied, and thus their fluxes, are already known, and basic physical questions already evident define certain instrumental parameters. For example, numerous QSOs are known from *Einstein Observatory* observations to have fluxes of 10^{-13} ergs cm^{-2} s^{-1} in the soft X-ray band. If we are to study the X-ray spectra of these objects with sophistication comparable to the optical spectroscopy done routinely today, and address such fundamental issues as emission line profiles, we will require spectral resolution $E/\Delta E \sim 1000$. Such resolution at these flux levels will require X-ray optics with $> 2\,\mathrm{m}^2$ collecting area, far beyond the capabilities of AXAF. Technology development, described in more detail later in this report, will be required in several areas of optics and detectors for projects such as these.

Similarly, clear instrumental improvements are needed in the γ-ray regime. GRO instruments all have a basic angular resolution of one to a few degrees. While this may be sufficient, for example, for currently achievable sensitivities of 10^{-7} photons cm^{-2} s^{-1} keV^{-1} at 1 MeV, with the order-of-magnitude improvement in sensitivity expected for post-GRO instruments, angular resolution of considerably better than 1° is needed to avoid severe source confusion. Similarly, while 1° resolution is sufficient for determining the spectra of bright γ-ray emitting objects, detection and identification of fainter objects will require arc minute resolution or better. Order-of-magnitude improvements in both sensitivity and angular resolution are expected given technology development in areas such as position-sensitive detectors and large-scale structures.

What is as yet unclear is precisely which technologies will prove most successful and cost effective, and their timescales for flight readiness for these very large missions or new instruments. It is thus premature to define such missions in detail at this point. Instead, we recommend a vigorous, major technology development program in certain key areas discussed later in this report.

A New Program of Moderate Missions

Highest Priority

Our panel believes strongly that specific moderate/Explorer missions should be selected competitively through the established peer review process. We have identified two particularly important scientific problems which can be well addressed by moderate class missions in the coming decade. These areas, which we regard of equal priority, are hard X-ray imaging and γ-ray spectroscopy.

Hard X-ray Imaging A serious programmatic gap is developing in American high energy astrophysics in the 10–250 keV range. During the coming decade, AXAF and GRO will almost surely revolutionize our understanding of high energy phenomena above and below these boundaries. Yet this intermediate energy range remains poorly explored. There are indications that it may be particularly fruitful. This range is high enough that the emission is dominated by non-thermal processes, thereby promising results quite different from those obtained by AXAF and its predecessors; yet it is sufficiently low that the expected photon fluxes from a large number of both galactic and extragalactic objects are respectably high, and these objects are thereby amenable to study in great detail.

Among the key scientific objectives of a hard X-ray imaging mission are: determination of the log N–log S relationship for hard X-ray emitting AGN out to $\sim$ 100 keV; determination of the accretion processes and particle energy sources occurring in the immediate environment of the central engines of AGN; determination of the physical parameters of accreting neutron star systems, including magnetic field strength and geometry; study of possible signatures of accretion onto stellar-mass black holes; and study of the physics of the soft γ-ray repeaters. To study the high energy processes associated with galactic compact objects and AGN, a mission with broad-band energy response, extending over two decades of energy, is required: from the Fe K-shell line at low energies ($\sim$5 keV) to the positron annihilation line at high energies ($\gtrsim$ 500 keV). Such a mission would probe a very wide range of accretion physics over a wide range of central object mass scales. Many of the same physical emission processes expected to be operative in accretion onto stellar mass systems are also expected to occur in massive AGN. Study of Fe K-shell emission and comptonized photon spectra above 15 keV will constrain the physical parameters of accretion disks, while non-thermal power-law spectra in the range 50–500 keV and pair-plasma signatures above 500 keV will help determine the key characteristics of the physical environment in the immediate vicinity of the central compact object. In the case of accretion onto magnetic neutron stars, phase-resolved spectroscopy of cyclotron line emission will be critical in determining the magnetic field and plasma characteristics of the accretion column.

No high sensitivity survey of hard X-ray and soft γ-ray emitting objects currently exists. Detailed studies such as those described above should take place within the context of a proper understanding of the classes and number densities of galactic and extragalactic hard X-ray and soft γ-ray emitting objects. For this reason, survey objectives are a critical part of a new hard X-ray mission, and absolutely require imaging capability for the identification of objects and the elimination of source confusion. The technology of coded aperture masks is sufficiently mature that experiments with angular resolutions of one to several arc minutes seem achievable. Experience with soft X-ray astronomy indicates that this is the spatial resolution threshold where optical identifications suddenly become feasible for a large fraction of detected sources, thereby vastly increasing our understanding of each class of source detected. Spectral resolutions of 5–10% FWHM also seem straightforward to achieve, assuring accurate measurement of continuum slopes, and excellent data on the cyclotron emission features expected to dominate at these wavelengths for many objects.

A hard X-ray Explorer-class imager with $1\,\mathrm{m}^2$ of effective area will be two orders of magnitude more sensitive than the HEAO-1 A-4 experiment, the only recent sky survey, at 100 keV, and 10× more sensitive than XTE at this energy. In a long (10^6 s) pointing, a 100 keV limiting sensitivity of 2×10^{-7} ph cm^{-2} s^{-1} keV^{-1} will be achieved, which will easily detect Her X-1 (or a source 10× fainter than Cygnus X-1) in the LMC. Although Cygnus X-1 is not quite detectable in M 31, an Eddington-limited $1\,\mathrm{M}_\odot$ source would be. For extragalactic work, this long pointing could detect 3C 273 at 100 keV at 15× its true distance. A survey with 3×10^4 s per pointing can cover the entire galactic plane in 6 months, yielding 10^3 sources, or cover the entire celestial sphere in 2 yr, yielding 10^3 AGN to a limiting flux of 2% of 3C 273. In all the above cases, source confusion is never approached, and positions sufficiently accurate for optical identifications are obtained, except at very low latitude. Figure 1 shows the sensitivity of a typical 1 m^2 hard X-ray imager mission concept (EXOSS).

We strongly endorse an Explorer-class satellite to achieve these imaging and spectroscopic goals.

γ-ray Spectroscopy   GRO will be exceptionally effective at mapping the sky at a variety γ-ray wavelengths, and providing initial glimpses at spectral features in a large number of sources. It is clear that substantially higher resolution γ-ray spectroscopy, at flux levels far beneath those attainable by the GRO instrument complement, will be the next step to build on GRO results. Sensitivity of at least 10^{-6} ph cm^{-2} s^{-1} (100× superior to HEAO C-1, and 10× superior to GRO OSSE) is needed to address the next set of interesting problems. To measure line profiles and Doppler shifts, spectral resolution of $E/\Delta E \sim 1000$ is necessary. Both of these goals appear feasible with technology available during the coming decade. Similarly, angular resolution of a few degrees or less will be needed, again an achievable goal.

Among the key scientific objectives of such a high resolution γ-ray spectroscopy mission are determination of the sites and rates of recent nucleosynthetic activity in the Galaxy; exploration of the physics of Type I and Type II supernovae, including nucleosynthetic processes and the characteristics of the explosion; determination of the characteristics of low energy cosmic rays and the interstellar medium; study of the physical environment of collapsed objects, in particular the relativistic pair-plasmas; and measurement of the magnetic fields of neutron stars, including studies of cyclotron line emission and positron production.

Maps of the Galaxy in the light of several nuclear lines (e.g., $e^{\pm}$, ^{44}Ti, ^{26}Al, ^{22}Na, ^{60}Fe) will identify the nature and general sites of nucleosynthetic activity over the past 10^6 yr, including possibly Type I and Type II supernovae, novae, and Wolf-Rayet stars. The rates of these processes and the mixing of nucleosynthetic material into the interstellar medium will be measured. It is likely that several unknown supernova remnants less than 500 years old will be discovered. The profile of the $e^{\pm}$ line will be mapped to obtain a determination of the temperature and density phase where annihilation occurs in the interstellar medium. Nuclear lines from the interactions of low energy cosmic rays (E<100 MeV) with the interstellar gas and dust may allow a determination of the intensity and distribution of these cosmic rays, as well as the dust fraction and size, and the elemental abundances in the interstellar medium.

Discrete source observations will allow a number of Type I supernovae to be studied at distances approaching the Virgo cluster and their nucleosynthetic yield, energetics and expansion dynamics to be determined through measurements of their ^{56}Ni and ^{56}Co decay γ-rays. Detailed study of $e^{\pm}$ annihilation radiation from relativistic pair-plasmas in the vicinity of AGN will determine important physical parameters of the environment of the central engine and provide clues to its nature. The detection of nuclear lines, if accomplished, would provide significant new information on the energetic particle environment in AGN. Similar studies of galactic compact objects, in particular neutron stars and stellar-mass black holes, would yield correspondingly important measurements of the plasma and energetic particle environment around these objects. Observations of cyclotron, $e^{\pm}$ annihilation, and possibly nuclear lines can give new insights into the nature of γ-ray bursts, and the detection of gravitationally redshifted lines from the surface of neutron stars can provide direct information on the neutron star equation of state.

A spectroscopic mission of the this class would profoundly impact virtually every problem in γ-ray astronomy discussed earlier in this report. Recent advances in detector and cryogenic technology imply that a very powerful experiment of this type can still fit within the Explorer envelope; the sensitivity of such a mission (NAE) is shown in Figure 2. Many of the relevant technologies are already under test in balloon-borne spectrometers, although such experiments of course fall orders of magnitude short in the desired integrated exposure times. We thus enthusiastically endorse an Explorer-class high resolution γ-ray spectroscopy mission.

Programmatic Considerations There are moderate-mission concepts in relatively advanced design stages for both the hard X-ray imaging and the γ-ray spectroscopy missions discussed above, namely EXOSS and NAE, respectively. These missions address the scientific goals described here, goals which our Panel strongly supports. Other innovative technical approaches to these goals may also be possible.

Additional Mission Concepts

In X-ray astronomy, there are several additional important scientific areas which are also amenable to Explorer-class missions. One example is high resolution ($E/\Delta E \sim 10^4$) spectroscopy, to obtain for the first time data on line profiles and Doppler shifts of comparable sophistication to that of optical spectroscopy, albeit on a relatively small number of bright sources. There is also interest in an X-ray component to a "panchromatic" facility, which obtains simultaneous observations at UV and visible wavelengths as well.

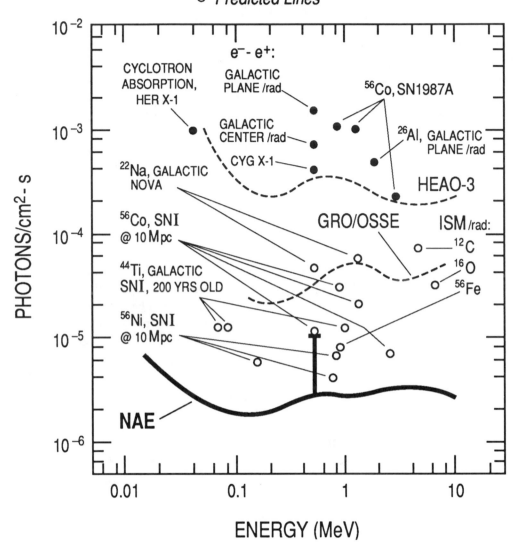

FIGURE 2   The line flux from a variety of known and predicted galactic and extragalactic γ-ray sources, together with the sensitivity (10^6 s exposure, 3σ significance) of several past/approved (HEAO-3, GRO/OSSE) missions and mission concepts (NAE).

Following the GRO mission, there will be a need for high-energy γ-ray observations with sufficient sensitivity and angular resolution to accurately locate sources, and to define detailed spatial features of emission regions such as molecular clouds, Galactic arms, and nearby galaxies, as well as to measure variations of compact sources. From a threshold of approximately 3×10^7 eV, the possibility of extending the energy range to approach 10^{11} eV is a desirable goal, requiring a sensitivity about five times that of the GRO high energy γ-ray telescope. Source location approaching $1'$ should be attempted, in part by detector improvements and in part by greater sensitivity for the high energy γ-rays whose character permits inherently better individual angular resolution. The possibility of measuring polarization may be considered. Additional areas of interest include the spectra of γ-burst sources at very high resolution (presumably with Ge detectors), and extrasolar applications of pinhole subarcsecond resolution imagers designed primarily for solar work.

New Opportunities for Small Missions

Our Panel has identified a number of exciting scientific problems that may be addressed in the next decade via small missions. In this context, we use "small" as a cost <$30 M, a payload mass <500 kg, and a launch vehicle smaller than a Delta. We discuss here several examples of such missions, *not with the* goal of establishing scientific or schedule priorities, which we believe should be left to the Announcement of Opportunity/peer review process, but rather to illustrate the diversity of important problems in high energy astrophysics that are accessible by these relatively low-cost projects.

A panchromatic attack on γ-ray bursts is available through HETE, the mission described previously above. It is remarkable that for a total mass of 100 kg, including both instrumentation and platform, one can cover the UV, X-ray, and γ-ray bands simultaneously, with sensitivity and angular resolution almost surely sufficient to make very significant, and quite possibly definitive, progress on the issue of the nature of γ-ray burst sources.

As a specific example of future possibilities, we note that recent innovations in X-ray optical design raise the possibility of a Wide Field X-ray Telescope in this small mission category. With an estimated payload weight of <300 kg (half of which is the optics), such a mission could survey 10^3 deg^2 in the 0.4–3 keV band, yielding $>10^3$ clusters (to z$\sim$2) and $>10^4$ AGN to limiting fluxes of $\sim 3 \times 10^{-14}$ ergs cm^{-2} s^{-1}. Such an instrument, specifically optimized for surveys, could be a splendid complement to AXAF, which is not designed for such work. As noted earlier, X-ray observations of clusters at this level of sensitivity may open entire new avenues for research in the large scale structure of the Universe.

We have identified a number of other important scientific problems which we believe are potentially well-suited to these small missions. Examples include application of multi-layer optics technology to produce a wide-field soft X-ray telescope with quite small energy bandpass, but very large field of view and effective area; very high resolution X-ray nebular spectrometers; all sky X- and γ-ray monitors to routinely monitor transients; and certain approaches to X-ray polarimetry.

We see a bright future to experimental X- and γ-ray astronomy in the small mission category, and we urge vigorous exploitation of these flight opportunities in the coming decade.

Technology Development Issues

X-ray Astronomy

A vigorous flight program in X-ray astrophysics should be accompanied by an equally vigorous program in technology development which will enable the design and testing of new experimental tools to be incorporated in future missions. There has been a virtual revolution in this field over the past few years owing to the discovery of a number of promising technologies which have not previously been utilized in X-ray astronomical instrumentation. Below we discuss developments in several key areas which are especially deserving of further study:

Large Area Telescopes (0.1 - 10 keV) Recent discussions of scientific requirements for the next phase of X-ray astronomical facilities have pointed to the need for large ($\geq 2\,\mathrm{m}^2$) collecting area with moderate-to-high angular resolution. A reasonable goal is $\sim 2 - 3''$, which will avoid source confusion problems and still be adequate to resolve galaxy clusters and even galaxies at high redshift. Collecting areas should approach 10 m^2 at reasonable cost. The primary driver will be to achieve these parameters in the 0.1–10 keV range, but we should also continue to explore ways to push up to 20 or 30 keV with very low graze angle mirror arrays.

A number of techniques have been suggested to achieve these goals. Several approaches, including the use of thin metal foils, electroforming, epoxy replication onto thin carriers, and flat mirror plates with long focal length, appear especially promising. All of these concepts, and quite possibly others, are deserving of further development. Normal incidence optics may also play a role, especially if the multilayer technique can be pushed to shorter wavelengths. Much of this effort can be accomplished via small grants to universities and research institutions; however, some larger-scale coordination with industry may be warranted in this area.

Focal Plane Imaging Arrays The past few years have witnessed the rise o f the CCD as the "workhorse" imaging detector for X-ray astronomy. CCDs certainly present an attractive option for future missions because of the combination of very high spatial resolution coupled with moderate energy resolution that they provide. The use of CCDs for X-ray astronomy has consequently received much attention in this field, both in the US and abroad. Nevertheless, there is much yet to be achieved. In particular, large area telescopes will inevitably require larger focal plane detectors — the present technology is already being "pushed" in this regard for AXAF. Present CCDs are roughly the size of a flattened ping-pong ball, whereas detectors the size of ping-pong tables may be required in the future. Larger and/or smaller pixel sizes are also necessary for some applications. Deeper depletion regions are needed to enhance the high energy efficiency, and there is continuing concern about CCD susceptibility to long-term radiation damage.

At present, most CCD fabrication is performed in large industrial firms where the effort is driven primarily by commercial and defense interests. It is likely that the astronomical devices of the future will need to be "customized" in ways that are not necessarily consistent with the requirements of these other, more lucrative applications. NASA should take account of this problem in planning for future CCD development.

Non-Dispersive High-Resolution Spectrometers An extremely exciting recent development in X-ray instrumentation has been the introduction of non-dispersive, high resolution spectrometers which rely on cryogenic technology. These devices combine the high spectral resolution characteristically achieved with dispersive systems with the high quantum efficiency of conventional lower resolution detectors. A reasonable scientific goal for these detectors would be a spectral resolution of 0.5 eV, so that one could achieve resolving powers better than 1000 at the oxygen Kα line and the iron L lines. Ideally, these detectors should be position sensitive, at least over a limited field. Logistics of the space program in the foreseeable future require detector lifetimes of at least five years, which puts strong demands on the cryogenic systems.

There are several viable technological approaches under study in this category. These include resistive calorimetry, dielectric calorimetry, kinetic inductance read-out schemes, and superconducting tunnel junction arrays. The tunnel junction arrays, in particular, appear especially promising; however, all are deserving of further support. Fabrication of some of these kinds of devices may be outside the scope of what can be achieved by a university group with a typical SR&T grant, so NASA should consider some other funding options in this area. In addition, nearly all of these designs require cryogenic systems capable of holding stable temperatures at the sub-Kelvin level. NASA should assure additional funding for cryogenics technology development to guarantee that the special needs of these devices can be met with practical flight systems.

Other Basic Technology In identifying the major areas of basic technology, it is important that we also emphasize the importance of continuing a variety of other, more specific development activities. In effect, we must maintain a balance between large and small research programs just as we must maintain a balance between large and small missions. It is clear that remarkable returns have been achieved from modest investments in the SR&T program. During the long dry spell of the 1980s in terms of flight opportunities, the SR&T program has delivered an impressive number of successes in technology development. That effort has to be continued while we proceed with the missions of the 1990s. A partial list of areas for further investigation includes:

- high pressure gas counters and liquid noble gas detectors
- synthetic multilayers (the goal should be 2-D spacings of ~ 10Å, both as Bragg diffractors and normal

incidence mirrors)
- ultra-thin windows
- improved reflection gratings
- improved transmission gratings
- various types of polarimeters

γ-ray Astronomy

It is likely that the scientific objectives of γ-ray astrophysics in the 21st century will be addressed by multiple missions, involving a mix of platforms for scientific observations. Examples include: a broad-band high-sensitivity, high-spectral resolution mission in the 5 keV to 10 MeV energy range with angular resolution of 1' or better, a high-sensitivity, high-energy mission in the energy range 30 MeV to 100 GeV, also with angular resolution better than 1', and a mission capable of high-sensitivity, high-spectral resolution observations of γ-ray bursts. Several technology thrusts are readily identifiable to enable such missions. Significant improvements in angular resolution, energy resolution, and sensitivity are both required by the scientific goals, and feasible given anticipations for improvements in technology during the next decade. Specifically, investment is needed in the development of advanced detector technologies.

A common characteristic of the most promising new detector technologies is detailed event visualization, *i.e.*, excellent spatial resolution for identifying the primary interaction of the γ-ray and reconstruction of the geometry of secondary photon and particle interactions. Such event visualization is critical over the entire hard X-ray and γ-ray energy range for determining the photon arrival direction accurately and for improved sensitivity due to enhanced background rejection. Typically, sensitivity is correlated with size of the detector systems, so development of cost-effective, large-area detector technologies are important. Candidates for future technology development include high-pressure gas detectors, liquid Xe and Ar detectors, imaging scintillation detectors, position sensitive Ge and Si detectors, superconducting transition detectors, and high-energy bolometer systems. These are typically the same types of detector systems under development in high-energy and nuclear particle physics. It is difficult to currently assess which of the emerging technologies will be most attractive for 21st-century space missions. A prudent approach for technology investment would be diversification, with support both for investigating the feasibility of new detector concepts, and bringing to maturity those concepts already demonstrated to be feasible for space mission application.

In addition to detector technology, investment is required in various support technologies. These are likely to include large structures for imaging approaches such as coded-aperture and Fourier transform imaging, cooling technology for several of the promising detector systems, and advanced electronic and optical readout technologies.

Policy Issues

Changes in NASA Management Style

NASA science management has been responsive to the astronomy community in formulating a program whose content reflects community scientific priorities. NASA has also been successful in providing a continually growing pool of monetary resources for science, which has been especially enhanced in the past few years in the areas of data analysis and theory. One area in which there has been a growing problem, however, is the decreasing frequency with which instrumentation can be placed in space. For major missions, the timescale from initial study to fruition is now approximately a generation. For moderate missions of the Delta-class Explorer variety, the timescale is only a factor of two smaller, at 10–15 years. It is not clear that even small missions can be mounted much faster than a decade with standard NASA management practices.

The unfortunate policy error that resulted in the suspension of expendable launch vehicles in favor of the Space Shuttle also resulted in a management style that stresses formal safety, reliability and accountability over innovation and sensible risk. The *Challenger* tragedy is at least partially responsible for this cautionary approach to the development of flight hardware. The administrative burden of properly documenting even the smallest space mission results in stretching out the development schedule and therefore increasing cost. Furthermore, the formal reporting procedures may not even be the most effective way of providing the level of reliability that NASA desires. The cost/benefit ratio of this approach is not at all clear, as instrument development costs per pound of payload typically exceed launch costs by a factor of 10 or more.

NASA has already recognized some aspects of this problem and has begun to address it. A "mixed fleet" policy is now in effect, so that expendable launch vehicles can be used for at least some missions. The SMEX (SMall EXplorer) Program is an attempt to alter the management style of the smallest missions by establishing an in-house project team to produce standard spacecraft systems.

We encourage NASA to continue these efforts and to guard against their inevitable tendency to slip back toward the more cumbersome and costly approach. We also urge investigation of other changes in management style for space missions, such as less concentration on formal documentation and management oversight, and more direct management responsibility for involved scientists. The very successful example of the Japanese X-ray astronomy program, which features a fixed budget on a fixed schedule, where scientists and their small management teams make all the tradeoff decisions, is an interesting paradigm. We would like to see NASA investigate the utilization of similar management practices, and we suggest experimenting with such innovative management approaches on the "low" end, with the missions that have smallest costs. NASA is already making a first step in this direction with the HETE project.

The important issue here is flight frequency. Timescales of more than a decade for most missions dissuade much of the community from participating in the development of space missions; most university scientists are now extensive users of NASA data, but only a few contribute to the development of new missions. The negative implications for the training of the next generation of graduate students is obvious and could be disastrous to the future health of space science.

Use of Expendable Launch Vehicles Versus Manned Missions

We support the current NASA policy that manned vehicles should be used only for those launches where the presence of a man is mandatory for completion of the experiment. We therefore strongly recommend that all future high energy astrophysics experiments be launched on Expendable Launch Vehicles (ELVs), except in those cases so far into development that a change in plans would cause severe financial disruption. (Those experiments well matched to Space Station *Freedom* obviously also require manned support). A corollary of this recommendation is that, lacking some new and compelling rationale to the contrary, the concept of recoverable/interchangeable buses should be abandoned.

Barriers to Mixing Ground-Based/Space-Based Funding

Multi-wavelength studies of high-energy phenomena in the Universe are becoming increasingly fruitful, and are crucial for solving a significant subset of problems. These studies are greatly facilitated when sponsoring agencies fund problem-oriented, as opposed to wavelength-oriented, research. We encourage continued examination of the structure and policies of funding agencies with these issues in mind.

Lunar Base

In the time interval considered by this report, we can see no compelling experiments which can be performed only from the lunar surface. On the other hand, the moon might on longer timescales provide an effective site for future high energy astrophysics missions that require large, stable structures (such as long focal-length telescopes, coded apertures, large detector arrays), or that can make good use of the lunar soil for shielding (such as γ-ray spectrometers). The technology developments described above for earth-orbiting experiments will be a prerequisite to exploit future lunar experiments in X- and γ-astronomy. At the same time, planners of lunar initiatives should keep several strawman high energy instruments in mind as they plan the capabilities of the future lunar bases.

Mission Operations and Data Analysis Funding

It is hardly novel to emphasize that Mission Operations and Data Analysis (MODA), rather than construction of spacecraft and instrumentation, are the true goals of scientific research from space. MODA funds are invariably a small fraction of the capital costs of most projects; unfortunately there has often been severe pressure on the NASA MODA line item during times of budget difficulties in the past decade. If the scientific program advocated in this report is to be implemented, MODA funds must not only be successfully protected from such cuts, but expanded in a manner commensurate with the expansion of flight opportunities. The survival of graduate students and postdoctoral fellows is closely tied to the health of MODA funding, making this issue central to the question of the next generation of space scientists as well. We hope that in the next decade, high energy astrophysicists can work together with their sponsoring agencies to heighten Congressional awareness of the critical importance of MODA funding.

Line Item for International Instrument Opportunities

The opportunity to develop and fly individual instruments on foreign spacecraft has emerged as an important and very cost effective way for US investigators to build hardware, and for the US community at large to obtain data. In the past, each of these initiatives has been handled as a special case, often with extraordinary effort required from NASA Headquarters and the investigators. The NASA Astrophysics Division has attempted unsuccessfully to obtain a regularly recurring budget line item specifically to fund US participation in foreign space missions. We strongly commend this approach to financing these missions, and urge its implementation immediately. Each such opportunity should be reviewed by the appropriate NASA scientific advisory committee. Instrumentation for these opportunities should be selected by open competition and peer review wherever possible, and the needs for incremental MODA funds created by each such opportunity must be kept in mind.

Smaller NASA Programs

Sounding rockets We perceive that a substantial fraction of the funds currently expended in NASA-supported sounding rocket programs goes towards detector development, rather than to acquisition of unique scientific data. We therefore believe that consideration should be given to more overtly redirecting a portion of funds from the sounding rocket program to detector/technology development, where they may be more efficiently expended and are likely to produce a greater return.

Balloons We note that the γ-ray balloon program has been greatly oversubscribed recently, and is producing exciting science as well as serving as an essential testbed for instrumentation development. We recommend doubling the current scope of this modest program.

Support of young scientists The key problems of training the next generation of space scientists are well-known and have been discussed above: even if funds are available to support the best graduate students, the time required to complete most exciting projects has grown so long that such projects are hopelessly incommensurate with the graduate education programs. As a result, few if any students can see a space research project through completely, from the initial intellectual formulation of the problem to publication of results. Shortening the relevant timescales is a complex policy issue, but providing funding for support of junior scientists is not. We laud NASA for its recently-initiated Graduate Student Researchers Program, which, although limited in scale, represents a gradual return to the more halcyon days of graduate support of the 1960s and 70s. Likewise, the Hubble Fellow program is a welcome acknowledgement of the problems of supporting the best young scientists in our field. Programs such as these must be nurtured, expanded, and seeded to other agencies that support research in high energy astrophysics. A prerequisite for adequate support of graduate students is the maintenance of a healthy research base at universities where these students are trained. A careful balance of research support must exist between universities and other centers.

Theoretical Programs Theoretical research in support of space missions has always been crucial for advances in high energy astrophysics, and there are firm grounds to believe it will become yet more vital to the success of the missions of the 90s. As both X- and γ-ray astronomy shift an increasing fraction of their observations from imaging to spectroscopy, far more sophisticated theoretical models will be required to interpret the data. The cost of these indispensable modeling efforts is very small compared with the capital costs of virtually any mission, and even with the operations costs of most projects. Again we commend NASA for the recently instituted Astrophysical Theory Program, and urge its maintenance and vigorous expansion.

Laboratory Astrophysics The preceding comments on the rapidly increasing importance of spectroscopy in X- and γ-ray astronomy also impact laboratory astrophysics. It would be regrettable if data from elegant space-borne experiments cannot be adequately interpreted due to lack of fundamental laboratory data, especially given the sophistication of current techniques in laboratory plasma physics. We recommend that NASA and NSF support experiments in laboratory astrophysics, especially where there is promise of close connections to analysis of AXAF and GRO observations, or significant impact on the design of future instrumentation.

Two moderate missions	$300 M
Four SmEx	$100 M
Technology development	$50 M
International budget line	$50 M
Theory, suborbital	$40 M

Table 1. Incremental Costs of New Initiatives for the Decade

Conclusion

High energy observations from space by American investigators have undergone a relatively quiescent phase during the 1980s, a time when important scientific problems have been amply clear, but opportunities for presence in space have been all too rare. The launches of AXAF and GRO, as well as a variety of smaller currently approved projects described earlier, give the United States the opportunity to regain much of the momentum it built in X- and γ-ray astronomy in the 1970s. The new program for the 1990s which we have described here is of modest incremental cost, as can be seen from the estimates in Table 1, but we believe will lead the discipline into the new millenium with an exciting array of new data, a technology base needed for future experiments, and an infrastructure of both scientists and facilities essential for leadership in the field.

PARTICLE ASTROPHYSICS PANEL

BERNARD SADOULET,* University of California, Berkeley, *Chair*
JAMES CRONIN,* University of Chicago, *Vice-Chair*

ELENA APRILE, Columbia University
BARRY C. BARISH,* California Institute of Technology
EUGENE W. BEIER,* University of Pennsylvania
ROBERT BRANDENBERGER, Brown University
BLAS CABRERA, Stanford University
DAVID CALDWELL, University of California, Santa Barbara
GEORGE CASSIDAY, University of Utah
DAVID B. CLINE, University of California, Los Angeles
RAYMOND DAVIS, JR., Blue Point, New York
ANDREJ DRUKIER, Applied Research Corporation
WILLIAM F. FRY, University of Wisconsin, Madison
MARY K. GAILLARD, University of California, Berkeley
THOMAS K. GAISSER,* University of Delaware
JORDAN GOODMAN, University of Maryland
LAWRENCE J. HALL, University of California, Berkeley
CYRUS M. HOFFMAN, Los Alamos National Laboratory
EDWARD KOLB, Fermi National Accelerator Laboratory
LAWRENCE M. KRAUSS, Yale University
RICHARD C. LAMB, Iowa State University
KENNETH LANDE, University of Pennsylvania
ROBERT EUGENE LANOU, JR., Brown University
JOHN LEARNED, University of Hawaii
ADRIAN C. MELISSINOS, University of Rochester
DIETRICH MULLER, University of Chicago
DARRAGH E. NAGLE, Los Alamos National Laboratory
FRANK NEZRICK, Fermi National Accelerator Laboratory
P. JAMES E. PEEBLES, Princeton University
P. BUFORD PRICE, University of California, Berkeley
JOEL PRIMACK, University of California, Santa Cruz
REUVEN RAMATY, NASA Goddard Space Flight Center
MALVIN A. RUDERMAN, Columbia University
HENRY SOBEL, University of California, Irvine
DAVID SPERGEL,* Princeton University
GREGORY TARLE, University of Michigan, Ann Arbor
MICHAEL S. TURNER,* Fermi National Accelerator Laboratory
JOHN VAN DER VELDE, University of Michigan, Ann Arbor
TREVOR WEEKES, Harvard-Smithsonian Astrophysical Observatory
MARK E. WIEDENBECK,* University of Chicago
LINCOLN WOLFENSTEIN, Carnegie-Mellon University
STANFORD E. WOOSLEY, University of California, Santa Cruz
GAURANG YODH,* University of California, Irvine

PARTICLE ASTROPHYSICS

EXECUTIVE SUMMARY

In the last few years, particle astrophysics has emerged as a new field at the frontier between high energy astrophysics, cosmology, and particle physics. A spectacular achievement of this new field in the last decade has been the establishment of neutrino astronomy with the detection of solar neutrinos by two independent experiments and the spectacular observation of the neutrinos from the supernova SN1987A. In addition, the field has produced tantalizing hints of new physics beyond the standard models of astrophysics and particle physics, generating enthusiastic attempts to confirm these potential effects.

The next decade promises to be even more productive. Extrapolating within the present conceptual framework, we expect the next ten years to bring fascinating results on the following issues:

• The elucidation of the nature of dark matter, especially if it is made of nonbaryonic particles.

• The explanation of the solar neutrino puzzle and if a supernova explodes in our own galaxy, a detailed account of the explosion mechanism by the analysis of the emerging neutrinos.

• The possible confirmation of the existence of point sources of energetic particles leading to the production of gamma rays, neutrinos and maybe new particles at energies as high as 10^{14} eV.

• The understanding of the origin of cosmic rays, including the physical processes responsible for their synthesis and acceleration on a wide variety of scales.

It is even more likely that as yet unsuspected phenomena will be discovered, that entirely new concepts will be tested and that the deep link between particle physics, the early universe and the high energy astrophysical processes will be extended beyond what we can imagine today!

In contrast with more mature observational fields, particle astrophysics is likely in the next decade to still be based on the succession of experiments which will increasingly sharpen their scientific focus. We see, therefore, the evolution of the field as governed by a number of decision points occurring in the next decade when the information from previous experiments or technological development becomes available:

• The highest priority at the moment is the rapid implementation of the presently approved program of observations on the ground and in space. Within this program, existing neutrino detectors should be coordinated and maintained to provide an efficient supernova watch.

• The technology is now available to tackle the question of the extragalactic origin of cosmic rays with the study of their spectrum at 10^{20} eV. We recommend that the high resolution Fly's Eye be supported, contingent on a favorable detailed technical review.

• We foresee important funding decisions to be made in a few years when the required information will be available. If the feasibility of the cryogenic technologies can be demonstrated, it is clear that a full scale search for nonbaryonic dark matter will have very high priority. If the large extensive air shower detectors currently being deployed confirm the claims for localized gamma ray sources above 10^{14} eV, understanding these unsuspected acceleration mechanisms (and the new particle physics if the claimed anomalous muon content is substantiated) would require more powerful detectors. The results of current solar neutrino detectors may steer the field into new observational directions. Finally, if the early promise of gamma ray astronomy at 10^{12} eV is fulfilled, then larger installations will surely be required.

The preparation for these decisions appearing on the horizon demands a strong development of new detection techniques: we emphasize in particular the cryogenic particle detectors for dark matter searches, the new solar neutrino schemes, and the test of the water Cerenkov technique for the detection of extensive air showers.

Such a complex and fundamentally multidisciplinary enterprise requires a strong theoretical activity and a variety of coordination mechanisms. Our main recommendation here is the establishment of a particle astrophysics

Advisory Panel which could advise the federal agencies involved in particle astrophysics (NSF, DOE and NASA) on the relative scientific priorities and help them set up long term policies.

The emergence of a new scientific field

Particle detection techniques have been essential to the development of high energy astrophysics since the first cosmic ray investigations in the early years of this century. Their role has been greatly expanded during the past decade as a variety of large experiments have begun to address fundamental astrophysical questions.

Two of the most recent and important advances in astronomy have been made by the direct detection of neutrinos. Two independent experiments measured the solar neutrino flux using totally different techniques, and neutrinos from the supernova 1987A have been directly detected by experiments designed to observe proton decay. Moreover, observation of the electromagnetic spectrum from astrophysical objects has been extended above 10^{11} eV using ground based Cerenkov detectors. Evidence obtained with air shower arrays suggests that gamma radiation in excess of 10^{14} eV may also have been observed from astrophysical objects. Cosmic rays have been detected up to 10^{20} eV but their nature and origin at these energies remains a mystery, as does the means of their acceleration.

In parallel, progress in the understanding of particle physics has suggested that the missing matter in the universe may consist of as yet undiscovered elementary particles, which are relics of the very earliest phases of the formation of the universe. It also appears that quantum fluctuations and topological singularities generated in phase transitions occurring at very high temperature in the early universe could have played a fundamental role in the formation of large scale structure. In addition, it is now well understood how the properties of neutrinos could be responsible for the solar neutrino puzzle, and the powerful acceleration mechanisms evidenced by the highest energy cosmic rays may require new particle physics.

The discoveries and activities described above have been mostly carried out by an unconventional breed of "astronomers" whose backgrounds have been in particle or nuclear physics. The nature of the research ranges from solid experiments with well defined systematic goals to investigations which test speculative ideas or follow on experimental hints. Therefore, such a field is a vital and exciting one, with new ideas, new practitioners, and the certainty of scientific progress. It may well be that the cosmos is providing us with the first evidences for physics beyond the standard models of astrophysics and particle physics.

In this new field, which we may call particle astrophysics, we can distinguish four interrelated areas dealing respectively with cosmology and particle physics, stellar physics and particles, high energy gamma and neutrino astrophysics ($>10^{12}$ eV) and cosmic ray astrophysics.

We review first the essential scientific questions being tackled and the present experimental program, before turning to priorities and institutional questions.

Cosmology and Particle Physics

The physics of the early universe is intimately related to particle physics at the very highest energies and it is not possible to distinguish them in the quest for the answer to the fundamental questions of cosmology: What is the nature of the ubiquitous dark matter? What is the origin of the predominance of matter over antimatter? What is the explanation for the smoothness, flatness, and old age of the universe? What is the origin of the primeval inhomogeneities that triggered the formation of structure and eventually galaxies in the universe? Conversely, the cosmological observations provide essential constraints in the construction of unified theories of particle interactions and may be the only source of information on physics at the very highest energies (up to the Planck scale - 10^{19} GeV). Physics at these energies is difficult to probe in terrestrial laboratories and so, at the same time the early universe provides a natural laboratory in which physics at the most fundamental level can be studied.

Particle Physics and the Early Universe

The past decade has seen the consolidation of two standard models: the SU(3)xSU(2)xU(1) gauge theory of particle interactions and the Hot Big Bang model. The former provides a fundamental theory of the elementary particles and their interactions at distances down to 10^{-17} cm (energies up to 10^3 GeV), while the latter provides an accurate accounting of the history of the Universe from about 10^{-2} sec after the origin of the universe. Encouraged by these impressive successes, particle physicists and cosmologists have begun to extrapolate to earlier times and attempted to answer the fundamental questions outlined above. The origin of the matter-antimatter asymmetry seems to involve forces that violate both CP and baryon number conservation. At very early times (10^{-34} sec?) phase transitions may have played an important role. For instance, inflation provides a very attractive explanation for the flatness and old age of the universe and of the primeval density inhomogeneities. In addition, many theories that go beyond the particle physics standard model, such as supersymmetry, predict the existence of stable relic elementary particles that are left over from the early moments after the Big Bang and that may constitute dark matter. Combining such cold dark matter (which could also be made of condensed astrophysical objects or primordial black holes) with the Harrison-Zel'dovitch spectrum of adiabatic density fluctuations predicted by the simplest models of inflation, it has been possible to obtain a fair first approximation to a theory of the formation of galaxies and clusters of galaxies. Alternatively, it is possible that cosmic strings or other topological defects associated with an early phase transition or with inflation, may be at least partly responsible for structure formation; in this case, there are indications that hot dark matter (e.g. relativistic particles) rather than cold dark matter may produce structures like the ones we see.

After more than a decade of intense theoretical work which has produced these fascinating ideas, the time is ripe for strengthened experimentation and observation. The main problems are easily identified:

- *Determination of the basic cosmological parameters.* We are still lacking a definitive determination of the Hubble parameter which enters in most cosmology calculations. A reliable measurement of the ratio Ω of the average universe density to the critical density, for instance by probing the geometry of the universe on large scale, is central for solving the problem of dark matter. The age of the universe is still uncertain. The combination of these three parameters would allow us to determine the spatial curvature of the universe (testing therefore the inflation paradigm) and the value of the cosmological constant, the small value of which remains a mystery.

- *Measurement of primordial abundances*: This will allow us to test in more detail the standard Hot Big Bang model of primordial nucleosynthesis and determine the average density of ordinary (baryonic) matter.

- *Study of diffuse backgrounds*: The primordial fluctuations that presumably lead to the formation of structure in the universe must be reflected in small anisotropies of the 2.7 K cosmic microwave background. When these are finally detected, they will provide a firm foundation for theories of structure formation and will point back to the processes that gave rise to these fluctuations in the early universe. The X ray and γ ray diffuse backgrounds are not yet fully understood but are potentially of cosmological origin. Timing measurements of the millisecond pulsars place the most stringent bounds on the density of the stochastic gravitational wave background and help to constrain models such as cosmic strings.

- *Mapping of the universe.* The systematic measurement of the galaxy density and velocity fields on increasingly larger scales will help to reveal the initial conditions and basic evolutionary processes in the universe.

- *Search for dark matter.* The nature of dark matter remains a mystery but begins to be accessible to observations: condensed astrophysical baryonic objects may be detectable by micro-lensing, relic particles by direct detection via elastic scattering in the laboratory and primordial black holes by the gamma ray bursts they should generate.

Although the needed observational evidence involves a wide variety of observational fields of astrophysics, most of which are not covered by this panel, it is important to stress the importance of all these observations in order to understand the role of particle physics in the early universe.

Dark Matter

Dark matter currently is probably the best example of the interpenetration of particle physics and cosmology. Based upon decades of astronomical and cosmological observations we are certain that most of the matter in the Universe is nonluminous and transparent. Various cosmological and astrophysical arguments suggest that the dark matter may not be ordinary matter (baryons), and that the most likely candidate is a relic elementary particle. The abundance of the candidate relics, their properties, and means of detecting them have been studied, and we are now ready to undertake the most important step: the experimental testing of the particle dark matter hypothesis.

Three types of dark matter particles are particularly well motivated: light neutrinos, axions and weakly interacting massive particles .

Light neutrinos of mass of about 30 eV would solve the dark matter problem, although they may complicate the explanation of the formation of large scale structure of the universe. Unfortunately, no experimentally viable method has yet been proposed to detect the cosmological neutrinos directly and we will require laboratory measurements of the neutrino mass for the three neutrino generations to fully test this hypothesis. The detection of a large number of neutrinos from distant supernovae could eventually provide interesting direct mass limits.

The axion, a very light (mass of about 10^{-3} eV to 10^{-6} eV) pseudoscalar particle still represents the best solution to a fundamental problem of the standard Particle Physics model (the strong CP problem of QCD). Unfortunately the axion interactions are expected to be very weak, not much stronger than gravitational, and it is a testimony to the talents of the experimentalist involved that the sensitivities of the first generation experiments employing resonant microwave cavities were within about a factor of 300 of that required. It may be possible to improve the sensitivity by this factor with large cavities in the lower mass region. The fundamental experimental problem remains that present detection schemes are narrow band, and, since the axion mass which leads to closure density is only known to a factor of about 100, scanning the entire region requires a long time and a large effort. Recent theoretical work suggests that the mass uncertainty may even be larger as radiation from global strings may be the dominant source of axions.

Weakly interacting massive particles (WIMPs) are another general class of candidates which correspond to the case where heavy dark matter particles were in thermal equilibrium with the rest of matter in the early universe. Their interaction cross sections can then be estimated from their current density and turn out to be of the order expected for "Weak Interactions" (in the technical sense). This coincidence may be purely accidental or may be a very precious hint that physics at the W and Z^0 scale (e.g., Supersymmetry) may also be responsible for dark matter! For instance the lightest supersymmetric particle may constitute dark matter: this particle is the neutralino, sometimes referred to as the photino or higgsino, which are special cases. In order to test this fairly general hypothesis, we could attempt to detect directly the interaction of halo particles with a laboratory target, in a way complementary to the new particle searches at accelerators (LEP, Tevatron and the SSC). This requires, unfortunately, very sensitive detectors with a good rejection of the radioactive background. Current detectors using ionization techniques have set interesting limits, for instance excluding the possibility that dark matter is made of heavy Dirac neutrinos (Fig. 1) and severely limiting the existence of cosmions which could simultaneously explain the deficit of ^{8}B solar neutrinos. But before the neutralino model can be probed, the rejection of backgrounds must be improved by two or three orders of magnitude. This factor could eventually be reached with emerging technologies based on the detection of phonons or quasiparticles in superconductors which should allow better energy sensitivity and much higher redundancy. Although not sufficient to establish the feasibility of a definitive experiment, the results obtained currently with cryogenic detectors of a few tens of grams are encouraging. It may also be possible to detect indirectly the presence of these dark matter particles in our galaxy, by their annihilation products: positrons, antiprotons, γ rays and neutrinos. Recent balloon measurements indeed indicate an excess of antiprotons and positrons but the interpretation in terms of dark matter annihilation products now seems less likely than some other possible explanations. Of these indirect methods, neutrinos originating from the sun may be the least model dependent.

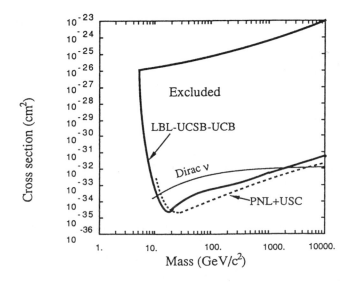

Constraints on the mass and the cross section of Dark Matter particles, obtained by the LBL-UCSB-UCB and the PNL-USC collaborations (as of December 1989).

Figure 1

Other Relics

Other relic particles may be significant in cosmology even if they do not constitute dark matter (recall that the cosmic microwave background contributes less that 10^{-4} of the critical density). The observation of relic cosmological neutrinos would of course be of exceptional interest but no good ideas for their detection have been proposed. Significant relics include superheavy magnetic monopoles, decaying neutrinos, a neutralino species, or decaying axions. Monopoles are one of the two fundamental new predictions of grand unified theories (the other being proton decay). They may be detectable by a large, football field-sized detector, MACRO, which is well on its way toward completion. It will, for the first time, probe the flux range below the "Parker bound", 10^{-15} /cm^2/sec/ster. Should relic monopoles be discovered they would both confirm the ideas of grand unification and probe the universe at an age of about 10^{-34} sec.

Stellar Physics and Particles

Neutrinos play an important role in the physics of stars and there again particle physics and astrophysics are intricately intertwined. Neutrinos are elementary particles with very small interaction cross sections which, along with photons and positrons, are emitted in the nuclear processes which power the stars. Photons diffuse out of the stars with time scales of the order of millions of years, losing all information about the reactions which created them. Most of the neutrinos traverse the interior of the star without interaction and thus in times measured in seconds. Therefore, they carry direct information about the processes which created them in the stellar interior. To date, neutrinos have been detected from a middle-aged star, namely the Sun, and from one star at the end of its life, supernova SN1987A.

The solar neutrino Chlorine experiment in the Homestake mine.

Figure 2

Solar neutrinos

It is believed that the processes which produce energy in the solar interior are sufficiently well understood that the rates of the different nuclear reactions contributing to this energy release can be calculated, provided the empirical inputs to the calculation are known sufficiently well. Thus, measurement of the flux and spectra of solar neutrinos should confirm in detail our understanding of processes in stellar interiors. The well documented and recently confirmed disagreement between the calculated solar neutrino flux and the measured flux indicates that processes in the solar interior may be different from our expectations, or that the propagation of the neutrinos from the interior of the sun to a terrestrial detector may be influenced by fundamental properties of the neutrinos themselves. If neutrinos have a mass, the observed deficit of high energy neutrinos may be due to the transformation of electron type neutrinos into another neutrino type either in the sun (matter oscillations), or during propagation to the earth (vacuum oscillations). Whatever the solution of the puzzle, an imperfect modeling of the sun or fundamental properties of the neutrinos, it will be important for astrophysics, cosmology and particle physics.

The original solar neutrino experiment (Fig 2) uses a subterranean radiochemical detector in which neutrinos above a threshold of 814 keV transform ^{37}Cl nuclei into ^{37}Ar nuclei through inverse beta decay. It is mainly sensitive to the neutrinos resulting from the decay of ^{8}B, an end product of one of the reaction chains in the sun. Over a twenty year exposure, the average response of this detector is less than one half of the best calculated lower limit of the flux. This deficit of high energy solar neutrinos has recently been confirmed by the Kamiokande II experiment operated by a Japanese - U.S. collaboration. Their technique is completely different and is based on the detection of the recoil electrons scattered elastically by the ^{8}B neutrinos in a large water Cerenkov detector. The electron direction is kinematically correlated with the neutrino direction, so that the angular distribution of electrons with respect to the direction from the sun shows, for the first time, that the sun is the source of the neutrinos (Fig 3).

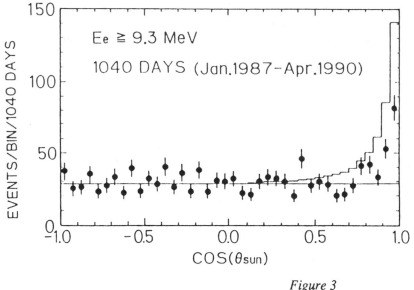

Angular distribution with respect to the sun of electrons produced by solar neutrinos in Kamiokande II.

Figure 3

As 1990 begins, the ^{37}Cl detector has observed an apparent time dependence in the detection rate which appears to be anticorrelated with sunspot number, or solar activity. Since solar activity is a surface phenomenon, and the neutrinos are born in fusion reactions deep in the solar interior, a connection between the two phenomena would be surprising. Such a flux variation is not confirmed by the Kamiokande II experiment but the small counting rates in both experiments precludes any definitive conclusion.

Because of their relatively high threshold, neither the ^{37}Cl experiment nor the Kamiokande II experiment are sensitive to the copious neutrinos from the main pp reaction chain which is responsible for most of the energy production of the sun. It is clear that it is advantageous to detect those lower energy neutrinos since the uncertainties associated with the calculation of their flux are much smaller than for the ^{8}B neutrinos. The necessary technology involving a radiochemical method using inverse beta decay on a ^{71}Ga target was developed in the United States in the early 1970's. Unfortunately, partially because the funding structure was not flexible (we will come back to the problem of particle astrophysics experiments falling between the cracks in section 4), funding could not be secured for such an experiment in the United States. Gallium experiments have been implemented by the Soviet Union in the Baksan Laboratory and by Europe at the Gran Sasso Laboratory in Italy. Both experiments, which will produce results during the early 1990's, are international collaborations, with a significant participation by United States scientists, especially in the Baksan experiment. The latter experiment recently presented tantalizing preliminary results, possibly indicating an even larger neutrino deficit at low energy. It is too early to draw any firm conclusion but it is clear that these gallium experiments with their sensitivity to the main energy generation mechanism in the sun are a critical component of the total solar neutrino program.

A second generation experiment dedicated to the measurement of the ^{8}B neutrino flux has just been approved. The Sudbury Neutrino Observatory collaboration (Canada, U.K., and U.S.) plans to use for this purpose a kiloton of heavy water, D_2O, as a solar neutrino target in a Cerenkov detector. In addition to measuring the elastic scattering of neutrinos by atomic electrons as in the Kamiokande II experiment, this experiment will measure the energy spectrum of electron type neutrinos through the inverse beta decay of deuterium, and the total flux of neutrinos, independent of type, through the neutral current disintegration of deuterium by neutrinos above the 2.2 MeV reaction threshold. This neutral current capability will be essential if it turns out that the deficit of solar neutrinos is due to oscillation phenomena. While the electron neutrino component may be depressed, the total flux of neutrinos should be constant. Furthermore, the counting rate in this experiment will be an order of magnitude higher than in the ^{37}Cl experiment, an important advantage when trying to calibrate out systematics or to understand the time dependence in the signal.

The solar neutrino program appears therefore to be well directed. The results of the ^{37}Cl and ^{71}Ga experiments, and the findings of the Sudbury Neutrino Observatory will very likely clarify the overall situation within a few years. Follow up experiments may then become necessary to confirm the insight thus gained. If the anti-correlation with the solar activity is confirmed, high rate experiments based for instance on ^{127}I may become necessary. The observation of a low rate in the gallium experiments would focus the attention on the low (pp) and intermediate (^{7}Be) energy region, and the need for a spectrum measurement, for instance with ultra-low radioactive background scintillators, possibly loaded with ^{115}In or ^{11}B, or cryogenic methods in ^{4}He.

Supernovae

The most significant event in establishing neutrino astronomy was the observation of a burst of neutrinos from supernova SN1987A in coincidence in the Kamiokande II and IMB water Cerenkov detectors in February 1987. Because the supernova was in the Large Magellanic Cloud at a distance of 52 kpc, smaller scintillation detectors in the U.S., Europe, and the Soviet Union did not see it. The observation of a flux compatible with the expected energy release provided an impressive confirmation of our understanding of the basic supernova mechanisms.

Should a supernova occur in our galaxy, the existing detectors and detectors under construction will be able to detect orders of magnitude more events. This will allow both important studies of the emission of neutrinos from the collapsing core including the time development and energy spectrum of electron, mu and tau type neutrinos and antineutrinos and detailed measurements of properties of neutrinos. It may be, for instance, possible to achieve a mass limit of 100 eV for the mu and tau type neutrinos. The array of detectors capable of detecting signals from a galactic supernova will increase during the 1990's. These detectors include solar neutrino detectors, dedicated supernova search detectors, and deep underground scintillation detectors searching for magnetic monopoles and other "cosmic ray" phenomena. Given the importance of this information for refining current understanding of the mechanism of stellar core collapse and of the neutrino sector, all such detectors should be instrumented to maximize supernova detection capability including burst handling and absolute timing.

Unconventional Particle Physics and Stellar Physics

In addition to the neutrinos, there may be other weakly interacting particles (e.g. axions or neutralinos). However these additional hypothetical particles may perturb supernova explosions and the stellar energy generation mechanisms. Their properties have recently been severely constrained by the SN1987A observations and by "conventional" stellar physics (e.g. population of the Hertzprung-Russel diagram and lifetimes of stars). We expect these cross-disciplinary analyses to expand in the coming decade. More generally, many particle physics theories have implications for astrophysics. For example, if strange matter is the ground state of matter at high density and/or atomic number, then some (or all) neutron stars may be strange stars: This would have important implications for astrophysics which in turn will provide one of the few probes of the strange matter hypothesis.

High Energy Gamma Ray and Neutrino Astronomy

When astronomers developed techniques which extended observations of the electromagnetic spectrum beyond the optical range, many new discoveries were made. It is natural to want to extend observations of astrophysical objects in as broad a wavelength range as possible. Thus, at the upper end of the electromagnetic spectrum, space-borne instruments have been used to observe gamma rays up to about 10^9 eV. Higher energy observations are difficult because of rapidly falling fluxes. The Gamma Ray Observatory, GRO, which will soon be launched, will extend space-borne observations to their practical limit, a few 10^{10} eV.

At about 10^{11} eV, gamma rays begin to produce showers in the earth's atmosphere which can be detected by ground-based detectors. In addition, it is possible to build detectors on earth that have large collection areas and good sensitivity for very low fluxes. Attempts to observe astrophysical objects by detection of air showers have been made for more than twenty years. For the most part, these efforts have been an offshoot of cosmic ray studies. In the energy range 10^{11} - 10^{13} eV, ground-based optical detectors can observe the Cerenkov light produced by the shower in the upper atmosphere. In the U.S., such detectors have been developed at the Whipple observatory, Haleakala and Albuquerque. For energies greater than 5×10^{12} eV, a sufficient number of shower particles reaches the ground so that they can also be directly observed by particle detectors. Sparse arrays of such detectors at moderate altitude are typically sensitive to shower above 5×10^{13} eV.

However, both techniques, the atmospheric Cerenkov technique, and the air shower technique, have a problem with severe backgrounds from ordinary cosmic rays, which arrive isotropically. The ordinary extensive air showers are initiated in the upper atmosphere by protons and heavier nuclei, but the air showers they produce are not easily distinguishable from showers produced by the gamma rays being sought. The technique which can separate the signal from the background has primarily been the identification of an excess of showers from a given source direction. If the source is identified to have a well-established periodic modulation in another wavelength region, one can search for the same periodicity in the gamma ray signal. Other methods to suppress cosmic ray background depend on the technique. For the atmospheric Cerenkov technique, the Cerenkov light is imaged on the focal plane of a mirror which follows the source. The Cerenkov image is quite different for a gamma ray shower from that of a hadronic shower. For the air shower technique, one exploits the fact that a gamma ray shower is expected to have at least twenty times fewer muons accompanying it than a proton-induced shower of similar size. The separation techniques are statistical in nature and encounter all the dangers of statistical fluctuations when the signals are weak. Only recently has a steady flow of gamma rays in the energy range $(0.3 - 2.0) \times 10^{12}$ eV been convincingly observed from the Crab nebula with the 10 meter telescope at the Whipple Observatory. Figure 4 shows a view of the telescope. The imaging algorithms used improved the signal to background ratio by a factor of 50.

The 10 meter Air Cerenkov detector of the Whipple Observatory.

Figure 4

Air shower arrays have reported point sources of showers at energies >10^{14} eV from X-ray binaries such as Cygnus X-3 and Hercules X-1. A statistically significant observation of Cygnus X-3 over a four-year period was reported by the University of Kiel in 1983 and confirmed by Haverah Park, and the CYGNUS experiment at Los Alamos has reported a short burst from Hercules X-1 in 1987. Both results showed an excess of showers from the source as well as a periodicity that was characteristic of the source. In the case of Hercules X-1, the period is slightly different from that of X-rays. The most curious aspect of both observations was the fact that the muons accompanying the shower were not reduced as would be expected for a gamma-induced shower. Imaging of the Cerenkov light from lower energy showers coming apparently from Hercules X-1 and showing the same slightly displaced period, leads to the same puzzling conclusion that showers seem hadronic. Such observations are inconsistent with the expectations of particle physics for γ rays and, if confirmed, would require the existence of new, light, and strongly interacting particles or of dramatic new interaction thresholds. Furthermore, these results imply new compact acceleration mechanisms at the source. While radiation at 10^{12} eV from the Crab nebula can be understood with conventional astrophysical mechanisms, radiation at 10^{14} eV from compact objects, if it is gamma radiation, must come from neutral pions produced at the source. Given the exciting consequences which follow from the air shower observations, there are many groups throughout the world which are engaged in the search for point sources emitting radiation at energies greater than 10^{14} eV. So far there have been no compelling confirmations of the Cygnus X-3 or Hercules X-1 results, a consequence, perhaps, of source variability. In the United States, two major installations at Los Alamos and Dugway, Utah are now starting operation. The expanded CYGNUS array is operating with 200 detectors covering an area of 0.8×10^5 m^2. By the end of 1990 an array built by the Universities of Utah, Michigan, and Chicago will be operating with 1089 detectors covering 2.3×10^5 m^2. These two installations are located at similar longitudes so that even short bursts of an object will observable by both detectors. Figure 5 shows a view of the CYGNUS array. Whether a genuine astronomy will develop at energies greater than 10^{14} eV depends on confirmation of the results discussed above. Detectors of sufficient sensitivity exist and it is a matter of time and patience before it will be known whether new instruments are required.

The Cygnus extensive air shower detectors. The photo shows the LAMPF accelerator in the background with the beam stop area in the middle right. The CYGNUS counters can be seen on the roof of the beam-stop building and deployed in the surrounding area. Several counters of the expanded array can be seen in the immediate foreground.

Figure 5

On the other hand, astronomy at 10^{12} eV is truly beginning. At present, a second 10m telescope (GRANITE) is being constructed at the Whipple Observatory, which will greatly improve the sensitivity so that sources other than the Crab are likely to be seen. There are new technologies using large coverage or tracking detectors whereby the threshold of the air shower technique may be lowered to nearly overlap the atmospheric Cerenkov technique, especially if the detector is located at high altitude. One of these techniques involves the instrumentation of more than 10^4 m^2 of a clear lake with photomultipliers so that the observation of an entire air shower by Cerenkov light in the water is possible. The advantage of such a method is that its operation is not restricted to dark moonless nights.

It should be remarked that the American teams are currently mostly involved in arrays located in the northern hemisphere. Since most of the galaxy is visible only at southern latitude, it is likely that the need for southern instruments will arise. Because of its altitude and the fact that the same patch of the sky is always visible, the South pole may be particularly interesting. Currently a modest American-British array, SPASE, is pioneering this approach.

High Energy Neutrinos

High energy neutrinos may provide another window onto acceleration mechanisms in the astrophysical environment, and if observed from point sources, would give a clear proof that hadronic processes play an important role. Such neutrinos are best detected by their production of upward-going muons. The neutrinos that pass through the earth may produce muons in the ground just below a large detector. One possibility is a water Cerenkov detector installed at the bottom of the ocean. Such experiments will have negligible background but the signal may be very small. The Dumand experiment recently approved by DOE represents a first attempt at exploring this virgin territory at a sensitivity sufficient to observe a few neutrinos from Cygnus X-3 if the continuous gamma fluxes are at the level reported by Kiel. Alternative techniques using the lake Cerenkov concept or the polar ice (as a Cerenkov light or microwave radiating medium) have been proposed.

If the existence of point sources radiating 10^{14} eV gamma rays is confirmed, then this effort will become even more interesting. If a compact source produces neutral pions, it must produce charged pions as well. Decays of these pions will produce a powerful source of neutrinos. Depending on the details of the source, the flux of these neutrinos can greatly exceed the gamma ray flux since the neutrino absorption cross section is so small. Observation of both neutrinos and gamma rays from a source would provide important information about the mechanism of the compact accelerator.

It would seem prudent, however, to await either observation by DUMAND (or possibly MACRO or the LVD) of neutrinos from point sources or results from the Dugway and Los Alamos gamma ray detectors before proceeding with neutrino detectors of larger scale.

Cosmic Rays

Cosmic rays which can be observed near Earth cover a large range of energies, from the MeV region to 10^{20} eV, and comprise the nuclei of all known elements, as well as electrons, positrons, and antiprotons. The study of this tenuous plasma of relativistic particles addresses scientific questions closely related to the themes discussed above: high energy particle acceleration, stellar and galactic astrophysics, and cosmology and particle physics. However, contrary to the observational techniques reviewed thus far, cosmic ray measurements up to energies around 10^{14} eV should be conducted in space or on high altitude balloons. It is for that reason that cosmic ray research in space is summarized separately in the following. We conclude the discussion with an account of the present status of air shower measurements from the ground which cover the range 10^{14}-10^{20}eV.

Cosmic Rays from Space

Particle acceleration is perhaps the most immediate scientific theme of cosmic ray studies. It is ubiquitous in nature, occurring in solar and stellar flares, interplanetary and interstellar shocks, pulsars, supernova explosions, and perhaps in shocks on the scale of entire galaxies. A major goal is the understanding of these cosmic accelerators and the determination of their energy sources and the physical mechanisms by which they are able to efficiently extract a small number of highly energetic particles from a nearly thermal distribution. The main observational tools available are studies of the energy spectra of the individual cosmic ray components over a wide energy range. Important progress was made during the 1980's by extending the energy spectra of major nuclei with measurements on balloons and from the Space Shuttle up to energies around 10^{13}-10^{14}eV/particle. The persistence of the power-law spectra of cosmic ray primaries over a large energy range lends support to the assumption that first-order Fermi acceleration in supernova-driven shocks acts as the prevailing acceleration mechanism. The spectral measurements also provided indications that during their propagation through the galaxy, the cosmic rays become enriched in heavy elements relative to protons at energies above approximatively 10 TeV. Significant progress in understanding the mechanism by which particles are selected for acceleration resulted from the discovery that cosmic rays undergo fractionation which is strongly correlated with the particle first ionization potentials. The great similarity of this fractionation to that occurring during the separation of the solar corona from the photosphere (Fig. 6) suggests stellar coronae may provide the pool of hot matter from which cosmic rays are extracted.

The origin and evolution of matter in the Galaxy is the major theme addressed by studies of the elemental and isotopic composition of cosmic rays. Since cosmic rays are much younger ($\sim 10^7$ yr, as deduced from the abundances of radioactive nuclei such as ^{10}Be and ^{26}Al) than the galaxy and the solar system ($>10^9$ yr), the comparison between cosmic ray and solar system abundances elucidates the chemical evolution of the galaxy due to ongoing stellar nucleosynthesis processes. Measurements on balloons, followed by more precise data from spacecraft, have led to the first high resolution observations of the isotopic composition of heavy primary cosmic ray nuclei, and to the discovery of excesses of the neutron rich isotopes of Ne, Mg, and possibly Si (relative to solar system composition), indicating differences in the nucleosynthesis history of these two samples of matter. The first reliable measurements of abundances of ultraheavy nuclei above the iron group ($Z \geq 30$) demonstrated that both slow and rapid neutron capture nucleosynthesis contribute to these elements, and that the heaviest nuclei are not produced solely by recent explosive nucleosynthesis as might occur if cosmic rays were both synthesized and accelerated in a supernova.

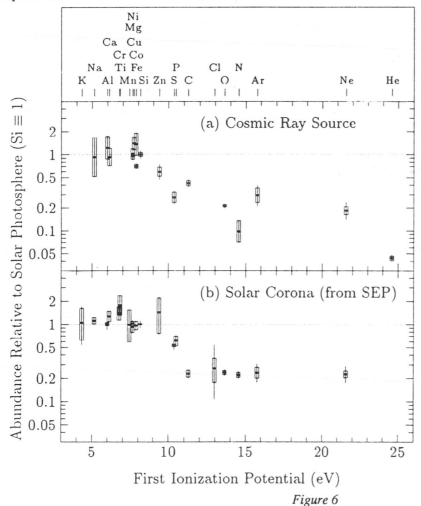

Heavy element abundances derived from energetic particle measurements are plotted relative to abundances in the solar photosphere (with the Si ratio defined to be 1), as a function of the element's first ionization potential (FIP). Panel (a) shows the composition of the galactic cosmic ray source, while panel (b) shows the compostion of the solar corona, as derived from measurements of solar energetic particles (SEP).

Figure 6

Observational tests of cosmology constitute another active area for cosmic ray investigations. For instance, the determination of the contribution of cosmic ray-induced spallation reactions to the production of light isotopes, particularly ^{7}Li, is essential for deriving the cosmogenic yields of these species and establishing the baryonic density in the universe. In the search for antiparticles in the cosmic rays, recent balloon measurements have placed stringent limits on possible contributions of annihilations of supersymmetric dark matter particles to cosmic ray antiprotons at low energies. However, in the GeV region, there are indications of fluxes of p̄'s and positrons that are larger than those expected to arise from interstellar nuclear interactions of cosmic rays. It remains to be determined whether these excesses are indicative of a cosmological source, or are due to peculiar phenomena in the galaxy, for instance $e^{\pm}$ pair production in the magnetospheres of pulsars.

The interactions of cosmic rays in interstellar and interplanetary space are probed by particles of the lowest energies, below a GeV/nucleon. Particle acceleration can be studied in the heliosphere in much more detail than in more remote regions of the galaxy, and will provide invaluable tests for theoretical models that are applicable at larger scales and higher energies. In order to establish the role of low energy cosmic rays in the heating and ionization of the interstellar medium, *in situ* measurement of these particles will be required in local interstellar space, as pointed out in the 1980 Astronomy and Astrophysics Survey Report (Field Report). An "Interstellar Probe" mission to carry out such measurements remains a high priority objective in this field, but will not be discussed in detail here since it falls outside the scope of the present study.

This brief discussion illustrates the importance of cosmic ray observations in space for the understanding of a variety of astrophysical phenomena relevant to fields ranging from cosmology to radio and gamma ray astronomy. In the coming decade, major advances are expected in a number of areas:

Antimatter studies. High precision observations of antiprotons and positrons over a wide range of energies with magnetic spectrometers on balloons and in Space ("Astromag") should make it possible to determine conclusively whether there is a significant contribution of these particles from a source other than interstellar interactions of cosmic ray protons. This will lead to improved limits on the production of $\bar{p}$ and e^{+}'s by the annihilations of candidate dark matter particles. In addition, the sensitivity of searches for heavy antinuclei should be improved by 2 to 3 orders of magnitude.

Isotopic composition studies. The abundances of essentially all stable and long lived isotopes of elements with $Z \leq 30$ should be measured, and exploratory isotope observations should be extended up to $Z \approx 40$. These data should clarify the pattern of isotopic anomalies, making it possible to determine the dominant nucleosynthesis process contributing to the production of cosmic ray source material. The observations made in particular with the Advanced Composition Explorer should establish the time between nucleosynthesis and acceleration, using the abundances of primary electron capture nuclides such as ^{57}Co. With the Astromag magnetic spectrometer, isotopic separation will be extended to GeV energies, making it possible, for instance, to use radioactive "clock" isotopes in a regime where their half lives are increased by relativistic time dilation.

Ultraheavy element studies. Abundances of individual elements with $Z > 50$ should be measured with passive track detectors (by the Heavy Nucleus Collector) with good statistical accuracy and resolution sufficient to separate adjacent elements. These will make possible studies of neutron capture nucleosynthesis and determination of the relative importance of steady state and explosive processes in the production of cosmic ray source material.

Studies of high energy spectra and composition. Direct measurements of the spectra of major elements at high energies can be achieved through exposure of large instruments in space. By exposing detectors of several m^2 ster in space for at least a year, direct composition measurements can be extended to 10^{14}-10^{15}eV/particle. Such measurements will not only test models for particle acceleration at high energies, but will also provide an essential overlap and calibration for the indirect measurements by ground based air shower arrays that provide information up to the highest particle energies known in nature. This could be accomplished by a one-year flight of the transition radiation detector system that was successfully tested on the shuttle (Spacelab-2) in 1985, or by new instrumentation to be developed for attachment to the Space Station.

Studies of interaction with the solar system. A number of space missions selected for flight in the 1990's will measure the nuclear composition and atomic charge states of solar energetic particles and anomalous cosmic rays. These should significantly improve our knowledge of the composition of the solar surface, and should conclusively establish the origin of the anomalous component. Information on the three dimensional structure of the heliosphere and its role in the modulation of galactic cosmic rays will be greatly improved through particle observations at high heliographic latitude by the Ulysses mission, and by ongoing studies using the Pioneer and Voyager space probes in the outer heliosphere. Investigation of differences in the modulation of particles of opposite charge sign, particularly electrons and positrons, will also contribute to understanding the physical processes responsible for the solar modulation of cosmic rays.

Cosmic Rays - Ground Observations

While cosmic rays with energies below about 10^{14} eV/particle may be best studied by experiments in space, the low flux at greater energies limits studies to the observation of the showers they produce in the atmosphere.

Their energy spectrum is rather well known, falling from $10^{-4}/m^2/sec$ for energies greater than 10^{14} eV to ~1/Km²/year for energies above 10^{19} eV.

The highest energy region is particularly interesting. The mere existence of cosmic rays up to 10^{20} eV is rather surprising and there is at present no understanding of a process which could accelerate or produce directly particles of such energy. Moreover, at energies greater than 10^{19} eV there are sufficiently distinctive phenomena which should permit the identification of the source of the radiation (galactic or extragalactic) and the identity of the radiation, protons, heavy nuclei, or gamma rays. The radius of curvature of a proton of energy greater than 10^{19} eV in the galactic magnetic field is larger than 10 Kpc, which is comparable to the size of galactic disk. Heavier nuclei will have correspondingly smaller radii of curvature. In addition, if the cosmic rays at this energy consist of extragalactic protons, their spectrum should have a sharp cutoff at 10^{20} eV because of photon-pion production on the 2.7 K background radiation (Greisen cutoff). Below this cutoff, the spectrum is expected to flatten because of the pile-up of particles produced in these interactions.

The Fly's Eye experiment of the University of Utah has collected the largest number of events (200) with energy greater than 10^{19} eV. Their technique of observing the cosmic rays by their fluorescence in the atmosphere allows many details of the induced shower to be observed. The direction, energy, and longitudinal development of the shower can be measured. The longitudinal development permits incident gamma rays, protons, and heavier nuclei to be distinguished on a statistical basis. The energy spectrum of the upper end of the spectrum is shown in Fig. 7. It suggests that there is a flattening of the spectrum above 10^{19} eV and a possible cutoff based on the absence of events at energy higher than 10^{20} eV. These cosmic rays arrive isotropically and the longitudinal development suggests that they are protons. Thus, there is some evidence that these cosmic rays come from outside our galaxy. Other groups using conventional extensive air shower techniques dispute these findings but their sample is also small and their energy resolution worse.

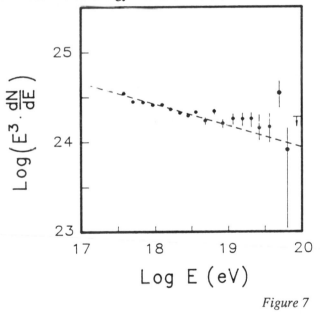

Differential energy distribution multiplied by the cube of the energy, for the highest energy cosmic rays as observed by the Fly's Eye experiment.

Figure 7

A new proposal to build a Fly's Eye with ten times the sensitivity has been made to the NSF. This instrument would have much finer resolution for the measurement of the longitudinal development of the shower. Measurement of the depth of maximum of the shower and its longitudinal extent would permit a clear separation between incident protons, heavy nuclei such as iron, and gamma rays. Gamma rays with energy greater than 5 x 10^{19} eV could be observed by the Landau, Migdal, Pomeranchuk effect whereby at these high energies the mean free path for pair production and Bremsstrahlung is greatly increased. Thus, by a combination of the measurement of the spectrum, composition and anisotropy it would be possible to establish the galactic or extra-galactic nature of these cosmic rays. The ability to collect over several years 2000 events with energy greater than 10^{19} eV would make this determination possible.

Cosmic rays in the 10^{15}-10^{19} eV region also present mysteries. They may be galactic in origin but there is no established galactic acceleration mechanisms that can produce protons with energies greater than about 10^{15}

eV. The existing air shower arrays referred to in the gamma ray section, or the combination of surface arrays with underground detectors should be able to make crude composition measurements which will help answer these questions. As mentioned above, cross calibration with space instruments will be essential. One would also like to know whether the galactic cosmic rays are produced in localized compact sources, or are accelerated slowly over much larger regions in the galaxy. The observation of discrete gamma ray sources at energy greater than 10^{14} eV may resolve this question, since a discrete source implies the production of neutral pions by considerably higher energy primary particles.

Highest Scientific Priorities

From the discussion above, we can extract the most likely themes in the next decade:
• The nature of dark matter and the possibility that it is made of nonbaryonic particles produced in the early universe. More generally, the imprints left by physics at ultrahigh energy on the universe are fundamental to the understanding of both cosmology and particle physics.
• The solar neutrino problem and the use of neutrinos to understand supernovae. In the process, stellar models may be refined and the intrinsic properties of neutrinos will also be better delineated.
• The nature of particle acceleration mechanisms, especially if gamma rays are indeed observed as high as 10^{14} eV. Confirmation of a muon anomaly would require major revisions of particle physics. In any case, detection of gamma rays in the 10^{12} eV region will complement observations at longer wave length.
• The origin of cosmic rays, their composition, the nature of their acceleration mechanisms and the possibility that they may partially be extragalactic.
We can be assured that other concepts and fascinating questions will be generated along the way, but these problems are useful tools to define the highest scientific priorities.

Implementation of the current program

The highest priority is to implement and strengthen the current program. As explained above, very important experiments are coming on line or have just been approved which will shed critical light on the scientific problems tackled, while small developments explore the feasibility of new techniques. It is to the credit of the funding agencies and to the community that this represents a good first approximation for a balanced program and *it is critical that these experiments and developments be vigorously supported at the fastest rate feasible.* Concurrently a *strong theoretical effort* should be continued. In order to expand on these points, we reiterate the fields considered above:

Particles and Cosmology

• Although they may not yet have the required sensitivity to probe the entire available parameter space, first generation experiments attempting to detect dark matter particles (axions and weakly interacting massive particles) are essential to explore an already large class of models, develop the technology and master the experimental problems. The support of the present low cost efforts both by NSF and DOE should be continued and expanded. For axions, the current technological development should be aggressively pursued and a very interesting next step would be to extend the sensitivity of the search to the low mass region. It is also important to continue the weakly interacting massive particle searches with improved ionization detectors (possibly with isotopically enriched materials to improve the sensitivity to Majorana particles) and eventually with the transformation of these set ups into pilot experiments using cryogenic detectors of a few tens of grams. Indirect searches for dark matter particles with existing neutrino detectors [mostly DOE] and the antiproton, positron and gamma ray observations [NASA] should also be actively pursued.
• MACRO [supported by DOE] is expected to reach an important milestone in the search for monopoles with a sensitivity below the Parker bound and to improve significantly our searches for high energy neutrinos potentially produced by annihilation of dark matter particles in the sun or the earth. More generally, its large volume and area allows this instrument to detect supernova neutrinos and high energy muons from cosmic rays. Similarly the LVD [supported by NSF and DOE] will begin in late 1991 to contribute to these studies.
• In the past ten years, theory has led the way in establishing the bridge between particle physics and cosmology. It is critical for this effort to be expanded at the time when experiments begin to be implemented.

Particles and Stellar Physics

• The highest priority for the United States solar neutrino program [DOE] is the participation in the Sudbury Neutrino Observatory (SNO). This experiment will produce a high counting rate study of ^{8}B solar neutrinos, including the energy spectrum of the electron type neutrinos and a measurement of the total flux of neutrinos independent of their type.

• Participation of the US teams in the gallium experiments [DOE] is important, since their results will determine a portion of the overall picture of the solar neutrino problem.

• It is important to keep the ^{37}Cl experiment running over the long term, until a larger ^{37}Cl detector planned by the Soviet Union becomes operational.

• It is critical to organize a well coordinated supernova watch. The detectors currently running or being constructed [Kamiokande, IMB, MACRO, and SNO all funded by DOE, and LVD funded by NSF and DOE] have a mass well suited for the detection of neutrinos from supernovae in our own galaxy. It is important to maintain IMB in operation at least until MACRO fully comes on line.

• Support of the theoretical efforts in understanding the sun and the supernova explosions are obviously an important complement of this experimental program.

High Energy Gamma and Neutrino Astrophysics

• The installation at the Whipple Observation of a second atmospheric Cerenkov detector [GRANITE supported by DOE] is an important step in the observation of gamma rays at 10^{12} eV. This region is particularly interesting, with the only unambiguous observation of a steady flux from a high energy source, the Crab nebula, and the expectation that many other sources may be observable in this energy region where the flux is expected to be larger than at higher energies. Moreover such studies will complement nicely the observations in space made by the Gamma Ray Observatory, which will be sensitive up to 10^{10} eV.

• Two large extensive air shower arrays are currently being put in operation: the Chicago-Michigan-Utah array [NSF-DOE] and the expanded Cygnus array in Los Alamos [NSF and Los Alamos (DOE) discretionary funds]. They will be critical in clarifying the observational situation in the 10^{14} eV region: existence of pulsed sources and muon content anomaly. Since they are located at approximately the same longitude, the two experiments will be able to cross check each other.

• DUMAND II is a useful first exploration of the virgin field of high energy neutrino astrophysics and an important pilot implementation of a technology which, if successful, could be extended to a larger size, if this is required for the study of neutrino sources.

Cosmic Rays

High priority should be given to carry out expeditiously the space missions recently selected for cosmic ray investigations:

• The Advanced Composition Explorer (ACE) will measure isotopic, elemental and charge state composition over six decades of energy (0.5 keV/nucleon to 0.5 GeV/nucleon).

• The Astromag facility selected for the Space Station Freedom, will be a unique facility with a large magnetic spectrometer which should allow major advances in the study of cosmic antimatter components and isotopic composition over a wide range of energies.

• The Heavy Nucleus Collector (HNC), another Space Station payload, will make the first high resolution determinations of the abundances of individual ultra-heavy elements with $Z \geq 50$.

• The POEMs experiment on the Earth Observation System, will measure spectra of electrons and positrons below 1 GeV, and determine the relative contributions of primary and secondary production.

These space observations have to be complemented at the highest energy by observations from the ground:

• Composition studies in the 10^{14} - 10^{16} eV region are natural by-products of the gamma ray experiments mentioned above and of the MACRO [DOE], LVD [NSF-DOE], Homestake and Soudan 2 [DOE] underground experiments which are operated in connection with surface arrays.

• In addition, important new information is being collected at very high energy by the Fly's Eye detector [NSF] which has been a recognized experimental success and should continue to be supported.

New Initiatives in the coming decade

The program described above is well balanced with a number of new experiments. However, the scientifically essential *supernova watch* requires a better coordination, and the proper exploration of the *highest energy cosmic rays* and the elucidation of the old puzzle of their origin requires a new instrument. In addition it is clear that the experiments currently implemented and the present technological developments will lead to major new initiatives, the character of which will greatly depend on the results obtained in the intervening time. In charting the future of particle astrophysics it is then essential to take into account *decision points* that we see occurring naturally in a few years when the following information is obtained: feasibility of cryogenic detection of dark matter particles, nature of the solar neutrino problem, existence of sources at 10^{14} eV. This is summarized in table I.

Table I

Potential New Initiatives

Potential New Initiative	When	Cost ($M)	Information Necessary	Technological Developments
Supernova Neutrino Watch	Now	1/yr		Long term maintenance
10^{20} eV cosmic ray facility	Now	15		High resolution Fly's Eye
Full Size Dark Matter Experiment	1993	5-15	Feasibility of cryogenic detection techniques	Cryogenic Detectors
New Generation of Solar Neutrino Detector	1993	10-30	Results of Gallium experiments	Cryogenic and Scintillator Techn. High Rate
Gamma ray facility in the 10^{11}-10^{13} eV region	1993	10	Abundance of sources	Stereo operation of air Cerenkov
Gamma ray facility in the 10^{12}-10^{16} eV region	1994	30	Finding 10^{14} eV sources (with present large arrays)	Lake technology

Essential Technological Developments

Technology	Recommended funding levels ($M)
Cryogenic Detectors for Dark Matter Searches	1/yr in addition to present ~ 1/yr
Cryogenic and Scintillator Techniques for Solar Neutrinos	1-1.5/yr
Lake Cerenkov Technology for Extensive Air Shower Detectors	2.5-5 (1 prototype)

Immediate recommendations

The two projects that we recommend for immediate action are the systematic organization of a supernova watch and the construction of the High Resolution Fly's Eye, currently proposed to the NSF, provided the technical review of the instrumentation and the institutional arrangements is favorable. The first project involves mainly the continuation of the present support to existing instruments at least until new ones of similar supernova detecting capabilities come on line. The cost involved in this transfer period will be minimal. In the long run, however, the support for this activity may have to increase since the low rate of galactic supernovae may require the long term maintenance and regular upgrading of detectors not designed initially to be operated for more than a few years. We will come back to the institutional problems involved. The second project is technically ready to go, has a capability of collecting ten times more events than the present instrument, and a multi-university collaboration has been established. Prototype mirrors and detectors have been built. The cost of the instrument is about $15M. It has the capability with several years of operation to observe a Greisen cut off, to determine the extragalactic origin of the particles, and to establish whether they are protons or heavier nuclei. The instrument will be capable of observing cosmic rays down to 10^{16} eV, and it is likely that at a later date ancillary detectors will be added to enhance the capability of the instrument.

For space-borne investigations, the frontiers will remain in the areas of the rare ultraheavy cosmic ray nuclei (up through the actinides), and of direct composition measurements well above TeV energies. The former are essential for understanding the roles of steady state and explosive nucleosynthesis processes in the galaxy, and the latter are necessary to unravel finally the mystery of cosmic ray acceleration. It will be important that NASA develop both an Ultraheavy Nuclei Explorer Mission and a Trans-TeV Particle Composition Mission.

Future Initiatives

The other initiatives which are presently being considered in order to tackle some fundamental scientific issues of particle astrophysics require more scientific or technical information before a responsible decision can be made. We can see major regions where such initiatives are likely.

• The nature of dark matter is one of the most pressing scientific problems today, and the discovery that it is made of nonbaryonic particles would be of extremely fundamental significance. Unfortunately the level of the present technology for the *direct detection of dark matter particles* prevents us from launching the definitive experiments now which would fully test the particle hypothesis. We have to rely on a step-by-step strategy based on the use of existing technologies to probe part of the parameter space while developing the new techniques necessary for a full exploration. This applies both for the axion search where new methods are needed to explore the high mass region, for instance using ferrite loaded detectors, and for the weakly interacting massive particles. In this case, the feasibility of building phonon or quasi-particle detectors with sufficient energy sensitivity and powerful background rejection and the practicalities of reaching very low radioactive background in low temperature refrigerators must be proved by the pilot experiments mentioned above before a full size experiment could be implemented. However, if and when the technology can be firmly demonstrated, it will be important to move rapidly to the implementation of such experiments. The cost may be in the 3 to 5 million dollars range per experiment. In addition, if present underground laboratories appeared unsuitable, a modest, dedicated ultra-low background facility might have to be built at a cost not expected to exceed $10M.

• The gallium experiments and the measurement of the neutrino flux as we enter a period of enhanced solar activity are likely to tell us a great deal about the *nature of the solar neutrino problem*. It is quite natural that a few years from now new proposals will be made which attempt to check the insights thus gained. There again, it will be important to answer rapidly the new round of questions, whether they concern particle physics or astrophysics. We could imagine detectors in the range of 10 to 30 million dollars but their precise type will depend on whether high rate, energy resolution or threshold will be the most important required feature. We estimate that a design for the next generation of detectors may be ready by 1993.

• The third area where a major initiative is likely is that of *gamma ray observations in the 10^{11}-10^{13} eV region*. The scientific interest of this region has already been pointed out. The atmospheric Cerenkov technique has demonstrated the necessary background rejection capability at 10^{12} eV, but it remains to be seen with the second telescope being installed at Whipple Observatory how a multiple mirror system behaves and how rich the sky is at TeV energies. A decision for a large array of Cerenkov mirrors can be taken as early as 1993. The cost may be of the order of 10 million dollars.

• If the current investigations at 10^{14} eV show that new phenomena are indeed occurring at this energy, a *careful study of the region 10^{13}-10^{16} eV with more powerful instruments will become essential*. For that purpose a major facility using the lake water Cerenkov technique may be particularly interesting as it would allow full analysis of the extensive air shower. It may be advantageous to locate it at high altitude in order to extend its sensitivity down to the lowest energy possible, to overlap with the Cerenkov technique and map in detail a potential transition region. However, *this panel believes that a decision to construct such a facility will have to wait for the confirmation of the existence of sources in the 10^{14} eV region and the demonstration at the prototype scale* *of the water Cerenkov technology applied to extensive air shower measurements.* In any case, it is very likely that a powerful ultra-high energy gamma ray facility will become essential. A decision may be made in 1994. Depending on its exact scope and on the technology or combination of technologies used, the cost may be of the order of 30 million dollars.

• With the approval of DUMAND II, there is no immediate need for an additional high energy neutrino instrument. However another type of detector (such as a deep lake detector) may have to be considered in a few years in the case the deployment of a few phototubes strings by the DUMAND II team fails to confirm the practicality of operating a detector in the deep ocean. By that time, a confirmation of 10^{14} eV point sources showing that hadronic processes are important may have been obtained and would provide a better estimate of the size needed for a neutrino observatory.

• We should acknowledge that the last two points are not accepted by the team which is proposing the GRANDE detector, including several of its members who are serving on this Panel. The proposed instrument is a deep lake water Cerenkov detector, located at low altitude in Arkansas. It would simultaneously study γ rays between a few TeV and 10^{15} eV with excellent muon coverage, angular resolution and background rejection and neutrinos above 6 GeV in a more hospitable environment than the deep ocean. The proposers argue that whatever the results of the large arrays currently being set up, an instrument like GRANDE would have to be built, and that in the general interest of the field, no time should be lost. After thorough discussions and in spite of the power of the proposed detector and the breadth of the physics tackled, this panel concluded that it is preferable to wait for the results of the current large arrays before starting the construction of a new major instrument, which could then be optimized to follow up on the findings of the previous generation detectors.

Longer Term

In the longer term (beyond this decade), it is of course more difficult to predict the evolution of the field. The future depends so much on what happens during the next ten years that it is impossible to guess the discoveries and what will be the most exciting in the long term. These are some of the items discussed:

• Cosmology is likely to remain an extremely active field of study. But ideas are evolving rapidly, and radically new windows may be opened onto the early universe. It is therefore not very useful to attempt to guess in which direction we will be going. However, the existence of a solid theoretical effort and a strong experimental community with powerful techniques at its disposal, spanning the various observational fields from radioastronomy and optical methods to particle physics techniques, is the best guarantee that any breakthrough will be exploited rapidly.

• In the field of particles and stellar physics, the detailed study with neutrinos from the interior of the closest star will certainly continue. In 10 years we may not yet have seen a supernova from our own galaxy but it will still be vital that observations continue so that such a rare and important event would not be missed. The greatest challenge that we can presently conceive in this line of research is probably the development of techniques capable of reaching the tens of megatons necessary for supernova neutrino detectors to be able to detect supernovae in the Virgo cluster. Such instruments would permit us both to study many collapses and to put precise limits on the mass of all types of neutrinos, an enterprise extremely important for cosmology and particle physics.

• The future of high energy gamma ray and neutrino astronomy depends to a large extent on the confirmation in the present decade of the existence of very high energy sources. If they do indeed exist, we can envision a long program of powerful gamma ray and neutrino observatories, probably with some installations in the southern hemisphere and the support of large international collaborations.

• A dedicated Interstellar Probe mission that would rapidly leave the heliosphere would provide invaluable *in situ* measurements of the nearby galactic environment.

• The deployment of very large orbiting detectors, 100-1000 m^2 ster, with exposure of several years, may allow us to observe directly very high energy primaries (charged and neutral) and allow in particular a direct study of the composition of cosmic rays near the spectral knee (10^{16} eV). Measurement at these very high energies may be done with massive calorimeters recording the cascade generated by interactions. One promising possibility is to develop such an instrument at a lunar base, where most of the required mass would come from the moon itself.

Essential Technological Developments

With the results of present developments and experiments, major branch points in the scientific strategy of particle astrophysics will present themselves in the coming three to five years. It is essential to prepare the technologies likely to be needed and to study now the necessary facilities so that we can proceed with the next generation of detectors as soon as the scientific questions are clarified. Following the discussion of the last section, the items of highest priorities are:

Cryogenic Detectors of Particles.

As argued above, answering the central question of the nature of dark matter requires the development of new detection methods. For the specific case of weakly interacting massive particles, the technique would have presumably to be based on the breaking of Cooper pairs in superconductors or the production of phonons in materials at very low temperature. Reviewing the level of support of this activity in this country (around \$1M for massive detectors shared nearly equally between DOE and NSF), this Panel is struck by the disparity with the European development funded at a level approximately 4 times higher. Moreover some development areas such as that of tunnel junctions are nearly absent in the US, while they are fairly successful in Europe. *We recommend multiplying the US funding level by at least a factor of two and involving more institutions* so that within 2 to 3 years, pilot experiments could effectively attempt to demonstrate the radioactive background rejection and the redundancy of these methods and compare various technologies. This is an essential element if we are to install full size experiments featuring the 5 to 10 kg detectors necessary to probe the existence of supersymmetric particles as a significant dark matter component.

New Solar Neutrino Techniques

As the nature of the solar neutrino problem is being unravelled by the present generation of detectors, it will be important to have at our disposal technologies able to tackle rapidly whichever aspect emerges as the most important then: high rate, energy spectrum, low or high energy region.
Two directions need to be actively pursued:
    • *Cryogenic techniques* similar to those developed for dark matter. An interesting possibility, given the large mass of detector necessary, is the use of ^{4}He, presently being developed by Brown University, in particular if it is important to measure the spectrum of pp neutrinos. This effort must be sustained.
    • *Scintillator techniques* utilizing as target electrons ^{11}B or possibly ^{115}In in the scintillating material. The main problem appears to be the control of their radioactive background and a strong development effort should be directed to that question.

New Extensive Air Shower Detectors

The lake water Cerenkov detector technology is a radically new technique which exploits the experience gained with water Cerenkov proton decay detectors and potentially offers several advantages with: full calorimetry which may allow excellent hadronic shower rejection, or if necessary the study of the muon anomaly; excellent angular resolution; and low threshold especially at high altitude. This last feature could make this technology competitive with the atmospheric Cerenkov technique, with the additional advantages of an "all" sky sensitivity, continuous operation and the possibility of detecting sources with short duty cycles (if such exist).
    We believe, however, that this very interesting technique initially proposed by the GRANDE collaboration should be tested at a small scale before being fully deployed. We therefore recommend *operation of at least one prototype instrument (if possible at high altitude), at a modest funding level* in order to evaluate the potential of the technique. This technology may become critical at the decision point which is likely to appear within 3 to 4 years, when the situation with 10^{14} eV gamma ray sources is clarified by instruments that are currently being constructed. The experience gained with smaller scale water detectors may be invaluable for the optimization of a full size lake facility. In addition, modest developments of other potentially useful technologies such as tracking detectors should continue.

An Active Balloon Program

The NASA balloon program produces important scientific results and supports the development of new instrumentation for future space flights. Balloon-borne experiments play a vital role in the education and training of students and young scientists. Recent improvements in the reliability of heavy-lift balloons and the initiation of efforts to provide long-duration flights around the globe or at the South Pole promise to further enhance the scientific output. An adequate level of support of the balloon program and its technical staff as well as enhanced funding for state of the art detector development are essential.

Critical Institutional Issues

Particle Astrophysics attracts a growing number of physicists, reaching 200 experimentalists in the United States alone, while the cosmic ray astrophysicists working in space represent a well structured and organized community. Increasingly these two communities overlap. A certain number of institutional issues will become critical in the coming decade.

The Funding of Particle Astrophysics

Particle Astrophysics happens to be in a very peculiar situation. If one excludes traditional cosmology observations funded mainly by NSF (Astronomy) and NASA, and the cosmic ray studies in space which have been a traditional element of the NASA scientific program, the other observational fields covered by this panel have been supported by other divisions of NSF (Physics, the Polar Program, and the Science and Technology Office--with the Center for Particle Astrophysics in Berkeley), and by two divisions of DOE (High Energy Physics mainly for high energy astrophysics, and Nuclear Sciences for solar neutrinos).

These new funding sources have helpfully substituted for the traditional channels which appeared saturated, and they have supported instruments which have made some of the most significant astrophysical discoveries of the decade (e.g., the direct measurement of neutrinos coming from the sun and the neutrino burst from SN1987A). They have facilitated naturally the evolution towards astrophysics of a different physics community (mainly particle physicists). Moreover agencies such as DOE are particularly well equipped to handle large facilities.

In spite of these strengths, this funding situation has significant weaknesses. There is a clear danger for multidisciplinary efforts such as particle astrophysics to "fall between the cracks". Part of the problem may come from ambiguities in the perceived mission of the funding agencies. In particular DOE has historically a clear responsibility in particle physics but its involvement in astrophysics does not appear to have been fully accepted. As the distinction between particle physics, cosmology or astrophysics becomes increasingly blurred, this Panel feels strongly that *too narrow an interpretation of particle physics would be detrimental to scientific productivity; we suggest that the responsibility of DOE in particle astrophysics should be clearly affirmed.* Funding decisions of proposals should then be based on their scientific merit, the techniques used and the scientific priorities recommended by the advisory bodies, without attempting to define too rigidly what is particle physics and what is astrophysics. For instance, it will not be productive to argue that a supernova watch is a purely astrophysics problem; most of the current neutrino detectors are DOE supported, and it is clear that the detection of a supernova in our galaxy would also provide crucial information on neutrino properties.

Coordination between the various funding decisions is also difficult in this new field. High Energy and Nuclear Sciences decisions at DOE have traditionally relied heavily on the accelerator laboratories which, with their respective program advisory committees, filter proposals and set up priorities and long term policies. Such an intermediate structure does not exist in particle astrophysics. Small to medium cost projects are judged through conventional mail review while larger ones are evaluated by ad-hoc committees. But in most cases, the reviewers do not have access to the broader picture, where it is necessary to maintain the balance between experiments of widely different costs but similarly fundamental interests or between short term endeavors and long range technical development.

Recommended Funding Mechanisms

This analysis has led this Panel to make the following recommendations:

• *A Particle Astrophysics Advisory Structure.*

There is first a clear need for the establishment of a *Particle Astrophysics Advisory Panel* which could advise the federal agencies involved in Particle Astrophysics (NSF, DOE and maybe NASA) on the relative scientific priorities and help them set up long term policies. Such a body will certainly help proposals not to fall between the cracks, and will be critical in the decisions at the decision points that we identified above. Consultation with the community and discussions with some of the agency officials have shown a surprising level of consensus on the need for such a committee!

In order to guarantee coordination of priorities, this specific advisory structure should be organically linked to any general advisory structure for astrophysics. For example, its chair may be a member of a general Astronomy and Astrophysics Committee, if such a committee is formed. Similarly, official links to the High Energy Physics Advisory Panel (HEPAP) and to the Nuclear Science Advisory Committee (NSAC) and the relevant NSF and NASA advisory committees will have to be established. However, the specificity of the Particle Astrophysics community, the operational methods of the agencies involved, the multidisciplinary aspects of the program and the cost of large observational facilities are strong reasons for organizing a specific advisory channel.

• *A multidisciplinary approach to cosmology.*

Funding of multidisciplinary research is always difficult and particle astrophysics offers a typical example. The problem has been particularly acute for cosmology which appears to have been underfunded in the last decade both for experiments and theory. This suggests that the present funding institutions were unable to adapt to the emergence of a new discipline which cuts across many observational fields (including particle astrophysics). The scientific importance of the field, its philosophical implications and its appeal to the public at large require a much better coordination (among the various observational fields) and an improved general level of funding. This will require innovative institutional schemes bringing together astrophysicists of several observational fields and particle physicists, theorists and experimentalists alike. The theoretical astrophysics group at Fermilab is a recognized success and we encourage the creation of an experimental counterpart. It will be interesting to watch the evolution of the new NSF Center for Particle Astrophysics at Berkeley in its attempts to focus resources and expertise from many disciplines, institutions and funding origins on the problem of dark matter. It is conceivable that several such multidisciplinary center-like cosmology institutions nation-wide, complementary to individual PI grants rather than in competition with them, could significantly contribute to the implementation of the high scientific priority of cosmology and the coordination advocated above.

• *Balance between long range projects and short term scientific opportunities.*

In our above description of the future of the field we have focused on major facilities. It is important however to maintain a balance between a long range program, and the rapid exploitation of scientific opportunities when they emerge. This requires in particular a large enough financial "reserve" to be able to react rapidly to a promising idea or a breakthrough.

Along a similar line, the Panel recognizes the importance of frequent small to moderate mission opportunities within the NASA program. The rapid access to space that these could provide for new, innovative experiments is essential for the vitality of space science. Of particular importance in that respect is the Explorer program, including the new series of "Small Explorers" as well as balloon programs.

Recommended Facilities

Experimental particle astrophysics has been developing so far mainly at universities, although the Lawrence Berkeley Laboratory and the Los Alamos National Laboratory have been among the pioneers of the field and the Fermi National Laboratory is seriously considering the possibility of getting involved in experimental aspects. The absence of national structures such as accelerators which naturally regroup the physicists and support their activity leads this panel to recommend the following actions.

Establishing high energy astrophysics observational facilities on nearby sites sharing common facilities is in many cases quite natural scientifically and has already been informally implemented at sites such as Dugway in Utah and institutionalized at Gran Sasso in Italy. But the Dugway experience may demonstrate the need for significant support facilities that only a formal structure can fully provide at such remote sites. Such regrouping of facilities should not preclude experiments at other sites, when this is clearly preferable for the specific purpose or technique proposed. An obvious example is the use of the existing infrastructure of astronomical observatories.

This panel also studied in some detail the need for a national underground facility. It concluded that a national facility of the size of Gran Sasso was not justified, since large experiments could easily use this Italian facility. However the specific needs of searches for weakly interacting dark matter particles and the cryogenic detection of solar neutrinos (difficulty of transporting ultra low radioactivity detectors, low temperature refrigerators, need for ultra-quiet environment in terms of radioactivity, electric and mechanical noise) does warrant the *immediate study of the necessary ultra-low background facility or facilities*. Their size will probably be modest, they may well be installed in underground laboratories existing on the American continent (e.g., Soudan, Sudbury) or be built at shallower depths. In any case, a clear understanding of the requirements and of the possibilities has to be obtained in parallel with the technological developments, so that if and when the necessary detector technologies have been demonstrated, no unnecessary delay is encountered in implementing significant experiments.

The necessity of a continuous supernova watch over possibly several decades will require innovative institutional mechanisms. Although a straightforward coordination of the down periods of the various underground experiments will be sufficient in the short run, the operation and maintenance of instruments for tens of years cannot easily be accommodated in the present research structure. National laboratories may have to take the responsibility for this long endeavor.

More generally, this Panel strongly favors *the expansion of the involvement of National Laboratories* (including SLAC and SSC) in particle astrophysics. While helping this new field with their technical expertise, they may benefit from the intellectual stimulation and the diversification provided by a fundamentally multidisciplinary discipline.

International Collaborations

The field of particle astrophysics is already strongly benefiting from international collaborations, in part because of the tradition in Particle Physics. In addition to the continuation of the use of the deep underground facilities in Europe and Soviet Union (Gran Sasso, Frejus and Baksan), we see the potential for new collaborations around large observational facilities, in particular for supernova watch and for high energy gamma ray, neutrino and cosmic ray astrophysics. The probable need for a major extensive atmospheric Cerenkov telescope and air shower array in the southern hemisphere may provide an excellent opportunity for the involvement of developing countries.

Education and Technology

Particle astrophysics is in a unique position to help in the improvement of the scientific and technical education in the United States. Its scientific focus combines astrophysics which has always been a source of fascination and particle physics which enjoys a very prestigious position at the frontier of knowledge. It is therefore particularly easy to explain some of the problems addressed by the field to the public at large and the number of recent articles in the general press on dark matter or neutrinos, for instance, demonstrates a real interest. This certainly contributes in the long run to a better general understanding of science and technology.

For the same reasons, particle astrophysics attracts an increasing number of undergraduate and graduate students, and postdoctoral researchers. The experimentalists are in particular excited by the possibility of tackling very fundamental problems with relatively small instruments, modest sized teams, and reasonable time scales. Such a combination allows them to have a complete grasp of all aspects of the experiment and probably provides the best training for a scientist. Moreover, the diversity of the science involved offers experimentalists and theorists a marvellous multidisciplinary education.

Particle astrophysics also contributes directly to the technological base of this nation. Its progress depends critically on the sensitivity of its sensors, the power of its electronics, and the efficiency of its computer codes. In a way similar to particle physics, such performance is achieved by a combination of innovative in-house development, adoption of the most advanced technologies developed elsewhere and specific collaboration with industry to adapt manufacturing processes. Some of the developments specifically recommended in this report may be particularly beneficial for other fields. For instance, it is likely that the development of cryogenic detectors will have spin-offs not only in astrophysics (e.g., X and γ ray astrophysics) but in particle and nuclear physics, high resolution X and γ ray spectroscopy, and biomedical imaging. Ultra-low background techniques may also have interesting applications in biology (e.g., high sensitivity radioactive tracing) or in electronics (e.g., soft errors in large scale integrated circuits).

Therefore, far from being an esoteric abstract field, particle astrophysics can contribute at a modest but real level to the solution of some of the fundamental causes of recent economic difficulties.

THEORY AND
LABORATORY ASTROPHYSICS PANEL

DAVID N. SCHRAMM, University of Chicago, *Chair*
CHRISTOPHER F. McKEE, University of California, Berkeley, *Vice-Chair*

CHARLES ALCOCK, Lawrence Livermore National Laboratory
LOU ALLAMANDOLA, NASA Ames Research Center
ROGER A. CHEVALIER, University of Virginia
DAVID B. CLINE,* University of California, Los Angeles
ALEXANDER DALGARNO,* Harvard-Smithsonian Center for Astrophysics
BRUCE G. ELMEGREEN, IBM T.J. Watson Research Center
S. MICHAEL FALL,* Space Telescope Science Institute
GARY J. FERLAND, Ohio State University
BRADLEY W. FILIPPONE,* California Institute of Technology
MARGARET J. GELLER, Harvard-Smithsonian Center for Astrophysics
PETER GOLDREICH, California Institute of Technology
ALAN H. GUTH, Massachusetts Institute of Technology
WICK HAXTON, University of Washington
DAVID G. HUMMER, University of Colorado, Boulder
DIMITRI MIHALAS, University of Illinois at Urbana-Champaign
MICHAEL J. MUMMA, NASA Goddard Space Flight Center
PETER PARKER, Yale University
P. JAMES E. PEEBLES,* Princeton University
MALVIN A. RUDERMAN, Columbia University
GREGORY A. SHIELDS, University of Texas, Austin
PETER L. SMITH, Harvard-Smithsonian Center for Astrophysics
SAUL A. TEUKOLSKY, Cornell University
PATRICK THADDEUS, Harvard-Smithsonian Center for Astrophysics
SCOTT D. TREMAINE,* Canadian Institute for Theoretical Astrophysics
JAMES W. TRURAN, JR., University of Illinois at Urbana-Champaign
JOHN WEFEL, Louisiana State University, Baton Rouge
J. CRAIG WHEELER,* University of Texas, Austin
STANFORD E. WOOSLEY, University of California, Santa Cruz
ELLEN G. ZWEIBEL,* University of Colorado, Boulder

THEORY AND LABORATORY ASTROPHYSICS

Executive Summary

• Theory provides the basic paradigms within which observations are planned and interpreted, and without which observations degenerate into catalogs of meaningless data. A vigorous effort in theoretical astrophysics is necessary to realize the benefits from the wealth of new astronomical data expected in the 90's.

• Laboratory Astrophysics provides the basic data required to infer the intrinsic properties of astronomical sources from astronomical observations. Enhanced support for Laboratory Astrophysics is necessary if we are to take full advantage of the Great Observatories and of the new initiatives planned for ground–based astronomy.

• Commensurate support for Theory: NSF and NASA should support Theory at a reasonable fraction of their support for observational astronomy. DOE should support Theory insofar as it is relevant to its mission.

• Laboratory Astrophysics initiative: NASA should establish a long–term program in Laboratory Astrophysics to support major missions. NSF Astronomy should find new funds to establish a viable program in atomic and molecular laboratory astrophysics. DOE should support Laboratory Astrophysics that is relevant to its core programs.

Table of Contents

I - INTRODUCTION

For traditional reasons Theory and Laboratory Astrophysics have been coupled together in this report. However, since they are rather separate areas with different communities and different impact, we have tried to separate clearly the science and the needs of these two communities. The vitality of both Theory and Laboratory Astrophysics is critical to the success of essentially all future astronomy.

While Arthur Eddington's famous statement that no astronomical observation can be believed until confirmed by theory may be a bit excessive, there is no question that branches of science progress most rapidly when there is a close interplay of theory, observation and experiment. When theory runs too far ahead of what can be measured, a field becomes more philosophy than science, and when data taking yields huge archives without understanding, fields go through intellectual stagnation. The success of modern astrophysics illustrates the close interdependence of observation, experiment and theory. To maintain a vital science requires a strong theoretical community commensurate with a strong experimental/observational community. A strong theoretical community does not only attempt to explain data and establish frameworks with which to analyze, but it also makes predictions about what should eventually be seen. Furthermore, theory can provide a deep and satisfying understanding of how things fit together into a coherent view of the universe as a whole. In the two science sections of this report, opportunities for the 90's and successes of the 80's, we see both the impressive opportunities that lie ahead as well as recent successes upon which we can build. In particular, we note that theory has been an important driver of our subject in many areas.

Laboratory Astrophysics plays a very different role than Theory. It provides the firm laboratory base of atomic, molecular and plasma data necessary to understand and direct observations in space. It also provides the nuclear data necessary to carry out calculations of cosmological and stellar nucleosynthesis as well as energy generation and other nuclear processes. Furthermore, high energy accelerators and other particle experiments are now not only providing the particle data necessary for calculations, but also in some cases even testing cosmological predictions. Another area of Laboratory activity has been in the determination of meteoritic abundances which play a key role in the interpretation of nucleosynthetic ideas. In the science sections of this report, each of these wide-ranging sub-areas will be discussed.

The recommendations are divided into two sections. The first, immediately following the science opportunities, will discuss funding needs. The second, at the end, will discuss policy and procedural questions. Since Theory receives significant funding from three separate agencies, NSF, DOE and NASA, while Laboratory Astrophysics receives funding from those three plus NIST, and since each receives support from different subsections within those agencies, it is obvious that policy and interagency co-operation questions are non-trivial. In particular, the argument is made that both Theory and Laboratory Astrophysics require that funding be commensurate with the levels of funding received throughout astrophysics. Furthermore, since most of this funding is for relatively small individual research grants, it is important that such individual programs are not allowed to be lost or overlooked in the zeal for large projects.

II - SCIENCE OPPORTUNITIES FOR THE 90'S

Theory in the 90's

Experience suggests that many of the most exciting theoretical developments in astrophysics in the 1990's will be in directions whose significance was at best only poorly appreciated in the 1980's. However, it is an encouraging measure of the growing maturity of our subject that there is a considerable catalog of puzzles that we know how to address and whose answers will teach us something new. The following is meant to be a representative but certainly not complete sampler of such topics.

The Large-Scale Structure of the Universe

A major occupation for theoretical astronomy in the 1990's will be the search for a world picture consistent with the observations of the large-scale structure of the Universe, as it is now and as it was in the distant past. Observations are yielding new pieces to the puzzle, advances in computer hardware and algorithms are improving our ability to explore how the pieces of the puzzle might fit together, and new ideas from particle physics are offering a variety of world pictures that the pieces might fit into. The feeling among many workers is that the theory will be so tightly constrained by new observations in the coming decade that only one plausible picture of the cosmos will emerge. On the other hand, it is also possible that it will become clear that new physics will have to be invoked in order to achieve such a solution.

The large-scale structure of the Universe is characterized by the fractional departure of the mass density from homogeneity and the mean peculiar streaming velocity relative to the general expansion. If non-gravitational forces can be neglected, the standard cosmological model predicts a definite relation between inhomogeneity and streaming velocities. The deep galaxy redshift surveys in progress and planned will contain the information from which direct and statistical measures of the large-scale galaxy space distribution and peculiar velocity field can be derived. It will be a fascinating task to devise ways to extract this information and learn how to compare this information to the large-scale fluctuations in the galaxy space distribution, and to compare these fluctuations to those predicted by theoretical models.

Particularly exciting to theorists are the rapid advances in observations of galaxies and quasars at redshifts greater than unity, since these objects open a window onto the Universe when it was young. What is the significance of the observation that the intergalactic medium, as probed by the absorption lines in quasar spectra, was cleared of the bulk of the neutral hydrogen as early as redshift $z \sim 5$? Why do high redshift galaxies show indications of youth, such as alignment of optical and radio images, along with symptoms of age, such as small scatter in the Hubble diagram?

The cosmic background radiation (CBR) is the thermal radiation left over from the very early hot, dense phase of the expanding Universe. Its presence was predicted many years before its discovery, and extensive theoretical studies have shown that the structure and evolution of the very young Universe in principle can be inferred from anisotropies of the CBR. Convincing small-scale anisotropies have yet to be detected, but the observational situation is rapidly improving with the launch of the COBE satellite and with the development of sensitive detectors which can be sent aloft in balloons and rockets or used in the best ground–based sites, such as Antarctica. By conventional estimates, departures from an ideal heat bath originate in the early Universe, at redshifts $z > 100$. Here the theorists' tools have already been sharpened and oiled, and we are well-prepared to incorporate the anisotropy data into our growing fund of information about the Universe as it is now and at more modest redshifts.

The diffuse X-ray, γ-ray, and very high energy cosmic ray backgrounds are more enigmatic than the thermal 2.7 K background radiation (the CBR), but the observational constraints on sources are growing increasingly tight, and we may hope that in the 1990's we will understand the origins of these backgrounds, whether active galactic nuclei, events in young galaxies, or something completely new.

Because groups and clusters of galaxies are held together by gravity, it seems plausible that gravity was the dominant force in the final assembly of these systems. In the 1980's there was considerable progress in the development of analytic and numerical studies of the evolution of mass clustering in an expanding Universe. This work depends heavily on large-scale computation; this is one area where progress has been directly limited by scarce computing resources. Thus, we may expect to see considerably more progress in the 1990's as the available hardware and software continue to improve at a rapid rate. Non–gravitational forces may also play a role in the formation of groups and clusters, and the initial studies of these effects in the 80's will be greatly expanded in the 90's as computer power grows.

There is finally the great problem of understanding what the Universe is made of. The straightforward interpretation of the bulk of the observational evidence is that the mass is dominated by baryons with mean density about one tenth the critical Einstein-de Sitter value. However, since most of the mass has to be dark, it is easy to adduce theoretical arguments for exotic matter in an amount consistent with the theoretically preferred relativistic cosmology, the Einstein-de Sitter model. The problem of having baryonic matter in amounts much greater that 1/10 the critical density hinges on cosmological nucleosynthesis arguments; given the importance of such arguments, exploration of the standard theory as well as possible loopholes will continue to be important. Furthermore, new measurements of light element abundances and their evolution will further test the standard model predictions. It would be hard to overstate the impact on our subject of an unambiguous detection of dark matter, exotic or baryonic, either astronomically or in the terrestrial laboratory. And of course, a believable picture for the origin of the structure of the Universe will require convincing evidence, direct or indirect, on the nature of the dark matter.

Where will all this activity lead us? We may learn that the evolution of structure on the scale of galaxies and larger is consistent with the gravitational growth of gaussian mass density fluctuations that were present in the extremely early Universe, as predicted in the simplest inflation scenarios. However, it is also possible that structure grew out of non-gaussian fluctuations in the mass distribution, as the result, for example, of the stress of a primeval cosmic magnetic field, or of explosions, or of the gravity of the fields of phase transitions, or of primeval isocurvature fluctuations during baryosynthesis. And it certainly is possible that none of these popular ideas from the 1980's will be found to fit the observations. The exciting point is that theorists in the 1990's will have a broad variety of scenarios to evaluate in the light of an increasingly rich network of observational clues. Furthermore, the interface with particle physics in the study of the early Universe has proven rich, and the use of cosmological arguments to constrain fundamental physics can be expected to continue to influence particle physics experiments.

Galaxies

Here, as in the study of large-scale structure, much of the theoretical activity will interact strongly with new observations. The space-based observatories planned for the 1990's, such as HST, ROSAT, GRO, EUVE, XTE, SOFIA, SIRTF, FUSE, AXAF, and ISO, are justified in large part by the theoretical concepts developed in the past two decades, and the data they will provide, together with those from the planned powerful ground–based observatories, should lead to major advances in our theoretical understanding of the Universe.

Much of the theoretical work on galaxies in the 1990's will be related to a single fundamental puzzle: Can we elucidate a coherent picture of the structure and evolutionary history of galaxies such as the one we live in? This basic question has many aspects. For example, is the dark mass in the disk and halo of our galaxy comprised of low mass gas spheres (brown dwarfs, Jupiters, gravitationally bound comets), or the remnants of massive stars, or something exotic, like axions or black holes? Did our galaxy form out of a single primeval gas cloud, or by merging of gas-rich dwarfs, or by steady or episodic accretion of matter, or by something altogether different? Did our galaxy form at high redshift when the mean density was high, or at low redshift by a large compression of gas? When did the disk of our Galaxy form? Is the galactic magnetic field primordial, or is it sustained by a dynamo? When and how did the nucleus of our galaxy form? How did globular clusters form, and how do they evolve? Can we adduce evidence that there really are black holes in nuclei of galaxies? What is the connection between quasars and galaxies? Have there been violent mergers in our Galaxy's past? What determines the Hubble type of a galaxy? How do the processes of mass exchange between stars and the interstellar medium of a galaxy on the one hand, and the accretion of intergalactic matter on the other, combine to determine the abundances of the elements in galaxies? Initial steps toward answering all these questions have been taken in the last decade, and the availability of data of higher spatial resolution and at greater redshifts in the coming decade should permit enormous strides in our understanding.

The relative brevity of this section in comparison to the previous one reflects a fundamental change in our understanding of galaxies: more and more, it is becoming clear that the study of galaxy *structure* is inseparable from the study of galaxy *formation*. There are many examples: spiral structure in galaxies is likely to be strongly influenced by recent mergers or close encounters or by infall of intergalactic gas; the structure of the orbits in and shape of the galactic spheroid and halo is likely to reflect a fossil record of the formation process rather than subsequent internal relaxation; the warps in galaxies reflect the interaction of the disk with a misaligned halo or asymmetric infall rather than an internal mode, and so forth. It is possible that the 90's may see the theoretical subject of galactic structure largely subsumed within cosmology.

Star Formation and the Interstellar Medium

The problem of star formation is central to a broad variety of problems in galactic astronomy and cosmology. For example, what does a young galaxy look like? Does the dark mass around galaxies consist of Jupiters or brown dwarfs, or does it consist of the remnants of an early generation of massive stars? Substantial progress has been made in the past decade in elucidating the process by which isolated, low-mass stars form, generally with a surrounding disk and an energetic wind. This is only a beginning, however, and many fundamental questions remain: How do the molecular clouds, out of which stars are born, form and evolve, both in structure and in chemical composition? How do massive stars form? What determines the initial mass function? How does star formation in one part of a cloud affect that in other regions of the cloud? How does star formation proceed in starburst galaxies? The goal of this research is to develop a predictive theory of star formation, and substantial progress is anticipated. The results of this research will have a substantial impact on the studies of galaxy formation described above and on the theory of planetary system formation discussed below.

Most of the volume of the ISM is occupied by gas too tenuous to be molecular. This gas exists in several different states: cold ($T \sim 10^2 K$) and neutral; warm ($T \sim 10^4 K$), both neutral and ionized; and hot ($T \sim 10^6 K$) and ionized. The warm gas is observed to extend a kiloparsec away from the Galactic plane, and the hot gas is presumably more extensive, yet, forming a Galactic halo. The weight of this gas determines the pressure in the disk and thus determines a boundary condition for molecular clouds. Over the past decade, a variety of theories, involving supernovae, cosmic rays, and magnetic fields, have been advanced to account for the pressurization of the ISM; the increasing sophistication of the models together with the anticipated influx of new data should lead to a resolution of this problem in the 90's. The source of ionization of this gas—stars, shocks, or the cosmic EUV background—should be clarified as well. Diffuse interstellar gas has structure on scales ranging from much less than an AU to kiloparsecs, and a more coherent picture for the origin of this structure should begin to emerge.

Magnetic fields have long been recognized as playing a central role in the dynamics of the ISM, a role confirmed by recent measurements of field strengths, both in molecular gas and in diffuse atomic gas. The ISM is observed to be highly turbulent. It thus poses a complex problem in non–linear MHD (magneto-hydrodynamics), and numerical simulation is an essential complement to basic theory. The 90's should see a revolution in our theoretical understanding of the effects of the interstellar magnetic field as increasingly sophisticated algorithms and computer hardware combine to permit accurate, large scale, three–dimensional simulations of interstellar MHD processes. The fundamental question of the origin of interstellar magnetic fields will also be addressed.

The chemical evolution of the interstellar medium is intimately connected with interstellar dust, which contains most of the refractory elements in the interstellar medium. At present, the origin of this dust in not known: How much comes from evolved stars? How much from the debris of supernovae? How much from accreted interstellar gas? Laboratory studies of dust growth and destruction can aid theoretical studies of this issue.

The relativistic component of the ISM, the cosmic rays, has an important effect on the dynamics of the interstellar medium as well as on the ionization and thermal equilibrium of interstellar gas. The coming decade should see important advances in our understanding of the acceleration and propagation of the cosmic rays, driven in part by measurements of the cosmic ray composition and of the gamma ray emission associated with cosmic rays.

Stars

Perhaps the major unsolved problem in stellar astrophysics remains the solar neutrino problem. Whether this problem will be resolved with new neutrino physics, nuclear physics or stellar structure remains a mystery. However, over the next decade, much activity is bound to occur due to the new experiments, such as the gallium experiments (SAGE and GALLEX) and the D_2O experiment of SNO, as well as continued operation of Kamiokande, and, of course, the chlorine experiment of Homestake.

Another central problem in stellar structure is the development of predictive theories of convection and of stellar dynamos. The last decade has seen two observational advances which suggest that considerably more progress will be made on these problems in the next decade than in the last. First, the level of stellar activity, which is the observational manifestation of stellar convection and magnetic fields, is far better characterized now than a decade ago. Second, and more important, is the advent of helioseismology, which permits the inference of the internal structure of the Sun through observations of oscillations of its surface. Extension of this technique to other stars—astroseismology—is producing valuable further clues. Finally, increasingly realistic simulations of compressible magnetoconvection are making it possible to address problems such as convective overshoot and mixing more quantitatively than ever before, and substantial progress is anticipated.

A substantial fraction of all stars are known to occur in binary systems, a factor which introduces many interesting complications. Significant problems to be addressed in this area include the role of binaries in the gravothermal collpase of dense star clusters, the evolution of the binary progenitors of Type I supernovae, the possible relationship between low mass X-ray binaries and millisecond pulsar binaries, and the general implications of magnetic fields for diverse phenomena associated with the evolution of close binary systems. It will be important in the 90's to seek to build on the progress which has been achieved in these areas in the 80's.

Following the observational stimulus and theoretical interpretation of SN 1987A, the brightest supernova in almost four centuries, we can expect a renaissance of supernova studies in the next decade. Supernova search programs are being mounted in both northern and southern hemispheres, and facilities are being planned, both on the ground and in space, that will follow supernovae at all wavelengths. The kind of detailed data that enables accurate theoretical models to be constructed for SN 1987A will be obtained for dozens of other more distant supernovae. The 1980's saw the first steps towards providing a critical analytic link between theoretical models and observation—the development of realistic synthetic spectral codes. These efforts will reach fruition in the 1990's and be applied to a data set of increasing diversity and complexity. Many pressing issues remain to be addressed, and we can expect substantial progress on all fronts. Type I supernovae (those without hydrogen) and Type II (those with) probably have very different origins (though Type Ib may closely resemble Type II in mechanism). Understanding both is of major importance.

Pressing issues for Type Ia's include (1) the nature of the progenitor star (cataclysmic variable, white dwarfs merging by gravitational radiation?); (2) whether the explosion propagates as a subsonic deflagration or supersonic detonation; (3) the isotopic composition of the ejecta and the role of Ia's in galactic chemical evolution; (4) the γ-ray line signature of ^{56}Co that might be observed by GRO or NAE and how that constrains the models; and (5) an adequate understanding of the explosion and radiation transport to use the objects as cosmological standard candles.

For Type II's, some issues are: (1) the explosion mechanism including possible asymmetries or jets; (2) the immediate pre-explosive history of the star, its mass loss rate, radius, and composition; (3) the nucleosynthesis; (4) the nature and magnitude of mixing during and immediately following the explosion; (5) the nature of the compact remnant, how frequently is a pulsar formed and how fast does it rotate at birth? can a supernova explosion leave a black hole behind? and (6) possible restrictions that can be placed upon the μ and τ-neutrino masses from a galactic supernova.

In all cases we would like better diagnostics of the explosion energy in order to constrain the explosion mechanism. For SN 1987A this quantity was determined because we could estimate the mass of the helium core (from the presupernova star) and thus could obtain the mass of the hydrogen envelope by when the light curve peaked and from hydrodynamic calculations. In the more general case we will not know the progenitor star and will need accurate models of the time histories of spectral line profiles in order to obtain the velocity distribution of the ejecta.

We would also like to know the frequency with which supernovae of all types occur in galaxies of all types (both a theoretical and observational issue). This is related to the more general question of how stars terminate their existence as optically luminous objects, whether by winds, novae, supernovae, or possibly silent collapse. All of these terminations are challenging problems with importance for the chemical evolution of the Galaxy.

High-Energy Astrophysics

A significant portion of the energetic radiation in the Universe is believed to originate from the accretion of matter onto collapsed objects—neutron stars, stellar black holes, and massive black holes in galactic nuclei. A major task for theorists in the 90's is to provide observers with the tools necessary to interpret observations of galactic nuclei and to determine whether massive black holes indeed lie there. Physical processes in the high–energy environment of galactic nuclei and quasars—such as relativistic shocks, particle acceleration, and e^-e^- pair creation—could be diagnostics of the existence and properties of a central black hole. A fundamental task is to understand the accretion process itself. If the accreting gas has even a small amount of angular momentum, it will form an accretion disk. Such a disk is like a two–dimensional star: it has an interior in which energy is generated, a photosphere in which the continuum is generated, a chromosphere in which emission lines are generated, and a corona which can produce a wind. However, in contrast to stars, accretion disks are energized not by nuclear reactions but by the release of gravitational energy. Understanding how this energy is released—the physics of the effective "viscosity" in the disk—is one of the major problems in theoretical astrophysics. Some steps toward the solution of this problem have been taken in the 80's, and the 90's should see significant advances.

Highly collimated outflows, or jets, often accompany accretion disks. Although much progress has been made in modeling these jets, there are still many unresolved questions associated with their origin, collimation, and stability that must be answered. The study of the interaction of such jets with the surrounding interstellar medium is a promising area for research. High resolution observations combined with increasingly realistic simulations should significantly advance our understanding of astrophysical jets.

It is worthwhile noting that almost 50% of the COS-B γ-ray sources are still not identified with known objects. Furthermore, cosmic γ-ray bursts remain a major research area. Recent observations by the Ginga satellite of the hard x-ray emission accompanying γ-ray bursts shows evidence for both a cyclotron feature and the first harmonic. Thus, evidence is increasing that at least some of these events originate from Galactic neutron stars. During the 90's several space missions, both U.S. and foreign, will address γ-ray bursts (GRANAT, GRO/BATSE, NAE, HETE). These bursts may be the chief detectable emission of old neutron stars in our Galaxy. There is obviously room for exciting new developments. It is also worth noting that the combination of rapid rotation and high magnetic fields thought to exist in many of the high energy sources may lead to particle acceleration beyond any energies so far achievable in the terrestrial lab.

The Solar System

The long-term stability of the solar system is one of the oldest unsolved problems in physics. The 1980's saw a rekindling of interest in this problem, sparked by at least three separate developments. First, the availability of inexpensive computing encouraged long numerical orbit integrations, both on supercomputers and special-purpose machines. Second, the tools developed by dynamicists in other areas (resonance overlap, analog mappings, etc.) began to be applied to the solar system. And perhaps most important, work on the Kirkwood gaps in the asteroid belt demonstrated that dynamical evolution over timescales of 10^6 years or longer was important for the present-day structure of the solar system. With this encouragement that the solar system is not boring on very long timescales, the 1990's should see a broadly based attack on the long-term dynamical stability of the solar system, with the ultimate goal of understanding to what extent the present structure of the system is determined by the requirement of dynamical longevity.

The 1990's should also see great advances in our understanding of the formation of the solar system and of the possibility of the formation of planetary systems around other stars. Much of the recent activity here is centered on the interface with the theory of star formation, which can now be used to constrain theories of formation of the planets, as well as on the application of tools developed in the study of other astrophysical disks (disk galaxies, accretion disks, etc.) to the protoplanetary disk. In the 1980's, the *Voyager* spacecraft offered the first close look at the structure of the outer planets, and the incorporation of this data into theories of planet formation will only be fully realized in the 1990's.

The questions of isotopic anomalies in solar system material relate solar system formation questions to nucleosynthesis and galactic evolution as well as the overall question of star formation.

Laboratory Astrophysics in the 90's

We anticipate that the 1990's will provide several significant, new windows on our universe. With the Great Observatories, we can expect an enormous amount of new data in the infra-red, optical, X-ray, and γ-ray regions of the electromagnetic spectrum. In addition, new underground detectors and ultra-high energy cosmic ray experiments will provide information on neutrinos as well as attempt to shed light on the nature of dark matter in the universe. In order to interpret this large amount of new data, laboratory measurements in molecular, atomic, nuclear, and particle physics will become increasingly important.

Molecular, Atomic and Optical Physics

Because almost all our knowledge of the Universe reaches us in the form of photons, atomic, molecular and optical physics is an essential component of research in astronomy and astrophysics. An extensive data base containing reliable values of the parameters which characterize atomic, molecular and optical processes is an integral part of quantitative theories describing astronomical phenomena. The 1990's will see the deployment of an array of powerful new instruments for astronomical spectroscopy and an unprecedented growth in the quality and range of astronomical spectroscopic data. The increasing sophistication, precision and range of observational techniques and of theoretical models create new demands for more and better data on atomic and molecular properties. There is a still greater need for a deeper understanding of atomic, molecular and optical physics so that those processes that are relevant to the interpretation and guidance of the observations are identified, subjected to laboratory investigations, and incorporated into astronomical theories.

Quantitative analyses of the spectra of astronomical sources and of the processes that populate the atomic and molecular energy levels that give rise to emission and absorption require accurate data on transition frequencies, oscillator strengths, transition probabilities, electron impact excitation, deactivation and ionization cross sections, photoionization and photodetachment cross sections, radiative and dielectronic recombination and radiative attachment rate coefficients, and cross sections for heavy particle collisions involving charge transfer, excitation, and ionization and hyperfine and fine-structure transitions. Data on slow and on fast collisions are needed. If molecules are present, processes such as radiative association, rotational and vibrational excitation, ion-molecule and neutral particle chemical reactions, dissociative recombination, photodissociation and collision-induced absorption take place and must be quantitatively described. For example, even for the most simple and fundamental molecule, H_2, there are great uncertainties regarding the cross sections for collisional excitation of the vibration-rotation levels from which we observe emission.

Most of the mid to far-infrared radiation in the universe originates from interstellar dust grains. Both laboratory and theoretical work is needed on the physical properties of candidate grain materials, in order to interpret absorption, emission, and polarization measurements. The study of small grains, whose sizes (5-30 Å) put them in the transition regions between large molecules and bulk grains, seems especially fruitful. For example, the optical properties of polycyclic aromatic hydrocarbon (PAH) clusters or of hydrogenated amorphous carbon (HAC) require careful laboratory study in order to interpret the infrared continuum and emission features seen from interstellar dust clouds. One of the most important problems in laboratory astrophysics is the identification of the diffuse interstellar bands. The laboratory study of small organic grains/large molecules in the 1990's could solve this long-standing problem. Laboratory spectroscopy of mixed molecular ices of astrophysical interest is another area of great potential which will help exploit the data of such missions as SOFIA and SIRTF, revealing the origin and evolution of solid matter in interstellar clouds and in the solar system. In the near and mid-IR, laboratory studies have often driven observations leading to discoveries such as frozen CO on interstellar dust.

A combination of experimental and theoretical research is required to recognize which processes are significant to astrophysical environments and to provide the data base for detailed analysis. In the past, much of the research activity in atomic and molecular physics provided spectroscopic data needed by astronomers, but that is no longer the case. The experimental and theoretical research relevant to astronomy will not be pursued in the normal progress of research in atomic, molecular and optical physics. That research is driven by more fundamental considerations which will only incidentally produce data useful to astronomy. Our ability to carry out research in laboratory astrophysics has diminished severely over the past decade and scientists with the appropriate skills and interests are becoming rare, and the decline in quality of existing laboratory equipment is a grave concern. Because the problems raised by astronomers are almost never at the forefront of research in atomic and molecular physics, funding cannot be obtained from sources that ordinarily support atomic and molecular physics. Therefore, in order to obtain the atomic, molecular and optical data that underpin many astrophysical studies, astronomy will have to provide support in a more systematic way and with a longer term perspective than it has in the past. Another important consequence of the divergence between the research directions of atomic physics and astrophysics is the need to train and provide academic positions for people active at the boundaries between the two fields. Prominent practitioners of "astro-atomic physics" are approaching retirement. Very few of their highly qualified students have found secure academic homes, as physics and chemistry departments (who, in any event, are reluctant to recognize atomic and molecular physics as high priorities) regard such people as astrophysicists, i.e., not engaged in cutting-edge physics. It is as crucial for the community to provide secure positions for such people as it is for the funding agencies to support them.

Nuclear Physics

Nuclear physics continues to provide key input to astrophysical calculations. From solar neutrinos, to stellar nucleosynthesis and energy generaton, to big bang nucleosynthesis, to understanding the mechanism for supernovae, to black hole formation, to cosmic ray propagation, nuclear physics has played an important role in the last ten years that will continue into the next decade. The 1990's are likely to see continued research in nuclear physics with application to astrophysics. Likely developments include: utilization of new techniques to measure important reaction cross sections, the measurement of key reactions in inhomogeneous big bang models, and direct and indirect measurements of reactions involving radioactive nuclei important for understanding explosive burning.

While there is strong interest from the nuclear physics community in astrophysical problems, there is a clear change of direction in nuclear physics that is likely to have an important impact on nuclear astrophysics. This new direction is towards a small number of very large facilities which focus on the frontiers of nuclear physics. In order to proceed with these new facilities in an era of nearly constant budgets, many of the older cornerstones of nuclear physics (small tandems, dynamitrons, cyclotrons) have had to be shut down. This winding down of small facilities can only be expected to continue over the next decade. It is, of course, these small machines that have been the workhorses of nuclear astrophysics.

The last decade has also seen a significant increase in the use of large higher energy facilities in attacking astrophysical problems. The next ten years will certainly present new opportunities with the construction of the Relativistic Heavy-Ion Collider (RHIC) searching for the quark-gluon phase transition (crucial to our understanding of the big bang and potentially important in neutron stars as well), along with expanded use of present facilities.

Particle Physics

Fundamental research in particle physics has recently spawned whole subfields in astrophysics. In particular, the physics of the early universe is closely coupled to particle physics. There has also been significant interplay between particle physics and astrophysics with respect to the properties and detection of neutrinos. We can expect that the 1990's will see a wealth of new information relevant to astrophysics as new accelerators come on line. Important data has already arrived from the e^-e^- colliders, SLC and LEP, and new data from these will continue during the 1990's. New data at even higher energies (and therefore back in time closer to the big bang) is expected in the coming decade from the new hadron colliders — SSC in the US and LHC at CERN.

In addition, small scale experiments probing the properties of elementary particles will certainly play an important role for astrophysics. Such research, by better defining the scope and limitations of the "Standard Model," will have immediate feedback to astrophysics and should be strongly supported. As in other fields, there is the concern that in advancing to the forefront of particle physics, some facilities/capabilities that are important in addressing astrophysical questions will be forced out of operation. One important example in this context is the future availability of medium and high energy neutrino beams, which can play an important role in probing neutrino interactions and searching for possible new phenomena such as neutrino oscillations.

III - FUNDING RECOMMENDATIONS

We begin with two overall funding recommendations, and then give specific recommendations for each agency, broken down into theory and laboratory astrophysics.

1. Commensurate support for theory. NSF should establish a separate theory program funded at a level commensurate with that of other federally funded basic research in the physical sciences, at about 15% of the University grants program. NASA's support for theory, already strong, should grow with the increase in the science portion of the astrophysics budget. DOE should support theoretical astrophysics at universities and DOE laboratories insofar as it is relevant to its mission.

2. Laboratory astrophysics initiative. NASA should establish a long–term program in laboratory astrophysics to support major missions such as the Great Observatories and CRAF–Cassini. NSF Astronomy should find new funds to establish a viable program in laboratory astrophysics, to be coordinated with the Physics and Chemistry Divisions. DOE should support laboratory astrophysics, particularly atomic and low energy nuclear, insofar as it is relevant to its core programs.

Recommendations by Agency: Theory

NSF

Let us first discuss the recommendation regarding theoretical funding. In particular, we noted from interviews with physics grants officers at NSF and DOE that physics programs tend to put 15% to 22% of the operations and university grant funds into physics theory, whereas at NSF astronomy, theory was at a level of about 9% of university grants or about 3% of the overall AST annual budget. Even when capital equipment costs are removed from the total budget and only operations and university grant funds are examined, the amount (about 5%) is obviously well below that allocated for theory in physics. Furthermore, this amount has fluctuated considerably over the years in fact as well as in the interpretation of what is defined as theory. A third case is NASA which has made dramatic strides in theory support following the field committee report. Now, out of all astrophysics at NASA, approximately 10% is theory (see Table 1 and Figure 1).

Grants to individual investigators have traditionally been one of the greatest strengths of U.S. science, particularly at NSF. Theory is one area where such individual grants not tied to "major missions" are crucial. Therefore, the strength of NSF support in theoretical astrophysics is vital to our future.

*1. *Theory Program.* Theory is by its nature often both multidisciplinary and interdisciplinary, and therefore does not readily fit into the object-oriented classification of the Astronomy Directorate at NSF. A theorist might use the same tools to study radiative transfer in molecular clouds as in intergalactic clouds; to analyze the dynamics of accretion disks around protostars as in active galactic nuclei; or to simulate the formation of a star as of a galaxy. The best theorists often apply a variety of theoretical tools to a wide range of problems, and this broad, non-mission oriented research is particularly well–suited for NSF support. We therefore repeat the recommendation in the Field Committee Report that a separate program officer be appointed for theoretical astrophysics. We note that the separate theory program in the Physics Division of NSF has been highly successful. In order for the theory program in the Astronomy Division to be equally successful, it must be adequately funded: a level of at least 15% of the university grants program would begin to approach that in the Physics Division, and would be consistent with the recommendation of the Field Committee (which was never implemented) that the level of funding for theory be increased by about 50%.

*2. *Postdoctoral Program.* A postdoctoral position has long been an essential component of the training of a theoretical astrophysicist, but the small size of typical NSF grants makes it difficult for an individual investigator to hire a postdoctoral fellow. Only a very small fraction of the postdoctoral postions sought by Ph.D. graduates in theory are funded by the NSF; some go to the NASA theory centers, while many other go abroad. The problem is particularly acute in theory, for several reasons. One is that the desirable goal of allowing the fellow to work on a variety of problems, often not specified in advance, is incompatible with the usual project-oriented grant; a second is that because theorists are less closely tied to major telescopes or spacecraft, there is less chance of obtaining funding through a component of a large-budget project. We therefore recommend that the Theory Program provide funds for 5 to 10 three-year postdoctoral positions per year through grants to individual investigators and groups of investigators. Funding for these positions should be based on the scientific accomplishments of the proposers and, for those investigators who have supervised postdocs in the past, the success of the postdocs in their subsequent careers; funding should not be based on a detailed research plan for the proposed postdoctoral fellow. We also note the attractiveness of the reduced overhead rates employed in certain programs such as NSF's Research Experience for Undergraduates (REU) which are awarded as low overhead supplements to existing grants for the support of undergraduates. It would be highly desirable if a similar style program could be developed for postdoctoral researchers.

*3. *Theory Programs in the National Astronomy Centers.* Theoretical astrophysics provides the underlying concepts and the basic models with which data are interpreted, and it is thus an essential part of any astronomy program. We therefore recommend that the National Astronomy Centers (NOAO, NRAO and NATC) establish strong theory programs to support their user communities. The theory program should involve both a visitors program, to bring in theorists working on problems of relevance to the Center for long term visits, and joint support of permanent theoretical staff with local universities.

NASA

NASA has responded commendably to the Field Committee's recommendation that it establish a strong, broad program in theoretical astrophysics. Many of the arguments made by the Field Committee still apply. New observations from missions operating in the 90's, including COBE, HST, ROSAT, GRO, EUVE, XTE, SOFIA, AXAF, NAE, SIRTF, HETE and FUSE, are likely to "consume" a great deal of existing theory and point the way to more sophisticated modeling and interpretation; indeed, if this were not true, the missions themselves would be scientific failures. The complete wavelength coverage of these missions, from radio to γ–rays, demands the integrative effects of theory if the full scientific benefits of these missions are to be achieved. It is therefore crucial that NASA maintain and even expand its support for theoretical astrophysics in the 1990's.

NASA GRANTS
In Constant FY90 Millions of Dollars

Fiscal Year	'87	'88	'89	'90	'91	'92	'93	'94	'95
Theory in R & A Program									
Astrophysics Theory Program	2.0	2.0	2.0	2.0	4.0	4.0	4.0	4.0	4.0
Wavelength-Specific Theory	3.3	3.2	3.2	3.0	3.0	3.0	3.2	3.3	3.5
Total R & A Theory	5.3	5.2	5.2	5.0	7.0	7.0	7.2	7.3	7.5
Theory in MO & DA Grants	1.1	1.5	1.6	3.3	6.9	7.4	8.5	9.7	10.3
Total Theory	6.4	6.7	6.8	8.3	13.9	14.4	15.7	17.0	17.8
Research & Analysis	24	23	23	22	24	24	25	26	27
MO & DA Grants	12	15	17	31	58	65	77	85	91
Total Astrophysics Science Grants	36	38	40	53	82	89	102	111	118

Note: Based on data provided by NASA. Figures for FY '91 and beyond are projections prior to input from this report.

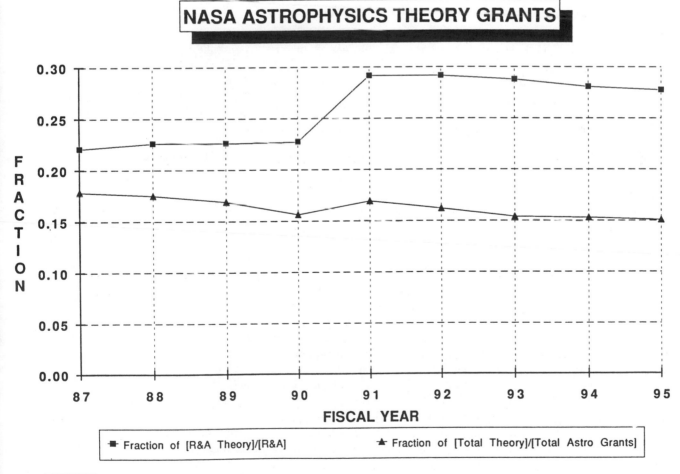

FIGURE 1

Table 1 and Figure 1 show the NASA budget for research grants to universities and national centers for FY87 through FY90. Also shown are the projections for FY91 through FY95. Note that the budget for the Astrophysics Theory Program is expected to double in FY91 and to remain constant thereafter, although this projection is not yet definite. The support for wavelength-specific theory (i.e., not including the Astrophysics Theory Program) is about 15% of the Research & Analysis budget. Funding for theory of both types is provided by the High-Energy, UV/Optical, and IR/Radio Branches. There is also support for theoretical astrophysics from the Mission Operations and Data Analysis Program. As a result, the total support for astrophysics theory by NASA somewhat exceeds 15% of the total budget for astrophysics science grants.

The specific recommendations for NASA by the Panel are as follows:

*1. NASA should expand the Astrophysics Theory Program to $4 million in FY91 (as planned) and progressively to $8 million/year by the end of the decade (in constant FY90 dollars). A mixture of large and small groups should be supported. As the funding increases, the announcements of opportunity should be issued more frequently. Exceptionally meritorious proposals should be funded for five year terms.

*2. NASA should continue to support wavelength- and mission-specific theory in both the Research & Analysis Program and the Mission Operations and Data Analysis Program so that the total support for theory (including the Theory Program) is at least 15% of the total support for astrophysics science grants. The solar and planetary divisions should continue their strong suport of theory as well.

*3. NASA should continue to support theoretical astrophysics at the Space Telescope Science Institute and at NASA Centers such as Ames and Goddard. There should be at least comparable theoretical presence at any new organizations responsible for the scientific oversight of the Great Observatories (GRO, AXAF and SIRTF).

DOE

* DOE should support theoretical astrophysics at universities and the DOE laboratories insofar as it is relevant to DOE programs (e.g. particle astrophysics, nuclear astrophysics, atomic and molecular astrophysics, plasma astrophysics, computational astrophysics). The existing theoretical programs at Fermilab, Lawrence Livermore National Lab, and Los Alamos National Lab are successful examples which should be strongly supported and should be emulated at other DOE labs, such as RHIC and the SSC as they develop.

Recommendations by Agency: Laboratory Astrophysics

Theoretical and experimental laboratory astrophysics require special attention because of some unique problems.

Laboratory astrophysics is interdisciplinary. This asset can, unfortunately, lead to bureaucratic confusion and inaction when no branch, division, agency, etc. will take responsibility for reasearch projects with applicability to several fields of astronomy.

Much of the atomic, molecular and nuclear laboratory astrophysics that is vital to astronomy is no longer at the forefront of basic research in physics and chemistry. Consequently, data for astronomy can no longer be expected as spinoffs from fundamental research supported by funding agency divisions that are not dedicated to astronomy. Laboratory astrophysics must be supported by astronomy divisions.

In the particular case of experimental laboratory astrophysics, a decade of reduced support has left the laboratory infrastructure in decay and the practitioners demoralized and, thus, leaving the field. Increased and consistent funding is needed to maintain the national capability to provide fundamental atomic, molecular and nuclear parameters required for effective use of the vast amounts of data that will be produced by the new, very expensive, ground-based and space observatories of the 1990's.

The creation of new funds expressly for laboratory astrophysics, with the goal of dealing with specific astrophysical problems, will attract more researchers in the related fields. In addition, if some of these funds are used to improve the communication between astrophysics and the related fields, such as by cross disciplinary workshops/conferences, the astrophysical problems can be brought to the attention of those most competent to do the research.

Mechanisms are needed to insure long-term stable funding for laboratory and theoretical groups interested in pursuing nuclear, atomic, molecular, and optical research related to astronomy and astrophysics.

If the decline in capabilities in laboratory astrophysics is to be reversed and the research strengthened so that the present and future needs of astronomy can be met, new funding will have to be found, dedicated to laboratory astrophysics. New researchers will be attracted to the field if sufficient long-term support can be assured and steps are taken to stimulate laboratory activities by connecting them to significant astronomical questions.

NSF

* To ensure the continued viability of laboratory astrophysics, we recommend that the NSF provide new funds in laboratory astrophysics in the divisions of astronomy, physics and chemistry with a particular emphasis on upgrading laboratory equipment and enlarging the number of graduate students entering laboratory astrophysics, including its theoretical component. Laboratory astrophysics cuts across the sub-divisions of astronomy, and the NSF should look to coordinate its support across astronomy and with physics and chemistry so that the funding of individual proposals can be provided at an adequate level.

NASA

* NASA's missions in the coming decade, in both astrophysics and in planetary sciences, will provide data of a range and quality far beyond our ability to interpret with our existing knowledge of the physics (and chemistry) of atoms, molecules, and dust grains. We recommend that NASA recognize laboratory astrophysics as a significant part of data analysis and allocate to the support of laboratory astrophysics an appropriate fraction of the funds for data analysis.

DOE

1. The primary mission of the DOE is not astrophysics. Nonetheless, much of laboratory astrophysics is related to the mission of the DOE, and the cross–fertilization of ideas between energy research and astrophysics is beneficial to both. We recommend that the DOE include the astrophysical implications of proposed research in its assessment of its scientific merit; laboratory astrophysics proposals which are related to the mission of the DOE and which are highly rated scientifically should be funded on the same basis as other highly rated DOE proposals.

2. The continued use of large higher energy facilities and the employment of new facilities like the Relativistic Heavy-Ion Collider (RHIC) for nuclear astrophysics are to be encouraged. Much of the relevant work in nuclear astrophysics, however, needs to be done on machines that are no longer regarded as being on the cutting edge of nuclear physics research. The Panel recommends that some of these more modest machines be maintained in operation and that astrophysically-oriented experiments be regarded as a key component of the research done on the machines. These machines should be available to outside users through collaborations, and would be used for nuclear physics, atomic physics, and applied physics, as well as astrophysics.

NIST

* NIST is world-renowned for the quality of the atomic, molecular, and optical research carried out in its laboratories in pursuit of its mission to maintain physical standards. Astronomers, too, are deeply involved in standards of measurement, and their ability to measure time (with the aid of millisecond pulsars) now rivals that of laboratory scientists at NIST. Both astronomers and scientists at NIST have a strong interest in the measurement of the properties of atoms and molecules, and in maintaining compilations of these data. We therefore recommend that NIST strengthen its ties to laboratory astrophysics; the astrophysical implications of the research carried out in its laboratories should be accepted as a valid subsidiary justification for the research.

All agencies

* A major need is the establishment of efficient Data Centers in the various branches of laboratory astrophysics. Until now, heroic efforts on inadequate funding have been made by a small number of people at a small number of institutions to assemble, assess and catalogue laboratory data. The work has been invaluable, but needs to be extended with a greater emphasis in ease of access and the use of electronic means of communication. This is already done in particle physics at LBL in a way that is compatible with astrophysics needs. Similarly, the Kellogg Lab at Caltech has carried out this function to date in nuclear astrophysics, but it is unclear

how long this can continue. Atomic and molecular data are not so well archived from an astrophysical viewpoint, and this needs to be remedied. We recommend that small, efficient data centers be established in atomic, molecular, and nuclear astrophysics by the appropriate agencies through the normal peer review process. The agencies should coordinate so as to avoid duplication of effort.

IV. BASIS - THE SUCCESSES OF THE 80'S

Theory in the 80's

Theory performs a number of functions in the overall enterprise of astronomy. It provides the basic paradigm within which observations are framed and without which they degenerate into a catalog of uninterpreted data. At its most satisfying, theory makes predictions that are later verified by observation. Theory can also have dramatic impact in a "post-dictive" mode by explaining previously observed phenomena. It catalyzes specific observations, which in turn stimulate new model building in a mutually interactive enterprise that drives progress in the field. Finally, theory provides much of the conceptual stimulation that invests astronomy with excitement.

In assessing the impact of theory in the decade of the 1980's, it is important to bear in mind that the big payoff of successful predictions, explanations, or guiding frameworks may be a long term process, so that the successes of the 1980's often had seeds planted earlier. Likewise, the general activity in theory in the 1980's may not bear fruit until the 1990's or beyond. In the following discussion, paragraphs marked with "*" highlight verified predictions.

The Solar System

In the solar system, the *Voyager* spacecraft provided dramatic proof or illustrations of a number of theoretical concepts, including tidal heating as a cause of vulcanism, and density waves and shepherd satellites in planetary rings. Further impressive accomplishments were compelling explanations of Jupiter's great red spot, the Kirkwood gaps, and the apparently chaotic rotation of Saturn's satellite Hyperion.

Observations of the vibrational spectrum of the Sun led to the new science of helioseismology, with a firm theoretical basis that was rapidly developed. Models of the rotation of the solar interior based on the observed rotational splitting in the spectrum overturned previous conceptions of rotation on cylinders, but a fully self-consistent model that will account for the 11 (22) year solar cycle is still being sought in ongoing work. Continuing studies of solar neutrinos included new suggestions of induced mixing of neutrino types and the possible effects of hypothetical new particles. This work continues within a fundamental theoretical framework laid down in the 60's and 70's. The Sun is predicted to have a magnetic heliosphere, and Pioneer 10 and the Voyagers continue to search for the heliopause.

The conceptual notion that impacts of massive extraterrestrial bodies may have led to major biological extinctions illustrates the strong linkages that can develop between apparently unrelated disciplines, and the importance that apparently arcane theoretical studies (for example, the structure and dynamics of the Oort cometary cloud) can suddenly have for other subject areas (for example, the environmental consequences of a major impact may be similar to the effects of nuclear winter). In the broadest context, work in this area has led to the recognition that the solar system is a dangerous and unpredictable place, and that astronomical catastrophes are likely to have had a profound influence on the developments and survival of life on Earth.

The Interstellar Medium and Star Formation

Significant progress was made in the study of the interstellar medium (ISM) and star formation. Large structures in the ISM—supershells and superbubbles—with diameters of hundreds of parsecs were successfully modelled as being due to correlated supernovae in stellar associations, although for some structures—particularly the largest ones—this model has some difficulty. Observations of the diffuse ISM continue to be interpreted in light of a three–phase model developed in the 70's, in which the ISM is divided into cold ($T \sim 10^2 K$), warm ($T \sim 10^4 K$), and hot ($T \sim 10^6 K$) phases in approximate pressure equilibrium. In the last decade, debate raged as to the pervasiveness of the hot component of the ISM and on the relative importance of magnetic fields in determining the observed structure of the ISM. The discovery of small dust grains and polycyclic aromatic hydrocarbons through infrared observations led to extensive theoretical study of their role in the thermal and chemical balance of interstellar clouds. The theory of interstellar shocks grew to encompass shocks in which the dissipation is effected by ion–neutral collisions, of particular importance in molecular gas. Observations of H_2 fluorescence confirmed theoretical calculations of the response of molecular clouds to UV radiation. The theory of acceleration of cosmic rays by shocks was put on a firmer footing, though many questions remain. A fundamental understanding of the fluctuation of pulsar radiation in terms of interstellar scintillation was developed.

A major accomplishment of the 80's was the development of a coherent picture for the formation of low–mass stars: Beginning with dense molecular cores of several $M_\odot$ observed in many molecular clouds, the theory shows how ambipolar diffusion of weakly ionized gas leads to a collapse which commences at the center of the core and which leads to the formation of a young stellar object with properties similar to those actually observed. The magnetic field is crucial in reducing the angular momentum of the gas, but the residual angular momentum leads to formation of a disk, in agreement with infrared and submillimeter observations. Young stellar objects are observed to rotate far more slowly than would be expected if they formed from a disk, indicating that a substantial fraction of the binding energy must be dissipated. It was suggested that this dissipation is accomplished through a magnetically driven wind, thereby producing the bipolar flows which are a ubiquitous feature of low–mass star formation. The magnetic field is an essential ingredient in this entire picture, and models of magnetically supported clouds received striking confirmation at the end of the decade when direct observations of the magnetic field in molecular clouds showed approximate equipartition between the magnetic energy and the turbulent energy over a range of densities.

Stars

*The area of stellar evolution gave the most spectacular confirmation of a long-standing theoretical prediction. Supernova 1987A produced a flux of neutrinos, providing the first detection of extra–solar neutrinos. The information in those 19 detected events was not as detailed as would have been wished, but nevertheless yielded data on the time of arrival and energy spectrum of the neutrinos that was completely consistent with decades of work on the problem of gravitational collapse and neutron star formation. This observation and associated theory put important limits on the lifetime, charge, and magnetic moment of the electron neutrino. Another important prediction was that of the existence and properties of the gamma ray flux associated with the production and decay of radioactive ^{56}Ni to ^{56}Co and then to ^{56}Fe. Not only were gamma rays and X-rays from SN 1987A observed, and their properties used to study the nature of the explosion and subsequent mixing, but freshly synthesized Ni, Co, and Fe were directly observed, especially in the infrared. The mass of radioactive nickel produced by the explosion was predicted ahead of time to within a factor of two. The successful interplay between theory and observation was particularly apparent in the planning of the NASA missions to observe the supernova.

Theoretical study of SN 1987A provided basic explanations of the evolution of the light curve and spectra and led to a growing confidence that the supernovae will become valid distance indicators. Estimates of the distance to the LMC based solely on SN 1987A agree with classical estimates to within about 10 percent.

A great deal of work was done on the evolution of massive stars with an attempt to incorporate mass loss from radiative driven winds self-consistently. Such work was able to account for the observed properties of a wide range of hot stars, from Wolf-Rayet stars to the nuclei of planetary nebulae. The theoretical Hertzsprung-Russell diagram was refined for comparison with the observations of globular clusters. The theoretical framework of evolution of asymptotic giant branch stars and various stages of "dredge-ups" continued to guide observations of physical and chemical properties of red-giant stars. In the context of binary star evolution, the framework associated with the notion of accretion disks provides a crucial guiding principle for both theory and observations.

* Theoretical studies of oscillating white dwarfs rejuvenated the study of white dwarf stars. Models incorporating a thin layer of hydrogen provided the first explanation of the ZZ Ceti phenomenon of variable DA white dwarfs and also predicted that an analog should exist among the helium-rich DB white dwarfs. This unexpected prediction was then verified observationally. This work has been extended to hot post-planetary nebulae, and the study of the variability along the white dwarf sequence has given rise to the subject of white dwarf seismology, which is giving information on the masses, composition, and structure of white dwarfs as well as their evolution. The latter provides a unique and independent way to determine the age of the Galactic disk.

High Energy Astrophysics

The 80's have seen major quantitative advances in the capability to compute the properties of classical nova explosions. Many aspects of dwarf novae can now be successfully accounted for in terms of a limit-cycle thermal instability associated with partial ionization of hydrogen in accretion disks. This theory is now being applied to some classes of X-ray transients, as well as to models for AGN. The observation of cyclotron lines in some gamma ray bursters was interpreted on the basis of theoretical models to mean that these bursters are magnetized neutron stars.

Theory continues to provide a major impetus to the study of a variety of collapsed objects. The general theory of gravitational collapse received support from SN 1987A, but it remains a major theoretical effort to understand the physical mechanism of supernova explosions. The 80's saw a great deal of work on accretion-induced collapse to make compact objects in close binaries. Continued study of the binary pulsar has led to more and more accurate confirmation of Einstein's theory of relativity, including the first evidence for gravito-magnetic effects in the form of geodesic precession of the pulsar spin axis, and strong new limits on the time variability of the gravitational constant.

In the study of galactic nuclei, the theoretical notion of a massive black hole with an accretion disk embedded in a dense star cluster provided the framework that continues to guide work on active galactic nuclei and quasars. Numerical and analytic models of jets had some success in accounting for the radio jets observed in extragalactic radio sources. Observations of polarized line emission in some Seyfert 2 galaxies led to a "unified" model for Seyfert galaxies in which the central nucleus is surrounded by an opaque torus of gas, and the object is classified as a Seyfert 1 or 2 depending on whether the line of sight to the nucleus misses the torus or not. This model was substantiated by X-ray observations.

Galaxies and Cosmology

The theory of galaxy mergers provided a framework to study a variety of issues involving the formation and evolution of galaxies. The shells observed in some elliptical galaxies were successfully modelled as the result of mergers. Cooling flows in clusters of galaxies were a major topic of theoretical work, with one of the central issues being the effects of the flows on the evolution of the central galaxies in the clusters. The 80's also saw a great deal of theoretical work in the area of N-body calculations aimed at understanding the dynamics of stellar clusters and galaxies. These studies provided insight into core collapse in globular clusters and the beginnings of a basis for understanding the tri-axial structure of elliptical galaxies.

* The discovery of gravitational lenses provides an excellent example of the role of theory in astronomical discoveries. Essentially all the key features of lenses were predicted theoretically, on the basis of general relativity, in classic papers by Einstein, Russell, and Zwicky in the 1930's. After the discovery of lensed quasars early in the last decade, theorists refined their calculations to the point that there is now detailed agreement between theory and observation in a number of cases. The unexpected discovery of luminous arcs in some clusters of galaxies has been successfully accounted for in terms of gravitational lensing. The cosmological implications of the observed lenses have also been worked out.

The redshift debate was all but settled in the flood of observations that verified the basic theoretical result that the Hubble flow is the only viable source of the systematic redshifts of distant galaxies and quasars. The Lyman alpha clouds along the line of sight toward quasars were studied as indicators of the intensity of the intergalactic radiation field, as probes of large-scale structure, and as sources of baryonic material. Theoretical study of the intergalactic medium showed that, with current counts, quasars are unable to ionize the intergalactic medium.

One of the most interesting challenges facing theorists in the last decade was the attempt to understand how galaxies evolved in a Universe filled with such a uniform cosmic microwave background radiation. Searches were made for ever fainter galaxies in an attempt to see if voids were filled with small galaxies, as predicted by theories of cold dark matter. The continuing null results of the search for fluctuations in the background radiation were made critically interesting by the importance attached to them by theory.

*The Sunyaev-Zeldovich effect—the distortion of the microwave background by Compton scattering in clusters of galaxies—represents a major theoretical prediction which is receiving substantial support from observation.

*The 80's saw the basic big bang model predictions further refined and verified. Of particular importance have been the verification of predictions of standard homogenous Big Bang Nucleosynthesis (BBN). The early focus of BBN was on ^4He, but during the 70's, with the theoretical work showing how deuterium could be made only in the big bang, the focus turned to deuterium and its concomitant limit on the baryon density of the Universe. The end of the 70's saw the deuterium and ^3He constraints combined to predict that consistency could occur only if the abundance of primordial lithium was $Li/H \sim 10^{-10}$ and if the density of baryonic matter is about 1/10 the critical density. Observations in the early 80's appear to have confirmed the predicted lithium abundance, thereby providing a strong argument that most of the Universe is made up of some form of non-baryonic dark matter. (An alternative, less favored explanation is that the density of the Universe is only 1/10 critical.) The 80's also saw a flurry of activity in examining inhomogeneous non-standard BBN models. By the end of the decade, the robustness of the standard model was demonstrated by the failure of the efforts made to circumvent it.

*One of the most decisive verifications of an astrophysical prediction came when particle accelerator experiments at CERN in Geneva, Switzerland and at SLAC in Palo Alto confirmed that there were only three generations of light neutrinos. Theoretical calculations of the light element abundances (particularly ^4He) due to BBN are inconsistent with observation if there were a large number of types of light neutrinos.

The concept of inflation was one of the most important advances in cosmology in the 80's. Inflation raised the stakes in the competition between cosmological models, by showing that at least one plausible model based on Grand Unified Theories (GUTs) could explain the observed isotropy of the universe, provide a natural explanation for why the density is near the critical value, and suggest a natural spectrum for primeval density perturbations. While it remains to be seen whether inflation proves to be correct in detail, it has greatly expanded the horizons of what cosmologists can hope to explain, and has provided the framework for much of the work on the early universe that took place in the 1980's.

The interconnection of particle physics and cosmology that developed during the 80's is illustrated by other examples as well. Astrophysical constraints set limits on the existence and properties of various particles. The physics of baryon non-conservation in the GUT theories gave the first models for the origin of the excess of matter over antimatter and for the ratio of baryons to photons in the Universe. Phase transitions predicted from particle physics played a role in the study of a variety of effects from magnetic monopoles to seeds for galaxy formation, to inhomogeneous Big Bang Nucleosynthesis. The ideas of cosmic strings, both normal and superconducting, and other topological defects launched a major theoretical industry to explore their properties and to seek connections to observed large scale structure. All this work pushes the frontiers of physics further back toward the Planck scale, the ultimate problems of quantum gravity, and the origin of the Universe.

Laboratory Astrophysics in the 80's

Atomic and Molecular Physics

Spectroscopic studies of a diverse range of atomic and molecular species at all wavelengths were carried out at the greatly enhanced level of precision made possible by lasers and they assisted in the identification of emission and absorption lines of a diverse variety of astrophysical objects. Spectroscopic measurements on carbon monoxide yielded information on CO that removed a major uncertainty in the description of the photodissociation of CO in circumstellar shells and interstellar clouds. Theoretical calculations and laboratory measurements successfully identified the source of infrared emission lines in the solar atmosphere as excited Rydberg levels of magnesium, silicon and aluminum.

Charge transfer was recognized as an important mechanism for redistributing the charge in a high excitation nebulae. Emission lines resulting from the charge transfer of multiply-charged oxygen and neon in hydrogen were predicted and found in the spectra of planetary nebulae. Experiments to measure charge transfer at low energies were designed.

A productive, mutually stimulating collaboration occurred between observers of molecular emission and absorption lines and theorists carrying out basic quantum-mechanical calculations in joint studies of the spectroscopy of possible candidate molecules. The collaboration led to the discovery and identification of interstellar cyanoethyl, butadinyl, cyclopropenylidene, dicarbon sulfide, tricarbon sulfide and protonated hydrogen cyanide and of circumstellar silicon dicarbide, silicon carbide and deuterated butadinyl.

The molecular ion H_3^+, an ion of central importance in the ion–molecule theory of interstellar chemistry, was observed in emission in the atmosphere of Jupiter. Quantum-mechanical evaluation of the potential energy surface followed by the calculation of all the energy levels and transition probabilities provided a prediction of the emission line frequencies and intensities, whose accuracy could be assessed by experimental measurements of selected transitions. On the basis of the laboratory studies, emission lines appearing in the spectrum of Jupiter, originally labelled unidentified, were definitely attributed to H_3^+. The detection of H_3^+ is the first observational evidence for extraterrestrial H_3^+. The interplay of theory, experiment and observation was further demonstrated by searches for interstellar H_3^+ by observations of absorption lines looking towards infrared sources.

Major strides were made in laboratory studies of chemical reactions and the first measurements were made at temperatures below 50K. The expected enhancement of the rate coefficients of many reactions of positive ions with heteronuclear molecules at low temperatures was established experimentally and incorporated into models of interstellar chemistry. A beginning was made on the identification of the products of dissociative recombination. Progress was made experimentally and theoretically in understanding the role of internal energy modes in chemical reactions.

Another area in which the three pronged attack of experiments, observations and theory has resulted in very significant gains in our understanding is that of interstellar solids. Interstellar dust and ice composition went from a field rich in speculation to one in which laboratory analog studies provided strong constraints on observations and theories. For example, these studies predicted the presence of, and guided the subsequent detection of, important ice constituents such as carbon monoxide and methanol, revealing the complex interplay between the gaseous and solid phases in molecular clouds.

Experimental studies of excitation and ionization due to electron impact were improved in accuracy and were extended to many more systems in states of high ionization. They were complemented by increasingly elaborate theoretical models which made evident the importance of resonance structures in electron impact processes. A significant contribution of theory was the recognition of the importance of dielectronic recombination at nebular temperatures. The first reliable field-free measurements of dielectronic recombination were carried out. Transition probabilities of intersystem lines of light-atomic ions of astrophysical importance were measured by using ion traps to confine the ions and detecting the exponential decay of the emission intensity from metastable levels populated by laser radiation or by electron impact.

The Opacity Project was initiated under the guidance of M. J. Seaton. The project is an international effort to exploit the development of powerful theories by applying them systematically, using fast large computers, to calculate atomic and ionic parameters such as oscillator strengths, transition probabilities, radiative and dielectronic recombination coefficients, photoionization cross sections and line-broadening parameters. The activity, though of inestimable value to a broad range of applied physics, was stimulated primarily by the demands of astrophysics. The first results from the project are appearing.

Plasma Physics

A deepening understanding of the physical processes in plasmas was developed. Particular advances were achieved in aspects of thermal conduction, plasma relaxation and beam-plasma interactions. The concept of a saturated heat flux, which has been important for evaporation of interstellar clouds, was observed and explained in laser-irradiated pellets. Inhibition of thermal conductivity by small scale fluctuations has been seen in confined plasmas and invoked in stellar coronae and galaxy cluster cooling flows. The idea that magnetically dominated plasmas relax to a minimum energy state consistent with constant total magnetic helicity was developed to interpret reversed-field-pinch devices and has been used to model extragalactic radio jets and solar coronal loops. Beam-plasma interaction experiments were instructive in interpreting observations of radio emission from pulsars and from solar and stellar flares.

Nuclear and Particle Physics

The last decade has seen significant advances in both nuclear and particle physics, which have contributed to our understanding of astrophysical phenomenon. New information on nuclear cross sections has contributed to our understanding of solar neutrino production and big bang nucleosynthesis, along with stellar evolution and nucleosynthesis. New measurements of fundamental particle properties have also provided key input to our astrophysical understanding. More detailed measurements of the nuclear reactions important to the p-p chain have solidified our confidence in the nuclear input to the standard solar model. This new work thus makes a nuclear physics solution to the solar neutrino problem less likely. Also of relevance to the solar neutrino problem, medium energy measurements of (p,n) reactions and the extracted Gamow-Teller matrix elements have provided information on neutrino capture cross sections for several of the terrestrial solar neutrino detectors. New measurements of the key reactions of the standard big bang have improved our confidence in the predictions of the abundances of light elements produced during the big bang. While significant work has also been done on the reactions of helium burning, considerable uncertainty remains in some of the key reactions. New work with radioactive targets has also provided information on several reactions that occur during explosive burning where unstable nuclei can play a role.

Measurements of certain isotopic anomalies have also shed light on the origin of the solar system and the role of certain nuclear processes during nucleosynthesis. Detailed measurements of beta-decay strength functions for nuclei far from stability has allowed a better understanding of the path of nucleosynthesis as it approaches both the proton and neutron drip line.

In the particle physics realm, constraints on the possible modifications to the standard model of the strong and electroweak interactions, as well as better understanding of particle properties, have contributed significantly to astrophysics in the last decade. New data from SLAC and CERN has provided strong confirmation of the standard model, with the discovery of the W and Z^0 particles. And very recently, the detailed measurements of the Z^0 decay have constrained the number of possible generations (limiting any additional new quarks and leptons) to three, in excellent agreement with the bounds obtained from big bang nucleosynthesis. New high precision measurements of the free neutron lifetime have significantly reduced several key astrophysical uncertainties. These new measurements (some using advanced neutron storage-bottle technology) have direct bearing on the predictions from the standard big bang and on the rate of the p-p reaction in the solar model.

V - POLICY ISSUES

Agency Responses to Field Committee Report

Some of the recommendations of the Field Committee Report related to Theory and Laboratory Astrophysics have been implemented, but there has been little response to others. In general, we give higher marks to NASA than to the NSF in effectively responding to the Field Committee recommendations.

NASA has significantly increased its support of theoretical astrophysics through the Astrophysical Theory Program. This program has strengthened or catalyzed the formation of several research groups and supported a significant number of postdoctoral research associates and graduate students. The NASA Graduate Student Researchers Program has also supported a number of students working in theoretical or laboratory astrophysics. We consider this program to be highly successful.

The most significant positive development related to NSF funding is the Presidential Young Investigators program. The recipients of these awards seem to be able to ramp up to a high level of activity more quickly than investigators without these awards, and they are competitive in seeking NSF support once their PYI funding has ended.

Unfortunately, theoretical and laboratory astrophysics have not fared well within the general NSF astronomy grants program. Far from the 50% increase recommended by the Field Committee, these areas have actually lost ground. The fraction of the NSF astronomy budget going into Theory is about 1/3 the fraction going into Theory in NSF Physics. Despite the recommendation of the Theory subpanel that Theoretical Astrophysics be set up as a separate program in the Astronomy Division, similar to Theoretical Physics in the Physics Division, theory remains balkanized among different programs, to the perceived detriment of researchers who are oriented towards general physical processes rather than towards specific astronomical objects.

The DOE laboratories have made some progress in supporting work related to astrophysics, but much remains to be done in the DOE supported areas of atomic and molecular, nuclear, particle, and plasma physics as it relates to astrophysics. Support for theoretical astrophysics at the National Astronomy Centers is essentially nil, with the laudable exception of the NASA-funded Space Telescope Science Institute. The NSF Astronomy Centers seem to have ignored the Field Committee recommendations.

Advanced computing has been supported by the NSF and DOE, and most researchers now have greatly improved opportunities for large scale scientific computing. However, the strong support for supercomputer facilities has not been matched by support for more modest facilities at the individual, group or departmental level. This is unfortunate for several reasons: first, the smaller machines are far more cost-effective for many tasks than supercomputers, and in some cases now actually approach the supercomputers in speed; a second reason is that the solutions of many problems in astrophysics are more limited by lack of imagination than lack of computation, so that it is especially important to provide flexible, friendly and accessible local computing that permits working theorists to solve small and medium scale computational problems with a minimum of effort and time.

Finally, we consider funding of institutes and workshops. The ITP at Santa Barbara has sponsored some excellent programs, but its service to the community is limited by lack of a permanent staff in astrophysics. NASA and NSF (Physics) support Aspen and other summer workshops, but more travel money must be made available if these workshops are to be made available to a broader section of the community.

Policy Recommendations for the 90's: Theory

The committee examined many possible recommendations, including such things as a new national center for Theoretical Astrophysics. However, the committee felt that at the present time the real needs in theory are for increased stable individual grant support and postdoctoral support, as mentioned under funding needs. Policy recommendations to individual agencies for Theory are given below.

NSF

*1. Establish a separate Theory Program in the Astronomy Division, as mentioned before.

*2. We recommend that the directorate of NSF responsible for education increase its funding for graduate fellowships and traineeships; this would benefit all science.

*3. The use of umbrella grants, where several theorists at a given institution combine their grants, should be an acceptable but not mandatory option in order to reduce paperwork and permit the pooling of resources. The resources to be devoted to each of the investigators on such a grant should be clearly spelled out.

*4. An "adiabatic turnoff"–i.e., a year of reduced funding– should be provided for the termination of long-term programs.

*5. Accomplishment-based proposals should be actively encouraged for senior theorists. Again, this reduces paperwork and, for the case of senior theorists, provides adequate information for evaluating the proposal.

NASA

* NASA is to be congratulated for recognizing the importance of theory in carrying out its mission. Continued support for the Theory Program, for mission-specific theory, and for theory at the NASA centers is essential if the benefits of the missions planned for the 90's are to be realized.

DOE

* DOE should support theoretical astrophysics insofar as it is relevant to its mission. The Panel notes that the DOE has used astrophysics to help justify new accelerators such as RHIC and SSC, but to date its funding of astrophysically related projects has not been enlightened. Particle astrophysics, nuclear astrophysics, plasma astrophysics and computational astrophysics are all of direct relevance to DOE's mission.

All agencies

* A policy issue which extends across the boundaries of all agencies is computational facilities. The development of major super-computing facilities through NSF and DOE has provided a major improvement in the computational environment in the last decade. However, these facilities do not satisfy all of the computational needs of the theoretical community. Moreover, recent advances in small scale computing now mean that many tasks previously requiring supercomputers can be handled on much smaller machines at a fraction of the cost with only modest losses in speed. Yet these smaller machines are often difficult for individuals or groups to acquire through individual grants. The Panel recommends that all agencies should recognize the importance of computers of a range of sizes as powerful tools for theoretical research, and that funding for powerful machines dedicated to the use of individuals, groups or departments should be increased.

Policy Recommendations for the 90's: Laboratory Astrophysics

NSF

    * A coordination across directorates is essential for laboratory astrophysics because of its inter-disciplinary nature. Too often we hear of important laboratory projects falling through the NSF cracks because they are "not of prime interest to the nuclear or atomic or ... communities," while at the same time Astronomy says that "this is nuclear or atomic or , not astronomy."

NASA

    * The need for laboratory data for NASA missions goes beyond individual missions. NASA should recognize this broad requirement for laboratory data as they have done for theory.

DOE and NIST

    * These agencies should recognize that acquiring laboratory atomic, molecular, optical and nuclear data relevant to astronomy is appropriate to their missions and should be supported.

VI. SUMMARY

The future of astronomy is dependent on maintaining an active and dynamic theoretical community and on making the critical laboratory measurements necessary for future observations. In this report we have attempted to show how to achieve this. The funding agencies should provide a level of funding for Theory which is commensurate with their funding for observational and experimental astronomy. Laboratory astrophysics requires new funds, particularly from NSF and NASA, in order to obtain the data that is essential for the interpretation of astronomical observations. The successes of the 1980's and the opportunities for the 1990's show that support for these programs is necessary if astronomy is to achieve its goals in the coming decade.

SOLAR ASTRONOMY PANEL

Solar Astronomy

Table of Contents

0. EXECUTIVE SUMMARY

The following is an abbreviated overview of our Panel recommendations, including the priority ranking of major missions, brief descriptions of these major projects, and a summary of recommendations regarding small-scale science, theory, and programmatics.

0.1 Strongly-supported Major Ongoing Projects

- Global Oscillations Network (GONG); cost: $15M; start: 1987
- Solar Heliospheric Observatory (SoHO); cost: $211M; start: 1988

- *GONG.* The Global Oscillations Network Group will produce definitive observations of solar global oscillations (p-modes), based on an Earth-encircling network of automated observing stations. The major aim is to reduce the effects of sidelobes (by dramatically increasing the time during which the Sun is observed), thus increasing the S/N, and making mode measurements much easier (and more reliable). These measurements, in conjunction with the on-going and planned solar neutrino experiments, allow us to peer into the solar interior, and thus allow us to test our understanding of how a star like the Sun is constructed.

- *SoHO.* The Solar Heliospheric Observatory contains coronal imagers and spectrometers, and a helioseismic instrument, and will be placed at the L1 Lagrangian point. It is a European spacecraft, but has U.S. participation in the form of instruments, ground support, and subsequent data analysis. This observatory has probably the best "shot" at observing g-modes (which will allow us to probe the deep interior), as well as the lowest-degree p-modes. In addition, the solar imaging instruments on SoHO are designed to look in great detail at a totally different aspect of solar activity than is normally examined, namely the mass outflows. They will also be able to look at the "closed" corona, and here their principal virtue is their spectroscopic capabilities, combined with imaging, which also allows detailed density and temperature diagnostics to be performed. SoHO will be able to measure temperature and density of coronal material out to several solar radii; to measure outflow velocities; and possibly to measure motions with sufficient accuracy that one could detect with waves in the atmosphere. Thus, these instruments will directly address the recalcitrant problem of accounting for the acceleration of the solar wind.

0.2 New Ground-based Solar Projects

1. High Resolution Optical Imaging (LEST); cost: $15M; start: 1991
2. Large-aperture Solar IR Telescope; cost: $10M; start: 1996

- *LEST.* Our highest priority ground-based project is a new moderate-to-large-aperture adaptive-optics telescope in the visible region. This entails:

    (a) a vigorous development program in adaptive optics, based on the existing facilities;

    (b) the development of a moderate to large-aperture high-resolution telescope, either separately or jointly with other countries. The United States is now a participant in the Large Earthbased Solar Telescope

(LEST) consortium, which plans to build such a telescope starting in 1991. Contingent upon progress in the adaptive optics area, we propose to contribute up to 1/3 of the total cost of that telescope. The main improvements over existing telescopes will be in resolution (down to 0.1"), intrinsic polarization (less than 1%), and improved detectors. LEST will thus function as a high angular resolution ground-based optical imaging tool for examining flow-magnetic field interactions; for example, its polarimetric capabilities will enable it to measure Stokes parameters with higher accuracy than OSL, though over a much smaller field of view.

• *LARGE-APERTURE SOLAR INFRARED TELESCOPE.* The infrared solar spectrum offers several innovative ways to study solar magnetism. The opacity minimum at 1.6 microns gives us the deepest look into the solar photosphere. At 12 microns, emission lines arising from highly excited states of magnesium give us our first good look at magnetic regions that are difficult or impossible to probe in the visible. A further powerful advantage for magnetic field research in the infrared is the quadratic wavelength variation of Zeeman splitting. Our proposal again entails two distinct objectives:

(a) Acceleration of existing programs for the development of focal plane instrumentation suited to measure magnetic fields in the infrared;

(b) Building a larger telescope than the 1.5m McMath (because of the improvement of seeing that occurs in the infrared, combined with the decreasing resolution of a fixed aperture at increasing wavelength, and the decreasing photon flux per spectrum line doppler width in the infrared), which means one of the following three options: (i) join with our nighttime colleagues in construction of a 10m class telescope capable of doing solar infrared daytime work (with suitable protection); or (ii) construct a new, large-aperture solar infrared telescope, possibly in combination with a large-aperture reflecting coronagraph; or (iii) upgrade the McMath telescope to a 4m aperture.

This facility will allow us to probe the 3-D structure of magnetic flux tubes in the solar surface layers, and thereby permit us to understand the processes which lead to these highly fragmented structures in the solar photosphere.

0.3 New Space-based Solar Projects

   0. Orbiting Solar Lab (OSL); cost: $500M; start: 1992
   1. High-energy Solar Physics mission (HESP); cost: $200M; start: 1995
   2. High-resolution Coronal Imager (Context); cost: $250M; start: 1998

• *OSL.* The Orbiting Solar Laboratory represents a marriage of a high angular resolution optical telescope with a high angular resolution UV spectrometer and X-ray imager on a free-flying polar orbiter. The main objective of OSL is to image the solar surface layers with a sufficiently high angular resolution (0.1-0.3") so as to observe directly the interactions between surface motions and the magnetic fields on the solar surface. To date, we have only been able to do this from the ground (through the highly obscuring terrestrial atmosphere) and on one brief Shuttle flight; OSL will allow us to view the interactions we believe give rise to virtually all of solar activity over time scales comparable to the duration of the solar activity cycle. The physical processes we want to understand, and which will be observed with this mission, are: (a) The formation of magnetic flux bundles — we want to understand why solar magnetic fields are so highly spatially intermittent; (b) the evolution of surface magnetic fields — we want to understand how the distortions of surface magnetic fields by the surface convection can lead to plasma heating, including the creation of a multi-million degree corona enveloping the Sun; and how the process of magnetic field evolution proceeds (we know that solar magnetic fields must decay, but the observed decay rate is many orders of magnitude larger than what classical theory would predict.)

• *HESP.* The High Energy Solar Physics mission will carry out high-resolution spectroscopy and imaging of solar hard X-rays and gamma-rays, and will detect neutrons. It will image X- and gamma-ray sources with resolution better than 1", and will resolve spectral features with $E/\Delta E \sim 1,000$ up to energies of 10 MeV. HESP will study the processes of nonthermal energy release in the active solar atmosphere by observing the X-rays and gamma-rays produced by particles accelerated concomitantly with flare energy release. We know that in solar flares the release of nonthermal energy is mediated by efficient particle acceleration, that energy is transported rapidly by the accelerated particles, and that the acceleration and transport must result from complex interactions of plasma with shocks, turbulence, and rapidly varying electric fields. However, we do not understand the detailed nature of these processes, not can we reliably differentiate among different mechanisms. We must determine the location of the site or sites of particle acceleration observationally, we must define the properties of the particles directly from the observations, and we must have a clear picture of the relevant magnetic configurations. Observations with high spatial resolution will allow the localization of the sources of the particles and the tracing of their transport paths. Observations with high spectral resolution will allow the deciphering of the rich information encoded in gamma-ray lines, such as energy spectra, angular distributions, abundances, and energy content of the accelerated particles. Because of the proximity of the Sun, the study of the solar high-energy processes provides one the best techniques for investigating similar processes which are known to play a dominant role at many astrophysical sites. HESP has a definite time-critical need — it must by flown by the year 1999 to take advantage of the next solar maximum. For this reason, we aim at a modest mission (in the "intermediate class", in the context of NASA/Space Physics Division plans). HESP will consist of a single major high-energy instrument capable of carrying out high-resolution imaging and spectroscopy simultaneously. It is by virtue of this simplicity of design that a mission in orbit by 1999 is possible.

• *High-resolution Coronal Imager (Context)*. The *Context* mission addresses a major problem in astrophysics: How are hot plasmas in stellar atmospheric structures created? *Context* takes advantage of the opportunity created by the measurements of *Solar Probe*, which will fly to within 4 solar radii of the photosphere and conduct *in situ* measurements there. Through high-resolution imaging and spectroscopy, *Context* can identify the structures through which *Probe* has flown, in order to understand the environment of *Probe*'s particles-and-fields measurements. Such a combination of *in situ* and remote sensing will be unprecedented in astrophysics.

0.4 Strongly-supported Major Interdisciplinary Projects

  • Solar Neutrino Experiments; no costs available
  • mm Array; cost: $115M; start: 1991

• *Neutrino experiments* refer here to the panoply of solar neutrino experiments supported by DOE and NSF, many carried out in collaboration with foreign scientists, including the Homestake mine experiment, the Japan/US collaboration Kamiokande II, the Canadian/US collaboration in the Sudbury experiment, the Gallium experiments, and so forth. The long-standing solar neutrino problem involves a remarkably broad range of physical issues, from stellar structure and opacity calculations to the physics of neutrinos themselves. More recent data have only deepened the mystery of these particles, as the possibility of correlations of the solar neutrino flux with the solar activity cycle has been raised. The discrepancy between what we observe and what can be explained by theory, is so substantial, and is so basic to physics, that the new experiments must be assured of continued funding.

• The *mm array* is a facility which solar radio astronomers are enthusiastically looking forward to sharing with their astrophysics colleagues. Its main capability will be to allow imaging of the solar outer atmosphere at radio wavelengths with a resolution comparable to the what can be done at other wavelengths (from the optical to X-rays), and to do so in a wavelength window which has been essentially unexplored. A particular feature of observations at these wavelengths is that one can look "deeper" into the atmosphere than at longer wavelengths.

0.5. OTHER RECOMMENDATIONS AND POLICY CONCERNS

The Solar Panel also focussed on a number of issues not tied to the larger projects just discussed:

0.5.1 *The role of small-scale projects.* The Solar Panel strongly encourages the continued vigorous support of small-scale scientific projects, such as balloon and sub-orbital rocket projects; small PI-class space missions; and strong individual investigator grants programs at both NSF and NASA.

0.5.2 *University revitalization.* Solar physics is poorly represented at universities, a demographic fact which bodes ill for the future of the discipline. We therefore strongly support efforts to fund graduate students, such as the NASA Graduate Student Research Fellowship program; encourage more flexibility on the funding time period for NSF fellowships; and advocate stronger research ties between the universities and the national research centers.

0.5.3 *Funding flexibility.* We urge NASA and NSF to allow greater flexibility in the duration of funding periods.

0.5.4 *Balanced support of experiments and theory.* Balanced development of solar physics calls for consistent and realistic support of both theory and experimental efforts, including the support of theoretical work which is not directly tied to observational programs.

0.5.5 *Interdisciplinary research.* Neither the NSF nor NASA have a regular mechanism in place for handling interdisciplinary research programs. This is a particularly deeply felt problem in solar physics because solar physics research by its very nature tends to cross the discipline boundaries defined by the agencies. We therefore urge that provisions be made to allow such research programs access to funding.

0.5.6 *Access to large-scale computing.* Despite the vastly improved access to supercomputing afforded by the NSF Supercomputer Centers, it remains difficult to obtain large (> 200 hrs) contiguous allocations of supercomputer time. We therefore urge that exploratory programs initiated by NASA/Ames, NASA/Goddard, NCAR, and others to address this problem be strongly encouraged.

0.6. ENABLING TECHNOLOGIES

The following examples are just illustrative of the range of technologies with which solar physics is involved, and in which it plays a leading role. We strongly endorse a continued vigorous program in technological developments.

• Active control of large structures in space. This will be needed in order to carry out the Pinhole Occulter Facility (P/OF), a major imaging facility at hard X-ray and gamma-ray energies with sub-arcsecond capabilities.

• Adaptive optics, optimized for relatively low-contrast, extended images.

• Capabilities for analysis of extremely large data sets (>1 Terabyte).

• Improved, new-generation detectors for high energies (e.g., wide-band gap semiconductors, gamma-ray channeling detectors, etc.); and at infrared wavelengths (large format CCDs).

• Further development of normal incidence EUV and X-ray optics for high angular resolution studies at these wavelengths.

1. AN OVERVIEW OF MODERN SOLAR PHYSICS

1.1 General Perspectives

Recent theory and observation have established that the Sun is a complex dynamical structure, whose interior represents an active and mysterious universe of its own. There is no reason to doubt the basic features of stellar structure models, but it must be remembered that the ideal standard stellar model contains many arbitrary assumptions. The Sun is the only star that has been studied in detail, and the only detailed information we have is from scrutinizing its more or less inscrutable exterior. Its interior possesses internal degrees of freedom that are only gradually being discovered and described, and, once described, are only gradually being understood. Present knowledge of the interior of the Sun and stars is based largely on simplified static models constructed from the theoretical properties of particles and radiation as we presently understand them. Parameters can be adjusted to provide a static solar model whose radius and surface temperature agree with observation, and which represents a starting position for the next phase of the inquiry into stellar physics. This next phase concerns the dynamic and magnetic aspects of a star, a phase which is already well underway, and which is the primary focus of our report.

Now the dynamical effects ignored in the static models are already suggested by these models. Thus, for instance, the calculated temperature gradients indicate the existence of the convection zone, extending from the surface down a distance of about 0.3 solar radii. The gas continually overturns and operates as a heat engine whose work output is not subject to the usual thermodynamic limitations that apply to thermal energy. The activity at the surface of the Sun is a direct manifestation of this convective heat engine, which produces such diverse phenomena as sunspots, flares, coronal transients, the X-ray corona, and the solar wind, largely through the magnetic field as an intermediary. It seems not to be generally recognized in the astronomical community and elsewhere that the precise causes of solar activity are not yet reduced to hard science. For instance, it cannot be stated why the Sun, or any other solitary star, is compelled to emit X-rays, nor is it understood why a star like the Sun is subject to a mass loss of 10^{12} gm/sec. Indeed, it is not altogether clear why the Sun chooses to operate a 22-year magnetic cycle, producing the other aspects of the activity largely as a by-product. This means, then, that we do not understand the origins of stellar X-ray emissions; this branch of X-ray astronomy, with its remarkable powers of penetration into the active component of the universe, is for the present limited largely to phenomenological interpretation. Indeed, the present ignorance of the Sun reflects the general lack of progress in understanding stellar activity of all kinds. We cannot fully interpret nuances of the surface emissions of the distant stars until we understand the physics of surface activity through close scrutiny of the Sun.

However, the problems are deeper than the puzzles of surface activity. There are the mysteries posed by the different surface abundances of Li, Be, and B and in more stable elements such as Ca and Fe in some F and G dwarfs. There is the mystery posed by the theoretical evolutionary brightening implying that the Sun was 30 percent fainter 3 or 4 $\times$ 10^9 years ago, while over the same period of time, mean temperatures on the terrestrial equator have not varied more than a few degrees.

Turning to more direct problems, observations of the solar neutrino emission have failed to corroborate the conventional theoretical models of the Sun. The failure to do so has stimulated a careful review of the theoretical complexities and uncertainties of the model, leading to significant downward revisions of the predicted neutrino flux. Nonetheless the present discrepancy between the observed and predicted neutrino emission seems to be stuck at a factor of three. Until that dilemma is resolved, we cannot be sure of the rest mass of the neutrino or of the amount of the dark matter in the universe. We cannot be sure of the theoretical evolution of a star on the main sequence. We cannot be sure of the age of globular clusters or of the age of the galaxy. We cannot be sure of any theoretical interpretation of anomalous abundances in main sequence stars.

Helioseismology shows promise of providing a detailed and quantitative probe of the physical conditions (temperature, density, mean molecular weight, angular velocity and magnetic field) throughout the entire Sun. Complete success depends upon suitably long unbroken runs of data and on the detection and identification of g-modes. The analysis of the data currently available points to peculiar and puzzling effects, including interior velocity fields inconsistent with detailed hydrodynamic simulations and with classical dynamo theory. The frequencies of the p-modes differ by many sigmas from those calculated from the standard models of the solar interior. How drastic will be the necessary revisions to the standard solar model is a matter of conjecture. The present rapid development of seismological probing of other stars is an exciting and important adjunct to the exploration of the interior of the Sun.

It should be emphasized that there is far more at stake than the standard model of the solar interior. Our knowledge of the static structure of most stars is founded on the success of the solar model, and it is on the theoretical static structure of stars that our ideas of the age and evolution of the galaxy are based. From the broadest perspective, one of the fundamental tasks of solar physics is to develop independent observational checks on this central bastion of astrophysical knowledge.

Now, the remarkably active state of the solar periphery, driven by the convective heat engine, has been studied with increasing angular resolution, spectral resolution and wavelength range for several decades. Knowledge has expanded enormously, without, however, bringing immediate theoretical understanding. Solar activity, like the Earth's weather, is a tempestuous and complicated affair. To obtain some measure of the possible theoretical complexity, note that the Reynolds number N_R of the convective heat engine is of the order of $10^{12} - 10^{13}$, which means that the fluid is active on all scales from one solar radius R down to the fraction $10^2/N_R$ of R, or ≈ 10 cm. Hence, the convection has approximately $(N_R/10^2)^3 = 10^{30} - 10^{33}$ degrees of freedom, and for complete numerical simulation would require a grid with $N_R/10 = 10^{11} - 10^{12}$ intervals in each of three dimensions. What is more, the magnetohydrodynamic Reynolds number N_M is 10^{10}, whereas the terrestrial laboratory can achieve no more than 10^2 or 10^3, so there is no general body of knowledge from which the subtleties of solar magnetic activity can be interpreted. The enormous heat flux in the convective zone producing the superadiabatic temperature gradient and driving the convective heat engine on all scales, and the extreme magnetohydrodynamic effects of solar activity combine to provide a dynamical scenario of exotic character that can be understood only after it is described and studied in detail – it is far too complex for *a priori* predictions. First, detailed observations are required to describe the situation. Then, numerical modeling and theoretical studies of individual dynamical effects can be brought to bear. That is the nature of the activity of a star: There is no single effect, no single new principle that throws open the gates to a flood of understanding. The behavior of a convective, highly conducting fluid is a whole field of physics in its own right, which requires concerted close theoretical and observational study in its own right, progressing past dozens of "milestones" and enjoying dozens of "breakthroughs." The milestones and breakthroughs already add up to an impressive body of knowledge, but represent only a beginning.

A particularly important revelation occurred about two decades ago, with the realization from detailed observational and theoretical considerations that, outside sunspots, the magnetic field at the surface of the Sun, rather than being the expected continuum distribution, is effectively discontinuous, composed of small individual intense and widely separated magnetic flux tubes of $1 - 2 \times 10^3$ Gauss. The measured mean field in any region then depends mainly on the distance between the individual magnetic fibrils, because the individual fibrils or flux tubes are too small (about 200 km diameter), for the most part, to be resolved in the telescope. The crucial information for understanding the large-scale behavior of the magnetic fields on the Sun (which are, it must be remembered, the perpetrators of the Sun's magnetic activity) are: (a) the origin and structure of the individual fibrils and (b) their individual motions. So the pursuit of solar activity becomes solar "microscopy". This is a field in its infancy, although with great potential through the development of adaptive optics on ground-based telescopes and the development of diffraction limited telescope systems in space. Indeed, the high resolution UV observations from space, although nowhere near the ultimate necessary resolution of 50-100 km, have already established the general occurrence of myriads of tiny explosive events (nanoflares) and high speed jets in the solar corona, providing a clue about the heat input that causes the corona. The individual bursts of energy ($10^{24} - 10^{27}$ ergs per event), and indeed the entire supply of energy to the corona, is evidently a result of the motions of the individual magnetic fibrils in the photospheric convection. The motions undoubtedly involve both jitter and intermixing of the individual fibrils, producing Alfven waves and a general wrapping, respectively, of the lines of force in the fields in the corona. But at the present time, there is no direct measure of any aspect of the fibril motions, nor any direct detection of waves or wrapping in the coronal magnetic fields. Only the effects of the myriad of small explosive nanoflares can be seen. So the causes of the solar and stellar corona, although extensively developed theoretically, are still without a hard observational foundation.

Another important milestone was the *Skylab* discovery of frequent coronal transients, involving the eruption of matter from the low corona outward into space, often accompanied by flare activity at the surface of the Sun. Now many years later, we are beginning to realize that these mass ejection events apparently result from large-scale magnetic field eruptions – but why they occur is not clear. Further, the observations show clearly that coronal mass ejections and chromospheric eruptions (i.e., flares) can proceed quite independently. Thus, the coronal mass ejections reflect a form of solar activity not heretofore recognized. Their relation to the large scale evolution of the solar magnetic field – and to stellar magnetic changes – is not yet clear.

The remarkable X-ray photographs of the Sun, showing clearly the magnetic loop structure of the active corona and the interweaving coronal holes, have gone a long way toward formulating the problem posed by the existence of the active X-ray emission. Also, the high speed streams of solar wind issuing from the coronal holes demonstrate that the corona is active even outside of active regions. The X-ray and EUV studies of the solar corona, together with the discovery by the orbiting *Einstein* Observatory that essentially all stars emit X-rays, provide a profound scientific challenge as to why ordinary stars are compelled to such extreme suprathermal exploits.

The ability to release energy impulsively and accelerate particles is a common characteristic of cosmic plasmas at many sites throughout the universe, ranging from magnetospheres to active galaxies. Observations of gamma-rays and hard X-rays, radiations that can be unmistakably associated with accelerated particle interactions, as well as the direct detection of accelerated particles, such as cosmic rays, strongly suggest that at many sites a significant – and in some cases even a major – fraction of the available energy is converted into high-energy particles. Detailed understanding of the processes which accomplish this is one of the major goals of astrophysics.

Solar flares offer an excellent opportunity for achieving this goal. A large solar flare releases as much as 10^{32} ergs, and a significant fraction of this energy appears in the form of accelerated particles. It is believed that the flare energy comes from the dissipation of the non-potential components of strong magnetic fields in the solar atmosphere, possibly through magnetic reconnection. Immediate evidence for the presence of accelerated particles (electrons and ions) is provided by the gamma-ray and hard X-ray continuum emissions which result from electron bremsstrahlung, and by gamma-ray line and pion-decay emissions from nuclear interactions. Nuclear interactions also produce neutrons, which are likewise directly observable at Earth. The accelerated charged particles enter interplanetary space and arrive at Earth somewhat later, delayed by their circuitous paths of escape from the magnetic fields of the flare. The wide variation of the relative abundances of some isotopes and atomic species among the accelerated particles provides an indirect but detailed view of the physics operating within one of the flare particle-acceleration processes. These high energy emissions are the best tools for studying acceleration processes in astrophysics. Solar flares are one of the very few astrophysical sites where it has been possible to study simultaneously the acceleration of electrons and protons, and only for solar flares can the escaping accelerated particles be directly detected and correlated with the electromagnetic radiations produced by the interaction particles. In addition, lower-energy emissions (soft X-ray, EUV, UV, and radio emissions), which are also observed from flares, reveal much of the detailed properties of the ambient plasma (e.g., temperature, density, and magnetic configuration) before, during, and after the flare.

This is the broad view of the mystery of the Sun and the stars. The specifics are complex, but it is essential, if we are to grasp the scope of the problem posed by the Sun, to spell out these complexities in somewhat more detail. The next section, then, details some of the specific problems, measurements, observations, and theoretical studies that are necessary along the way to probe the mystery.

1.2 The Frontiers and Goals for the 1990s

By the year 2000 we hope to have a fairly detailed picture of the structure and dynamics of the solar interior, a better understanding of how the Sun generates magnetic fields, considerable measurements of how magnetic fields modulate the smooth outward flow of energy, a better description of the morphology of flaring plasmas, and some predictive capability for the flow of non-radiative energy through the heliosphere to Earth. These and other advances will be achieved by observational improvements in spatial resolution, temporal coverage, and new exploitation of radio, infrared, EUV, and X-ray spectral regions. Up to the present moment, the Sun has become increasingly mysterious the more we have studied it; by 2000 AD, we may reasonably hope to begin closing in on some of the more important mysteries.

As in any subfield of astronomy, there are more solar problems ripe for attack, and hence more projects in this report, than can be accomplished in a decade. Uncompleted tasks will form the basis of a program for 2000 and beyond. There is no likelihood that access to space will become rapid, cheap, frequent, or reliable enough to replace the need for a strong ground-based program. Hence, both space-based and ground-based components will be required in solar physics for several decades into the future.

In this section we focus on the principal specific goals for solar research which we believe are most important and also realizable in the decade of the '90s or shortly thereafter. This list will then determine the specific initiatives we will recommend in Sections 2 through 5. For convenience, we deal with the three principal components of the Sun – interior, surface layers, and outer layers – in sequence.

1.2.1 The Solar Interior

Two powerful and complementary techniques are available for probing the solar interior: neutrino spectroscopy and helioseismology. The flux of high energy neutrinos has been measured since the late 1960's. A discrepancy of about a factor of three between observations and model predictions gives rise to the well-known solar neutrino problem. Significant progress toward solving this problem can be made by a new generation of detectors that will allow the spectrum of solar neutrinos to be measured. An appropriate goal is to measure, by the end of the decade, the flux of solar neutrinos and its possible time variation as a function of energy, from the low energies associated with the neutrinos resulting from the main p-p reactions up through the energies of the ^{8}B neutrinos long-studied by the chlorine experiment. In addition, the fluxes (and their possible time variations) should be understood in terms of the structure of the interior (and its possible time variations), as well as the particle physics of the neutrino.

Helioseismology uses the frequencies of millions of normal modes of oscillation of the Sun as probes of the structure and dynamics of the interior. Since 1975, exploration of much of the solar interior has been done using this technique. Better measurements are required to explore both the deep and shallow interior regions as well as to refine the present fairly crude picture of the middle regions. A major goal is to discover why the theory of stellar structure and evolution fails to correctly yield either the structure or the dynamic picture now emerging from helioseismology. Some of the required measurements can be done only from space but others can be done from the ground more effectively. A newly emerging technique is seismic imaging of local regions on the Sun. This promises to give us the first subsurface views of solar activity, which should revolutionize our understanding of enigmatic features such as sunspots. A realistic goal is, by the end of the decade, to have made accurate measurements of the entire spectrum of p-modes; to have developed a new solar model consistent with these data within their errors; to have determined empirically the internal rotation as a function of depth and latitude; to have obtained a quantitative theoretical understanding of the physical processes giving rise to this differential rotation; to have detected the elusive large-scale circulation flows in the deep convection zone; to have mapped the three-dimensional structure of surface active regions and sunspots; and (if they penetrate to the surface with observable amplitude) to have made the initial detection of g-modes generated in the deep interior. This ambitious set of goals is made realistic because of the extraordinary observational and theoretical progress in helioseismology made in the '80s, coupled with realistic expectations for observational and theoretical capabilities to be developed in the '90s.

The cycle of solar magnetic activity is believed to arise in the interior by a combination of differential rotation and cyclonic convection acting on magnetic fields. More accurate and higher resolution measurements of surface magnetic fields are required to understand how magnetic flux evolves and is dissipated from the surface as part of the solar cycle. Also needed are accurate measurements of mass flows both on small and large scales to gain a better understanding of the role of magnetoconvection. Such observations, combined with some of the helioseismic results described above, should go a long way toward reaching the goal of an accurate model of the solar dynamo. A realistic goal for the '90s is one which has been sought for decades: finally to obtain a real physical understanding of the origin and nature of magnetic activity in the Sun and stars.

1.2.2 The Solar Surface

Observations have shown that the solar surface layers – from the chromosphere to the corona – are permeated, heated and controlled by the magnetic field rooted in the photosphere. The physics of this region is complicated by large density and magnetic variations and violent mass motions. A major observational problem is the difficulty in making accurate physical measurements in the face of spatially blurred observations. Physical quantities deduced from such blurred measurements may apply to an average within the measured volume of the quantity, but because of extreme nonlinearities they may apply to no physically realizable state at all. It is little wonder that important problems such as heating of the upper solar atmosphere or storage of magnetic energy and its violent release in flares have not been solved. The frontier in this research is very much controlled by how small a volume can be measured accurately. While space offers the most certain route to improvement, the development of adaptive optics promises significant benefits using ground-based telescopes. A realistic goal for the '90s is to obtain a clear physical understanding of the interaction of magnetic fields and convective motions at and immediately beneath the surface, and specifically to understand the surprising shredding of the field into spatially intermittent "flux knots," which appears fundamental both to the evolution of surface magnetic fields and to their consequences for atmospheric heating.

New measurement techniques will also be needed to characterize accurately the physics of the lower atmosphere. Particularly important will be observations of the magnetic field as a vector varying with height and time, along with corresponding vector mass motion measurements. The technology to make such measurements is under active development at several observatories. Advantages of diverse spectral regions, from the extreme ultraviolet through the infrared to mm wavelengths, are being exploited. Extremely important is the capability to relate physical conditions measured in the surface layers (through visible and infrared data) to conditions in the overlying heated chromosphere (through UV, extreme UV, and mm data); the data must have adequate spatial and temporal resolution to isolate physically near-homogeneous regions and to establish cause-and-effect relations between phenomena at the various levels. A reasonable goal for the end of the decade is to make substantial progress in developing and testing a specific physical description of the magnetohydrodynamic processes giving rise to atmospheric heating in these layers.

1.2.3 The Outer Solar Atmosphere: Corona and Heliosphere

Our understanding of the corona and heliosphere was revolutionized by space observations during the last three decades. While much was learned, we still do not have a good understanding of what compels the Sun to produce a hot, X-ray emitting corona. Evidently most other stars also have coronas. Observations have demonstrated that magnetic fields play a controlling role in the morphology and large-scale dynamics of the solar corona. There is also evidence, but not proof, that magnetic processes are responsible for supplying the energy to heat the corona and to produce violent events such as flares and coronal mass ejections. The key to further progress lies in obtaining improved observations on all accessible spatial and temporal scales, but particularly on the smallest size scales. Progress in X-ray and gamma-ray imaging promises to allow

ASTRONOMY AND ASTROPHYSICS PANEL REPORT

major advances in our observational understanding of the corona's structure, and of the relation between heated plasma and the sites of energy release.

The outer corona merges into a heliosphere that dominates interplanetary space. The heliosphere is Earth's non-radiative connection to the Sun, and events within it are of considerable practical as well as scientific interest. Much has been learned about the structure, dynamics and physics of the heliosphere by combining in situ and indirect measurements with correlative studies of driving phenomena at the solar surface and in the lower atmosphere and corona. The same cannot be said for violent events such as coronal mass ejections that occur frequently during active times of the solar cycle — better observations from the photosphere through the heliosphere, involving in particular coordinated in situ and remote sensing observations, are required for significant further progress.

Appropriate goals for the decade of the 1990s include localizing coronal energy release sites, understanding in detail the mechanism of solar wind acceleration and fixing the height above the surface at which it occurs, and determining how the speed of the resultant wind depends on magnetic geometry. Finally, an extremely important goal is to take advantage of the activity maximum around 1998-2002, using state-of-the-art imaging instruments at X-ray and gamma-ray energies, in order to achieve a major advance in our understanding of the basic mechanisms of solar flares, including very high energy particle acceleration.

1.2.4 The Solar-Stellar Connection

It should be emphasized in this overview of solar physics that the solar-stellar connection is an important part of the physics of the Sun and the physics of stars in general. For we may safely assume that most, if not all, rotating and convecting stars would prove as active and mysterious as the Sun if we could observe them as closely. These stars do not fail to exhibit great complexity in those aspects that can be studied. As already noted, it is astonishing to see that some stars support gigantic flares and starspots. Some exhibit mass loss enormously greater than the Sun. Essentially all of them exhibit X-ray coronae, from which we may infer that their coronal gas expands along the more extended lines of force, carrying the field into space to form a stellar wind much like the solar wind. The general existence of X-ray coronae implies the same nanoflares and microflares and the same coronal transients as can now be observed on the Sun, although there is no foreseeable means for observing them individually on the distant stars. The same complex magnetohydrodynamic and plasma processes must occur. The same puzzles concerning their internal structure, their internal rotation and their dynamo confront us, except that it is not possible to come so directly to grips with these puzzles as it is for the Sun. The best that can be foreseen is to understand the Sun and then perhaps to infer the solutions for the other stars. It is essential, therefore, to study the oscillations and seismology of the other stars, to monitor their activity cycles over long terms, and to make precise measurements of their rotation rates. Only in this way can we discover their individual quirks as well as determine the "average" behavior of each class of star. The deviation of the individual from the average provides insight into the variable conditions under which stars are formed, which then helps to understand the idiosyncrasies of the Sun. Other stars of different ages may provide an idea of the Sun in its youth, to be compared with the geological record for clues to the effects on the planetary environment. The spindown of the Sun at an early age may have involved profoundly different conditions from those that obtain today. In a similar vein, it appears that the Sun occasionally passes through centuries of suppressed activity (e.g., the Maunder Minimum), and centuries of enhanced activity. The human research program cannot encompass such fundamental long-term shifts in the nature of the activity, so one must turn to the hundreds of solar-type stars to provide a record of the many different moods of the Sun in the span of a human lifetime.

Thus, as a direct by-product of obtaining the goals described above, we may anticipate corresponding great advances in our understanding of many long-standing problems of stellar physics. But beyond this "spin-off" result, it is reasonable to adopt the goal of making far more detailed studies of stellar phenomena related to those studied on the Sun, through emerging capabilities of stellar seismology, through stellar observations with new advanced ground-based and space instruments expected to be operational during the decade, and through continued monitoring of time-varying stellar magnetic activity of existing observatories.

In concluding this general appraisal of current problems in the physics of a star like the Sun, it is appropriate to make some general comments on the future beyond the listing of specific research goals as we perceive them today. Even though solar physics is sometimes thought of as a mature field, in the coming decade it may be as unpredictable and full of surprises as any astronomical discipline. The observational techniques available in this closing decade of the 20th century have opened up entirely new horizons to solar research. It is too soon to guess where the neutrino observations will lead, but whatever the results of the present gallium detectors, the astronomical implications will be profound. Helioseismology may be expected to play an essential role in removing the ambiguities of anomalous neutrino fluxes, unless, of course, the discrepancy is entirely a matter of neutrino oscillations between three or more states, which would have deep cosmological implications. What is more, we can be sure that the investigation of the solar surface and the solar interior on so broad a front will provide surprises, perhaps of a fundamental nature. The present writing is based only on contemporary knowledge, and cannot anticipate what lies ahead when we probe into the unknown realm of the solar interior and the small-scale phenomena at its surface.

2. GROUND-BASED SOLAR PHYSICS

2.1 Introduction

For nearly 400 years the physics of the Sun has been studied from the ground. While much has been learned about the Sun – and by most astronomical standards it is well understood – the fact is that the Sun confronts astronomers with many unsolved puzzles, both old and new – a point made in some detail in the preceding discussion. Observations from the ground continue to play a major role in confronting these puzzles, many of which have consequences far beyond solar physics.

Within the general framework just discussed, we see the 1990s as the era in which ongoing key initiatives – discussed in Section 2.2 below – will be augmented by a major new initiative, which is needed to make really significant progress in our understanding of solar activity. This new "solar magnetohydrodynamics" initiative depends upon a concerted development program in high angular resolution optical observations and precision polarimetry, using existing ground-based telescopes; and aims for the establishment within this decade of a large-aperture ground-based optical facility – the Large Earth-based Solar Telescope – using adaptive optics techniques to image the solar surface in the subarcsecond range.

2.2 Status of Major Projects and Facilities

One major ground-based solar project is in progress: the Global Oscillation Network Group (GONG) project. It is aimed at a ten-fold improvement in the accuracy of intermediate spatial-scale helioseismology measurements for studies of most of the solar interior. The project is a community effort led by the National Solar Observatory. A prototype instrument will be completed early in 1991 and the next phase of the project is to build and install six identical instruments at selected sites around the world. Given timely funding, this network should be operational in late 1993. Observations and data reduction will continue for at least three years, to be followed by an intensive analysis effort by the helioseismology community. GONG and the helioseismology instruments to be flown on the SoHO mission in 1995 were designed to be complementary and interdependent: While GONG emphasizes intermediate spatial scale observations with a high duty cycle, the SoHO instruments emphasize large and small spatial scales difficult to observe from the ground. Thus, both projects are essential for the advance of helioseismology, and together can attain the helioseismology goals discussed in Section 1.2.1.

The U.S. program of solar physics includes a wide range of ground-based observational facilities operated by national observatories, federal agencies and individual universities. The national observatories with solar observing capabilities (the National Solar Observatory and the National Radio Astronomy Observatory) provide facilities that are publicly available to qualified scientists by peer review. These facilities generally have a scale that is not appropriate for a single university. Federal agencies (NASA, Air Force, Department of Commerce) operate solar facilities in support of various mission goals. These facilities are not generally available to the wider research community. The federally-funded High Altitude Observatory operates solar facilities for its own research programs, but also provides for the use of facilities to the community. Several universities operate solar observing facilities in support of the research of their faculties and students. The scope of these ranges from major, multi-purpose equipment to modest, single purpose instruments. The observational solar programs with two or more faculty members include the California Institute of Technology, California State University at Northridge, Michigan State University, Stanford University, the University of Hawaii, and the University of Maryland. Smaller programs (one faculty member) are found at Penn State University, University of Arizona, University of California at Los Angeles, the University of Chicago, and the University of Southern California.

The U.S. ground-based observational solar program led the world for most of the 20th century. This leadership role is now being challenged by the decline in funding in the US, and by other countries that are developing strong observational programs, most notably various European countries and Japan. The existing U. S. telescopes were built for the most part before 1970 and are equipped with focal plane instrumentation that has been upgraded but frequently dates to the 1970s or earlier. Nonetheless, these telescopes are important resources; with more modern focal plane instrumentation, many of the ground-based frontier projects for solar physics in the 1990s could be carried out with existing telescope facilities.

2.3 A Prioritized Ground-Based Program for the 1990s

2.3.1 Prerequisites

To keep the United States at the forefront of international solar research, the ground-based U.S. solar program for the 1990s must:

 • focus clearly on the most important scientific problems which can be attacked with available technology, and the technology now under development;

 • be able to translate the specific plans developed by the community into ongoing projects with some assurance of funding continuity;

 • provide for a balance between the large-scale, community-based programs and smaller-scale research programs led by individual investigators;

 • assure a balance between innovative new projects and the fulfillment of commitments to ongoing projects, including those which by their very nature are carried out over long spans of time, based on scientific merit;

 • assure the training of solar scientists at a level commensurate with the anticipated program beyond the year 2000;

 • encourage a symbiotic relationship between ground-based and space-based observers, and theoreticians;

 • provide for the infrastructure needed to support the anticipated program, including an effective capability for reducing and analyzing the extremely large data sets which new generations of solar instruments (and numerical simulations) will provide.

2.3.2 Prioritization

With specific regard to our priorities for the U.S. solar research program in the 1990s, several factors must be kept in mind in addition to the primary constraint, namely that of merit:

 • the priorities of ground and space-based research are interdependent;

 • the priorities of the U.S. program depend on international programs and programs of other nations;

 • the program is motivated by a mixture of long term and short term needs;

 • the program balances major, moderate, small, and interdisciplinary initiatives;

 • "major solar initiatives", as defined by the AASC, require multi-year commitments at the NSF division level, not just at the program level;

 • "moderate solar initiatives", as defined by the AASC, require commitments at the NSF program level, typically for 2-3 years;

 • "interdisciplinary initiatives", often very large projects, may be of interest to organizational units of NSF other than the usual solar funding sources;

 • operational systems, such as the U.S. Air Force Solar Observing Optical Network (*SOON*), must be recognized as programs whose principal responsibilities are to monitor and report solar activity, rather than to support solar physics research. Nevertheless, in the past, solar research has benefited significantly from USAF willingness to provide, for example, *SOON* Hα images for research purposes (e.g., Solar Maximum Year).

2.3.3 The Major New Initiative: Solar Magnetohydrodynamics, and the LEST.

A major scientific goal for ground-based solar research in the 1990s is to understand the physics of solar magnetic fields in the regions of the Sun that are observable from the ground. This is an ambitious and critical goal, and we propose that the major step be taken to reach it by building, together with international partners, a Large Earth-based Solar Telescope – the LEST.

Research in solar magnetohydrodynamics has the potential to revolutionize our understanding of solar and stellar activity, the outer solar atmosphere, and how the Sun affects Earth. The Sun is the only star for which we can investigate the small-scale phenomena that dramatically affect the transport of energy and momentum through stellar atmospheres. One of the most important discoveries of solar physics in the last two decades has been the recognition that the solar magnetic flux outside sunspots is highly structured in the form of intense, isolated magnetic flux tubes. This structuring of the field evidently plays a key role in the heating of the outer solar atmosphere, the occurrence of flares, and the driving of the solar wind. Therefore, the small-scale structure of the magnetized atmosphere of a star is in large measure responsible for many of the large-scale phenomena that have long been observed but not understood. The Sun is the only star for which one can achieve the spatial and spectral resolution and photon flux necessary to measure all components of the magnetic field vector, from which the structure and density of magnetic fields can be observed, and their role in solar activity understood. In this sense, the Sun acts as a Rosetta stone for interpreting the ubiquitous stellar activity observed by *IUE*, *Einstein*, *EXOSAT*, and now *ROSAT* throughout the H-R diagram.

The Solar Panel has identified as its highest priority for ground-based solar research a major initiative in magnetohydrodynamic (MHD) studies of the solar photosphere and overlying atmosphere. The time is right for this MHD initiative in this decade because recent advances in telescopes, polarimeters, detectors, data processing, and radiative transfer theory offer ways to break through barriers that up to now have seriously impeded progress. The major impediments have been inadequate spatial resolution for extended time periods, incompletely compensated telescope polarization, difficulty in obtaining measurements of the full vector magnetic field over a range of heights, and oversimplified interpretation of measurements. High resolution should be achievable through the use of adaptive optics systems with existing telescopes and with new, large-aperture solar telescopes located at excellent seeing sites. Telescope polarization can now be compensated to the 0.1% level at existing telescopes and to the 0.01% or better level in new, low polarization designs. Highly efficient modulators and detectors offer orders-of-magnitude improvements over most existing equipment for the measurement of the full vector magnetic field. Additional layers of the solar atmosphere have been made accessible to robust diagnosis in the infrared portion of the spectrum. Modern computing equipment offers orders-of-magnitude improvements in both the quantity and quality of measurement interpretation. Finally, recent advances in computer simulations of compressible convection now permit us to address these observations from a quantitative point-of-view. These advances in theory, coupled with the advances in observations, will allow a major improvement in our understanding of phenomena such as the turbulent diffusion of magnetic fields by confronting theory – such as the results of simulations – directly with high spatial resolution observations.

In addition to excellent angular resolution, it is necessary to be able to make accurate polarimetric measurements of the Zeeman effect across profiles of spectral lines formed over a range of heights in the solar atmosphere. In the past, progress in this area has been frustrated by the large amount of polarization produced within the existing solar telescopes at Kitt Peak and Sacramento Peak and by the technical demands of precision polarimetry through the Earth's atmosphere. Compensation techniques have recently been developed to allow measurements to be made at a level of about 0.1% using existing groundbased telescopes, and parallel advances in polarization analysis techniques and data acquisition will soon allow quantitative measures of the vector magnetic field with angular resolution of one arcsecond, or better. Substantial efforts are now underway at a number of US institutions to advance high-resolution solar polarimetry. Such work needs continued support at high priority. It will form an essential part of the magnetohydrodynamics initiative.

As for facilities, this ambitious initiative cannot be fully addressed with a single ground-based approach. First, we give highest priority to a new moderate to large aperture adaptive-optics telescope in the visible region and near infrared. However, this initiative must not preclude continued progress in the far infrared, where rapidly advancing technology, combined with intrinsic advantages of the far infrared (5 to 20μ), may make a large-aperture infrared telescope a high priority. Second, this magnetohydrodynamics initiative includes moderate projects that are already started, as well as new ones. Third, the initiative will be greatly strengthened by, and will greatly strengthen, the Orbiting Solar Laboratory (OSL), presently NASA's top priority new start mission, whose UV telescope-spectrograph and X-ray imaging instruments provide powerful observational tools for studies that cannot be accomplished from the ground, and whose optical imaging capability will provide "space truth" for the much longer-term ground-based observations from LEST, and will help define evolving instrumentation for the LEST focal plane.

2.3.3.1 The Large Earthbased Solar Telescope (LEST)

Existing telescopes cannot provide the 0.1" resolution needed to measure directly such basic quantities as the strength of the highly concentrated and fragmented photospheric magnetic field. Both sufficient aperture and adaptive optics, in combination, are required to reach this goal from even the best ground-based sites. The United States, through a partnership between the National Solar Observatory, industry, and the Defense Department, has developed a world leadership position in the field of solar adaptive optics. Recently we have seen a significant advance in our ability to image the solar surface at close to the diffraction limit of moderate-aperture telescopes using this technique. This work, currently underway at a low level at the National Solar Observatory and Lockheed, should be vigorously pressed to an operational system at existing telescopes. Funding at a level of about $1M per annum for a few years should allow the speedy completion of an adaptive optics system on an existing telescope which is demonstrably adequate for a large-aperture solar telescope.

Following the successful completion of these developments, the United States should develop a moderate to large-aperture high-resolution telescope, jointly with other countries. The United States is a participant in the *Large Earthbased Solar Telescope* (*LEST*) consortium, which plans to build such a telescope starting in 1992, and pending the results of the adaptive optics program and of the *LEST* final design study, the United States should plan to contribute up to 1/3 of the cost of that telescope (expected cost for 1/3 participation: $1M design, $15M construction, $1M annual operation).

The *LEST* will be an ideal complement and long-lived extension to the *Orbiting Solar Laboratory*, a space mission for high-resolution solar studies in the optical, ultraviolet, and X-ray regions. (*OSL* is discussed in Section 3.3.1 below.) Whereas the hallmarks of *OSL* are uniquely high and uniform angular resolution, and unique access to the ultraviolet and X-ray regions, the hallmarks of *LEST* are large aperture, precision polarimetric capability, longevity, access to the near infrared (routinely out to 2.2 microns), and flexibility of instrumentation. The *OSL* will obtain visible light images and spectra in selected spectral lines, and continua with angular resolution and stability that could not be matched by any conceivable ground-based telescope. Because *OSL* is a free-flying telescope in a polar orbit, however, it will have no flexibility for changing its observing capabilities as new scientific and instrumental opportunities arise, and its lifetime (nominally three years, possibly longer) will of course not approach the 25-or-more year useful life of *LEST*. *OSL's* design precludes the aperture and extremely low polarization required for the most sensitive measurements of three-dimensional magnetic field structures: although *OSL* will obtain uniquely highly resolved data on relatively strong magnetic fields, a telescope like *LEST*, with its large aperture (2.4 times the aperture of *OSL*) and extremely low polarization, is necessary to measure the full vector magnetic field of weak features. The *LEST* will build upon results from *OSL* characterizing the fine-scaled magnetic flux distribution and evolution at the solar surface, to investigate the full (non-potential) vector magnetic fields and their specific implications for energy transport and heating of the outer atmosphere. With the capability to probe deep in the photosphere at the 1.6 micron opacity minimum, *LEST* will also provide complementary data to *OSL* about the nature of convection and its overshoot, and the magnetic fields entrained therein. It may be able to provide important helioseismology information as a follow-on to the *GONG* (and in space, *SoHO* experiments), using adaptive optics to provide high-resolution over small fields of view for tomography of subsurface structure. Other key research areas abound, and will evolve over the long lifetime of *LEST*.

LEST will be the only new large solar telescope built with US participation since the mid-70's. It will complement and extend the aging US national facilities (McMath telescope complex on Kitt Peak, built in 1960, vacuum telescope feed added in 1973, and the Sac Peak facilities, the most modern of which dates to the mid '60's). Aside from these national facilities, there are few large university solar telescopes, so that the *LEST* will be of great importance across the community. In particular, *LEST* is expected to play a significant role in the training of advanced students in solar research, with great benefits to the vitality of the U.S. solar research community.

The *LEST* telescope will be the premier solar telescope on the face of the Earth. As befits such an endeavor, it is receiving scientific and financial support from around the world. The *LEST* foundation contains nine member countries: Australia, Germany, Israel, Italy, Norway, Spain, Sweden, Switzerland, and the United States. The international community of scientists represented by these nations provides a critical number of capable solar researchers able to utilize fully the capabilities of *LEST*. At the same time, the capital and operating costs will be jointly borne by the member nations. Thus, the United States will gain access to this unique instrument by providing no more than one-third of the total costs. As already mentioned, the US costs will be less than $20M, so that the *LEST* telescope will be an extremely cost-effective investment for the United States.

The *LEST* telescope is in a very mature state of design: It has been under study for nearly a decade, and more than 40 technical reports have been written on its design, scientific rationale, instrumentation, and site selection. A superb, proven, and developed site in the Canary Islands has now been selected.

The Solar Panel recommends in the strongest terms the rapid completion of the Adaptive Optics program, key to the success of *LEST*, followed by the funding of the US share of the *LEST* construction and annual operating costs.

2.3.3.2 Infrared Facility Development

The infrared solar spectrum offers several unique ways to study solar magnetohydrodynamics. The opacity minimum at 1.6 microns gives us the deepest look into the solar photosphere. At 12 microns, the emission lines arising from highly-excited states of magnesium, aluminum and other elements allow an unambiguous measurement of the strength of magnetic fields for most known features of the lower solar atmosphere. A powerful advantage for magnetic field research in the infrared is the quadratic wavelength variation of Zeeman splitting. This means that Zeeman components are cleanly separated at moderate field strengths and uncertain modeling of blended line profiles can be avoided. This advantage applies even with only moderate angular resolution. The McMath telescope of the National Solar Observatory is well suited to infrared solar research and, accordingly, several institutions are developing focal-plane instruments with modest resources to measure magnetic fields in the infrared. These efforts should be accelerated by increased funding to a level of about $150K per annum.

The promise of the solar infrared for magnetic and other investigations is so great that a larger telescope than the 1.5m McMath is urgently needed. The need follows from the improvement of seeing that occurs in the infrared, combined with the decreasing resolution of a fixed aperture at increasing wavelength. Another reason for a larger aperture is the decreasing photon flux per spectrum line doppler width in the infrared. Three options are available, and all should be investigated: (a) join with our nighttime colleagues in construction of a 10m class telescope capable (with suitable protection) of doing solar infrared daytime work, (b) construct a new, large-aperture solar infrared telescope, and (c) upgrade the McMath telescope to a 4m aperture. It is well-worth aggressively pursuing all of these options. We do note however that the first option has the attraction that it is likely to be cost-effective for both the night-time and solar communities, and it would foster much-needed scientific interaction; similar cooperation would also obtain for the McMath upgrade, as this telescope could also be used to great advantage for stellar observations, in particular for asteroseismology. The singular attraction of the second option is that the telescope could be designed to do an optimum job from the start, and it could be placed in an excellent infrared site. The third option offers a rapid intermediate solution (in both cost and capability) to the need; the main experimental compromise is that it falls short of the desired angular resolution in the 12-micron region. The cost estimate for a 4m primary and 6m tracking mirror system (using cooled, actively-supported aluminum mirrors) is $7M, including instrument upgrades and a 25% contingency. As already noted, this option also would provide an upgraded facility for solar-stellar research.

The Solar Panel recommends that detailed definition studies be carried out in the near future on the several approaches to development of infrared facilities, and that development of an appropriate facility be carried out later in the decade.

2.3.4 Moderate Initiatives

During this decade ground-based instrumentation offers the only practical approach to measurement of vector magnetic fields and electric fields in the photosphere and chromosphere and high-resolution imaging and spectroscopy of the corona, both during and outside of solar eclipses. These moderate initiatives are all of sufficiently small scale to allow them to be accomplished from a healthy grants program at NSF. Our goal here is simply to list examples without assigning priority order: The peer-review process, not this document, should determine when and how they are done.

In arbitrary order, the most important moderate initiatives include:

Global Solar Dynamics

The relationship between large-scale flow fields and long-term trends in solar activity is largely unexplored. The initial observations of "torsional oscillations" have been confirmed: The torsional oscillation represents a zone of enhanced latitudinal shear, and its migration from the poles to the equator is definitely associated with the location of magnetic activity. Since the solar dynamo depends in part upon the existence of shear flows, these oscillations may play some role in the magnetic field generation process; whether or not this is the case, and if it is, exactly what this role might be, remains largely unresolved. The amplitude of the circulation associated with the activity is much smaller than that of various other solar velocity fields, and is therefore difficult to measure reliably. Observations from a single site do not allow us to remove the non-circulatory velocities by time-averaging. The most effective filter requires that we sample the non-circulatory velocity pattern uniformly in time over the lifetime of the process we wish to eliminate. A suitable network of sites will be necessary to address this objective from the ground, perhaps similar to the GONG network. A study is required to specify fully the individual telescopes and network necessary to carry out this project.

Gravity Modes in the Solar Interior

Oscillations of the solar interior for which buoyancy is the restoring force are expected to be excited within the solar radiative zone. These "g modes" are confined to cavities which extend roughly from the core to the bottom of the convection zone. They should thus be far better indicators than p modes of conditions deep within the energy-generating solar interior. Many attempts have been made to observe g modes, but to date there is no consensus on a positive detection. The information that the g modes promise on the solar interior is unique and valuable, so efforts to overcome the observational difficulties must be pursued. The observation of g modes is challenging because their periods are long, their amplitudes are expected to be weak, and the spectrum is expected to be very crowded. The main observational problem is to suppress the large background noise produced by supergranulation and active regions. A program to successfully detect and measure g modes will require several phases. First, funding must be provided to explore methods of detecting g modes. This will involve developing new observational techniques to suppress the high background noise produced by the Sun itself. Second, after a promising technique is developed, funding will be required to produce a demonstration instrument capable of producing convincing preliminary results. Finally, the long periods and crowded spectrum of g modes mean that helioseismic-quality measurements will not be possible from a single site, and either a ground-based network or a space mission will be required. The former would involve funding of the magnitude of the GONG project.

A Solar-Dedicated Frequency-Agile Radio Array

In solar flares, brightness temperature spectra (as distinct from flux spectra from a single telescope) can provide important diagnostics of the flare plasma, particularly the electron energy distribution/temperature and magnetic field strength. In active regions, spectra provide unique measurements of the coronal magnetic field. Three small dishes are presently being added to a solar dedicated, frequency-agile (1 - 18 GHz) array at Owens Valley, California, to expand the array to five dedicated telescopes. The power of the array lies in its ability to obtain detailed microwave spectra on time scales of seconds. It is highly desirable to increase the number of telescopes in the Owens Valley array to 8, giving 1260 u-v plane measurements. It would then rival the VLA in its solar imaging capability, surpass it in its spectral coverage, but would be dedicated to solar work.

A Large-Aperture Reflecting Coronagraph (LARC)

Heating of solar and stellar coronae, acceleration of solar and stellar winds, condensation of coronal plasma above active regions, and instabilities leading to coronal transients and mass ejections are among the fundamental astrophysical problems that would benefit greatly from the improved observations of the solar corona offered by a large aperture reflecting coronagraph (LARC). This concept overcomes most of the limitations of conventional refracting coronagraphs. Such a coronagraph has no fundamental restriction on aperture size. Because it is achromatic, the LARC can simultaneously observe multiple wavelengths, and can also remove the dust component that constitutes a major portion of the sky background through rapid image cadencing. It provides extended spectral coverage from the UV into the IR, and operation in the IR would further reduce the sky background. Indeed, a large aperture reflecting coronagraph would be superbly suited for solar IR observations. Such a coronagraph simplifies heat-flux rejection, and can be designed to offer extremely small net polarization. Internally-scattered light, the historic drawback of reflecting coronagraphs, can now be suppressed by superpolishing techniques. Prototype reflecting coronagraphs of 5cm and 15cm aperture operated at NSO/SP have demonstrated their ability to measure coronal structures that are much fainter than would be measurable with conventional refracting coronagraphs of comparable aperture.

A 1-2m aperture LARC could measure magnetic field structure at many heights in the corona, providing thereby the key to understanding coronal plasma processes. It could measure processes associated with electric current dissipation, MHD wave generation and damping, magnetic field strength and direction, magnetic reconnection, flows and condensation in prominences, small spatial scale dynamical phenomena such as the events thought to be involved in high speed solar wind streams, and the mechanisms giving rise to coronal mass ejections. It could make excellent low-scattered-light photometric measurements of disk features, and could also make many nighttime observations requiring low levels of scattered light. The LARC, with its high spatial resolution, would neatly complement the high spectral resolution offered by the small reflecting coronagraph onboard SoHO.

Tomography of the Convection Zone

Early studies of the hydrodynamic structure of the solar interior through observations of pressure-driven (p) acoustic modes of oscillation provided a 1-dimensional description of the thermodynamic properties of the solar convection zone and interior. More recently the question of 2-dimensional structure has been opened, in the study of the solar rotation rate as a function of depth and latitude. The opportunity to study the 3-dimensional structure of subsurface layers – what has come to be called "tomography" – is now before us. Inhomogeneities produced by convection, large scale flows, and dynamo action all show promise of detection. There is reason to believe that tomography will provide the ability to probe the near-surface regions on the unseen hemisphere, leading to predictions of active centers that will rotate into the solar hemisphere that is visible from Earth.

Observational requirements include high spatial resolution (CCD formats $\approx$ 4096-square), full-disk field of view (small-aperture active-optics telescopes), high signal-to-noise ratio (CCDs with large full well in combination with suitable narrow-band filters), and near-continuous observations over several days (a campaign mode is adequate, since the phenomena are not long-lived).

Macroscopic Electric Fields

Remote sensing of both quasi-static and wave-related macroscopic electric fields opens up a new diagnostic technique for particle acceleration and energy dissipation in the solar chromosphere and corona. Quasi-static transverse fields of order 10-100 Volts per cm are expected in reconnection models of flares. We would expect these fields to be canceled by polarization of the moving plasma. However, they would be detectable in a non-comoving population of neutral hydrogen. Whether a sufficient emission measure of non-comoving HI exists in the possibly highly filamented neutral sheets of a flare volume is a matter that can be settled only by electric field measurements. Current discharge flare models, on the other hand, predict large parallel E-fields, whose intensity may also reach 100 V/cm or more. Again only sensitive E-field measurements can determine the emission measure of HI that might exist in such double layers. In addition, wave- or turbulence-related E-fields of comparable 10-100 V/cm intensities have also been predicted in some flare models.

A technique using the transverse Stark effect in the high Paschen lines has been shown to yield sensitivities in the 10V/cm range, and observations with such a CCD "electrograph" are now underway at the NSO/SP Evans facility. Several directions appear to be promising in future application of this new technique. One is development of an imaging electrograph using a narrow-band filter instead of a spectrograph to study the two-dimensional distribution of the transverse electric field component. Such an instrument might use a liquid crystal birefringent filter as a monochromator to achieve ease of tunability over several spectral lines. Another direction is to build an electrograph optimized for the Brackett and Pfund series lines. The same wavelength-squared advantage used in infrared magnetic measurements at 1.6 and 12 microns could be realized by such an instrument.

Precision Solar Photometric Telescopes

Progress in measuring and understanding the Sun's radiative outputs, stellar light variations, and the impact of solar variations on the Earth's climate and atmospheric chemistry requires high precision and assured continuity in several kinds of photometric and radiometric observations of the Sun and Sun-like stars. These data, obtained from the ground and from space, are essential for improved physical understanding of the Sun's present, past and future variations in the EUV, UV, and total radiative output.

The essential item of hardware for the solar component of these studies is a stable, well-characterized, low-scattering system for precision solar photometry. This system would include a small telescope, a monochromator, a large CCD detector and a modern data acquisition system. At least two of these instruments would be deployed at suitable sites, and manned by staff experienced in highly reproducible, carefully documented photometric observations.

2.3.5 Interdisciplinary Initiatives

In this section we describe several initiatives important for solar physics, but also important to other disciplines, and whose funding would come from outside the normal solar physics sources. It is thus not within the capability or purview of the traditional funding sources for solar physics to make any of the initiatives described in this section happen. Our intent here is simply to recognize that each and every one of these projects will allow us to learn something interesting and important about the Sun. On the basis of their interest to solar physics, they can be placed in rough priority as follows:

Neutrino Research

Neutrinos produced in the solar core, as a result of the nuclear reactions that give stars their long, stable lifetime, are predicted to have a mixture of line and continuum energy distributions. There are now two experiments which are sensitive only to the relatively rare, but high energy, ^{8}B neutrinos arising from a side-reaction of little energetic importance in the Sun; these experiments – the US Homestake mine experiment and the Japanese/US Kamiokande II collaboration – show consistent results, both at a level statistically significantly below that predicted by theory.

The next steps are to measure the energy spectrum of the electron-type ^{8}B neutrinos (as will be done by the Sudbury Solar Neutrino Observatory deuterium experiment, a Canadian/US/Great Britain collaboration, and by the liquid argon detector being developed by CERN and Italy), to measure the time-dependence of the neutrino flux (as a function of solar cycle as well as on the much shorter time scales associated with solar flares), and to determine the solar neutrino flux from the p-p reactions which play the central role in the energetics of the solar core. Unfortunately, the US participation in the two ^{71}Ga experiments now underway (the GALLEX experiment, principally funded by West Germany, Italy, France, and Israel, and the SAGE Soviet-American experiment) is relatively minor. Projects which are now under study will allow the energy spectrum of low-energy neutrinos to be mapped out (using, for example, low temperature detectors). Both the neutrino flux measurements and the neutrino spectroscopy are essential for progress in understanding the physics of the solar core (and hence the cores of other stars); indeed, it can be argued that such neutrino spectroscopy is as crucial to a thorough test and understanding of stellar structure and evolution as photon spectroscopy has been to fostering astrophysical developments over the last century. It is critical that these future developments be well supported.

NRAO Millimeter Array (MMA)

Exciting opportunities exist for high spatial resolution solar mm-wave studies using the array envisaged by the National Radio Astronomy Observatory. This wavelength domain is one of the last frontiers of radio astronomy. A key attribute of the MMA is that its spatial resolution is comparable to that now obtainable at other wavelengths, ranging from the optical to the X-ray domain – of order 1 second of arc; furthermore, at these short wavelengths, one can observe far deeper into the solar atmosphere than is possible at cm and longer wavelengths.

The science problems which can be attacked by the MMA are manifold. Consider, for example, mm-wave radiation from solar flares: At the lower chromospheric level, it likely to arise from thermal bremsstrahlung, allowing one to relate radio wave brightness to the density-temperature structure in the heated regions; relative timing of mm-wave vs. cm-wave bursts should then help distinguish among the possible causes of this heating. High time resolution studies of mm-wave flare continuum emission from 10-100 MeV electrons, and comparison with continuum and nuclear gamma-ray line observations, will constrain models for the as yet poorly understood near-simultaneous acceleration of electrons and protons to very high energies. Mapping of solar active regions, filaments and prominences which takes advantage of the partial polarization of mm-waves (resulting from the difference between x mode and o mode emissivities) will inform us about the magnetic field strength and topology at heights greater than the photosphere (where most magnetogram data apply). The arcsecond resolution of the MMA will help to understand why coronal holes are brighter than quiet regions in mm-waves, contrary to what is observed at almost all other wavelengths. Thus, there is wide recognition of the MMA's outstanding potential for solar research, but it is important that adequate time be devoted to solar studies, and that the technical challenge of providing the desired data be met, for significant progress will surely result from applications of this array to solar studies.

An Antarctic Observatory for Long-Duration Observations

South Pole observations avoid two significant disadvantages of low-latitude networks, i.e., the need to merge observations from different instruments and effects of diurnal fluctuations in observing conditions at each site. Additionally, the South Pole offers better than arc second seeing and extraordinary infrared observing conditions. The scientific benefit of a South Pole facility, which could be used for a number of different investigations by different groups, is significant for the study of solar activity variations over time scales of hours to weeks. Such a benefit has been realized for helioseismology and could be used to advance our understanding of solar activity in the same way.

The Canadian Compact Cm/dm Array

A study of a compact synthesis array for decimeter/centimeter waves is presently underway by Canadian radio astronomers; an international collaboration may be sought. The present concept has about 100 antennas, each about 12 m in diameter, occupying an area about 2 km in extent. This array would have a spatial resolution in arcseconds approximately equal to the wavelength in centimeters, and operate between about 0.4 and 5 GHz (possibly to 15 GHz). It would have a time resolution of 1 s or less. Its most important property for solar studies is the excellent u-v plane coverage in the snapshot mode (approximately 5,000 samples) that is needed to make high quality images (dynamic range > 100) in circumstances where the brightness distribution is changing. With such an array it would be possible to make definitive studies of the quiet Sun, active regions, and flares. It is particularly important to have flexibility in choosing observing frequency, as lower frequencies arise preferentially from greater heights, allowing changes of atmospheric structure with height to be measured.

Stellar Oscillations

The measurement of the stellar analogues of solar global p-mode oscillations will provide unique constraints on the physical structure of stellar convective zones and thereby test interior models. In addition, observations of the rotational splitting of such modes will yield the depth-averaged internal rotation rate. Comparison of the internal rotation rate with the surface rate, obtained from either line broadening measurements or modulation of the stellar flux by asymmetrically distributed surface active regions, will provide the first direct information on a fundamental parameter in dynamo theory, namely, differential rotation. The scaling of oscillation amplitudes with stellar parameters will elucidate the physics of the excitation process (e.g., stochastic excitation should scale in a different manner than other instability mechanisms). Finally, the observed mode frequencies can provide important new constraints on evolutionary processes. All of this information is central to understanding the structure, evolution, and internal dynamics of our own star, the Sun. The detection of stellar oscillations is the classic "photon-starved" problem where every increase in aperture yields gains. However, long uninterrupted blocks of time are required for success in asteroseismology. Hence, a facility that can be dedicated to programs of this kind is essential to progress further in the direct study of the interior structure of solar-type stars. A modest first step would be to construct several sets of identical focal plane instruments, each fed by an optical fiber. These could then be deployed at existing moderate-aperture telescopes (2-m or larger) at different longitudes in a campaign mode, to provide early measurements of key stellar seismic parameters on the brighter Sun-like stars.

Stellar Magnetic Activity

Stellar activity on both rotational and cyclic time scales is being investigated for a sample of solar-type stars and a sample of stars that spans a broad range of physical parameter space. Of particular interest are the various activity cycles and large luminosity variations that have appeared after only a couple of decades, foreshadowing occasional variations in the Sun at widely spaced and unpredictable intervals. Thus far, this fundamental and extremely important study has received but little attention and support from the federal funding agencies. The relevant stellar properties include mass, effective temperature and gravity, rotation rate, age, metallicity and fractional convection zone depth. An especially critical boundary condition for dynamo theory occurs in the region of the Hertzsprung-Russell diagram where stars become fully convective, i.e., for spectral types later than about dM5. It is in this region where the nature of the operative dynamo is expected to undergo a fundamental change. We have no information on the nature of activity cycles, if they exist at all, in stars with wholly convective interiors.

At the very least, funding is required to maintain the important, if long-term, synoptic studies underway with existing facilities. However, because of the intrinsic faintness of many stars of interest, a large aperture facility is required to obtain synoptic observations of key activity diagnostics such as the Ca II K line. A 4 m-class telescope with modern instrumentation will permit observations of stars 2 magnitudes fainter than is possible with existing smaller and older facilities, thus expanding the sample size by more than a factor of 10. In this way, meaningful statistics on the properties of the activity in the fully convective, intrinsically faint M dwarf stars can be obtained.

2.4 Conclusions and Summary

During the next decade, ground-based solar physics will enter a new observational frontier enabled by advances in adaptive optics, high precision polarimetry, and infrared imaging: direct observations of interactions between solar surface motions and solar magnetic fields on the sub-arcsecond spatial scales which theory predicts are relevant to the evolution of solar surface magnetic fields and energy input into the solar outer atmosphere. These new capabilities complement the new view of the solar interior which helioseismology has revealed over the past decade. As a starting point, we therefore strongly encourage the continued vigorous support of the *GONG* project (which will push the frontiers of exploring the solar interior from the ground); and we strongly recommend, first, the immediate development of adaptive optics necessary for high and uniform angular resolution optical observations, with the specific aim of building, together with international partners, the *LEST* facility to attack the frontier of high angular resolution solar surface observations; second, the vigorous development of infrared imaging and spectroscopy instrumentation, together with development of a detailed plan for a large-aperture infrared facility; third, a concerted effort to implement the moderate-scale initiatives listed in Section 2.3.4, and fourth, support for the interdisciplinary initiative listed in Section 2.3.5.

3. SPACE OBSERVATIONS FOR SOLAR PHYSICS

3.1 Introduction and Summary

Beginning with the early V-2 rocket flights, the observing capabilities opened up by access to space have led to a continual flow of discovery and understanding in many branches of solar physics. The new techniques made possible by space — new wavelengths, unparalleled "seeing", and photometric stability — have formed a large part of the modern renaissance of solar physics. The solar observational domain has at the same time broadened tremendously, now ranging from neutrinos from the solar core to resonance-scattered interstellar gas at the heliopause.

Much interest in solar physics is associated with magnetic activity: the dynamo, the surface magnetism in sunspots and faculae, solar flares and coronal mass ejections, and the solar wind itself. Observations during the last decade, with the *Solar Maximum Mission* (*SMM*) and other instruments, have provided new insights into the problems of explosive energy release and particle acceleration associated with solar activity. But in addition, surprisingly, the deep solar interior has stimulated flourishing new fields of investigation – led by the neutrino puzzle and the remarkable exploits of helioseismology. These new discoveries are just beginning to be reflected in the content of the U.S. space program in solar physics.

To meet the challenge of these new discoveries, we recommend the immediate development of approved missions such as the *Solar and Heliospheric Observatory* (*SoHO*) and the *Orbiting Solar Laboratory* (*OSL*), together with the support and encouragement of small observational programs (sub-orbital, international missions, small Explorers, and individual experiments on various spacecraft). Among the proposed new missions beyond *OSL*, the solar community strongly endorses the *High Energy Solar Physics* (*HESP*) mission, a small-to-moderate mission capable of studying magnetic active regions and flares by emphasizing hard X-ray, gamma-ray, and neutron observations, through the maximum of the forthcoming solar cycle (ca. 2000 A.D.).

3.2 Ongoing Programs

3.2.1 U.S. Programs

There are no currently approved U.S. spacecraft dedicated to solar physics. The only NASA programs in observational solar physics approved at present consist of one instrument on the Japanese (ISAS) *Solar-A* (1991 launch), instrumentation on board the European (ESA) *Solar Heliospheric Observatory* (1995 launch), and the remainder of the small suborbital program (including two Max 91 balloon payloads, some other rockets, balloons, and one Spartan mission). In addition there are important interdisciplinary instruments with potential solar applications, e.g. the planned launches of the *Gamma-Ray Observatory*, *Ulysses*, and *WIND* in the early 90's.

One flight instrument (the *Ultra High Resolution XUV Spectroheliograph*) and one concept study for a major mission (the *Pinhole/Occulter Facility*) have been approved for Space Station *Freedom*. Depending upon the rapidity of deployment of the Space Station, and of its utility for attached payloads, it may become a major factor in space observations for solar physics. The *Pinhole/Occulter Facility* is discussed further below.

The *Orbiting Solar Laboratory* is in a special category, in that its Shuttle-based predecessor was an approved mission in the past. It is very much an ongoing program and its approval to resume development for flight as a free-flyer in a polar orbit is anticipated in the very near future. We discuss it in detail in Section 3.3.1 below.

3.2.2 Non-U.S. Programs

At present there are a number of solar spacecraft already under development in all of the non-U.S. major space programs (Japan, Europe, U.S.S.R.). These consist of

• *Solar-A*. This Japanese spacecraft, to be launched in 1991, will observe high-energy phenomena in solar flares. It contains one major U.S.-supplied instrument, a soft X-ray telescope with a 1024 × 1024-pixel CCD camera (pixel size 2.5 arc seconds). This will obtain the first soft X-ray images of the Sun from orbit since the *Skylab* Observatory of the early '70's, and the first not using a film readout. *Solar-A* also carries instrumentation for hard X-ray imaging and non-imaging X-ray spectroscopy.

• *SoHO*. This ESA mission contains three U.S. experiments: a solar oscillations imager for helioseismology, and white-light and UV coronagraphs for studying coronal mass ejections and the solar wind. The spacecraft will be placed at the L1 Lagrangian point of the Earth-Sun system, offering continuous sunlight. In addition to these experiments, the *SoHO* instrumentation will carry out detailed observations of the chromosphere, transition region, and corona with EUV and XUV instruments.

• *CORONAS*. This Soviet project consists of a series of spacecraft launches to study solar phenomena simultaneously with their influence on near-Earth space. The measurements will be made from the *Automatic Universal Orbital Station* with solar orientation, to be launched into a quasi-Sun-synchronous orbit between 72 - 82 degrees at an altitude of 500 km. Measurements of particular interest will be made with a soft X-ray telescope (*TEREK*) for location of solar outbursts. Other instruments (*IRIS, SONG*) will determine the spectra of gamma-rays and neutrons. The spectra and composition of solar cosmic rays near the Earth will also be measured. Additional instrumentation includes a radiospectrometer for the frequency range 0.1-30 MHz, and photometers to study solar oscillations. There are no U.S. experiments aboard the *CORONAS* spacecraft. Currently two *CORONAS* launches (I and F) are planned between late 1990 and 1992.

3.3 New Missions

3.3.1 The *Orbiting Solar Laboratory*

The *Orbiting Solar Laboratory* (*OSL*) is the prerequisite space mission for solar physics in the 1990s. It will break through the barrier of poor spatial resolution which has severely retarded studies of magnetic activity on the Sun. Earlier solar space observations have primarily emphasized wavelength regimes not accessible from the ground. The *OSL* combines UV and soft X-ray observations of the chromosphere and corona, with photospheric measurements from a diffraction-limited 1-m telescope totally free from atmospheric blurring. This will permit observations of basic phenomena on spatial scales comparable to the density scale height of the solar photosphere. Accurate measurements of physical properties free from non-linear spatial averaging over a wide range of varying conditions will be possible for the first time.

As a well-instrumented, long-lived and readily available facility, the *OSL* will be capable of conducting a wide and varying range of solar research. A primary goal for *OSL* research is the nature of solar magnetic fields from the deepest observable layers of the photosphere upward to high temperature regions of the solar corona. A particularly important location is where the solar plasma changes from domination by radiative and convective processes to magnetic control; processes in this region are fundamental in creating solar activity. Thus the dynamic interaction of magnetic fields with mass motions is another key goal of *OSL* research. A third major goal is to study magnetic energy storage in the atmosphere, and the conversion of violently released energy to high temperature plasma in phenomena such as flares. Many of these processes are currently mysterious even for as well-observed a star as the Sun. Solving these problems will give us more confidence that our understanding of processes elsewhere in the Universe is well founded.

The *Orbiting Solar Laboratory* will resolve the individual flux tubes that have widths comparable to the intergranular lanes of the photospheric convection (about 200 km or less) and will track their migration, intermingling, and interaction. At the same time, it will measure the response in the overlying chromosphere and corona to these MHD interactions in the photosphere. This cause-and-effect observing capability is the core rationale for *OSL*, leading to knowledge of:

• How the interaction between the convection and the magnetic field drives the heating that sustains the chromosphere, corona, and solar wind.

• How nonpotential energy is built up in large-scale metastable configurations, how these configurations evolve (and possibly become unstable), and how the stored energy is released (including possibly in flares and coronal mass ejections).

• How magnetic flux in the photosphere is "processed" by turbulent convection, and caused to diffuse away from active regions. This apparently sows the seeds for a new magnetic cycle in the dynamo process that sustains the Sun's magnetic field.

OSL is an extremely well-defined and thoroughly-studied mission. Planning for the critical component, a large-aperture diffraction-limited visible-light telescope (the *Solar Optical Telescope*, or *SOT*), began 17 years ago. Indeed the *SOT* was approved for development as a *Spacelab* mission more than a decade ago, but the development was then postponed for programmatic reasons unrelated to the scientific merits of the mission. Since then, technical capabilities have increased greatly, so that today the *OSL* has scientific potential *far* transcending the original mission. *OSL* is the key mission for solar physics, and enjoys the highest possible endorsement of the community. The Solar Panel notes with satisfaction that the *OSL* is now at the top of NASA's Strategic Plan for Space Science, and urges in the strongest possible terms that its development be resumed as soon as possible.

3.3.2 The *High Energy Solar Physics* (*HESP*) Mission

The scientific goals of *HESP* – a new mission to study the active Sun and flares during the activity maximum toward the end of the decade – center on the mechanisms and processes of explosive energy release and particle acceleration associated with solar activity. These processes are at the core of the solar flare problem, and they play a major role in all of astrophysics, particularly in objects dominated by high energy processes. *HESP* will observe the high-energy radiations (hard X-rays, gamma-rays, and neutrons) which are the most unambiguous signatures of accelerated particle interactions. Observations of these emissions with high spatial resolution will allow the localization of the sources of the particles and the tracing of their transport paths. Observations with high spectral resolution will allow the deciphering of the rich information encoded in gamma-ray lines, such as abundances in both the ambient gas and the accelerated particles, beaming of the accelerated particles, temperatures and states of ionization of the ambient gas, and the structure of the magnetic fields. The combination of high spatial and energy resolution, by providing diagnostics which are qualitatively different from anything available so far, offers unique opportunities for resolving some of the most complex issues in solar physics (flare mechanisms, particle acceleration, coronal heating). Understanding of these issues will also be of great benefit to high energy astrophysics. In particular, *HESP* will provide information on:

• The nature of acceleration mechanisms, by determining the ratio of accelerated protons to electrons from observations of gamma-ray lines and continuum. An overabundance of protons would favor shock and stochastic mechanisms, while the overabundance of electrons would favor electric fields. The combination of high spatial resolution and energy resolution could distinguish sites where electron acceleration dominates from sites where ion acceleration is dominant.

• Angular distributions and magnetic field structures, from direct imaging of hard X-ray bremsstrahlung in the active flux tubes. The shapes of the nuclear deexcitation gamma-ray lines also reveal the angular distribution of the interacting ions, which in turn depend on the nature of the acceleration and transport of the fast particles and on the structure of the magnetic fields. With the combined angular and spectral resolving power of *HESP* it will be possible to determine angular distributions accurately as a function of position in the atmosphere.

• Abundances in both the ambient gas and in the accelerated particles, from the relative intensities of gamma-ray lines. These abundances include that of ^{3}He in the photosphere. The high energy resolution capability of *HESP* will allow the separation of many more lines than was possible with *SMM*, thereby qualitatively enhancing the power of the technique. Abundance variations shed light on mechanisms of mass motion in the atmosphere and turbulent mixing in the interior.

• Temperatures, densities and states of ionization of the ambient gas, from the shape of the 511 keV positron annihilation line. Recent galactic observations of this line with high energy resolution have yielded information on the interstellar medium. No comparable solar observations of the 511 keV line have yet been carried out. Based on the analogy with the galactic case, we are confident that much new qualitative information will be forthcoming from high resolution observations of annihilation radiation from solar flares.

• Heating of the corona, by observing bremsstrahlung from electron acceleration in microflares and other non-thermal processes. If even minimal efficiency for 10 MeV proton acceleration accompanies the energy deposition in the solar corona, then the corona can also be observed as a steady source of 2.223 MeV line emission; that is, it has been shown that even if one part in several thousand of the energy needed to accelerate the solar wind is deposited in the corona by protons of energies greater than 10 MeV, the Sun would be a steady source of gamma-ray line emission. Because of its great sensitivity to the very narrow 2.223 MeV neutron capture line, *HESP* could detect the presence of ion interactions which heat the nonflaring solar atmosphere. The question of the heating of the solar corona could therefore be investigated with *HESP* in a novel way.

The *HESP* payload will consist of a single instrument, a Ge spectrometer that combines high spectral resolution for hard X-rays and gamma-rays with high-resolution imaging from modulation-collimator optics. It will also have the capability for observing meson-decay gamma-rays and neutrons using its anticoincidence shield elements. The instrument will have angular resolution below one arc sec, energy resolution on the order of 1 keV (over an energy range up to 10 MeV), and time resolution of < 1 s. Observations with *HESP* will characterize the evolution of solar activity from the beginning and through the maximum of Cycle 23. To carry out its mission properly, *HESP* should be accompanied with diagnostic observations of parameters of solar plasma in the $10^4 - 10^7$ K temperature range, and with vector magnetic field measurements in the photosphere. Such observations could be provided by the instruments on *OSL* if *OSL* is operating during the next peak of solar activity. (If not, *HESP* should carry XUV, EUV, and enhanced visible instrumentation capable of imaging and spectroscopy of the solar atmosphere.) *HESP*, as defined here, lies in the small-to-moderate mission category, and is consistent with the scope of the Explorer program. The Solar Panel regards *HESP* as the highest priority *new* mission for solar physics in the 90's, and strongly recommends its rapid development, so that it is in orbit by the next rise to maximum solar activity around 1999.

3.3.3 "Quick" Opportunities in Space

In addition to *HESP* and the on-going missions, we feel strongly that a full spectrum of space flight opportunity should exist and be exploited properly for solar physics. The small end of the spectrum — to which Freeman Dyson's comment that "Quick is Beautiful" applies perfectly — includes Small Explorers, Shuttle-based experiments (including GAS payloads), the suborbital program (balloons and rockets), partial payloads on various U.S. and international spacecraft, and small attached payloads on the Space Station. Many of these smaller, short time-scale opportunities have a major positive effect on the nature of the solar physics program because of the involvement of students and the impetus towards innovation in instrumentation.

Balloons and rockets will continue to be at the forefront of instrument development activities for future space missions. These vehicles can also produce highly valuable data. Long-duration balloon flights, for example, can offer 7-20 days of data, thus providing essentially mini-spacecraft missions at low cost for certain wavelength ranges (gamma-rays, hard X-rays, optical). The current Max '91 Solar Balloon Program is a good example of an initiative exploiting this capability, but has been the subject of major cutbacks. The *SPARTAN* program and other limited-duration experiments on board the Space Shuttles also deserve notice, and indeed have already produced important data in several branches of solar physics. Suborbital and other "quick" opportunities are the ideal vehicles for student training in experimental work, essential for the future growth of the discipline.

The Space Station *Freedom*, with proper planning and adequate transportation back and forth to orbit, could meet many of the requirements for small solar space instrumentation (as has the *Spacelab* program, in fact). For example, precise radiometric instruments could be deployed, intercalibrated, exchanged with new versions, etc. Such measurements would benefit solar and stellar astronomy as well as radiometry in general. The *Ultra High Resolution XUV Spectroheliograph* (*UHRXS*), recently selected for flight on Freedom, should achieve spectrally resolved images with angular resolution of ≈ 0.1 arc seconds over the wavelength interval $\approx 6\text{Å} < \lambda < 1,550\text{Å}$, which is indicative of the potential that *Freedom* represents for solar physics.

3.3.4 Other missions

Many other important space missions have been proposed, each with unique capabilities for solar and related sciences. We do not prioritize these here and recognize that some of them will have to wait for some time, but we do think that their scientific content strongly warrants their eventual development. Some require deep space, and these will have increased priority in the eventuality of human travel to the Moon and the planets. Some of these missions have been well-studied, and others are just ideas at present.

• The *Pinhole/Occulter Facility*: The *P/OF* scientific objective is to provide sensitive, high-resolution observation of the solar corona, incorporating large-aperture UV and visible imagers as well as the crucial high-energy observations (hard X-rays and gamma-rays) at 0.1 arc sec resolution. The *HESP* mission defined above serves as a predecessor of the full-fledged Pinhole/Occulter high-energy experiments. The *P/OF* coronal observations would provide an ideal Earth-orbital complement to the *Solar Probe*, for its observations would record the "context" of the *Probe*'s *in situ* measurements. As mentioned, *P/OF* has been approved as a concept study for the Space Station *Freedom*, and its development schedule will depend upon that of the Station.

• *Solar Polar Orbiting Imager*: Solar observations from above the solar poles, including imaging and stereoscopy in conjunction with observations from near Earth. The polar regions of the Sun play a unique role in the evolution of solar activity, particularly on solar-cycle time scales. A solar polar orbit would let us look directly at these regions for the first time, without the major uncertainties caused by projection from ecliptic-plane observations. A polar orbiter will also give a global view of the mid-latitude regions of activity and permit us to follow their evolution on the crucial few-week time scales not accessible from a single perspective in the ecliptic.

• *Solar Probe/Context*: While not a solar remote-sensing mission in the classical sense of solar astronomy, *Solar Probe* can make *in situ* observations from its unique perspective, within a few solar radii of the solar surface. True scientific productivity from *Solar Probe* demands the existence of facilities for "context" observations of the corona from one A.U., with state-of-the-art coronal imaging and spectroscopy; the scientific return from *Solar Probe* can be amplified many times by conducting high-resolution coronal observations before, during, and after the perihelion passage – it may be possible to carry out true tomographic remote-sensing observations, capable of defining the three-dimensional structure through which *Solar Probe* would fly. This is the aim of the *High-resolution Coronal Imager (Context)*. The *Context* mission addresses a major problem in astrophysics: How are hot plasmas in stellar atmospheric structures created? Through high-resolution imaging and spectroscopy, *Context* can identify the structures through which *Probe* has flown, in order to understand the environment of *Probe*'s particles-and-fields measurements. Such a combination of *in situ* and remote sensing will be unprecedented in astrophysics.

• *Advanced Solar Observatory (ASO)*: The deployment of advanced instrumentation in all areas of observation, into a comprehensive facility for space observation of the Sun, represents a natural goal of the solar space program. Concepts for an *ASO* have been studied for the Space Station. The *Pinhole/Occulter Facility* and the *Ultra High Resolution XUV Spectroheliograph* represent the initial steps toward a comprehensive *Advanced Solar Observatory*.

• *Mercury Orbiter*: Unique solar physics observations can be made from a spacecraft orbiting Mercury. High-energy neutral emissions, such as hard X-rays, gamma-rays, and neutrons, will be considerably more intense than at 1 A.U. (about 10× for photons, but more than 1000x for neutrons). Solar energetic particles, when observed at 0.3 A.U., are less affected by the intervening interplanetary medium, and hence will more directly reflect the propagation of the particles accelerated in the flare region than they do at 1 A.U. Long-term observations of these flare emissions from Mercury could be carried out for a large part of Cycle 23 after the year 2000. Additional observations at other solar aspect angles can give valuable stereoscopic information bearing on the directivity of solar emissions or their height of origin. These observations can be made with small instruments (<25 kg) with modest power and telemetry requirements.

• *Solar-Stellar Activity Mission*: The stars teach us about solar activity by allowing the study of analogs to the solar mechanisms; they also give us rich fields for new discovery. Stellar magnetic activity can be sensitively studied with wide-field (one degree), high-resolution (few arc sec) soft X-ray observations, preferably arranged with observing sequences that allow broad ranges of time scales. A dedicated solar-stellar activity mission might also carry EUV instrumentation for monitoring stellar chromospheric activity, and visible-light instrumentation for stellar seismology.

• *Solar Watch*: a network of deep-space solar observing nodes in deep space can provide unique views of the Sun from different perspectives. Such stereoscopic viewing is essential scientifically, because of the globally variable structure that solar activity presents to us and because of its inherently three-dimensional structure. A global view is also essential for characterization of solar activity – on relevant time scales – at a level capable of supporting accurate predictions. This prediction capability is a prerequisite for human space travel in the future, for example to the Moon or to Mars, because of the threat posed by high-energy radiations. We envision a "Solar Watch" program beginning with fairly simple operational measurements, and building up to sophisticated research-class instrumentation in the future.

• *Solar Synoptic Observatory*: a permanent space facility for basic solar observations ranging from total irradiance to H-alpha flare patrol. This type of work is currently done from the ground using antiquated technology, as a service function for various commercial and government users (e.g., power distribution networks interested in service disruptions due to magnetic storms). Such data are often used for scientific purposes, because they uniquely cover the longer time scales. Both functions would improve dramatically – much improvement in precision would be possible, for example in sunspot records – if a permanent facility could be established, perhaps at geosynchronous orbit as a part of the NOAA meteorological satellite system.

• *Janus*: a mission to characterize both the origins and the results of the Sun-Earth interactions, largely by simultaneous remote sensing. The name Janus comes from the Roman god of doorways, represented with two faces for bidirectional viewing. The Janus mission would require observation both from deep space (e.g., the L1 point) and from a polar 24-hour platform. Janus would simultaneously tackle (a) the precise characterization of Sun-Earth relationships of all kinds, essential for the understanding of anthropogenic changes in the environment; and (b) the accumulation of solar data essential for prediction of solar and heliospheric activity – the "solar weather."

3.4 The Space Exploration Initiative

A Presidential initiative is now under consideration to return human beings to the Moon and eventually to send them to Mars as well. While the ultimate objectives of this endeavor are not primarily scientific, this enterprise will inevitably affect solar research in strong ways: "applied solar research" insofar as solar high-energy radiations pose a danger to astronauts' health; and "pure solar research" that can take advantage of uniquely valuable research platforms enabled by the initiative. We discuss these briefly here.

The Initiative presents both challenges and opportunities to solar physics. The challenges relate to the prediction of solar activity, a practical subject whose success is closely linked to our theoretical understanding of the underlying physics. We have addressed this prerequisite with the "Solar Watch" program described above, which will provide qualitative and quantitative improvements in the data base for solar activity forecasting. Deep-space observations of the invisible hemisphere of the Sun alone will appreciably improve activity forecasting via the early warning capability of active-region growth. It should be noted that the present solar maximum is (at least) one of the two largest on record, and that the "millennium maximum" will also produce large particle fluences.

Forecasts of space environment conditions will be more important than ever before when man returns to the Moon and starts for Mars. NASA made a huge effort during the Apollo program to keep aware of and avoid potential dangers from space disturbances. The Space Exploration Initiative will call for an even better effort because of the longer periods of astronaut exposure involved. The needs of society and those of solar physics coincide in research that will lead to a deep enough physical understanding of the Sun to permit accurate predictions of its magnetic activity, or to at least specify the limits of its predictability.

Opportunities for solar research are provided by the unique platforms made available under the Initiative.

The surface of the Moon has some explicit advantages for solar observations. It can provide a stable platform for extremely long focal-length instruments or interferometers, for example. A future Lunar Solar Observatory could therefore give us extraordinary capabilities, perhaps beyond those conceivable in free space. A particularly attractive idea is to build (on the far side of the Moon) a low-frequency radio interferometer. Such an instrument – in a crater far drier than Clark Lake or the Plains of Saint Augustine – would be in an ideal site because of the suppression of terrestrial radiation by the body of the Moon. The instrument would be capable of non-solar observations during its local nighttime.

The deep-space explorations will give us platforms for stereoscopic solar observations. Such observations (of the photosphere, chromosphere, and corona) will allow us to make true three-dimensional images of solar phenomena, probably a key qualitative improvement in many areas of solar physics. Also, the hidden hemisphere of the Sun carries unique information about global processes, especially on the time scales (crucial for active-region development) comparable to the rotation period. For these reasons, an interplanetary network of solar observing instruments would be an extremely desirable goal for solar physics in space.

Finally, there is a sociological point to be made: Permanent inhabitants of the Moon, at some future time, will rather naturally have a strong interest in the Sun and its activity because of its practical significance to their survival. Solar observing facilities should have priority, and this priority should include both the early and late phases of development of lunar habitation.

3.5 Solar-Terrestrial Physics

By the time of the next solar maximum, environmental issues will likely be much more the focus of the space program than they are today. Already, better understanding of our environment is the premier achievement and most challenging goal of the space program. The EOS mission is a recognition of this. EOS will include solar irradiance monitors, but solar physicists will need to complement EOS with a research satellite dedicated to the predictive foundations of solar magnetic activity and solar-forced geomagnetic activity. While we do not make extravagant promises, there is little question that a vector magnetograph, a full-disk-viewing X-ray telescope, a chromospheric filtergraph, a coronagraph and a complement of high energy burst detectors in Sun-synchronous orbit or at L1 would make strong contributions to our understanding of disturbances of the near-Earth space environment.

3.6 Conclusions and Summary

To meet the challenge of the next decade in space, we recommend the continued development of *OSL* and the other approved missions. Beyond these, our top priority for a new space project is a small-to-moderate mission capable of studying solar activity through the maximum of the forthcoming solar cycle (ca. 2000 A.D.); this *High-Energy Solar Physics* (*HESP*) mission will emphasize high-energy observations with an instrument of unprecedented spectral and spatial resolution, and sensitivity. We also strongly recommend the support and encouragement of small observational programs (sub-orbital, international missions, small Explorers, and individual experiments on various spacecraft), as well as ongoing space programs such as *Solar-A* and *SoHO*. Finally, we recommend expeditious development of the remaining missions listed in Section 3.3.4, each of which represents unique, first-class science.

4. TECHNOLOGY FOR SOLAR PHYSICS IN THE 1990s

4.1 Introduction

This section contains brief descriptions of technology development efforts that must be conducted during this decade to enable advanced research projects during the decade following AD 2000. Several of these efforts are now underway and should be continued. Some of the technology developments are common with other fields of astrophysics but others are unique to solar research. For those efforts that are well defined and already initiated, the implementation strategy is simple: finish the efforts as soon as possible and promptly convert the results of successful efforts into useful research tools. For less well defined activities and ones that this panel cannot foresee, the strategy for implementation is to let the peer review process assign priorities in the normal way. We note that although both NASA and the NSF have a mechanism for funding advanced technology initiatives, the scale of these two efforts is substantially different. Partly because of this difference, we list the specific efforts directed toward space and ground-based research separately, but note that there is often much commonality.

4.2 Ground-based solar physics

An observational science such as solar physics depends on improvements in technical capabilities for acquiring and reducing observational data. Two vital initiatives are required to insure a strong ground-based program both for the decade of the 1990s and beyond. The highest priority is a continuation of development of adaptive optical systems to allow existing and future telescopes to achieve high spatial resolution. The second priority is development of advanced data processing for handling the large data sets typical of solar observations.

4.2.1 Adaptive Optics

High spatial resolution measurements are needed for extended time intervals to investigate the detailed structures and dynamics associated with energy transport and storage, heating, and activity in the solar atmosphere. Image blurring and distortion make it difficult to obtain the needed data. Post-facto image reconstruction techniques in principle can be made to work for solar filtergrams, but real-time correction of the seeing is necessary for spectroscopic studies. A prototype solar adaptive optics system built by Lockheed demonstrated that a segmented active mirror could be made to function on the Sun, using high contrast features such as sunspots as targets. Current efforts are toward: (a) development of a more robust system that is easy to keep phased and that can track any arbitrary region on the Sun, locking on the low contrast granulation; and (b) interfacing such a system to an existing solar telescope in a user-friendly manner. Development of and experience with an operating adaptive optics system on an existing solar telescope is critical for the success of the *Large Earth-based Solar Telescope* (*LEST*) project.

4.2.2 Analysis of Extremely Large Data Sets

Progress in observational solar physics depends increasingly on analysis of large datasets of high dimensionality. Examples include helioseismology image time series (3-D), high speed flare spectroscopic imaging time series (4-D), and Stokes polarimetric multiband imaging time series (5-D). Data volumes range from a few gigabytes to a few terabytes. Current capability is only marginally adequate for the near term, and will be a major constraint later in the decade if not improved. Several requirements appear in handling large datasets. A need to monitor instrument performance and verify data quality requires quick access to small samples of the data spread throughout the set. Interactive processing of a subset is required to identify the systematic errors and biases, and the nature of gross defects. Finally, the need to reduce whole datasets to scientifically meaningful results depends on a pipeline analysis keeping pace with the data collection.

Raw computing power coupled with high data flow is necessary to maintain this currency. The development of suitable hardware will be driven by commercial needs far more than astronomical needs. Exceptions may be massively parallel processors or single language processors, which are still most common in research environments. It is critical to exploit useful developments, and to make available the most modern hardware practical for the needs of solar research. Solar physics has software needs that are seldom met by software developed for nighttime astrophysics. There is a need to develop specific solar research software and make it available to the solar community.

4.2.3 Instrumentation

A number of technological developments are important for the highest-priority ground-based solar research. Improvements in infrared narrow-band filters, polarizers, and photodetector arrays will contribute to continuing progress in measurement of magnetic and electric fields in the solar photosphere and chromosphere, as well as improved thermodynamic diagnostic capability in those same parts of the solar atmosphere. Optical photodetector arrays with shorter readout times than are presently available will produce an immediate benefit in any measurements such as the above, including development of adaptive optics. In some polarimetric applications, kilohertz rates would be ideal for processing of detected images in order to eliminate the undesirable effects of seeing. This need could be met by detectors with several (8) storage areas on the chip itself, along with efficient and rapid avenues to transfer the charge among the storage areas and the light-sensitive portion of the chip. Such devices would have to be developed in collaboration with the semiconductor industry. Superpolished mirrors are the key to the extension of coronagraph capability to the UV and IR regions of the spectrum, as well as to improved sensitivity and spatial resolution. Improvements in tunable birefringent filters using liquid crystals as electro-optic elements will bring direct benefit to all polarimetric observations, which encompasses virtually all of the above. Ongoing development of Fabry-Perot filter devices for very narrow-band imaging of the Sun should also be encouraged. Finally, for many applications, a device is urgently needed which allows spectral coverage simultaneously with imaging. This need would be met by an imaging Fourier transform spectrometer operating within a restricted bandwidth of the spectrum. This development will require a large advance in the speed with which the information from area detectors can be stored and processed.

4.3 Space-based Solar Physics

4.3.1 New Technology

The development of new technology for space observations, traditionally the domain of small space programs, has dwindled alarmingly with the lack of emphasis given by NASA to the suborbital programs and university groups. The recent successes of normal-incidence X-ray optics are a classic example of the kind of rewards that can be reaped by small programs.

Urgently needed technology areas include advanced detectors for all spectral ranges, mirror technology, filters, lightweight large structures and data systems. Many of these developments benefit more than one field of research, and should be energetically pursued in NASA's Office of Applications and Space Technology, in conjunction with OSSA.

4.3.2 Large Structures in Space

Most disciplines of astrophysics, including solar physics, envision large future instruments in space. These include telescopes of large aperture and/or focal length, interferometers for a variety of wavelengths, and large-scale occulters such as the *Pinhole/Occulter Facility*. Such structures can be made extremely lightweight and robust by using active servo control *of the structure itself*. This technology exists where needed for structures on the Earth's surface, for example in active-optics telescopes, but has never been utilized in space. NASA may also need this technology for its large engineering structures (for example, Space Station *Freedom* itself). We therefore encourage the orderly development of "Controls-Structures Interactions" techniques.

4.3.3 High-energy Instrumentation

A variety of new high energy (> 10 keV) radiation detection techniques which offer the potential advantages such as high spatial resolution, low background, polarization measurements, large detection area, and high spectal resolution at room temperature should be developed. These may include high pressure gas scintillation detectors, wide band-gap semiconductors such as $HgI2$ or $CdTe$, position-sensitive germanium detectors, and crystal diffraction and channeling techniques.

5. POLICY AND RELATED PROGRAMMATIC RECOMMENDATIONS

5.1. University Research and Education

The discipline of solar physics is currently poorly represented in universities, and is consequently hampered in the development of a strong theoretical and experimental personnel base for future research. There are several initiatives that would specifically address this problem. First, the NASA Graduate Student Research Fellowship (GSRA) program, which funds graduate students semi-independently of a primary advisor, should be greatly expanded. This very successful program has commendably increased its student enrollment beyond allocated levels by ad hoc additional funding supplied by various OSSA Divisions, but we urge that NASA instead consider simply increasing the scale of this program. Second, the NSF graduate fellowships should be awarded over a flexible time period that would allow support to be shifted into the latter part of a student's graduate education, when he or she is most likely to be doing research (NSF Fellowships are currently awarded for the first three years of study). Third, stronger connections between universities and national research centers should be encouraged. This could be done by establishing additional cooperative research programs, expanding visitor programs (at both the postdoctoral and faculty level), and increasing the size of student visitor programs. The latter could be done through the NASA GSRA program, which at present funds students only at universities or NASA centers.

5.2. Facilitating Solar Research

Two steps can be taken by NASA and NSF to facilitate solar research through minor restructuring of existing funding programs.

1. NASA and NSF contracts/grants should have flexible funding periods, ranging from one year to five years, with the length of the funding period being determined by the quality and the requirements of the proposal.

2. NSF grants should allow for release time for teaching faculty, which could, for example, provide one half year of salary every three and one half years to support dedicated research time either at the home university or at another institution.

5.3. Integrated Support of Solar Research

It is important that both NASA and NSF take an integrated approach to the support of solar physics, particularly with regard to theoretical and observational research and to solar and related astrophysical research.

1. Whenever possible, a balanced support of theoretical and observational research should be provided; in addition, it is wise to provide a balance between theorists and observers tied to large research groups and/or projects, and theorists and observers who operate independently of such research groups. At the same time, we strongly urge NASA to improve its past and present record of funding only minimal amounts of science in conjunction with instrument proposals; past funding practices have led to the unfortunate situation that theoreticians are recruited as co-investigators on instrument proposals, contributing to the credibility of the proposal during peer review, but then are not funded to provide substantive scientific input prior to launch.

2. NASA Space Physics and Astrophysics should develop a mechanism for supporting interdisciplinary solar/astrophysical research. Other agencies should consider establishing similar arrangements, as appropriate.

5.4. Computing

1. Supercomputers: Although we applaud the recent initiatives in supercomputing by the NSF and, more recently, by NASA, we have some concern that this new capability is not optimally used. In particular, supercomputers are a singular resource for very large simulations (viz., calculations which consume of the order of 200 or more Cray X/MP CPU hours), which cannot be replaced by machines of the minisupercomputer class. However, national resources devoted to supercomputing in fact generally do not allow realistic access to such use of supercomputers. We urge the NSF and NASA to encourage the provision of additional resources to such very large scale computing, as has in fact been done by the NASA Ames supercomputing facility and, in isolated instances, by some of the NSF Supercomputing Centers.

2. Workstations: Workstations are becoming essential tools for both experimentalists and theorists in the analysis of their "data". For this reason, we recommend that contracts and grants should continue to be able to provide for both the purchase and maintenance of such workstations as a matter of standard practice.

3. Networks: Efforts to provide high speed digital networking capability to large numbers of solar researchers should be enhanced. Frequently used data sets should be made readily available via these networks.

5.5. Theory Initiatives

The NASA Solar-Terrestrial Theory Program (STTP), and its successor, the Space Physics Theory Program (SPTP), have made a commendable start in the support of theoretical work which is not directly mission-oriented. As a number of previous National Academy reports have noted, provision of such support — in addition to more mission-related theoretical studies — is essential for the health of the space physics disciplines.

These reports also noted that theory needs to be supported on two distinct scales: first, at the individual investigator level; second, at the level of group efforts with significant "critical mass". The STTP program indeed was created initially with the specific intention of responding to this second need.

However:

1. The current typical grant size of the NASA SPTP is well below what is desirable for support of "critical mass" theory groups at universities; this desired mean support level has been discussed by NASA, and projected at the roughly $300K level, but has never been implemented. We recommend augmentation of the current program in order to increase the mean grant size to the previously-discussed support levels.

2. Much of current NASA theory grants to individuals is funded through the SR&T budget, in which there is substantial pressure to focus funding on directly mission-related work. This means that theoretical studies in solar physics, which are carried out by individuals but are not directly mission-related, are strongly discouraged, contrary to the recommendations of previous National Academy studies. We recommend that NASA modify the "ground-rules" so that such grants can be funded, based on the quality of the peer reviews.

3. Maintaining an appropriate balance between experimental programs, data analysis, and theory is a major challenge for both the NSF and NASA. We recommend that both agencies establish rough guidelines for balancing these programs, based on input from the appropriate advisory panels.

5.6 Recommendations by the National Academy of Sciences Study on Solar Physics from the Ground

Finally, we note that the National Academy of Sciences has recently published the report entitled "The Field of Solar Physics: Review and Recommendations for Ground-Based Solar Research" (National Academy Press, 1989), which contains the following recommendations which we endorse and incorporate into this report:

"1. Develop a coherent, well-defined infrastructure for solar physics within NSF, with that agency properly assuming the lead role in support of basic research in ground-based solar physics. Thus the committee recommends that the internal structure for funding of solar research within NSF be changed so that support for both grants and centers is administered by a single entity within NSF whose primary responsibility is solar physics. Such a reorganization will permit the development of appropriate advocacy within NSF, the definition of an overall coherent approach to the subject, a unified vision of the field's national facilities and university grants program — its scope and its development — and the implementation of new efforts. The directorate in which to place the recommended section could be either the Geosciences Directorate (the residence of support for solar-terrestrial sciences and the High Altitude Observatory) or the Mathematical and Physical Sciences Directorate (the residence of support for astronomy and the National Solar Observatory). Placement of the recommended section is a matter for NSF decision. The committee believes that such a section will benefit the nation's solar physics efforts.

2. Support and encourage university programs in experimental and observational solar physics and take steps to strengthen the partnership between, on the one hand, federally supported research centers and, on the other hand, universities. In particular, the committee recommends that specific programs to enhance education and training of students in solar instrumentation and observational techniques be established in the university community and that those universities willing to commit themselves to such programs receive support for the extended periods required to carry out such efforts. In addition, the committee recommends that more effective partnerships be forged between federally funded centers and universities—partnerships involving the exchange of faculty and technical staff, hardware and software, and workshops and short courses.

3. Protect newly funded initiatives in solar physics by ensuring their continued support until they are completed. Unless funding for such initiatives can be assured within the limits imposed by general federal budget restrictions, avoid pursuing additional new initiatives. The committee further recommends that NSF refrain from commingling funds targeted for new initiatives with base-level support funds in response to budget-cutting pressures."

The second and third points apply equally well to the solar programs within NASA, and we hence also recommend them to NASA.

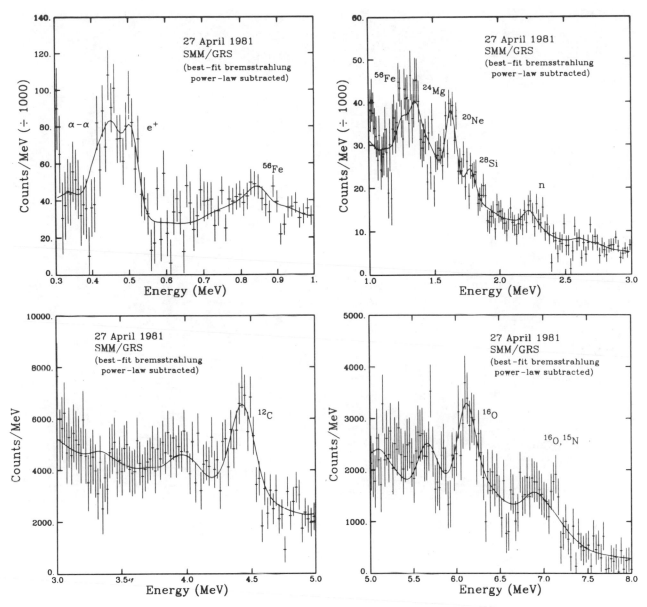

Figure 1. Observed gamma-ray spectrum of a solar flare. The comparison of the calculated spectrum (solid curve) with the data allows the determination of abundances of both the ambient gas in the gamma ray production region (thought to be the chromosphere) and the accelerated particles (thought to be accelerated in the corona). The aim of HESP is to improve our understanding of these gamma-ray spectra.

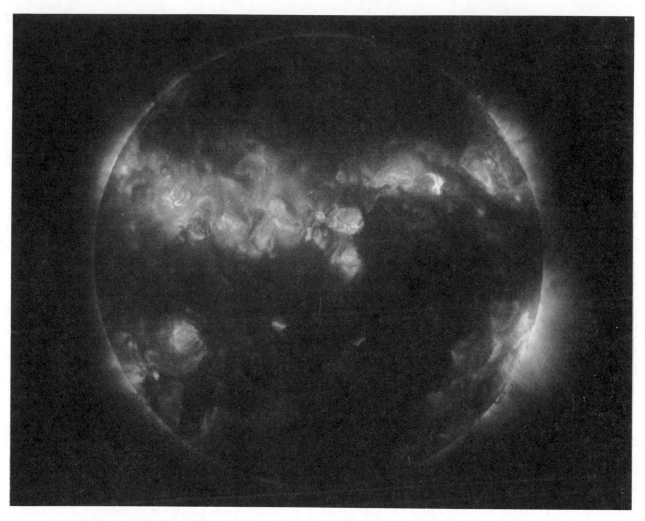

Figure 2. The solar X-ray corona photographed on September 11, 1989 from a rocket carrying a normal-incidence X-ray telescope. Normal-incidence optics represents a great advance over previous methods of solar X-ray imaging and will be the basis for high-resolution X-ray imaging on the Orbiting Solar Laboratory. Image courtesy of IBM Research and the Smithsonian Astrophysical Observatory.

Figure 3. The solar chromosphere, imaged in H-alpha light at the Sacramento Peak Observatory. Spicules are seen near the limb as dark thread-like features (bottom), and are seen in emission above the limb (top). Spicules may be very important to our understanding of the energy balance of the solar atmosphere. Unfortunately, they are only poorly resolved in existing ground-based telescopes, but they could be thoroughly investigated using the high-resolution capabilities of LEST on the ground, or OSL in space.

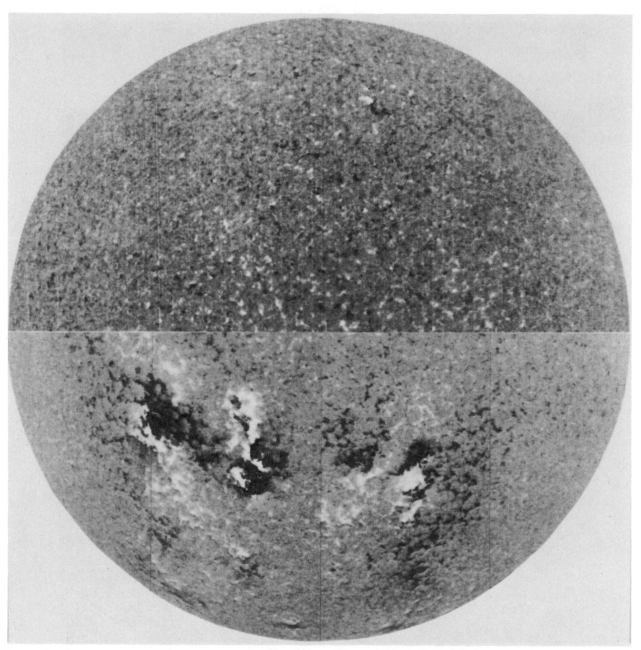

Figure 4. A composite magnetogram image, with the bottom half showing the magnetic pattern at maximum of the 11-year solar magnetic activity cycle and the top half the pattern at solar minimum. The dramatic change holds important clues to the nature of the magnetic cycle. It is important that new high-resolution observations, both from OSL and LEST, follow the changing magnetic patterns, and their consequences in the overlying atmosphere, throughout a large fraction of a solar cycle.

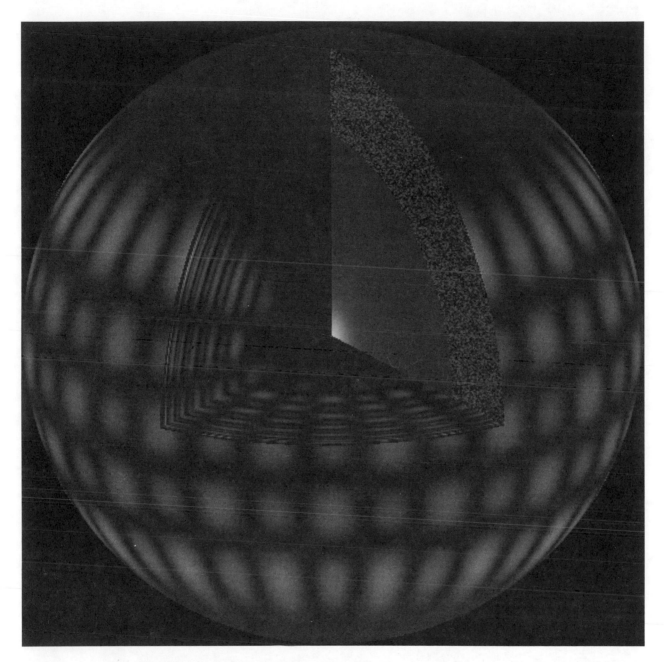

Figure 5. A computer representation of the two-dimensional spectrum of some of the 10 million solar oscillation modes. This covers the range of degrees from 0 to 150 (from left to right) and frequency from 2 to 5 mHz (bottom to top). These observations, made in a single day, do not have sufficient frequency resolution for accurate characterization of solar internal structure or rotation; needed are continuous observations for extended periods, such as are planned with the GONG experiment as well as helioseismology instruments on the SOHO mission.

PLANETARY ASTRONOMY PANEL

Planetary Astronomy

INTRODUCTION AND SUMMARY

Planetary astronomy - the study of planetary bodies and phenomena by astronomical remote sensing techniques - has represented a major element of astronomy since the invention of the telescope. Most of the discoveries by Galileo and his successors during the 17th and 18th centuries were in the field of planetary astronomy. In our own century, astronomical observations have provided the basic data, ranging from planetary masses and orbits to the properties of their atmospheres and surfaces, that have made possible the initial spacecraft reconnaissance of the solar system. Even today, in an era of numerous spacecraft missions, astronomical techniques still yield much of the information we have on the physical and chemical nature of planets, satellites, rings, comets, and asteroids. Furthermore, astronomical techniques are providing the opportunity to expand our perspective to embrace the discovery and study of planetary systems orbiting other stars.

The field of planetary astronomy is a subdiscipline of astronomy dealing with observations of the members of our own solar system carried out from ground-based, airborne, and orbiting observatories. It employs many of the same facilities and techniques that are used by other astronomers, spanning the electromagnetic spectrum and encompassing laboratory and computational tools as well as telescopic observations. This field is strengthened by support from two major federal agencies, the National Aeronautics and Space Administration (NASA) and the National Science Foundation (NSF).

A number of recent discoveries highlight the continuing capability of astronomical techniques to contribute to planetary studies. These include:

- Discovery of the ring systems of Uranus and Neptune in advance of Voyager, and continuing ability to monitor changes in the rings and thereby study ring dynamics and planetary structure.
- Radar mapping of the topography of Venus and identification of volcanoes and rift zones that bespeak an active geology beneath that planet's opaque clouds.
- Discovery of deuterium in the martian atmosphere and use of this isotope as a measure of loss of water from Mars in the past.
- Discovery of numerous atmospheric constituents of Jupiter and Saturn, including organic compounds that record the complex photochemistry in the upper atmospheres of these giant planets.
- Identification of the composition of the surfaces of the satellites of the outer planets, including water ice (frost) on many satellites, sulfur dioxide and hydrogen sulfide on Io, and methane and nitrogen gas on Triton.
- Determination of the radius, mass, density, and surface reflectances of Pluto and Charon, and determination of the structure of Pluto's atmosphere.
- Identification of the wide-spread presence of black organic material on the surfaces of many satellites, comets, and asteroids throughout the outer solar system.
- Identification of the compositions of many asteroids with well-known meteorite types (including the discovery of iron asteroids), and determination of a compositional gradient across the asteroid belt that is a remnant of the original compositional gradient in the solar nebula.

In this report we profile the field of planetary astronomy, identify some of the key scientific questions that can be addressed during the decade of the 1990's, and recommend several facilities that are critically important for answering these questions. The most important of these facilities for planetary astronomy, prioritized according to their cost within the "space-based" and "ground-based" categories, are:

Major Space-based:	1. Space Infrared Telescope Facility (SIRTF)
Moderate Space-based:	1. Stratospheric Observatory for Infrared Astronomy (SOFIA)
	2. Search for Other Planetary Systems from Earth Orbit
Small Space-based:	1. Orbiting Planetary Telescope/Planetenteleskop (OPT/PTEL)
Major Ground-based:	1. Infrared-Optimized 8-meter-Class Telescope
Moderate Ground-based:	1. none
Small Ground-based:	1. Arecibo Radar Upgrade
	2. Astrometric Facility for Planet Detection

These facilities, which will complement and extend those already available to the United States astronomical research community, will ensure U.S. leadership in astronomy into the next millennium and will enable planetary astronomy to continue as a major component of our effort to understand the universe and its origin and evolution.

STATE OF THE PROFESSION

Planetary astronomy has contributed greatly to our current understanding of the solar system. Today, both astronomical and spacecraft studies, together with laboratory research on meteorites and lunar and martian samples, constitute an essential element of our quest to understand the solar system.

During the first half of the 20th century, planetary astronomy in the United States declined to the level of a minor branch of astronomical research, but a strong resurgence in this field began in the early 1960's as NASA initiated its program of lunar and planetary exploration by spacecraft. Not only did the need exist to learn as much as possible about the potential targets of spacecraft missions, but NASA's exploration goals also rekindled scientific interest in the planets. NASA, with its charter to explore the planets, took the lead in stimulating and supporting planetary astronomy. In the 1960's, it established a grants program, supported graduate and postdoctoral students, and funded the construction of three large telescopes for planetary work at the Universities of Arizona, Texas, and Hawaii. In the 1970's, NASA built the 3-meter national Infrared Telescope Facility (IRTF) in Hawaii and contributed to the development and operation of the Arecibo planetary radar facility in Puerto Rico.

These efforts together with the exciting new research opportunities stimulated a resurgence of planetary astronomy. Today planetary studies represent a significant and healthy component of astronomical research, with between 200 and 300 active planetary research astronomers in the United States. The great majority of these individuals draw at least partial support from federal grants; about 100 are Principal Investigators (P.I.s) in NASA's Planetary Astronomy Program, and about 15 more are P.I.s in the NSF planetary program. During the 1980's, an average of between 5 and 10 students were granted doctoral degrees in this field each year, which represents a significant drop from the previous decade. This Panel estimates that planetary astronomers now represent approximately 15 percent of research astronomers in the U.S. and somewhat more than 25 percent of U.S. planetary scientists (most of the remaining planetary scientists in the U.S. have backgrounds in the Earth sciences or physics.)

The primary professional society representing planetary astronomy is the Division for Planetary Science (DPS) of the American Astronomical Society, with a membership of about 700, including nearly 100 members outside the U.S. Typically annual DPS meetings attract 400 registrants, and approximately 300 individual papers are presented. The U.S. planetary science community also has its own journal, *Icarus*, published in affiliation with the DPS.

Throughout the 1980's, the combined annual budget for NASA and NSF grants programs for planetary astronomy averaged about $7 million (not including observatory operations). Currently, approximately half of these research funds are expended for studies of primitive bodies such as comets and asteroids. The outer planets and their satellites account for another 25 percent, with the balance devoted to studies of the inner solar system, instrument development, and the search for other planetary systems.

Approximately half of the U.S. planetary astronomers are optical/infrared observers. We estimate that these observers are granted about 600 nights (6,000 hours) per year on the 15 or so U.S. telescopes with apertures of 2 m or larger. This includes 50 percent of the time on NASA's 3-meter IRTF, more than 20 percent each on the University of Hawaii's 2.2-meter and the University of Texas' 2.7-meter and 2.1-meter telescopes, and less than 5 percent on each of the other telescopes in this class. During the 1980's, less than 3 percent of the time on the major telescopes of the National Optical Astronomy Observatories was assigned to planetary work. Other planetary observers use the Kuiper Airborne Observatory (KAO), International Ultraviolet Explorer (IUE), Arecibo radar, or the Very Large Array (VLA) and various millimeter and submillimeter telescopes.

In order to improve our understanding of the planetary profession, the Panel carried out a demographic survey of members of the DPS in 1989. We found that about half the professional membership (student members were not included) of the DPS categorize themselves as "planetary astronomers", and that this group consists of about 90 percent males and 10 percent females. Typically a planetary scientist works primarily on planetary studies; more than 60 percent of respondents stated that they spend more than half their time on planetary research, and 40 percent stated that they devote all of their time to this activity. Half of the respondents work for universities, 30 percent for government labs (including JPL), and 20 percent for other organizations. About 75 percent draw a portion of their salaries from grants ("soft" money), with more than 25 percent fully dependent on "soft" money.

The soft-money planetary scientists are about equally distributed over all age groups. A surprising number of even the most senior people in the field draw all or most of their salary from grants. Consequently, planetary astronomers are highly vulnerable to fluctuations in federal funding, and many scientists have preferred to leave the field in the face of apparently arbitrary threats to their core funding. In general, these funding uncertainties have resulted in lowered morale and have discouraged young people from entering the field of planetary astronomy.

While the numbers of planetary astronomers and their funding have generally held level or even declined in the U.S. during the 1980's, this field has been growing elsewhere. During the past decade, planetary science has seen a dramatic resurgence in Europe, parallel to the evolution of this field in the U.S. during the 1970's. We may anticipate a similar growth within Japan during the 1990's. Largely as a byproduct of its own space exploration effort, the U.S.S.R. has maintained a small but highly capable cadre of planetary scientists.

SCIENTIFIC OPPORTUNITIES FOR THE 1990's

Planetary astronomers study a wide variety of objects and phenomena, using techniques that range from traditional telescopic studies of faint objects to laboratory studies of meteorites and cosmic dust. Generally, however, we can divide this work into efforts to answer two fundamental questions: How did the solar system form; and what can we learn about our own planet Earth by comparative studies of the processes on other planets? These two major themes provide the basis for the discussion that follows.

Origin and Evolution of the Solar System

Planetary Systems in Formation: Protoplanetary Disks

It is widely believed that the formation processes that generate stars from the condensation and collapse of interstellar material also are capable of forming planetary systems. The star formation process is discussed in some detail in the report of the Optical/Infrared Panel in Chapter III of this volume. As reported there, recent observational evidence points to the widespread association of disks of gas and dust orbiting young stellar objects. These disks may be the equivalent of the solar nebula out of which our own planetary system formed some 4.5 billion years ago.

An important step toward placing the origin and evolution of the solar system in the context of other planetary systems was the discovery by the Infrared Astronomy Satellite (IRAS) that disks persist beyond the era of star formation and can be found orbiting some main-sequence stars. Particle disks have now been discovered around 150 main sequence stars. While we cannot yet detect planets orbiting other stars, we can begin to study the related properties of these disks, which may be analogous to our own comet clouds, but more populous.

An important problem during the next decade will be to measure properties of the disks such as mass, temperature and albedo distributions, as well as the carbon monoxide/dust ratios. Collisions of planetesimals in orbit around other stars may be expected to generate extended circumstellar dust disks which may be observable in the infrared region of the spectrum. The characterization of circumstellar dust and gas disks around main sequence stars could provide important information on the statistics and evolution of planetary systems. The Space Infrared Telescope Facility (SIRTF) will be an especially powerful tool for the investigation of circumstellar disks.

Completing the Inventory of the Solar System

To understand the origin of our own solar system, it is essential to establish its limits and inventory its contents. The decades of the '70s and '80s were spent in intensive study of the major bodies of the solar system and have greatly enriched our knowledge of their nature. However, we still do not have a complete inventory of all of the bodies of the solar system, nor of the detailed characteristics of those that have been discovered. Over the past decade, estimates of the mass of cometary material have increased by an order of magnitude, and the prevalence of dark organic material has been established on numerous bodies throughout the system. Today we recognize that the Trojan asteroids (orbiting the Sun at the distance of Jupiter) are as numerous as the asteroids of the main belt, a fact that was unsuspected a few years ago. And there is much yet to be learned, even about the populations of objects in the inner solar system.

Our direct knowledge of the contents of our solar system stops at Pluto's orbit. Information on the mass distribution beyond Pluto is important for understanding both the processes which control planetary accumulation in protoplanetary disks and those which control the dispersal of the disks. Furthermore, the reality of a Kuiper belt of comets beyond Pluto, apparently required to explain the origin of short-period comets, has not yet been observationally established. On a completely different scale, we have identified less than 10 percent of the estimated population of Earth-approaching asteroids. We are unable yet to say whether the observed number is consistent with injection from the 3:1 resonance in the asteroid belt or requires that many such asteroids be extinct cometary nuclei. Thus systematic surveys continue to be of great value.

The technology is ripe to perform much more thorough surveys, particularly in the outer solar system. There have been several estimates of the number and total mass of cometary nuclei in the Kuiper belt, and plausible estimates of the size distribution indicate that the number of such objects detectable with current technology is large. The most complete current survey is to visual magnitude $V = 22.5$ covering 4.5 square degrees. An all-ecliptic survey to red magnitude $R = 23$ (and perhaps to $R = 24$) is feasible. This project should be of extremely high importance since the Oort cloud, if it really exists, represents a significant mass. It also validates the concept that comets may be true cosmo-thermometers useful for determining conditions in the

A distant analogue of our own planetary system is shown in this coronographic photo of Beta Pic. The star is accompanied by a flat disk of solid material with dimensions of hundreds of AU. Advanced infrared instruments (SIRTF, SOFIA, and large ground-based telescopes) will greatly expand our ability to study other planetary systems, especially during their formative periods. *Las Campanas image reprinted by permission from R.J. Terrile (JPL) and B.A. Smith (Univ. Arizona).*

early solar nebula. Such a survey would also place tight constraints on the existence of any planets as large as Pluto out to very large distances from the Sun.

Remnants of Creation: Primitive Material in the Solar System

The most fundamental property of the members of the solar system, once they are discovered and their orbits determined, is their composition. Many of the most important advances in planetary studies during the '70s and '80s involved the identification of ices and organic materials on many objects, ranging from the moons of Mars to the comets of the distant Oort cloud. These objects preserve relatively pristine material from the time that the solar system formed from the primordial solar nebula.

Preliminary evidence suggests that the low-albedo materials covering (or comprising) comets and some asteroids and planetary satellites consist of macromolecular carbon compounds of low volatility, similar to those materials found in the carbonaceous meteorites. It is of primary importance to establish in detail the connection between the low-albedo materials in the outer solar system and the carbonaceous meteorites, because the chemistry, mineralogy, and thermal histories of the meteorites are relatively well understood. At the same time, the connections between the organic contents of the comets and the interstellar medium require further elucidation, as the study of pre-solar material proceeds with samples of comets and asteroids (the meteorites and interplanetary dust particles) in the laboratory. These studies will be carried out primarily through infrared, millimeter, and submillimeter spectroscopy, with large aperture ground-based telescopes, supplemented by critical observations that can only be made by the Stratospheric Observatory for Infrared Astronomy (SOFIA) from above the terrestrial water vapor, or by a cryogenic optical system such as SIRTF in deep space.

The most primitive bodies are the comets. Our knowledge of the physical characteristics of cometary nuclei has advanced tremendously during the past decade. We now realize that these bodies are typically irregular in shape, very dark, and spectrally distinct from asteroids. Most are small with dimensions of only a few kilometers, but one much larger object (Chiron) also displays comet-like behavior. Several comets have been observed to be active at heliocentric distances greater than can be explained by a water ice model. Rotational periods are now known for a few comets, and at least one of these (Comet Halley) may be in a complex dynamical state. Most surprisingly, it has been learned that cometary activity is often confined to a small fraction of the total surface of the nucleus.

The chemical compositions of primitive bodies are key indicators of the processes that occurred during the formation of our planetary system, and it is now possible to characterize these compositions by astronomical means. Several key findings of the spacecraft missions to Comet Halley and of the extensive Earth-based observational program demonstrate the high degree of spatial and temporal variability of the composition of gases in the coma. The diversity of chemical species and the complexity of physical phenomena, both on the nucleus and in the coma, require extensive simultaneous measurements using several techniques and a variety of spatial scales in order to properly interpret individual comets and also to place individual comets in the context of the ensemble of comets. A major challenge of the 1990's will be to understand the physical processes that have produced the observed characteristics. In this regard, spectrophotometric observations of dormant comets at large heliocentric distances will be particularly valuable.

Major uncertainties in our current understanding are amenable to resolution in the next decade. For example, the variations from one comet to another in abundance of volatiles relative to water, particularly the dominant volatiles carbon monoxide and carbon dioxide, may be addressed with proposed orbital facilities. Similarly, the abundance of cosmogonically significant but minor constituents such as sulfur, formaldehyde, methane and ammonia, may also be addressed. Elemental and isotopic abundances must also be determined. Elemental abundances will allow us to assess the completeness of our inventory of parent species, while isotopic ratios for certain molecules bear direct cosmogonic significance. Additional new species must be sought, such as the noble gases and complex hydrocarbons, and the heterogeneity of all species within the nucleus must be assessed.

Radar observations have yielded a wealth of new information about the physical properties of several comets and dozens of asteroids. The first direct detection of a cometary nucleus in 1983 was followed by the discovery of large particle clouds associated with Comets IRAS-Araki-Alcock and Halley. The radar signatures of near-Earth asteroids are highly diverse and reveal that a number of these small objects have extremely irregular, non-convex shapes. Radar has also been used to establish the metallic nature of a few asteroids and thus to verify less direct compositional inferences from visible and infrared spectrophotometry. Radar refinement of orbits is important for maintaining the accuracy of both inner-planet and asteroidal ephemerides. This capability is critically important for newly discovered Earth-approaching asteroids, since the detection of radar echoes can guarantee the optical recovery of asteroids on subsequent apparitions.

Many asteroids are compositionally intermediate between the primitive comets and more highly-processed planets and satellites. Spectrophotometric studies have revealed a variety of mineralogical classes, interpreted as representing various degrees of metamorphic and aqueous modifications of original primitive organic material. We have established further that composition is correlated with distance from the Sun, and therefore that the current asteroid population preserves information on the spatial variation of conditions in the solar nebula. However, telescopic studies must be extended to smaller objects and greater distances from the Sun, and also supported by improved laboratory and theoretical developments, before this record of the past can be interpreted with confidence.

A major opportunity for detailed study of primitive bodies will be provided by the 1996 launch of a NASA spacecraft to explore asteroid Hamburga and Comet Kopff. After flying past the asteroid in December 1997, the Comet Rendezvous and Asteroid Flyby (CRAF) spacecraft will match orbits with the comet, which it will study at close range from its arrival in July 2000 until past perihelion passage in December 2002. The spacecraft results will be enhanced and extended to other comets and asteroids through detailed Earth-based observations carried out in parallel with the *in situ* measurements. A number of the new astronomical facilities being considered for the 1990's will directly support - and benefit from - the results of the CRAF mission.

An End-Member Planet: Pluto-Charon

As the most distant known planet, Pluto and its satellite Charon present a unique opportunity to test current models of the formation and evolution of the solar system. Since the Pluto-Charon system has not been visited by spacecraft (nor are there plans to do so), what we do understand about it is based on astronomical observations from the ground, aircraft, and Earth orbit. From these we have a crude albedo map of Pluto's surface and know that its atmosphere contains methane, as well as a heavier gas, probably carbon monoxide or nitrogen. We also know the bulk density of the system is 2.1 g/cm^3, which is high enough to imply that Pluto and Charon formed separately in the solar nebula, rather than in a circum-planetary nebula with a subsequent escape into a solar orbit.

But how did Pluto and Charon become gravitationally bound, forming a "double planet"? Are these two bodies composed of the same materials? How does Pluto's atmosphere change diurnally and seasonally, as its solar distance varies from 30 to 50 AU? These questions can be answered by the traditional techniques of planetary astronomy, but implemented with more sophisticated detectors, instruments, and telescopes. The observations needed are: 1) far ultraviolet spectroscopy to detect ionized molecular species in Pluto's upper atmosphere; 2) high resolution infrared spectroscopy to determine the state of the methane and perhaps find additional molecular lines; 3) thermal infrared observations to constrain atmospheric and surface properties; and, 4) stellar occultation observations for determination of Pluto's atmospheric structure. Learning how the observed quantities change as a function of solar distance will be a further diagnostic for inferring more about the system.

Are We Alone? Detection and Study of Other Planetary Systems

Great strides have been made in the past two decades in terms of understanding the conditions and processes during the formation of the solar system. These strides have come in part from continued studies of solar system material, in part from advances in theoretical modeling of key processes thought to have played a central role in the formation of regions around mature and forming stars. The significance of the latter contribution is that it heralds the dawning of a new era in planetary science, in which the bridge between the study of our solar system as an isolated phenomenon is merged with studies of other examples of the phenomenon. It is not overstating the case to say that we will never fully understand the origin of our own planetary system without results from a successful search for and characterization of other planetary systems. Only then will we be able to determine which, if any, of the features of the solar system are proto-typical of planetary systems in general, and therefore, which properties must emerge as general results from a theoretical framework. We will begin to see the development of a new discipline, that of planetary system science, which will not only lead to an understanding of how our solar system was formed, but which will also provide a means to check our views of the process by which stars are born.

A star with a single planetary companion executes a reflex orbit that is a much smaller replica, in its projection on the sky, of the orbit of the planet itself. The dimensions of this stellar orbit are scaled down by the ratio of the planet to stellar mass. For a multiple planet system, the reflex motions of the star are independent and additive. The two measurable aspects of that stellar reflex orbit are the apparent displacement of the star, which can be sensed by precision astrometry, and the variation in its radial velocity, which can be detected by high-resolution doppler spectroscopy. These techniques are complementary, and both are maturing instrumentally to the point that their expected sensitivities in searches for other planetary systems are limited only by systematic effects associated with the physical properties of the stars themselves.

The appearance of Pluto and its satellite, Charon, as modeled by D. Tholen (Univ. of Hawaii) and M. Buie (Space Telescope Science Inst.), based on observations of mutual occultations. Ground-based occultation data have provided a unique opportunity to study these distant worlds. *Reprinted by permission from D. Tholen (Univ. Hawaii) and M. Buie and K. Horne (Space Telescope Science Inst.).*

The doppler spectroscopy technique measures the radial (line-of-sight) component of the star's reflex motion to a precision of a few meters per second, using specialized spectroscopic instruments on large ground-based telescopes. This technique works most sensitively for low-mass stars, and it is especially sensitive to planets in smaller orbits. Doppler spectrographic surveys are under way at the University of Arizona, the University of Texas, and at the Canada-France-Hawaii telescope at Mauna Kea, where precisions of 10 m/s have been achieved over several years. The current state-of-the-art would permit the detection of Jupiter around the Sun at the 1-sigma level, and of a jovian planet in a smaller orbit around a star of 0.3 solar masses with much better confidence. The observers at Mauna Kea have accumulated a data base on about 20 nearby dwarf stars, and several of them show marginal evidence for planetary or stellar companions. The primary advantages of the doppler spectroscopic technique are that it can be implemented from the ground, and its power is explicitly independent of the distance to the star under study. In addition, doppler spectroscopy is in many ways complementary to astrometric searches, which are most sensitive to planets in larger orbits. The complementary nature and the cost-effectiveness of doppler spectroscopy programs commend them as an important part of the overall effort to search for planets.

An alternative approach to searching for other planetary systems is astrometry, in which the reflex orbit of the star is determined from precision measurements of its apparent position on the sky. Unlike doppler spectroscopy, the detection capability of astrometry is inversely proportional to distance. However, for nearby stars, astrometric precisions of 10 microarcsec should be achievable from Earth orbit, and this level of precision is sufficient to detect planets with masses comparable to that of Uranus, in orbits with radii comparable to that of Jupiter around any star within 30 LY of the sun. In space, either direct telescopes or interferometer systems can be designed to achieve the required level of precision.

Comparative Planetology: Understanding Planetary Processes

Not all of planetary astronomy is focused on questions of the origin and evolution of the solar system. The major planets, and even many smaller objects, retain little memory of their beginnings. Instead, these objects have been subject to a long history of thermal and chemical modification, enhanced by random impacts and other external influences. To understand these planetary histories, and the conditions that determine planetary environments today, we must address the current state of the members of the solar system. Much of the current effort is also directed toward the dynamical processes that we see at work on other planets. As we develop a better knowledge of these processes, we can enhance our understanding of the forces that shape planetary evolution, and we may also develop a deeper understanding of our own planet Earth.

The following sections describe a number of opportunities to advance our understanding of planetary processes during the decade of the 1990's. Many of these focus on dynamic and time-variable phenomena. Most such studies require Earth-based astronomical studies, often spanning a considerable time base. Even when a spacecraft encounter has yielded a much more detailed study of a target at one particular epoch, these Earth-based studies must be made to relate the spacecraft data to events on the longer time scales characteristic of planetary seasonal changes or the 11-year solar cycle.

Dynamics of Planetary Atmospheres

The Voyager encounters with Jupiter, Saturn, Uranus, and Neptune have revealed great diversity in the dynamics and circulation patterns of the atmospheres of the giant planets. The winds vary greatly with latitude in a pattern that is unique to each planet. Saturn's winds, for example, are almost entirely eastward relative to the rotation of the magnetic field, while those of Neptune within 60 degrees of the equator flow westward. On Jupiter, the banded appearance of the atmosphere shows that planetary rotation plays the dominant role in global circulation, with rising air in the white zones and descending air in the dark belts, but this relationship is not valid for Saturn. Uranus shows essentially no atmospheric banded structure. Jupiter and Neptune, both of which have internal heat sources, each exhibit a gigantic storm system (the Great Red Spot and Great Dark Spot, respectively) at the same southern latitude and of comparable size relative to the planet's dimensions.

The depth of penetration of the zonal winds into the atmospheres of the giant planets and the energy source that drives those winds are unknown. Eddy activity on spatial scales that are below the resolution of Earth-based telescopes have only been glimpsed during the Voyager flybys and are poorly understood. Thus, our understanding of the coupling of the atmospheres to the motions of the interiors of the giant planets is very weak.

Many more observations with high spatial resolution and a time base comparable to that of the known seasonal and non-periodic changes are required to decipher the dynamics of the atmosphere of Jupiter and the other giant planets. At a resolution of 0.1 arcsec, Jupiter and Saturn can be studied on appropriate spatial scales. Unfortunately, there is little to be seen on Uranus, and Neptune is so distant that another order of magnitude in spatial resolution is probably required to monitor seasonal variations effectively. Much higher spatial resolution can be achieved near the 10 μbar levels with stellar occultations, which can probe the structure of these atmospheres with a spatial resolution of a few kilometers.

The wind patterns on Mars and Venus are not yet completely known nor understood. What is the circulation pattern in Venus' mesosphere? The underlying cloud layers show strong retrograde winds, while the thermosphere is expected to have a day-to-night wind pattern. Planetary-scale winds on Mars have never been observed, due to the absence of large stable cloud features in the atmosphere. The unknown winds on both planets can be observed by doppler spectroscopy utilizing rotational translations of the carbon monoxide line at millimeter wavelengths.

Planetary Rings and Ring Dynamics

Observations of occultations of stars by solar objects will continue to be a major tool of planetary science. Apart from *in situ* measurements, there exists no other method for directly probing refractivity and optical depth.

Occultation observations yield spatial resolutions on the order of a few kilometer (submilliarcseconds in the outer solar system), while constraining positional models to even higher precision, by one-to-two orders of magnitude. Such precision rivals or even exceeds many measurements from spacecraft, and is ultimately attributable to our precise knowledge of the position and rotation of the Earth. Occultation data can be continuously accumulated over a temporal baseline exceeding decades, which leads to continuous refinement of ring precession rates and related variables, which in turn provide precise geophysical measurements of quantities related to planetary interior structure.

The Neptune system presents a challenge to ground-based planetary astronomy over the next decade. The rings of Neptune are currently accessible to ground-based observation only via stellar occultation, and only the denser portions, or condensations, can be detected reliably. With the Voyager encounter data, we have a few sparse observations spread over an eight-year period. The orbital periods and radii of the condensations are so poorly known that predictions of their orbital phases over a decade-long baseline are subject to uncertainty of 10 degrees or more. No viable theoretical model for the condensations currently exists. Basic information about eccentricities, inclinations, precession rates, widths, and optical depths can be obtained from stellar occultations, but will require concerted worldwide campaigns and accurate predictions. Such a data set will provide fundamental constraints on ring theories, and may also yield accurate data on the interior structure of Neptune.

The Neptune system presents a challenge to

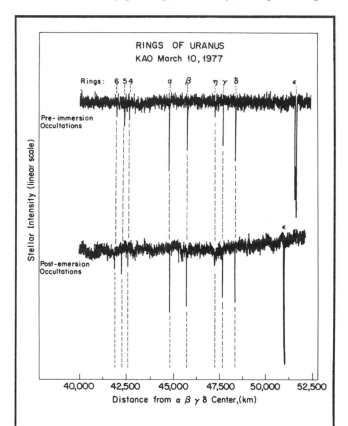

Discovery photometric data for the rings of Uranus, obtained by James Elliot (MIT) and colleagues from the Kuiper Airborne Observatory flying above the Indian ocean. The symmetry of the two traces demonstrates that the occultations are due to nearly circular rings and not to individual satellites. *KAO data reprinted by permission from J. Elliot (MIT).*

ground-based planetary astronomy over the next decade. The rings of Neptune are currently accessible to ground-based observation only via stellar occultation, and only the denser portions, or condensations, can be detected reliably. With the Voyager encounter data, we have a few sparse observations spread over an eight-year period. The orbital periods and radii of the condensations are so poorly known that predictions of their orbital phases over a decade-long baseline are subject to uncertainty of 10 degrees or more. No viable theoretical model for the condensations currently exists. Basic information about eccentricities, inclinations, precession rates, widths, and optical depths can be obtained from stellar occultations, but will require concerted worldwide campaigns and accurate predictions. Such a data set will provide fundamental constraints on ring theories, and may also yield accurate data on the interior structure of Neptune.

Other ring systems amenable to study at high spatial resolution afforded by the stellar occultation techniques are those of Uranus and Saturn. A highly sophisticated model for the orbits of the nine "classical" uranian rings has already been achieved by combining occultation observations obtained over the past 13 years. Even more precision with which to test dynamical theories of this system is available through continuing observations, as the aspect of Uranus changes from a polar to an equitorial view.

Until recently, Saturn's rings had not been studied with Earth-based occultation observations because of the prohibitively bright background of the rings themselves. However, occultation by Saturn's rings have begun being observed from the ground with new infrared techniques. Also, observations will be possible with the Hubble Space Telescope (HST), SIRTF, and SOFIA when they become operational. These data will yield immediate improvement in the precision of determinations of Saturn's gravity zonal harmonics, some first order information on the shape of the inner edges of the A, B, and C rings, and completely new information on the more opaque portions of the B ring.

Composition and Structure of the Atmospheres of Giant Planets

We are still learning much about the trace constituents of the atmospheres of the giant planets. Many important planetary molecular lines occur at wavelengths inaccessible from the ground, such as the 4 to 8 µm region. Studies from the KAO have revealed the presence of water, phosphine, carbon monoxide, germane, ethane, and arsine on Jupiter. The higher sensitivity provided by SOFIA's large aperture and higher flight altitude will permit numerous investigations not attempted from the KAO, such as high resolution spectroscopy of Uranus.

At radio wavelengths (millimeter to centimeter) one typically probes into and below the cloud layers. The main opacity is provided by gaseous ammonia. Observations at different wavelengths probe different depths, and hence provide information on the altitude distribution of the ammonia gas. High resolution images give information on the spatial distribution of this gas. Since ammonia is the primary constituent of the cloud layers composed of ammonia ice and solid ammonium hydrosulfide, radio observations provide direct information on these two cloud layers. The information can be tied in with atmospheric composition and thus planetary formations, as well as with cloud physics and dynamics.

Currently, excellent images at centimeter wavelengths can be obtained with the VLA. It would be worthwhile, however, to put more effort over the next decade into software to enable deconvolution techniques such that the images can be corrected for rotational smearing (images have integration times of at least a few hours), which would allow detection of longitudinal features in addition to the latitudinal structure.

At millimeter wavelengths, radio interferometry can image the giant planets on arcsecond scales and yield information on the ammonia gas and on trace constituents such as carbon monoxide, hydrogen sulfide, hydrogen cyanide, and phosphine at relatively deep levels of the atmosphere. Such observations imply that these constituents, if present, are brought up from deep levels in the atmosphere, as opposed to being formed at high altitudes or introduced from outside the atmosphere.

Volcanoes of Io

Jupiter's satellite Io is one of the most remarkable objects in the solar system. Its high level of volcanism, discovered by the Voyager spacecraft in 1979, is readily detectable from Earth, and observations made nearly a decade before the Voyager flyby can be interpreted in retrospect in terms of heat flow from volcanic hot spots. During the 15 years between Voyager and the arrival of the Galileo spacecraft at Jupiter, astronomical observations have permitted continuous monitoring of volcanic activity, and this ability to extend the time base of spacecraft data will continue into the future.

The volcanoes on Io channel the release of internal heat through a limited number of vents or hot spots that are observable from Earth in the thermal infrared region of the spectrum. The emission spectrum of the thermal regions is clearly apparent during eclipses, but even when not eclipsed the hot spots are detected, and with spatial resolutions of 0.1 to 0.3 arcsec individual volcanoes can be identified and studied in detail. Such studies have revealed the locations, sizes, and temperatures of the main hot spots, as well as estimates of the total heat flow. One interesting discovery is that about 50 percent of the total heat flow from Io is focused in one volcanic region, called Loki, which has been active for at least two decades. However, there are indications of variations in the temperature, area, and perhaps exact location of Loki on time scales as short as a few weeks.

The infrared spectrum of Io shows many bands diagnostic of the ices that lie on its surface. Sulfur dioxide (including several isotopes), hydrogen sulfide, and water ice have been found, but several other absorption bands are as yet unidentified. The ices on Io's surface are presumably emplaced locally by the active volcanoes, but they are largely unstable because of the temperature and the radiation environment. Sublimation and recondensation redistribute the materials on the surface, perhaps in patterns corresponding to their volatility. The compositions of Io's ices are a reflection of the chemistry of the volcanoes, and they are spatially and temporally variable. Spectroscopy of Io with higher spectral resolution and expanded spatial and time resolution will be valuable in the continued study of the volcanism on this satellite.

Ground-based observations of Jupiter's magnetosphere are obtained regularly. In fact, the presence of Jupiter's magnetosphere, its strength, tilt angle, and offset from center were determined from ground-based observations well before the spacecraft encounters. Radiation at decimetric wavelengths is synchotron radiation, emitted by high energy electrons near Jupiter's magnetic equatorial plane. The emission is a smooth function of central meridian longitude. Emission at decametric wavelengths is cyclotron radiation, emitted by electrons at or near their high latitude mirror-points. The emission appears in bursts, and is very irregular in character. The decametric emission is strongly modulated by Jupiter's central meridian longitude and the phase of Io in its orbit. More ground-based observations, with parallel modeling and theoretical investigations, are needed to add to the present knowledge of a number of unexplained effects such as a perplexing variety of types of bursts, polarization effects, and the presence of correlations with Io in some components and the lack of it in others. Long term monitoring is important to establish and investigate periodicities, to correlate the emissions with the solar wind or other solar phenomena, and to watch for effects that might reveal a non-periodic change in the jovian magnetic field structure or in its rotational period.

Mineralogy of the Martian Surface

Many studies of the surface of Mars can be carried out from the Earth. The 2 to 6 μm range is ideally suited for characterization of a variety of volatile-bearing minerals such as hydrates, hydroxides, water ice, carbon dioxide ice, carbonates, and sulfates. Some important features in this spectral range are accessible from the ground, while others require an airborne telescope. During the better oppositions the largest volcanoes are on the order of 1 arcsec in size, and polar caps are somewhat larger. Spectroscopy with comparable spatial resolution should isolate characteristic mineralogy, and permit geologic interpretation of the regional internal and surface processes which have been active throughout the planet's history.

Examination of photographs of the surface of Mars leads to the inescapable conclusion that there once must have been large quantities of fluid on the surface. Yet no obvious fluid sources are apparent today. Water vapor has been detected in the atmosphere of Mars and has been shown to be variable. The source of the water vapor is thought to be a permafrost trapped under the polar carbon dioxide ice caps. In support of the current Presidential and NASA initiative toward Mars, it is critical to continue to monitor the quantity of water vapor on Mars in order to quantify the atmospheric water cycle and to understand the seasonal cycles of the atmosphere.

High-Precision Dynamical Studies

We are used to thinking of many astrophysical measurements as being not very precise, i.e. as carrying no more than two or so significant figures. These are the type of measurements which are of most interest when we are in the discovery mode, when the mere existence of a phenomenon is in question. Much, but not all, of the work of ground-based planetary astronomy has been in this mode. But detailed comparison of theory and observation to several significant figures has also led to an elegant and profound syntheses. The planet Neptune was discovered in this fashion, and even in the twentieth century the careful analysis of orbits and spins has led to the discovery of subtle resonances. Most recently, careful and precise measurements of the mutual eclipses and occultations of the Pluto-Charon system has produced determinations of the sizes and orbital parameters to many significant figures.

Some examples of the precise measurements which are of interest follow: 1) The orbits of the Galilean satellites of Jupiter are evolving under the influence of tidal dissipation and exchange of angular momentum with Jupiter. Determination of the evolution of these orbits is possible, in principle, with high-precision observations of orbital phases over long temporal baselines. Some observations of this type may not be fully exploitable for many generations, but should be made now; 2) the shape parameters (both gravitational and geometrical) of planets are influenced by their spin states and by internal dynamics. It is feasible to measure these parameters to the level of a few kilometers in some cases, and over a period of years to build up a data base for detailed study of interior structure, as has been done for the Earth; and, 3) the jovian magnetic field, unlike the fields of the other giant planets, can be readily observed from the Earth, and its rotation period can be determined to a fraction of a second. According to dynamo theory, the geometry of the jovian field should be evolving, and one would expect to obtain information about such evolution if the current level of precision of observations were maintained for about a century.

CRITICAL TECHNOLOGY DEVELOPMENTS

The continuing success of planetary astronomy, even in the era of space missions, can to a large degree be attributed to technological advances in instruments, telescopes, and data reduction methods. New technology facilitates observations that have been previously impossible from Earth, and these open up new areas of scientific investigation. A recent example is the imaging of the Io volcanoes from the IRTF at 0.3 arcsec resolution. With this capability we can now watch new volcanoes emerging and old ones becoming dormant.

Obtaining these volcano images became possible just recently with the incorporation of an infrared array into a new instrument (the Proto Cam) at the IRTF and a better understanding of the seeing (degree of atmospheric interference) at Mauna Kea, which shows that images as small as 0.2 arcsec can be obtained in the near infrared region of the spectrum. A further bonus of this wavelength region is that the jovian planets have deep methane absorption bands, so that most, if not all, of the small satellites discovered by Voyager can now be studied from Earth.

Future technical advances will have similar impacts, since we are now in the midst of rapidly improving capabilities in several critical areas, including: 1) improvements to telescopes and the image quality that can be obtained from them; 2) increases in array detector size to cover an expanded wavelength region with lower background noise; 3) improvements to computer technology that are providing dramatic increases in speed, high capacity data storage, and worldwide network capabilities that will be needed to cope with the array data; and, 4) development of new, high-level software languages, some of which work with symbolic, as well as numeric input (such as IRAF, IDL, and Mathematica), thus dramatically reducing the time needed to develop and implement new methods for modeling and data analysis.

In the past decade planetary astronomy experienced a revolution brought on by the common use of the charge coupled device (CCD) as a detector. This device has had a particularly large impact on comet observations, which make full use of all four advantages that the CCD has over a photographic plate: 1) higher quantum efficiency; 2) larger dynamic range; 3) lower background noise; and, 4) output data already in digital form. At present, however, the greater number of pixels on a photographic plate still makes it the preferred detector for some survey programs, such as searches for new comets and Earth-approaching asteroids. This advantage will disappear as 2048 x 2048 arrays become available. Although infrared arrays are smaller (256 x 256 at present), they may have even greater impact on planetary astronomy than the optical CCD because of lower background fluxes from the jovian planets and the superior image quality in this spectral region.

As a result of these technological advances, we can expect not only higher quality data in the 1990's from Earth-based telescopes but the development of more efficient ways for planetary astronomers to work. The interconnection of the community and observatories through a common computer network will make it easier to share supported software. Another advantage of the ubiquitous computer network will be the facilitation of remote observing and data access. After an observing program has been set up and the observing techniques established, the program can be carried out by a local observer or through control of the telescope and instrument from a remote location. This will reduce travel time and costs and allow sharing nights between observing programs. The flexibility gained by remote observing could also be used to increase the effectiveness of certain observing programs.

Looking to the late 1990's and beyond, fundamental advances in the study of asteroids and other small solar system objects will be possible with the development of ground-based interferometers. Systems under development by several groups are expected to yield images of stellar sources with milliarcsecond resolution within a decade. Applications to solar system objects, with their lower surface brightness and irregular contours, may be more difficult, and they are expected to require both advanced adaptive optics and interferometer systems with extensive coverage of the U-V plane. However, eventually we hope to obtain asteroid images from the ground with resolution of a few kilometers throughout the main belt.

The extension of interferometric imaging to long baselines in space offers even more dramatic capability. A 10-kilometer baseline interferometer on the Moon could achieve resolution of a few tens of meters for the main belt asteroids. Such an instrument would also be expected to have astrometric accuracy of about 10 nanoarcseconds, opening a much larger volume of space to astrometric searches for extra-solar planets.

PROPOSED PROJECTS AND FACILITIES

The following are brief descriptions of the major new projects proposed for the 1990's that hold the greatest promise for new discoveries and deeper insight in planetary astronomy. They appear here prioritized according to their cost within the "space-based" and "ground-based" categories.

SIRTF: The Space Infrared Telescope Facility

SIRTF is a 1-meter class cryogenic telescope, designed to operate for five years in high Earth orbit (altitude approximately 100,000 km). SIRTF is the infrared component of NASA's family of Great Observatories, and it is currently planned for launch in the late 1990's. It will be operated as a facility for the entire scientific community, with over 80 percent of the observing time available to general observers. SIRTF's three instruments will provide images and spectra from 2 μm to beyond 700 μm, using an optical system which will provide diffraction-limited images at wavelengths longer than 3 μm over a 7-arcmin field of view. Using current infrared detector arrays, SIRTF will be natural background limited, and on a per-pixel or per-resolution element basis, it will be 1,000 to 10,000 times as sensitive as IRAS. SIRTF will be the first mission to combine the intrinsic sensitivity of a cryogenically-cooled telescope with the tremendous imaging and spectroscopic power of large-format detector arrays.

SIRTF will expand upon the important discoveries by IRAS of the disks and shells of particulate matter orbiting nearby stars. SIRTF will make detailed imaging studies of the most prominent disks to determine their shape and orientation, and to search for the dust-depleted inner regions predicted by models which may suggest the presence of planets close to the stars. Low resolution spectroscopy will be diagnostic of the composition of the dust in these systems, while high resolution spectra will trace gaseous constituents. SIRTF can detect less prominent dust systems around literally thousands of stars and determine how the prevalence and properties of such disks depend on the mass, luminosity, age, and other characteristics of the stars. These dust disks are spatially extended and thus demand the low background of a cryogenic telescope. SIRTF is the ideal platform for such studies, which can revolutionize cosmogony by extending the study of planetary system formation beyond the boundaries of our solar system.

The reconnaissance of the outer solar system accomplished by the Pioneer and Voyager spacecraft and the survey work of IRAS have revealed new aspects of the Sun's family that require further detailed study. Just as SIRTF can study material in the outer regions of other solar systems, it will extend the studies of our own solar system to the most distant planets and beyond. SIRTF will obtain spectra of comets, asteroids, and planetary satellites in the infrared region, where diagnostic molecular bands reveal the chemistry of these varied objects. The ices that condensed in the collapsing solar nebula further than 5 AU from the Sun and that are now locked in the comets and satellites carry the chemical history of the nebula in the zones where the outer planets formed. The chemistry of carbon, oxygen, and nitrogen in the solar nebula and in the protoplanetary clouds can be read in the spectral signatures of the comets and the relatively undisturbed planetary satellites. Organic solids and other forms of carbon, preserved from the interstellar medium and incorporated into the comets and asteroids, can be explored through their infrared spectral signatures in objects too faint and cold to be observed from ground-based telescopes. Infrared auroras on the giant planets will be observed spectroscopically, while spectra of the volcanically active Io will reveal the ever-changing surface and atmosphere. Far-infrared spectra of Pluto, Chiron, and Triton will, for the first time, give definitive information on the atmospheres and surface ices of these small and remote bodies. Dust from pulverized asteroids found by IRAS will be explored in high-resolution thermal infrared images to determine the sources and ages of specific asteroid collisions. SIRTF can also search for the hypothesized Kuiper belt, thought to represent a reservoir of comets beyond the orbit of Pluto. If this belt is as rich as is required by some recent dynamical estimates, one or two Kuiper belt comets should appear in a single SIRTF image taken in the zodiacal plane at 100 μm; these faint objects can readily be identified by their gradual motion relative to the galactic background.

SIRTF is the fourth and final of NASA's orbital Great Observatories, and as such it is the highest priority major new project of the Astrophysics Division. Assigned to JPL for implementation, it will receive a Phase B study in 1992. In the strategic plan of the NASA Office of Space Science and Applications, SIRTF is scheduled as a proposed FY93 new start, with launch anticipated for about 1998.

SOFIA: The Stratospheric Observatory for Infrared Astronomy

SOFIA will be an airborne observatory designed to address fundamental questions in galactic and extragalactic astronomy and in the origin and evolution of the solar system. Operating at altitudes from 12.5 to 14 km, SOFIA will provide routine access to wavelengths between 0.3 µm and 1.6 mm. Its 2.5-meter-diameter telescope will produce images ranging from roughly 4 arcsec at visible wavelengths down to about 1.5 arcsec in the near infrared region of the spectrum, and following the diffraction limit for wavelengths beyond about 10 µm. The planned schedule of 120 8-hour flights per year would support approximately 15 focal plane instruments and 40 P.I. teams annually. Many of these teams would involve one or more graduate students. The instruments would include photometers, polarimeters, and spectrometers developed and maintained largely by the interested P.I. teams. SOFIA's 20-year lifetime will provide a foundation of research, instrument development, and training of young scientists, bridging the gap between IRAS and SIRTF.

SOFIA will permit exciting new studies of the mineralogy of the martian surface. The 2 to 6 µm range is ideally suited for characterization of a variety of volatile-bearing minerals such as hydrates, hydroxides, water ice, carbon dioxide ice, carbonates, and sulfates. A number of important features in this spectral range are accessible from an airborne telescope. Spectroscopy from SOFIA should isolate characteristic mineralogy and permit geologic interpretation of the regional internal and surface processes which have been active throughout the planet's history.

Occultations of stars provide a powerful tool for the study of planetary atmospheres and rings. The altitudes of a planetary atmosphere probed by an Earth-based stellar occultation lie in the gap between spacecraft radio occultations and ultraviolet stellar and solar occultations. A series of stellar occultations by a planetary ring system can provide exceedingly precise orbital data, leading to an understanding of the age and evolution of the rings and to the internal structure of the planet through the harmonics of its gravitational field. SOFIA will readily accommodate the specialized instrumentation required for occultations, typically high time resolution, multispectral imaging systems. For example, the quality of an occultation observation is enhanced by working in wavelength bands where the planet is relatively dark. In the case of Saturn for example,

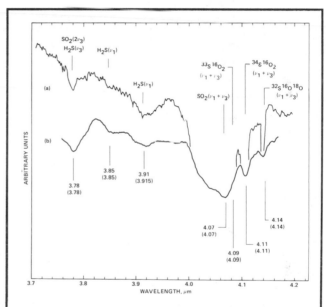

Infrared spectrum of Jupiter's satellite Io (solid curve) compared with laboratory spectra of possible analogue materials. Infrared spectroscopy is one of the most powerful tools for determining the compositions of both planetary atmospheres and solid surfaces. *Courtesy of F. Salama (NASA Ames Research Center).*

6.2 µm, a wavelength accessible from SOFIA but not from the ground, is an intensity minimum for both the disk and the rings. SOFIA will be a primary facility for observation of stellar occultations by Pluto, Triton, and Titan by virtue of its large aperture and mobility. Events involving these interesting objects will commonly involve faint stars, and will have uncertain predictions until a few days before the event occurs. Thus its sensitivity, mobility, scheduling flexibility, and insensitivity to weather assure SOFIA a unique role in the observation of occultations.

Spectroscopic studies of comets will make use of the spectral access available to a stratospheric telescope. The discovery of water vapor in Comet Halley from the KAO clearly demonstrated the value of airborne observations of comets. Ground-based measurements of Halley revealed organic material (the C-H stretch feature), but important

C-O, C-C, and C-N stretch bands of these species lie between 5 and 8 μm. Moreover, important cometary parent molecules have strong rotational transitions at submillimeter wavelengths. Radiation in these infrared and submillimeter bands does not reach the ground but is readily accessible from 12.5 km. Gaseous constituents require high resolution (roughly 1 km/sec) to discriminate between telluric and cometary features, to separate vibration-rotation lines in a band, and to permit kinematic analyses of cometary line profiles. SOFIA will be able to study a number of short period comets, observing their solid-state spectral features (such as olivine and water ice) as well as their gaseous features.

SOFIA will retain the major features of the Kuiper Airborne Observatory that have made the airborne astronomy program so successful, including flexibility of operations and broad access to investigators and their students. Particularly important are the opportunities offered to young scientists, who can accomplish complete astronomical investigations - from instrument concept and construction, through publication of results - within the time available to a graduate student or post-doctoral researcher. SOFIA will help bridge the gap between ground-based and space-based astronomy in the educational sense, as well as in its access to the electromagnetic spectrum. Extrapolating from experience with the KAO, one can project that SOFIA, over its 20 year lifetime, will generate nearly 100 Ph.D.s and involve participation by individuals from more than 150 institutions, about 70 percent of which will be universities.

SOFIA will be a joint project under U.S. leadership with the Federal Republic of Germany supplying a major part of the telescope system, supporting the operations at roughly the 20 percent level, and participating in the flight program at a similar level. NASA and the German Science Ministry completed Phase B studies of the project in 1989 and are currently continuing technology studies at a lower level. If approved for development in NASA's FY92 budget, SOFIA will be flying in the fall of 1996.

Search for Other Planetary Systems from Earth Orbit

The search for, and study of, other planetary systems can be a centerpiece of astronomy for the 1990's. Current ground-based studies using doppler spectroscopy and astrometry have the capability to detect Jupiter-mass planets orbiting the nearest stars. An improved astrometric telescope at an excellent site (discussed below, on page 17) could expand the search domain and extend to planets with the mass of Saturn. In addition, NASA's Search for Extraterrestrial Intelligence (SETI) project will begin operation in 1992, providing a radically different possibility for the discovery of other planets. If planetary systems are as prevalent as many astronomers believe, it seems likely that one or more of these surveys will achieve a positive result during the first half of the decade of the 1990's.

The discovery of one or more planets of jovian mass will be an important milestone, but it does not end the search for other planetary systems. The goal of a scientific survey must include the characterization of other planetary systems. To achieve this goal, we must use a search technique that is sensitive to smaller planets, at least down to the 10-Earth-mass level represented by Uranus, Neptune, and the refractory cores of Jupiter and Saturn. Further, we must anticipate that multiple-planet systems will be discovered, and that we will wish to determine the orbits and masses of these objects from their superimposed gravitational effects. The only technique capable of achieving the required accuracy is astrometry carried out at the 10-microarcsecond level for a period of a decade or more. Further, astrometry at the 10-microarcsecond level can provide definitive negative results in the search for other planetary systems; that is, failure to find any planets with such a search will challenge current ideas of the star and planet formation processes, and would therefore lead to a profound re-evaluation of current astrophysical thinking in these areas.

To achieve 10-microarcsecond precision, we must place an astrometric telescope beyond the Earth's atmosphere. In the long run, such surveys may best be carried out by interferometers on the surface of the Moon, and such systems should be carefully studied during the 1990's. However, it is also possible to pursue these goals with a free-flying astrometric observatory that could be launched and begin operation late in this decade. One example of the latter is the Astrometric Telescope Facility (ATF), a system capable, in principle, of relative astrometric measurements accurate to the several-microarcsecond level.

In addition to astrometry, which is an indirect method of detecting other planetary systems, there will be a need for systems capable of directly detecting other systems, or at least providing very high spatial and spectral resolution of disks which are likely precursors to planetary systems. In the region immediately adjacent to any bright point source lies a previously unexplored part of the universe. It is in this region that knowledge of the existence, origin, and formation of planetary systems lies hidden, masked by the scattered and diffracted light halo of the central source which blinds the optical system. The Astrometric Imaging Telescope (AIT) is a 1.5- to 2-meter-aperture orbital telescope which has unique capabilities to overcome this blinding light source and which combines direct imaging and astrometry to discover and characterize planetary and protoplanetary systems.

The astrometric function is implemented by a relatively wide field (20 arcmin) astrometric instrument, the Multichannel Astrometric Photometer (MAP). The imaging function combines a newly developed, highly efficient coronagraph (working against diffracted light) with a super-smooth optical system (working against scattered light) to form the Circumstellar Imaging Instrument (CII).

The imaging capability of the AIT is a powerful adjunct to all indirect methods. By design, it is extremely sensitive to circumstellar material such as zodiacal dust distributions or protoplanetary disks. Moreover, in the case of multiple planets, the image formation recorded from each planet is spatially distinct and detectability is not a function of the number of planets. In other words, all planets in a given system that are above the detectability threshold for that system are detected independently of one another within the same images. Even when no detections occur, upper limits can be placed on the absence of planets larger than a given diameter at a given orbital radius. This kind of information can be expected to help resolve any ambiguous indirect signals.

The high efficiency coronagraph and the super-smooth optical system of the AIT combine to produce a reduction of the light halo around a bright star by three orders of magnitude. Taking into account relative aperture sizes and optical figures, this amounts to a hundred-fold improvement in background level over comparable levels in the Hubble Space Telescope. This dramatic reduction in background is necessary to image extra-solar jovian planets, but it also implies that we will be able to do much more than learn that there is some nebulous region surrounding a given star; we will be able to image fine detail in that region and directly observe subtle features which would otherwise totally exhaust the dynamic range capabilities of typical imaging systems.

The AIT has been recommended to NASA's Solar System Exploration Division and is currently under study (Phase A) at JPL. If supported either through science funding for Space Station Freedom or as a free-flying spacecraft, it could begin operations before the end of the 1990's.

The Orbiting Planetary Telescope (OPT/PTEL)

The Orbiting Planetary Telescope/Planetenteleskop (OPT/PTEL) is a free flying 1-meter diffraction-limited telescope in high Earth orbit (geosynchronous). It is designed for an anticipated lifetime of 5 years and optimized for spectroscopic and high-resolution imaging of solar system objects. OPT/PTEL offers the following capabilities:

- Wide spectral coverage (115 to 5000 nm) free of atmospheric absorption.
- Fully multiplexed spectroscopy and imaging.
- Multiplexed ultraviolet-visible-infrared spectroscopic channels.
- Wide field of view (10 arcmin).
- On-target guidance and tracking to ±0.02 arcsec.
- Operation at small solar elongation angles.
- 17 hour/day operation above 40,000 km.

The long lifetime of OPT/PTEL will allow investigation of time-variable solar system phenomena and their response to changing input conditions (e.g. variable Extreme Ultraviolet (EUV) flux, changing heliocentric distance, climate, and/or episodic volcanism.) Applied to studies of Mars it should provide insight into the origin and development of global dust storms and could play an enabling role in the human exploration of Mars. The ability of OPT/PTEL to provide near-simultaneous measurements in widely separated spectral regions should allow it to provide data that will clarify the complex physical inter-relationships that occur in phenomena like the Jupiter-magnetosphere-torus-Io system, the Venus cloud-deck, and in active cometary atmospheres. It could be used to extend and clarify measurements made by orbiting or flyby planetary spacecraft by providing the measurements that define the large-scale context of *in situ* studies. For example, ultraviolet spectroscopy of comet Kopff will directly support the *in situ* spacecraft exploration of this comet in the first decade of the next century. It could also be employed to study the long-term evolution of phenomena in planetary atmospheres (e.g. outer planet atmospheric circulations), and on planetary surfaces (e.g. martian polar caps). Finally it will allow further geochemical mapping of the lunar and martian surfaces as well as a survey of the unexplored hemisphere of Mercury.

During its nominal lifetime the telescope would provide some 12,000 hours of prime observing time (equivalent, after taking into account typical weather factors and atmospheric seeing, to 17 years worth of prime-time (dark) observations on a ground-based telescope.) In its operation phase the project is conceived to operate in a mode similar to the highly successful IUE mission, but with the addition of a fully distributed data system that will allow fine targeting and data acquisition from the observers' home institution. It will provide rapid access to space for planetary scientists and therefore more frequent opportunities for novel solar system investigations in spectral regions unavailable from the ground.

OPT/PTEL is conceived as an international project between the Federal Republic of Germany, the United States, and, possibly, the European Space Agency. The United States' contribution would be the provision of the ultraviolet-visible-infrared spectrographic package, associated foreoptics, and detector cooling systems. The mission therefore provides opportunities to hardware development groups at U.S. universities and private institutions while maintaining a substantial role for a NASA center. NASA's Solar System Exploration Division has recently formed a Science Working Group to work with European partners in completing a Phase A study for the OPT/PTEL, which could be proposed for new start funding in 1994 and launched in 1998.

Infrared-Optimized 8-meter-Class Telescope

Astronomy in the U.S. is presently in a vigorous period of telescope construction; there are at least 11 projects under way or being planned for new ground-based telescopes in the 3- to 10-meter class. One of the current premier infrared telescopes in the world is the 3-meter IRTF on Mauna Kea, constructed and operated by the Solar System Exploration Division at NASA. Half of the time on the IRTF isdevoted to solar system studies, and this instrument has supported a significant fraction of all ground-based planetary astronomy during the 1980's. Now, however, it is opportune to take advantage of new technologies to construct a much larger infrared-optimized telescope in the 8-meter class. Such a facility would be an enormous improvement over any currently existing telescope, and it would provide the focus for much ground-based research in planetary astronomy and solar system formation.

Although several other 8-meter optical/infrared telescopes are planned or under construction, none of these is optimized for the infrared region of the spectrum. On an infrared-optimized telescope the mirror coating would be of the lowest emissivity possible (either gold or silver); however, such coatings are detrimental to the ultraviolet/optical performance of the telescope and would not be used on a general purpose telescope. Second, the infrared secondary mirror would be very small, producing a central obscuration of about 0.25 percent. Third, the secondary structure of the telescope would be of low emissivity. Fourth, the mirrors would be protected from dust, cleaned, and re-coated frequently to maintain the low emissivity of the mirrors. Finally, the telescope must incorporate active optics to ensure diffraction-limited resolution at near-infrared wavelengths. Coupled

Resolved IRTF infrared image of Jupiter's satellite Io. This image at a wavelength of 3 micrometers shows both reflected solar radiation from the full disk and bright thermal emission from the active volcanic hot spot called Loki. Resolution of the image is about 0.3 arcsec. *Reprinted by permission from J.R. Spencer (Univ. Hawaii).*

with state-of-the-art infrared instrumentation, this telescope would be the premier infrared telescope in the world. Significant studies that could be undertaken with this telescope include:

- Studies of the disks of young stars. With the 30 to 70 milliarcsec spatial resolution that is possible with diffraction-limited imaging at 1 to 2.2 µm, it is possible to obtain imaging at 5 to 10 AU on the nearest molecular clouds.
- Imaging of small bodies near bright planets. With an 8-meter telescope and advanced infrared imaging techniques, virtually all of the satellites discovered by Voyager could be routinely observed from Earth.
- Sensitivity to detection of faint planetary objects. A very deep survey of the composition, size, rotational periods, and density distribution of primitive objects could be undertaken with an 8-meter telescope.
- High spectral resolution observations. An 8-meter telescope provides sufficient aperture to carry out high spectral-resolution observations efficiently.

- Stellar occultation observations. With an 8-meter infrared telescope, the frequency of potentially observable occultation events becomes great enough that events can be chosen on the basis of systematic physical studies of atmospheres and ring systems.
- Detecting hot young super-planets, larger than Jupiter, around stars out to a distance of several hundred light years.

This telescope will also provide important support to NASA's CRAF mission, which will reach Comet Kopff in July 2000. It should be able to detect thermal radiation from the comet nucleus (nominally a few kilometer in diameter), near aphelion, thereby permitting a determination of its diameter in time to aid planning of spacecraft observing sequences. Additional observations of other Jupiter-family comets will help in extending results from this mission to a broader population of perhaps similar objects.

Studies of infrared-optimized telescopes are currently underway. One option is to optimize for infrared work one or more of the two 8-meter telescopes planned to be constructed by the National Optical Astronomy Observatories with support from the NSF. An alternative is to consider an 8-meter-class telescope as a possible replacement for the NASA-supported 3-meter infrared telescope (IRTF) built on Mauna Kea in the 1970's.

Arecibo Radar Facility Upgrade

Radar observations have yielded a wealth of new information about the physical properties of satellites, comets, asteroids, and the surfaces of the terrestrial planets. The most powerful radar facility in the world is the Arecibo radio-radar telescope in Puerto Rico, which is part of the National Astronomy and Ionosphere Center, operated by Cornell University under contract with the NSF. The instrument consists of a 305-meter-diameter fixed reflector, the surface of which is a section of a 265-meter-radius sphere. Movable line feeds suspended from a triangular platform some 130 m above the reflector correct for spherical aberration and can be aimed toward various positions on the reflector, enabling the telescope to point up to 20 degrees from the zenith. NASA support in the mid-1970's made possible the installation of the 2380-megahertz (13-centimeter) radar system, and annual support from NASA since then has proven essential to the continued operation of this unique facility. As a national center, Arecibo is accessible to the entire scientific community.

A recent proposal to NASA and the NSF for upgrading the Arecibo telescope calls for: 1) constructing a ground screen around the periphery of the dish; 2) replacing the higher frequency line feeds with a much more efficient Gregorian sub-reflector configuration; 3) doubling the output power of the 13-centimeter transmitter; and, 4) installing a fine-guidance pointing system. These upgrades will increase the instrument's average radar sensitivity by a factor of 20, more than doubling its range and reducing by nearly an order of magnitude the diameter of the smallest object detectable at any given distance. The quality, in terms of signal-to-noise ratio and/or spatial resolution, of all measurements performed routinely today would jump by more than an order of magnitude.

The impact of an upgraded Arecibo on planetary science will be fundamental and far-reaching, especially for studies of small bodies and planetary satellites. During its first decade of operation, the instrument will provide high resolution images of about 30 near-Earth asteroids and 100 main belt asteroids. Currently, Arecibo can barely skim the inner edge of the main asteroid belt, but an upgraded instrument will have access to asteroids throughout the belt. Short-period comets, which generally lie at the edge of the current detectability window, will become easy targets, letting us determine their nuclear characteristics and check for the presence of large-particle clouds. The asteroid flyby target of NASA's CRAF mission (Hamburga) will be accessible during its November 1993 apparition, when radar data can improve the orbit determination. A better orbit for Hamburga will be essential for ensuring maximum return from the 1997 spacecraft flyby, which otherwise will risk losing many of its close-up photographs due to pointing uncertainties. Similarly, asteroid Maja, targeted for a Cassini flyby in 1997, will be accessible to the upgraded Arecibo facility in February 1994.

Radar investigations of natural satellites will reap enormous benefits, especially for Io and Titan, whose near-surface physical properties and centimeter to kilometer structural properties will be readily discernible. The regoliths of the icy Galilean satellites could be probed to depths of 100 m or more and studied on a global scale, and the subsurfaces of Phobos and Deimos could be compared in detail. Iapetus will be detectable, and radar measurements could elucidate near-surface morphology and the disparate hemispheres of this unusual object.

Most elements of the Arecibo upgrade have been approved by NASA and the NSF. Barring unforeseen difficulties, the new capability should be fully in place by 1994.

Astrometric Telescope for Planet Detection

Doppler spectroscopy, which can be used to measure extremely small motions of stars in response to unseen orbiting companions, is a powerful technique for the detection of other planetary systems. The level of precision already achieved (10 m/s) opens an important segment of the phase space that might be occupied by other planetary systems; namely, those consisting of compact planetary systems accompanying low-mass stars. Present doppler spectroscopic observing programs, if extended to a sample of 100 or more G and K dwarfs and continued for a decade, should provide a definitive determination of whether such stars have compact planetary systems that include objects of jovian mass. Such a study constitutes an essential part of the systematic search for other planetary systems.

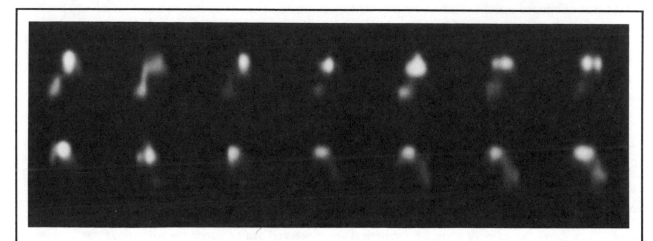

Arecibo radar images of the Earth-approaching asteroid 1989PB. The object, which is only about 400 m long, is resolved to show what appears to be a contact binary. The upgraded Arecibo radar will be able to reach hundreds of asteroids with this resolution. *Arecibo data, courtesy of S.J. Ostro (JPL).*

Astrometric searches for other planetary systems complement those carried out by doppler spectroscopy, in that they are more sensitive to extended rather than compact planetary systems. Current ground-based astrometry, at a precision of 2 to 4 milliarcsec, is capable of detecting a jovian-mass planet only for the nearest few dozen stars, and only if the stellar mass is less than about 0.3 solar masses. Thus, our own planetary system remains below the current detection threshold. However, the gains in astrometric precision that can be achieved with a dedicated modern astrometric telescope at a site with excellent seeing should permit studies with precision as good as 500 microarcsec, thereby bringing a wider variety of potential planetary systems within range.

Ground-based astrometry, with a dedicated telescope at an excellent site, will effectively complement the doppler spectroscopic method and will bring within our reach another significant segment of planetary system phase space. Such a telescope will be capable of detecting the reflex astrometric motion of G and K dwarf stars produced by planets of jovian or saturnian masses in orbits with radii greater than about 4 AU. The periods for such orbital motion are typically in the range of 5 to 10 years, so an extended survey is required. At least 100 stars are within range of such an astrometric telescope with modest aperture (1 to 2 m). An additional task for an astrometric telescope that would be important for planetary astronomy is its use for the accurate prediction of stellar occultations by bodies subtending small angular diameters, such as Triton, Charon, and Pluto.

The construction of a modern ground-based astrometric telescope is a next step in the search for other planetary systems. Such an instrument has been proposed to the NSF for construction on Mauna Kea. It should provide important fundamental data (e.g. proper motion, parallax) for stars and it will serve as a test bed for future astrometric instruments in space, as well as offering a good prospect of discovering one or more other planetary systems. Beginning the search now from the ground is both timely and prudent. The results promise to be exciting, and the experience will be invaluable as a precursor to the more advanced orbital facilities to follow, such as the system described above, on page 14.

RECOMMENDATIONS

Projects and Facilities: Prioritized Panel Recommendations

The Planetary Astronomy Panel finds that all of the projects and facilities discussed above are of the highest scientific merit. Each of them can be implemented during the 1990's (although not in all cases be fully operational before the end of the decade), and each will be of immense benefit to astronomers in addressing the exciting scientific challenges outlined in this report. We have categorized each of these in terms of cost and present the following recommendations, prioritized within each cost category. No effort has been made to prioritize one category against another, however.

Major Space-based:	1. Space Infrared Telescope Facility (SIRTF)
Moderate Space-based:	1. Stratospheric Observatory for Infrared Astronomy (SOFIA)
	2. Search for Other Planetary Systems from Earth Orbit
Small Space-based:	1. Orbiting Planetary Telescope/Planetenteleskop (OPT/PTEL)
Major Ground-based:	1. Infrared-Optimized 8-meter-Class Telescope
Moderate Ground-based:	1. none
Small Ground-based:	1. Arecibo Radar Upgrade
	2. Astrometric Facility for Planet Detection

Additional Recommendations

Research Support. Good science is a highly individualistic process that requires the participation of many researchers. Programs of individual research grants should be strengthened, and for these small grants the administrative system should impose a minimum of management or control beyond what is necessary to assure reasonable accountability. It is essential that inflationary erosion of the NASA and NSF grants programs be reversed and that funding levels permit new investigators to join the ranks of planetary astronomers.

Instrumentation. Progress in development of astronomical detectors and instrumentation, exemplified during the past decade by the CCD and 2-dimensional infrared arrays, has greatly increased the power of existing telescopes and has opened the way for spectacular advances during the 1990's. Many of the instruments built for the general support of astronomical research are also appropriate to planetary research. But planetary astronomy also has special requirements, for example in the need to study faint objects close to bright ones. It is important that funding be available to construct instruments optimized for planetary observations, if the promise of new observing facilities is to be realized.

Access to Telescopes. Planetary research often places unusual demands upon telescopes and scheduling committees. Many planetary events of high interest, such as occultations or the observations of newly discovered comets and asteroids, either make very specific requirement on time or are not predictable long in advance, or both. Some observations require the coordinated efforts of many telescopes working in concert. Observatory Directors and telescope scheduling committees are encouraged to consider these special needs and to retain sufficient flexibility to schedule planetary observations in spite of their sometimes complex requirements.

Telescope Support. Planetary astronomy has profited greatly from the construction and operation of ground-based facilities for planetary research, such as the IRTF in Hawaii or the planetary radar system at Arecibo. We commend NASA for its special efforts to recognize the unique needs of planetary astronomy, and we encourage the agency to continue to play a leading roll in the development of future facilities for planetary astronomy.

Surveys. Many important problems in planetary astronomy require a search for new objects (e.g. comets and Earth-approaching asteroids) or the systematic observations of properties (such as spectra or thermal emission) of large numbers of known objects. A variety of survey programs are therefore of great potential interest to planetary astronomers. However, these surveys must be designed so as not to exclude moving or variable targets. It is important that planetary astronomers participate in planning future survey programs, and that these programs be operated in such a way that important data on planetary sources are not discarded.

Interferometry. Although this report has not identified high-priority new interferometric facilities for the 1990's, we endorse continuing research in this field. Specifically, we support upgrades of the existing millimeter arrays in California (Hat Creek and Owens Valley), which have many research applications in planetary astronomy. Interferometry at many wavelengths is likely to be a central element of astronomical activity in the 2000's, and we must lay the ground work in this decade if the promise of these facilities is to be achieved in the next.

Laboratory Studies. The interpretation of data on solar system bodies frequently requires the acquisition and study of new laboratory data. The physical and chemical conditions in cometary comae, in planetary atmospheres, and on planetary surfaces are often outside the range normally studied by chemists, and new laboratory studies at the relevant temperatures, pressures, and radiation environment are needed. Major progress in understanding the chemistry of the interstellar medium, particularly its ice and organic content, is currently being achieved; many of these studies are directly related to the chemistry of the early solar system and those bodies which formed in it. In addition, the chemical and mineralogical *evolution* of the comets, planets, and asteroids can be deciphered with adequate laboratory data and simulations.

Access to Data. Astronomical data, often obtained with public funding, lose much of their potential value if not archived and made available to other investigators. Along with an archive, there is need for a catalog of the archive. We encourage efforts to apply new technology and establish uniform standards so that archiving of astronomical data can become routine before the end of the 1990's. In the case of planetary data, such an archive can be made available to the general community through the existing NASA-supported network of Planetary Data Centers.

Balance between NASA and the NSF. Planetary astronomy has profited by support from both agencies in the past, and we look forward to continuing this relationship in the future. However, we encourage the two agencies to work more closely together in coordinating their programs. The cost effective use of limited public research funds requires that NASA and the NSF complement, but do not duplicate, each others' programs.

COMPUTING AND DATA PROCESSING PANEL

LARRY SMARR, University of Illinois at Urbana-Champaign, *Chair*
WILLIAM PRESS, Harvard-Smithsonian Center for Astrophysics, *Vice-Chair*

DAVID W. ARNETT, University of Arizona
ALASTAIR G.W. CAMERON, Harvard-Smithsonian Center for Astrophysics
RICHARD M. CRUTCHER, University of Illinois at Urbana-Champaign
DAVID J. HELFAND, Columbia University
PAUL HOROWITZ, Harvard University
SUSAN G. KLEINMANN, University of Massachusetts, Amherst
JEFFREY L. LINSKY, University of Colorado, Boulder
BARRY F. MADORE, California Institute of Technology
RICHARD A. MATZNER, University of Texas, Austin
CLAIRE ELLEN MAX, Lawrence Livermore National Laboratory
DIMITRI MIHALAS, University of Illinois at Urbana-Champaign
RICHARD A. PERLEY, National Radio Astronomy Observatory, Socorro
THOMAS A. PRINCE, California Institute of Technology
CHRISTOPHER T. RUSSELL, University of California, Los Angeles
ETHAN J. SCHREIER, Space Telescope Science Institute
JOHN R. STAUFFER, NASA Ames Research Center
GERALD JAY SUSSMAN, The MIT Artificial Intelligence Laboratory
JOSEPH H. TAYLOR, JR., Princeton University
JURI TOOMRE, University of Colorado, Boulder
JOHN A. TYSON, AT&T Bell Laboratories
SIMON D.M. WHITE, University of Arizona
JAMES R. WILSON, Lawrence Livermore National Laboratory
KARL-HEINZ WINKLER, Los Alamos National Laboratory
STANFORD E. WOOSLEY, University of California, Santa Cruz

Computing and Data Processing

I. OVERVIEW

Computational technologies have been central to advances in astronomy and astrophysics for at least the last four decades. Martin Schwarzchild's stellar evolution codes used roughly half the cycles of John von Neumann's MANIAC "supercomputer" during the early 1950's. The 1960's saw the first detailed supernova computations by Colgate and White. The Einstein X-ray observatory and the VLA radio array, coming on-line in the 1970's, created images by the use of computers as intermediaries between the sensor and the observer. Theorists used supercomputers to model a wide variety of complex astrophysical phenomena in the 1980's.

Outside of astronomy, on the national scene, the strategic importance of high performance computing to the future competitiveness of broad sectors of the U.S. economy is coming to be widely recognized. As a consequence, a major national focus (the High Performance Computing Program) is emerging. The proposed components of such a program include high performance computing systems, advanced software technology and algorithms, a National Research and Education Network, and support of basic research and human resources.

A national initiative in computing, whether the one now proposed or a different one, will usher in a new context in which scientific research of all kinds will be practiced. Astronomy in particular stands poised, by virtue of its intrinsic data- and computation-intensive nature, its manageable size as a discipline, its past experience and future opportunity, to be the cutting-edge application discipline in a number of major aspects of a national program. Astronomy's task is to build its own internal computer infrastructure, in such a way as to maximize its leverage *vis a vis* the national program – and simultaneously to bring to astronomers the computational technologies that will be enabling of innovative astronomical discovery.

Modern astronomy and technology are often inter-related. New developments in technology have spawned qualitative advances in astronomy, and the promise of scientific discovery has often pushed technologies beyond their existing state-of-the-art. The charge coupled devices (CCD), new technology telescopes, active optics, and computing technology are examples of areas currently rich in this synergism. Analog devices are being replaced in new instruments with digital devices based on digital signal processors with greater precision and stability.

On the observational side, the scale of astronomical data that will be gathered in the 1990s, and which must be manipulated, communicated, and archived, will be on the order of many terabytes per year. (A terabyte per week is perhaps a reasonable figure.) Interposed between observation and actual understanding stands, increasingly, multiple stages of highly intensive data processing. Operation counts in the teraflop range (10^{12} floating point operations) per reduced data set will be increasingly common. Teraflop numerical simulations will be both helpful and practical in making the connections between astronomical observations, astrophysical theory and remote observing.

In the three complementary areas of digital data handling, intensive data processing, and theoretical modeling, astronomers are ready to take advantage of the expected technological advances of the 1990s: widespread of use of parallel computers, large increases in memory capacity, revolutionary improvements in data storage technologies, widespread use of graphics and visualization techniques, desktop high-performance workstations, high-speed networking, and powerful new algorithms.

The near future will see most researchers with access to powerful and flexible desktop computers linked over a national network, to each other as well as to high-value resources such as supercomputers and national observatories and data banks. Scientific visualization capabilities will be commonly available. The ability to bring together on the desktop the results of both complex simulations and detailed observations, and to be able to interact with each data set visually as well as quantitatively, could profoundly influence the progress of the astronomical sciences.

The Emerging National Information Infrastructure

In the 1960's, the Federal government provided the funds needed to set up first rate university computing centers. However, for fifteen years between 1970 and 1985, the Federal government removed itself from maintaining these facilities at the state-of-the-art. During that period few scientists had access to the newest computational technologies. Instead, shared departmental mini-supercomputers accessed by "dumb terminals" became the standard resource for most astronomers.

There was a radical reversal of this policy of "benign neglect" in 1985 when the National Science Foundation (NSF) formed the national supercomputer centers and began the national NSFNET network. These computational resources were financed from divisions of the NSF separate from disciplinary divisions. Access was not decided by money, but by peer review. Due to this democratization of access, in the last four years, over twenty thousand university scientists, engineers, social scientists, and humanists at over 250 universities and colleges have gained access to frontier computing technologies housed in the NSF supercomputer centers. There is a factor of 100 times the computing speed, memory, and storage capacity in the national centers as sits today on the desktop of the typical individual scientist. The National centers allow the benefits of substantial economies of scale with the cost of these facilities being borne across all fields of science and engineering. *We presume that the NSF, NASA, and DoE supercomputer centers, upgraded and enlarged, will continue to provide this resource to our community.*

During the same period, 1985–1990, individual workstations emerged which were as powerful as the previous departmental facilities. Most astronomers have managed to switch from "dumb terminals" to personal computers or workstations in the last five years. These desktop machines allow individualized control over one's computational research environment. The power and flexibility of these machines will continue to grow rapidly during the next decade. In addition, RISC (Reduced Instruction Set Computers) technologies have created a new version of the departmental computer which is near the speed and memory of a mini-supercomputer. The power of the departmental mini-supercomputers of the '90s will match or exceed those of the present generation of supercomputers. By the mid-1990's the computing power of the desktop computers, departmental minisupers and the central supercomputers will be at least 100 times what it is today.

The national network, which allows the researcher to "reach out" and grab that extra power when needed, has one thousand times the bandwidth compared to a user's access path just four years ago. The bandwidth of the national network will rise by yet another factor of 1000 during the coming decade. "Supernodes" arise naturally on the national network containing both specialized computational resources, and national digital archives of data - both from observations and from simulations. It is in computer networking that some of the greatest advances will come. As the gigabaud national network becomes a reality, there are three areas where revolutionary changes become possible. The first will be the use of facilities at the national centers from institutions all over the country. The second area is remote access to a distributed national digital library, which might contain scientific publications, previous observations, and results of theoretical simulations. Third is the remote control of "supertelescope systems" and the real time transport of the data to the astronomer.

We will also see a fundamental change in software. User interfaces are moving from command line, character oriented, single screens to menu driven, bit mapped, multiple windowed environments. It may come to pass that many important concepts in computer science (object oriented programming, distributed computing, and data structures) are becoming practical tools for computational scientists such as astronomers. A wide range of visualization technologies are changing the primary unit of information from number to image. Finally, large scale sharing of code is becoming accepted in at least the observational astronomy community, in the form of standard, portable, distributed data analysis systems.

The coupling of these elements in the next decade will transform astronomy and astrophysics into a *digital science*. The physical "glass and steel" of telescopes will be made orders of magnitude more powerful by addition of computing hardware and software. More researchers will "digitally observe" by accessing various national digital archives of all previously made observations, than will make new real observations on telescopes. Theoretical simulations of a complexity heretofore impossible will become commonplace. The ability to compare digital observation with digital theory will cross-fertilize both and lead to a much tighter mutual guidance than was possible in the past. Many of the aspects of this transformation will be shared with disciplines outside of astronomy, such as biology and environmental sciences.

Machines for the 1990's: Workstations to Supercomputers

Over the next decade, workstations will grow in performance to become comparable to present-day supercomputers. The present generation of RISC (Reduced Instruction Set Computers) -based machines are much more cost effective in terms of dollars per million floating point operations per second (MFLOP) than the current generation of supercomputers. These systems allow for affordable computing locally. Such systems also make possible a close coupling of analytic, numerical and visual computing.

Local systems offer several distinct advantages: First and foremost they offer the cheapest code cycle of any type of machine available. Second, the power and maintenance requirements (human and environmental) are much less. Third, and most important to the user, they offer i) instant response time, ii) little or no down time for maintenance, iii) local access for high resolution graphics, iv) sharing the resources with a much smaller group, as opposed to a central facility, where there are many times this number of users.

However, the advantages of supercomputers - large memory and disk capacity, vector or massively parallel processing, and extremely high input/output (I/O) rates, are crucial to a small fraction of computer users with large or demanding codes. A rough rule of thumb is the 80/20 rule: about 80% of users are performing small computations which can be supported effectively locally, *if* adequate funding resources are available. However, a small subset (perhaps 20%) of both the theoretical and observational community will attack problems whose computational requirements, in speed, memory, or storage, exceed those that can be reasonably provided locally. Such users will need access to national central facilities. The national centers will provide high cost technologies which will be available for experimentation by the community. These technologies will include vector processor and massively parallel supercomputers, very large memories, ultra high speed networks, large disk caches, the latest visualization technologies, all with teams of specialized experts.

The growth curves for desktop machines and for supercomputers are similar, so that with time, today's supercomputer capabilities will become affordable enough to be added to the local complement of machines. Of course, by then tomorrow's supercomputers will be more powerful too. Thus, the national supercomputer centers give researchers a chance to experiment with the future.

Essentially all active researchers need convenient access to good workstations, as well as "clear channel" coupling into the national network. In addition, small colleges and universities that have a history of training the students who become the future generation of scientists should be encouraged in their efforts to offer undergraduates exposure to modern computing. Support for high-performance workstations and mini-supercomputers would be a cost effective first step, by providing an accessible computational environment with modern software and graphics. It would also be a step toward geographical equality of resources and opportunity.

Lessons from the '80s

While certain subfields of astronomy (e.g., theoretical modeling) have always been demanding of forefront computer performance, astronomy as a discipline has not, in the past, been consistently out in front of other fields in the computing arena. However, during the 1980's the breadth of the computational base within astronomy has been expanding, a number of remarkable developments have begun, often from small beginnings, which are now poised to bring qualitative changes to the discipline.

A key example is the use of image processing algorithms in both optical and radio astronomy. These methods have long been used in radio astronomy, where synthesis arrays do not themselves form an image, but depend upon a digital computer as the image forming element of the telescope. However, much more than a simple Fourier transform of the measured visibilities is now standard practice. Powerful deconvolution algorithms have been developed which can greatly enhance the power of both radio and optical imaging telescopes. As discussed in the "Array Telescope Computing Plan", a conceptual proposal submitted by the National Radio Astronomy Observatory (NRAO) to the National Science Foundation (NSF) in September 1987 and resubmitted recently, the original design goal of the Very Large Array (VLA) of a dynamic range of 100:1 has now been increased routinely to 2,000:1, and a dynamic range of 100,000:1 is achieved for point sources. Thus, the VLA can be seen as an evolving telescope, with today's version being an instrument orders of magnitude more powerful and flexible than the one which was designed – all without hardware design modification. But the cost of this extra power is in the computing.

Unfortunately, the NRAO has insufficient computer power available to allow the full potential of the VLA to be realized. The computing problems with the VLA data have two origins - the first is the sheer volume and the second is the processing speed. To quote the above proposal: "The 10% of the expected proposals that generate 70% of the computing workload ... will be processed ... in supercomputers at national and regional centers. The rest will simply be deferred (i.e. users will not schedule the telescopes realizing that computational resources are not there or they do not reduce the data they have)..." .

In many cases, observations which comprise the most exciting and innovative of the possible radio synthesis projects cannot be carried out for lack of computing resources. This minority of projects is of great astronomical interest. We can summarize the types of scientific projects which lie within this category: a) All low-frequency imaging. In order to allow proper imaging at low declinations, VLA is designed as a 2-D instrument. This then requires a 3- D imaging for large fields of view. Due to special problems inherent in the computing requirements, all imaging at frequencies lower than 1 GHz, and much of the data taken at 21cm and 6cm must be processed with three- dimensional transforms. This is discussed in detail in section IV. b) All snapshot programs. One of the unique features of the VLA is its ability to make 2-dimensional images of bright, compact objects in only a few minutes of observing. This ability results in a speed enhancement of up to a factor of 200. But the cost is in computing. c) Studies of the interstellar medium of individual galaxies. These require extremely large images (up to 4096 × 4096 pixels) with 128 velocity channels. d) All Galactic absorption studies. These also require large images (comparable to the above) with high velocity resolution. e) OH and H2O maser emission. These are again, large images with high velocity resolution. The large VLBA computing needs are dominated by the same types of spectral line projects as listed above. All of the projects listed above except item b) require three-dimensional imaging from very large data bases. In some cases, four dimensional hyper-cubes may be required.

One can view the NRAO experience either as a great success – hugely multiplying *solely by processing* the peak capabilities of an expensive national instrument – or as a cautionary failure, a failure of vision (or national resource allocation) to provide the necessary computer power to a premier national astronomical facility. From either point of view, the conclusion is the same: astronomers are now sensitized to the importance of powerful computers and powerful algorithms, and they are determined that the negative aspects of the VLA experience will not be repeated.

The VLA experience is one of the best-documented examples of *computer starvation*. However, it is not the only example that we might offer. The Infrared Astronomy Satellite (IRAS) threw away a vast amount of information by binning its data too coarsely, due to computer hardware limitations. (This is now being redone by the Air Force!). While the Infrared Processing and Analysis Center (IPAC - the IRAS archive, administered by Jet Propulsion Laboratory on the CalTech campus) has been widely praised for its accessibility and servicability, it nonetheless has noteworthy limitations imposed on it by the computing power available to it. Another example involves the EINSTEIN observatory database. Although a unique resource for more than a decade, lack of adequate funding (until very recently) constrained the database and software to the mid-1970's technology on which it was developed. The forthcoming Gamma Ray

Observatory (GRO) data system also has significant computational shortcomings. The coming VLBA and the expanding millimeter-wave arrays such as the Berkeley-Illinois-Maryland Array will once again raise the threat of computational resources limiting and directing the science which can be done. On the positive side, the Hubble Space Telescope data analysis system (STSDAS) was designed to be exportable to a variety of computers. Furthermore, the budget for observers and archival researchers has been protected, and this funding can be used by successful, peer-reviewed proposers in part to acquire computer resources as appropriate. NASA has also recently encouraged work-station procurements in other projects, although lack of fundseasily compromises the program.

The technology trends of the next decade can dramatically improve this situation. It is important that our *national observatory system*, coupling multiple remote user analysts with data-gathering telescope facilities (both ground-based and space-based) have resources allocated to track the trends in computing technology. Emphasis must be placed both on a distributed computing environment, developing software to run efficiently on existing architechtures, including the difficult to program but extremely powerful massively parallel systems and making time available on existing supercomputers.

In light of the foregoing, our findings for astronomical computing are straightforward. Resources must be available for individual workstations, and for departmental or observatory mini-supercomputers. Networks must link the desktops of all investigators, all observatories, and all data archives. The development and maintenance of community software assets such as national data archives, data analysis programs, and theoretical simulation codes should be fostered. The allocation of computing resources is best carried out by peer review, but some oversight by the field is necessary to assure balance. The context for these findings is the assumption that current support from non-astronomical funding sources for the national network and the supercomputer centers continues throughout the decade.

Arrangement of This Report

Section II reviews some major challenges and technology trends encountered in facing the transformation to a digital astronomy - on both theoretical and observational grounds. Section III sets out in detail the need for a national data archive, and discusses some of its dimensions. Section IV consists of four "case studies" of high-performance data processing (both observational and theoretical), each one attempting quantitative estimates of what the requirements in the coming decade will be. Section V discusses how the transition from today's Megabit/sec national network to a 1990's Gigabit/sec fiber optic net will alter both observations and theory.

II. THE TRANSFORMATION TO A DIGITAL ASTRONOMY

In this section, we briefly review how astronomy and astrophysics will gain considerably from the technology trends and the implications of the national information infrastructure. With a discussion on real-time data processing, remote observing, theoretical simulations and community code development efforts, we show how our discipline is well poised to provide a leadership role in bringing the transformation and infrastructure into existence.

Supertelescopes

An emerging viewpoint is that all observational or laboratory instruments are *"smart sensors"* – a coupling of detectors to computers. The scientific power of a modern telescope is greatly leveraged by the amount and sophistication of the computing hardware and software applied to it. The lesson of the VLA is that a telescope is no longer a fixed capability instrument. Rather, it becomes a "supertelescope" which becomes more powerful with time, by virtue of its coupling to new generations of more capable digital computers. The balance of the "silicon to steel" tradeoff in designing a multiple decade national astronomical facility must be taken much more seriously in the 1990's than it was in the 1980's. Our report focuses attention on several examples of this, including large field CCD optical-IR arrays and radio telescopes. The increase in the sensitivity and resolution which computers will add to telescopes can be comparable in importance to the construction of large new telescopes now under construction or in the planning stages.

The NRAO, has proposed a distributed approach to meeting the computing challenges mentioned in the last section: 1) By writing and supporting flexible, all-purpose, exportable code (i.e., AIPS), much or even most of the projects scheduled on the VLA can be properly reduced at the observatory on mini-supercomputers or at the researcher's home institutions on desktop or departmental computers. This approach satisfies perhaps 85% of the individual observing runs, but falls far short of providing the capacity for the few demanding projects. 2) To provide the capacity for very large projects, the NRAO advocates a supercomputer access plan. The required software, perhaps specially coded to match various high performance architechtures, would be available on a few very large capacity machines. Fast data links must be available to allow real-time interaction of the user with the results. This is required since so much of radio astronomical data reduction is iterative in nature, and an experienced eye is required to judge the progress.

An idealized scenario for remote creation of VLA supermaps might work like this: The user physically or electronically sends his or her data sets to a designated contact at the national center. This person arranges to load the data onto disk. The user accesses the data on the supercomputer through their home workstation. The required commands can be issued from home, and the incremental results can be quickly transferred back to the workstation through fast data links for viewing by the user. After a number of iterations, which might take from hours to days to complete, the final results can be permanently archived, and the data deleted. Obviously, an efficient management structure will be needed to make this work. And, user-friendly, familiar code must be available to support the remote user.

A different high performance computing challenge faces optical/IR observers. Large charge-coupled device (CCD) focal plane imagers in the next generation of instrumentation for very large ground-based telescopes will require pre- processing in near real-time. Cameras with mosaic detectors larger than 5000x5000 pixels are now possible. The data rate from these detectors will overwhelm not only the traditional mini- or micro-computer or workstation, but also current array processors attached to mini- computers. About 1Gb of raw images (mostly calibration data) would be acquired in each 24hr period per instrument. Routine recording of such volumes of raw data for later reduction and analysis would create a data bottle-neck which would prevent the science programs from being carried out effectively. Real-time automated preprocessing and initial analysis of these data will be required. Special processors are now being built which can handle the high data rates from such large CCD detectors.

The scientist does not know precisely what is in the data, nor that it can be analyzed in one pass. What is necessary is real-time preprocessing of the raw data through all processing steps which are proven and which do not sacrifice other interesting scientific data. Reduction of data volume by a factor of at least 10 would result, for both imaging and spectroscopic CCD data.

Past examples of real-time array processing can be found in the fields of remote sensing, mail sorting, process inspection, radar signal processing, underwater topography, medical imaging and machine vision. Massively parallel real-time processing is constrained by the problem of transferring parallel data from a serial data stream at sufficiently high data rates. As in biological systems, analog image preprocessing at the detector becomes an advantage. Analog charge-coupled computing for focal plane image processing has been implemented in experiments. Neural networks, particularly analog VLSI preprocessors, have applications in real-time image processing. Digital computers have continued to keep pace with developing imager technology, so that most existing and planned astronomy instrumentation data systems are digital after the detector output and A/D converter. Eventually, optical computers are expected to increase array processing speed a thousand-fold.

Because optical telescopes form images directly, the optical community has been slower to experiment with deconvolution processing of their images than the radio community. However, the maximum entropy method (MEM) can be fruitfully used on CCD frames obtained under relatively poor seeing. Recent comparison of the results with frames obtained under better seeing, have shown that MEM deconvolution can improve the effective seeing substantially without creating any spurious features or structure. In addition to the effects of seeing, deterioration of images due to poor guiding and diffraction effects of secondary supports can be corrected. Routine MEM processing of all images taken with a particular telescope would require something like minisupercomputer power. The rapid rise in the processing power per dollar of available computers will within the next few years make it feasible to have a dedicated high-performance computer as an integral part of every optical telescope used for the acquisition of astronomical images.

Observing from your Desktop

Currently, many scientist's valuable times are inefficiently used in travel to remote sites to operate telescopes during observing runs. In addition, the optimal scheduling of a telescope is impossible, for the program of the astronomer on site always has the priority. If, instead, an astronomer at his or her home institution could monitor the data as it was obtained remotely with the same data rate as he or she would have at the site, and if telescopes could be dynamically scheduled as weather or other conditions change, the utilization of both the telescopes and the productivity of astronomers would increase.

The high performance national network will provide for *"teleobserving"* – the remote control of telescope systems and the real time transport of the data to the astronomer. In most cases the data will be obtained automatically, in several observing sessions, over a period of time. In other cases, observers will be notified if they wish to observe interactively. This will allow us to address the problems of optimizing the use of scarce national observing resources.

In this environment, the distinction between space and ground- based observing will begin to disappear. NASA's Great Observatories will automatically observe lists of targets under software control. New-technology ground-based telescopes, with their multiple fixed instruments, will use an optimum observing strategy, making rapid observing mode changes possible. Thus, for the optical-IR observer, a mode similar to that now used on the VLA may become common. Queue and "program" observing, taking optimum advantage of changing atmospheric conditions, will obtain the best possible data for all projects. These developments have the potential to change the way most observers work.

Electronic communication is also needed for real-time operations. For instance, planetary astronomy runs "campaigns", which may be multiple wavelength studies of the same object coordinated in time. They could be centered on a stellar occultation or mutual eclipses of a planet's moons. Such campaigns benefit from tight communication among the observers. Furthermore some planetary phenomena have time scales shorter than the terrestrial rotation period so that worldwide networks of telescopes are needed to properly characterize them.

Finally, the network can allow some synthesis radio telescopes to operate in near real-time. The present operational mode for synthesis radio telescopes is to acquire data and store the data on magnetic tapes for off-line processing at a later time. There is little opportunity for the astronomer to immediately see the results of his observations while the telescope is still available for follow-up observations - needed either because of poor data quality or to follow up an exciting, unexpected result. Particularly with a radio synthesis array telescope, the interval between data acquisition and working with the processed data ranges from weeks to infinity.

As a testbed and prototype of a tightly coupled telescope system, high speed network, and supercomputer, the Berkeley-Illinois-Maryland Array (BIMA) plans to implement a near real-time radio telescope. The BIMA will be a six-antenna millimeter-wave array with eight separate spectral windows of 256 channels each being available. In a typical eight hour tracking, the visibility function will be sampled sufficiently to make useful data cubes (8 cubes of 8 different spectral lines with right ascension, declination, and velocity axes) immediately. The telescope system, physically located at Hat Creek, California is completely under computer control and is accessible via computer networks, so an astronomer can monitor in real time the data acquisition process, edit the data, and set up command files for data processing. The data will be sent from Hat Creek to Berkeley to the supercomputer at the University of Illinois for immediate calibration, mapping, and deconvolution using an "expert system" controlled by the command files set up by the astronomer. An astronomer in Berkeley will then be able to display and begin the analysis of the data cubes on a local workstation by using a gigabaud national testbed network. If problems are found with the data, the project can be scheduled for immediate reobservation while the telescope system is still in the same configuration; if something unexpected is found, new observations can begin immediately.

This tightly coupled system of telescope, network, supercomputer, and workstation will very significantly raise the utilization of telescope systems and productivity of astronomers, and if successful, will be a prototype for all modern supertelescopes.

The NASA astronomy community is studying similar approaches, especially in the context of future space and lunar-based observing. Here, the need for automation in mission planning, expert systems for data analysis and experiment monitoring, space-borne data processing, advanced data compression and communications technology take on added significance. NASA is sponsoring studies in these areas, and is being encouraged to involve the user community in prototyping these technologies. Again, without direct user involvement in both the prototyping and all phases of the eventual implementation, the new capabilities run the risk of being inappropriate for the desired purposes.

Astrophysics in a Numerical Laboratory

Because of the nature of our (primarily) observational science, astronomers seldom can actively probe the objects of interest. Often these objects are complex in both form and temporal behavior, which hinders theoretical description even in cases in which we have the correct ideas regarding the underlying physics. From the beginning of the development of the digital computer, astrophysicists have been using this tool to simulate complex observed systems and to experiment numerically with new theoretical concepts.

Astrophysics depends on theory and modeling to a greater degree than other physical sciences, because astronomers can only observe remotely; active experimental intervention is not possible. Moreover, the observed phenomena are typically the result of usually complicated interactions among highly nonlinear processes occuring simultaneously. Therefore, it is necessary to construct rather elaborate models to achieve a satisfactory interpretation of the observations. As a consequence, the photons and fast particles which escape from astrophysical objects must be theoretically analyzed to the hilt, to extract meaningful physical information about the nature of their sources.

There is a long tradition of using analytic simplified models to capture the essence of a complicated astrophysical phenomena. Desktop computers are becoming increasingly important to support this work. Modern symbolic mathematics software allows the theorist to use more complex analytic formulations of the problem. Ordinary differential equations, which required a supercomputer to solve in the 1960's, are routinely evaluated and graphed by workstations and personal computers today. During the next decade the power of desktop machines will become so great, that many of the problems for which astronomers and astrophysicists are using supercomputers today, will also become soluble locally. Thus, we believe that desktop computers have become absolutely essential for theoretical astrophysicists.

As researchers build more and more complexity into their models, they outstrip the ability to compute locally in a reasonable turnaround time. This complexity arises for two fundamentally different reasons. First, the spatial dimensionality grows from one to two to three. Furthermore, as more and more realistic models are attempted, systems which are first studied as static, become time dependent. Typically, as the geometric complexity grows, so does the number of physical variables which must be solved for (e.g. from a radial velocity vector in spherical symmetry to all three components of the velocity vector in general). Second, one adds additional physics to the problem which increases both the number of equations and their coupling. For instance, one may add magnetic fields, nuclear or chemical reactions, radiation transport, or viscosity to an inviscid fluid flow code. In some cases, the introduction of new physics raises the effective dimensionality of the problem. For example, to describe the radiation flow in the most general case, one would add two angle variables and one frequency variable to the calculations. If there are scattering terms in the sources (Compton, Thomson, Rayleigh etc.) the system to be solved is a seven-dimensional integro-partial-differential equation. In addition, the calculation of realistic properties (opacities, equation of state) strain the resources of current machines almost to the breaking point. In short, both geometric complexity and additional physics can rapidly drive up total computational time and memory to values far exceeding today's fastest machines.

Real astrophysical systems are 3-dimensional objects evolving in time with extremely complex physics. Some aspects of these systems are currently being simulated on today's workstations, supermini computers, and supercomputers, subject to the restrictions on physical and geometrical realism which are imposed by the user's computer hardware and software. The goal of software designers is to make it possible to run codes transparently on any computer on the network, while retaining the interactivity and familiarity of local facilities.

The 1990's will be the decade where a number of long-standing astrophysical problems will be solved and computers will play an important role in these solutions. Areas which seem particularly ripe for rapid theoretical progress, and comparison with observations, can loosely be categorized as follows: large scale structure of the universe and cosmology, active galaxies and jets, star formation and the interstellar medium, dynamics of stars and stellar atmospheres, supernovae, accretion onto compact objects, generation of gravitational radiation, and the microphysics and magnetohydrodynamics of astrophysical plasmas. We will see significant advances particularly through modeling and numerical simulations approaching realistic complexity, which can be directly compared to observations.

As more powerful computers and community simulation codes become available, incorporating realistic physics and, where necessary, full three dimensional time-dependent geometry will greatly increase the ability of astrophysicists to directly compare their simulations with observational data. To illustrate the sort of progress we expect, consider a typical problem where interactions of radiation with matter are crucial. During this decade we will see the addition of nonequilibrium physics to hydrodynamics codes. A few codes have already taken a step in this direction with the introduction of two or three temperature systems comprising, say, electrons, nuclei, and radiation. But none yet allow for nonequilibrium effects in the excitation and ionization distributions. When this is done, radiation field and the state of the material become inextricably interwoven, making it impossible, even in principle, to calculate the thermodynamic properties of the material in terms of purely local variables. Rather, the system becomes fundamentally nonlocal, and we are forced to solve very large systems of globally interlocked equations, characterized by a wide range of characteristic spatial and time scales. These problems require the resources of massively parallel machines, and we should devote considerable efforts to algorithm development for such machines, forming an effective alliance with computer science experts working on such machines.

We estimate that nearly 10% of practicing astronomers are presently engaged in theoretical simulation of astrophysical phenomena. Some of this computational astrophysics is being done using local workstations and mini-supercomputers. Of the total allocations of time for all areas of academic science and engineering on the NSF Supercomputer Centers facilities, roughly 10% of the resources are being used by researchers in the field of Astronomy and Astrophysics. This is equivalent to about three processors of a current generation supercomputer which would cost around $20M to purchase. Those who are using supercomputers are trying to solve problems that push the system to the limits of software and hardware capabilities in existence today, and which could not be addressed using local computing resources. Some of these projects are also the ones attacking key problems in the discipline, and making seminal contributions that lead to major paradigm shifts in astrophysics. It has to be recognized that before the establishment of the national supercomputer centers, it was extremely difficult for astronomers to gain access to supercomputers. Consequently relatively few students were trained in the use of these machines, and the number of actual users remained very low. Now that the national centers exist, it becomes practical, for the first time, to train students in computational astrophysics. We just now have the first generation of these students receiving degrees and becoming professional astronomers. The percentage of professional astronomers who will be carrying out large computational simulations will grow rapidly over the next decade (assuming that the national supercomputer centers remain adequately supported).

In conclusion, the computing hardware needs of the theoretical astrophysics community can, with certain important exceptions, best be filled by a distributed system consisting of local mid-range computing facilities, including super-mini computers and graphics workstations, and upgraded national supercomputer facilities and high-speed network links.

Community Software

In order to effectively utilize the enormous advances in computer hardware expected in the next decade, we must have an accompanying development of scientific software. This is actually more costly and should be of at least equal concern with the computers themselves. Code development activities often require tens of man years, followed by a sizable budget and group for their maintenance.

The observational astronomy community has shown an admirable degree of coherence by developing systems like AIPS (Astronomical Image Processing System) and IRAF (Image Reduction and Analysis Facility), which have been adopted widely. These packages have saved an immense amount of time and duplicated effort. However it has often been difficult to identify adequate funding for ongoing efforts in the crucial areas of code maintenance and modernization. It is critical to augment efforts in the latter two areas.

From the perspective of theoretical simulations, code development efforts pertaining to multidimensional hydrodynamics codes, magnetohydrodynamic and particle-in-cell plasma codes, advanced stellar structure and supernova codes, N-body codes etc. should be encouraged and funded. At least part of this effort might fruitfully be located at a supercomputer center or a national laboratory, because this type of institution has the required broad infrastructure, and houses many related activities which are synergistic with the types of software development needed for astrophysics. Modest funding here might enjoy large leverage.

The astronomical community has an excellent record in the definition, maintainance, and distribution of uniform software platforms. Indeed, in the area of image processing, it appears that astronomy has already taken a technical leadership position relative to other scientific disciplines. Standard software development is a vital activity for the health of the commmunity: without the distribution of such software, and the use of standards, the handling of digital data is expensive and inefficient – or else doesn't get done at all.

One big success has been the FITS (Flexible Image Transport System) data format, now used internationally for the exchange and archiving of astronomical data. FITS was developed in 1979 by NOAO, NRAO, and NFRA (Netherlands Foundation of Radio Astronomy), and is an openly published standard. Since 1982, FITS has been the IAU standard for data interchange in astronomy. The FITS standard is maintained by regional committees in Europe and North America, which act under the authority of the FITS Working Group under Commission 5 (Astronomical Data) of the IAU. Recently NASA has established the FITS Support Office inside NSSDC at Goddard Space Flight Center.

The development by the NRAO of the Astronomical Imaging Processing Sysytem (AIPS) software illustrates both good and bad features of the software of the last decade. AIPS provides a very functional image processing system, with standards maintained by a national center. The general image processing capabilities of AIPS are sufficiently powerful that AIPS has been used extensively for optical and infrared image processing. An attempt was made during AIPS development, to isolate machine dependent features.

Software development for new large projects can learn from the experiences of the earlier models – a small core of people dedicated to the development of maximally transportable and evolvable software with the widest possible distribution in, and contributions from, the community.

The model introduced by NRAO for radio astronomy has also been adopted by other disciplines in the observational astronomy community, particularly by the large optical and x-ray groups. The National Optical Astronomy Observatories (NOAO) started developing (in about 1980) the Image Reduction and Analysis Facility (IRAF), a portable data analysis system designed to support their user community, and the European Southern Observatory (ESO) also started developing the Munich Interactive Data Analysis System (MIDAS). The development of AIPS, IRAF, and MIDAS provides the community with a limited number of very functional astronomical data analysis systems, which are portable to many computing platforms (ranging from PCs to minisupercomputers), and distributed widely, with standards maintained by national centers. The advantage of such a coordinated approach is demonstrated by the general willingness of the astronomy community to adopt these systems. STScI adopted IRAF as the environment for the HST data analysis system (STSDAS), and SAO adopted a similar approach for the ROSAT system (PROS). There are thus now several functioning groups, associated with national-level facilities, creating standardized software environments for data manipulation, analysis, and display. Grass-roots coordination has evolved among these groups, as well as via AAS and IAU working groups. The main impediment to further progress in this area would appear to be the lack of adequate funding, especially for maintenance of generic capabilities, not required for a specific project. Nonetheless, in the area of data analysis and image processing, it appears that astronomy has already taken a technical leadership position relative to other scientific disciplines.

The future will see more use of open systems, and standards, and very high speed networks. It will be essential that modern modular software standards be followed and that software be written to be portable to a variety of computers and usable over national high speed networks. Both the the national observatories and national supercomputer centers must lead in the astronomical software development effort. An example of this software effort is the development of a method for storing and transferring multi- object files across different machines on the network. One wishes to keep multi-dimensional floating point data arrays, palettes, images, and annotations together under one file name (for instance, in observational data sets such at FITS files, or theoretical simulation data sets). Further, one doesn't want to have to bundle and un-bundle these objects by hand. These computer science constructs should be discipline independent. An example is the Hierarchical Data Format (HDF), developed by NCSA. HDF allows different vendorsU computers on the network to automatically access combined files. The user's application code can read and write HDF files. NCSA is working with the NSF and NASA national observatories to create translators from the discipline specific file format FITS to the discipline independent file format HDF.

Careful attention to the "lifecycle" of software is also necessary. Major space missions such as the Great

Observatories are being designed for an operational lifetime of 15 years and for use by a large number of observers and archival researchers over a period of roughly 25 years. Given the typical 10 year development cycle, the ground data systems for these missions must function in a cost efficient manner over a 35 year period, despite rapid change in computing hardware, operating systems, and data analysis packages. Of even greater importance to the astronomical community is the very large cost for the ground data system of the first Great Observatory, the HST, and for subsequent missions unless fundamental changes are made. Over the lifetime of a long-lived Great Observatory, the ground data system may become as expensive at the construction costs for the observatory, and the fixed amount of money available for mission operations and data analysis may lead to a situation where rising costs of the ground data system will decrease the funding available for astronomical research.

One important way to reduce the costs of ground data systems is to design them from the start to accommodate rapid change in computing hardware and operating systems. Thus these systems should be built with evolvability and portability of software as requirements. In particular, layering to provide independence from specific operating systems and hardware is highly desirable. This design philosophy may be more expensive initially, but it will be very cost effective over the long term of these missions. Ground data systems should be portable to the major data analysis packages like IRAF and AIPS, which execute on a variety of vendor's platforms.

The HST Science Operations Ground System (SOGS) is often used as an example of the old methodology of developing ground systems, where a set of both operations and user requirements were implemented via a major formal procurement, and resulted in a large, monolithic, vendor-specific, hardware/software system. In fact, the portions of the ground system which were developed relatively late in the project, either with significantly greater user involvement, or by the users themselves, tend to be more in keeping with current ideas on evolvability and portability, as well as being more responsive to user needs. In the area of planning and scheduling, a portable expert system (SPIKE) has been added; in operations, workstations which run IRAF have been added to support off-line analyses and displays; and the pipeline calibration processing utilizes the identical algorithms available to any researcher through IRAF/STSDAS.

The end users of data from telescopes on the ground and in space are the people best able to determine sensible requirements for the software systems that will process the data. For this reason they should play a major role in formulating the requirements for such software systems, and they should closely monitor the development and testing of these systems.

This section has concentrated on observational astronomy community software. This is because such software is probably about one decade ahead of community codes for theoretical simulation. Although there has always been informal sharing among computational astrophysicists, there are not many truly national community astrophysics codes. Other theoretical fields such as chemistry, electron device simulation, plasma physics, and engineering have a rich history of the use of such community codes.

Efforts are underway at the national centers to develop, distribute, and support national users with new application software for astrophysical fluid dynamics research, incorporating the most accurate algorithms available for modeling astrophysical fluids. Versions will be developed for dynamics in 2- and 3-spatial dimensions, incorporating the important physical effects of self-gravity, magnetic fields, radiation, and thermodynamic properties of the gas. This software will incorporate the most accurate algorithms available for modeling astrophysical fluids in the Newtonian regime. The goal is an evolving software package implemented in a modern distributed UNIX operating system environment and optimized for high performance computers. Other environments needed include workstation tools with user-friendly interfaces for pre- and post-processing.

In addition, software for performing *"numerical observations"* of the simulations will also be developed. Numerical observations refer to the process whereby the fundamental physical variables of the simulated model are translated into observables (intensity, line widths, line shifts, polarization angle, etc.), including an assumed instrumental response, so that direct comparisons with observations can be made.

III. THE NEED FOR A NATIONAL ARCHIVE: SHOULD WE CONTINUE TO THROW AWAY DATA?

Historical Considerations

The first astronomical archives were just the written recordings of what astronomers saw in the sky. Since no detectors were involved, little was lost in the process, to the extent that the astronomer described his visual impressions in sufficient detail. These archives were quite useful to later astronomers. In general, these archives were not considered proprietary and were shared with other interested parties when that was requested. The archival medium proved quite durable; a variety of recent papers on sunspot numbers, supernovae, comets, etc. were based on observations many centuries in the past.

As photographic plates became suitable for astronomical purposes, and instrumentation developed to utilize this new type of detector, plate vaults became the new astronomical archive. Plates proved remarkably suited to astronomical demands, and remained the detector of choice for optical astronomy at least into the late 1960's. These plates have been, and will continue to be, extremely useful as a storehouse of information for the study of stellar motions, objects which vary in luminosity or morphology, objects whose spectra vary in time, etc. The degree of availability, or even records of the existence, of plates of a given object or region of sky varies greatly from observatory to observatory and observer to observer. However, a considerable fraction of the photographic database is available to the persistent astronomer.

Beginning in the 1960's, electronic detectors began to replace photographic plates for an increasing number of uses. By the present date, the main role remaining for photographic plates is wide field imaging, where the large format size of the plate compensates for the much lower quantum efficiency relative to two-dimensional electronic array detectors. The development of these electronic detectors, unfortunately, acted as a deterrent to maintaining a data archive. Problems with the data from the new detectors include: (1) the detectors have evolved rapidly in time, and proliferated in variety, making it difficult for an outsider from the future to understand what is available sufficiently well; (2) the data reduction required for the electronic detector data is often arcane, and specific to the detector and program being conducted, with an imperfect record of the required procedures; (3) the formats for storing data on tape or disk were generally not standardized, were in some cases machine dependent, and were often not well documented; and (4) the media on which the data were stored have, in many cases, become obsolete, been lost or overwritten, or deteriorated over time so that the data have been effectively lost. Thus, for some types of archival research, the recent electronic era may represent a regression - future astronomers will not be able to reconstruct the evolution of some phenomena because the relevant observations were not archived in a reliable fashion.

Problems and Opportunities for Archiving Ground-Based Observations

Some of the problems described above for data derived from electronic detectors are inherent to the field. Despite these problems, there are a number of reasons for optimism that now is a good time to seriously consider large scale archiving of ground- based data. It is important to note that space-based astronomy data is now routinely archived by NASA policy. Although subject to many of the aforementioned problems, it is nonetheless possible to retrieve most data obtained in the last 1.5 decades. In the case of the Great Observatories, standard data formats and standard data analysis systems are the norm. The development of the FITS format for data transfer and storage has allowed astronomers anywhere in the world, using a variety of computers and operating systems, to read and write tapes that are readable anywhere else. Given the will and the resources to smoothly transition to other storage media, FITS provides the means to create a long-lived astronomy archive so that data from the present will be accessible to future astronomers. Another important recent advance is the development of multipurpose, machine (and vendor) independent data analysis packages - such as AIPS, IRAF, and MIDAS. These packages provide standardized, well-documented methods to analyse a large percentage of the types of data produced with ground-based telescopes. In optical astronomy, CCD's have become the dominant detector for many purposes, and have thus become - at least temporarily - a de facto standard, further simplifying the task of archival research. In the planetary sciences community, with many of the problems similar to that of the astrophysics community, archiving is done with the Planetary Data System. Finally, but quite importantly, the recent development of computer networks and the imminent implementation of the NASA Astrophysics data archive system provides the means and the model for allowing widespread access to archives of ground-based observations.

Recent technological advances make this a good time to be considering development of an archive system. In the last few years, use of computer networks by astronomers has increased dramatically. The throughput of the system has improved considerably, and can be expected to improve much more in the near future. From a much different direction, the development of very large 2-D array detectors (2048^2 CCD's in the optical - soon to be mosaiced, and 256^2 arrays in the infrared - also soon to be mosaiced) makes data archiving more important for two reasons. First, the larger arrays mean that there is a greater chance for serendipitous discoveries - larger areas of the sky, or larger portions of the spectrum, make it more likely that data obtained for one purpose may contain information useful for another. Second, the greatly increased volumes of data make it much more likely that the data will not be fully utilized - valuable information relevant to a number of projects might be brought back from an observing run, but the PI only has the time and manpower to address the one question for which the data were obtained.

The fact that it is physically possible to develop an archive system for data from ground-based telescopes does not justify actually doing so. There must be an expectation that the scientific return for developing such a system is greater than the return for spending an equal amount of money to obtain new data. We believe that in many cases, the argument can be made in favor of archiving. CCD images obtained today provide an irreplaceable source of material for later proper motion studies. Similarly, the search for supernovae progenitors and other objects where one wishes to search for variability require access to archival data. During the next decade, a large number of US and international space missions will discover myriads of new and interesting sources - if we have had the foresight to be archiving ground-based data, much valuable information on these new sources may be obtained without using any new telescope time. Data archives that are well-documented and easily accessible over networks will provide fast "food for thought" following brain-storming sessions (instead of "... that's a great idea. Let's write a telescope time proposal, and wait six months ...", it can be "that's a great idea. Let's find out if anyone has obtained an image of that region in the last five years ...").

Impediments to Establishing a Data Archive for Ground-Based Astronomy

Data obtained from ground-based telescopes have traditionally been the property of the observer in perpetuity. Establishing proprietary periods and development of data archives would break this tradition. Many arguments could be advanced to keep the status quo: (1) the present system protects astronomers pursuing long-term projects; (2) graduate students should be allowed to finish their theses without having their data become public property; (3) astronomers with heavy teaching loads may not be able to reduce their data rapidly; (4) only the person who took the data fully understands it, and making it available to the masses would create more confusion than benefit; etc. We believe these are not reasons against archiving, but are instead just implementation questions - the rules determining propriety periods must be flexible enough to allow for differing circumstances, and the archival system must include calibration information and other relevant procedural data.

A more serious argument against archiving data from ground-based telescopes is simply that there is no money. This is particularly true for NSF funded national observatories such as NOAO and NRAO. The implementation of the archive system would require additional work. Archiving of data in Astronomy must be put into perspective with the other (time and resource) competitive aspects of doing our science. Observing with ground- and space-based telescopes, reducing the data, and publishing the results derived from them, all require time, expertise and money. Like everything else, archiving should be driven by the quality of the science that it delivers. Given limited financial resources, open competition and peer review - an explicit and case-by-case review for funds to archive the data is the most sensible approach. Our legacy is not just a stream of binary data, to be cared for and then preserved for future generations. Our legacy includes a distillation of those data into a perspective on our field that drives future thought, brings about a synthesis of new ideas and then precipitates innovation and additional observations. While we certainly cannot second-guess what future generations will find of interest in our presently gathered data, at the same time we cannot premise our science on that self-same lack of information; we cannot save everything out of ignorance. Justification as to the inherent value must be made competitively, proof that archiving will conform to the standards (of formatting, documentation, history etc., for raw data, calibrated data, and fully reduced data) must also be given and later verified. None of these additional tasks should be so onerous as to outweigh the potential benefits.

The contrast with NASA's outlook and expenditure for archiving and archival research is marked. NASA astrophysics currently spends of order $5 million per year strictly for archival research, primarily for data from IRAS, EINSTEIN and IUE. This amount is likely to grow considerably during the next decade as the archives for HST, ROSAT, and AXAF become available. STScI, working with other astronomy groups, has developed a prototype optical disk archiving facility which currently incorporates its digitized sky survey, and is being used to archive the early HST data. NASA is developing a major new large scale archive and distribution system for HST data which is estimated to cost over $20 million. A considerable amount of money will also be spent developing the Astrophysics Data System (and Master Directory) which will link the various NASA astronomy data archives together. Meanwhile, essentially no money is available to archive ground-based observations. The US observatories are lagging behind a number of major foreign observatories - archive programs have been started at the Anglo- Australian Telescope, the European Southern Observatory, and the La Palma Observatory.

Compatibility of Ground-Based and Space-Based Archives

Along with an archive there is a need for a catalog of the archive. This catalog should be as complete and easy to use as possible. Data and archiving systems for astrophysics space and ground-based missions should be compatible with, and ideally be a part of, the Master Directory Service of NASA's Astrophysics Data System. The data reduction and analysis software should be available at users' home institutions, not simply by remote access over networks to the mission data center. In particular, the available I/O bandwidth (over networks) would severely limit user access to the system. In contrast, the user community has (or soon will have) sufficient computing power locally to process the data. Finally, the scientific community should have free access to the archive.

It is desirable to develop cost-effective, useful archives of digital data from ground-based astronomical observations, available over a high-speed national network. Data are a legacy which we have a duty to bequeath to our successors. We note and commend NASA's important steps in this direction in the arena of space-based observation.

We recommend a similiar initiative for ground-based data with the following specific goals: (1) All major ground-based observatories, both public and private, should incorporate the capability for archiving of digital data. (2) The archived data should be accessible over the national network along with backup mechanisms such as rapid mail delivery of massive datasets on appropriate media. (3) This archive should consist of major homogeneous data sets with their requisite calibration information. Observatory Directors should establish and oversee appropriate criteria for the implementation of archives. We recommend that the funding agencies include the archiving program when determining the allocation of resources to an observatory. (4) The NSF should support high quality proposals for funding the capital costs of archiving data obtained at private observatories, and these data should become public after a proprietary period that extends no longer than 18 months after the last data in an observing program are obtained. (5) Archives must be designed to outlive any specific hardware, software, or media. (6) Archives should include all the raw data, calibration data, and information necessary to remove instrumental signatures from the data. (7) All observations obtained with large ground-based telescopes should be catalogued, whether or not the data are placed in an archive. (8) Original observers should get their data in the archive format so that both the original and subsequent analyzers of the data will start with the same data set. (9) A policy should be established to archive data described in papers published in the refereed literature. This policy should be enforced and implemented through journals, time-allocation committees, and the proposal reviewing process. The AAS Council should take actions to facilitate the archiving of processed digital data concurrently with publication in the Astrophysical Journal and the Astronomical Journal. (10) The standard for data interchange should be FITS or FITS extensions approved by the IAU and the NASA FITS Standards Office. Standards should be developed for data compression, archiving, catalogs, and user interfaces. (11) Peer review is essential for determining the allocation of resources for archiving in the context of all other competing requests for resources within astronomy.

We recommend a study of mechanisms for community input and user review of archive plans and operations; integrating ground-based archives with NASA's Astrophysical Data System; how further interagency collaboration on archiving between NASA and NSF should be implemented; and the impact of archiving plans in Europe and elsewhere.

The issue of international collaboration and coordination in archiving, and networks should be addressed. Reciprocal agreements must be reached with other international agencies with regard to access to and support of data archives. Astronomy on an international scale will develop much faster than in the past due to network access. Also a large percentage of the total useful astronomical database will be located outside the U.S.A. NASA has already begun to address this question with missions such as GINGA, EXOSAT and ROSAT, but no standard solutions exist; rather case-by-case solutions have been adopted. A goal should be to achieve a universal approach to archiving and data analysis. NSF should approach its foreign counterparts concerning ground based archiving and data access.

Giant model calculations are sometimes as valuable as observations. The numerical experiments together with their codes are directly analogous to major observational data together with the details of their acquisition techniques and calibration data. The implementation plans for archiving should include data from numerical computations with the source codes becoming public after an initial proprietary period.

Digital archiving will become increasingly important in sciences other than astronomy. Early serious study, and commitment of funds, can put astronomy community in a strong position as a testbed for other disciplines. This would allow NSF astronomy to seek joint funding with other Divisions, or from NSF management and Congress directly.

IV. HIGH-PERFORMANCE DATA PROCESSING: OBSERVATIONAL IMAGES AND THEORETICAL SIMULATIONS

This section presents four independent "case studies" – individual visions by panel members of the computational frontiers of astronomy in the 1990s. The first case study attempts to put a quantitative scope on the needs for realistic theoretical simulation. The second describes the specifics of one particular case, namely plasma astrophysics. The third case study documents the information explosion that is about to occur in optical data collection, with the advent of large CCD detectors and their attendant processing requirements. The fourth study is a corresponding analysis for the case of radio images.

Case Study A: Realistic Dynamical Simulations of Complex Systems

The "Fourth Dimension Supercomputer" is a system sufficiently powerful to calculate the evolution of complex, nonlinear systems in a fully three-dimensional space and in time. Presently large memory supercomputers are barely capable of providing this capacity, but only with crude resolution. Present trends in computer technology suggest that the next generations of machines will be serious "Fourth Dimension Supercomputers". To see what this implies for our science, let us examine what would be the parameters of a "Fourth Dimension Supercomputer", and what are the needs associated with its use.

Parameters needed for three dimensional simulations

Let us take a measure of our understanding to be how well our concepts match the images were observe. What is the quality of a "good" image? A variety of arguments and painful experiences with more meagre values suggest that a moderately good image requires about 500 x 500 pixels. The fact that in the personal computer market a popular screen resolution is about this same value is probably no accident (and 1000 x 1000 would be better). Our images are projections of 3D objects onto a 2D surface, so that a numerical description of the objects itself would imply $500^3 = 1.25 \times 10^8$ points. How many of these visualization points can be represented by a single computational point? This depends upon the power of our computational algorithm; significant advances are being made in algorithmic development. If several pixels are sufficient to describe a computational point, we have about 200 computational points per spatial dimension. Experience suggests this resolution is low but beginning to be interesting for a variety of problems involving simple physics, such as inviscid gas dynamics. We will assume $200 \times f$ to be required number of computational points per spatial dimension in subsequent discussion. This may be thought of as a minimum acceptable resolution for a variety of problems of current interest; $f = 2$ or $f = 3$ may often be appropriate. At each point a 3D calculation would require 4000 or more floating point operations (e. g., multiplications, additions, etc.), giving a value of $3.2 \times 10^{10} f^3$ "flops".

What about the resolution in time? Consideration of accuracy and stability of the numerical method often results in requiring a number of time steps which is of the order of the number of space steps in one dimension – in this case $200f$. For ten natural times, a short but not uncommon value, this implies $10 \times (200f)^4 = 1.6 \times 10^{10} f^4$ points, and $6.4 \times 10^{13} f^4$ floating point operations. A fast workstation may now provide 1 Megaflop performance, so that (if it would fit in memory) this task would take $6.4 \times 10^7 f^4$ seconds, or about $740 f^4$ days. The most efficient code on a four-cpu supercomputer might run at 400 Megaflops, so that the task would take $44 f^4$ hours. For twice the linear resolution, $f = 2$, this increases to 30 days.

Memory

How much memory is required? A floating point number of full precision requires 8 bytes. The physical state of a single point in the system may require 20 or more numbers. For two time slices (an old and a new state), this implies $3.2 \times 10^8 f^3$ numbers or $2.56 f^3$ Gigabytes. The largest memory presently available to astrophysicists at NSF supercomputer centers (1 Gigabyte) is about $2.5 f^3$ times smaller. Consequently on present supercomputersthe task would have to be paged in and out of memory, with attendant problems for speed and storage resources.

Storage

Using this value of $1.28 f^3$ Gigabytes per state (time slice), and assuming that 100 of the $2000f$ time steps are saved for analysis, the storage needed per project is $128 f^3$ Gigabytes. For 40 such projects, the storage requirement grows to 5 Terabytes. This is a major limitation: the extensive storage at the NSF supercomputer centers for researchers in all areas of science and engineering is estimated to be of the order of a few Terabytes. Data compaction and new storage technologies are needed to alleviate this bottleneck.

Communications

How do we get these data to the scientist for analysis? A reasonable estimate may be obtained by requiring that the time for data transfer of the results must be less than the time for computation — otherwise the data flow becomes the bottleneck. For one session of 100 time steps calculated, a 400 Megaflop supercomputer requires $8 \times 10^3 f^3$ seconds. It generates a new state needing $128 f^3$ Gigabytes of storage. This implies a data tranfer rate of greater than 1 Megabit/second, which is well above the actual performance of Ethernet, for example. Data compaction and upgrade of the NSF backbone to T3 are needed. It is particularly important to provide this national network resource to the wider community of scientists and students who do not reside at a supercomputer center site.

Speed

While low resolution projects are feasible on present supercomputers, higher resolution places serious demands upon computing speed. For 40 research groups per supercomputer, there are 200 hours per group per year (efficient parallel use of all cpus is assumed). At $44 f^4$ hours per project, this allows 4 projects per group per year, which is not really adequate for even moderate surveys of parameter dependence and testing. As progress is made on the constraints above, there will be demand for greater cpu speed, especially as observational comparisons make higher resolution necessary. Note that even this estimate is optimistic, as it assumes a level of vectorization and parallelization of code which is only occasionally obtained in practice.

Local Storage

Most scientists and students are not at the supercomputer sites. Upgrading the networking system will allow data to flow to them, but there must be local facilities to deal with it when it arrives. For example, a project of $2000f$ time steps, generating $20f$ saved states of $1.28 f^3$ Gigabytes each, would require $2.56 f^4$ Gigabytes of local storage. Factors of 2 and 4 could be saved by using 32-bit and 16-bit wordlengths.

It seems inescapable that analysis of 3D computations requires interactive 3D viewing, and that in turn requires extensive local storage capacity.

Viewing

Analysis of dynamical 3D systems requires looking at the time behavior of the system. This implies a requirement for local graphics capacity. It should be noted here that local RISC based machines are capable of high quality visualization. For example, suppose we want 60 seconds of images at 10 screens/second. For 3 Megabytes/screen (3 byte color on a 1000×1000 screen), this is 1.8 Gigabytes needed on a high speed disk to feed the graphics engine. This graphics pipeline must be fed at 30 Megabytes/second.

Algorithmic development in the area of 3D imaging is rapid, but needs support for standardization. Otherwise there will be a lot of redundant development of very similar software.

Case Study B: Plasma Astrophysics

Over the next decade, sophisticated numerical models and simulations will play a particularly critical role in the field of plasma astrophysics. The reason for this lies within the intellectual structure of the field itself.

It is widely supposed that plasma-physical mechanisms are responsible for many of the non-thermal processes observed in astrophysics, such as high-energy particle acceleration and the coherent emission of radiation. Similarly, non-classical transport mechanisms, such as anomalously large viscosity in accretion disks, or anomalously high resistivity in astrophysical dynamos, seem to be required by current astrophysical models. Plasma-based processes are at the heart of the micro-physics of these transport phenomena. An important goal for theoretical astrophysics is to develop quantitative calculations of the expected nature of these plasma processes, and of their observational consequences in relevant astrophysical situations.

But plasma processes both determine, and are determined by, their parent system's global configuration. Experience with laboratory and space plasmas has shown that a plasma's behavior is sensitive to the specific physical conditions and geometry in which it finds itself. At the same time, some knowledge of the plasma's behavior is often essential to constructing a credible large-scale model of the astrophysical system in question. Thus in order for astrophysical plasma physics to produce quantitative results that can be meaningfully related to astronomical data, an iteration must be performed between the microphysics (simulations of microscopic plasma processes) and the large-scale configuration which emits the photons observed by astronomers (simulations via hydrodynamic models, usually either radiation- hydrodynamical or magneto-hydrodynamical).

The field of plasma physics has been a pioneer in the development of successful computational models, including descriptions at the kinetic, magnetohydrodynamic (MHD), hybrid, and fluid levels. Indeed this development has been a necessity, due to the nonlinearity and geometrical complexity inherent in collective plasma behavior. Fusion research, of both the magnetic and laser-driven variety, has made extensive use of computational simulations in the interpretation of data from laboratory experiments, helped by the facilities of the National Magnetic Fusion Energy Computer Center and at national laboratories. Likewise, NASA's support of large-scale computing within the solar- terrestrial theory program has made computational simulation a regular tool for interpreting in situ space-physics data from NASA's solar- system probes. Driven by the fusion and space-physics communities, the computational simulation of microscopic plasma processes has shown considerable success over the past decade.

Unlike laboratory or space plasmas, one cannot probe the conditions in astrophysical plasmas directly. Thus astrophysical plasma physics research must take the additional step of integrating microphysics models with appropriate large-scale system models, so as to arrive at a quantitative prediction of the observed photon output. A start in the direction of large-scale astrophysical models has already been made. In the field of solar physics, MHD studies of turbulent convection and fluid-magnetic-field interactions will allow detailed comparison with the next generation of high-resolution solar instruments. Similarly, the first generation of MHD models of astrophysical jets has reached a sophisticated level, allowing comparison with high-resolution radio data. In addition to the further development of these two areas, over the next decade one can anticipate the development of MHD models for the large-scale structure of accretion disks, supernova remnants, pulsar magnetospheres, solar active regions, and planetary magnetospheric structure.

An important feature of the next generation of macroscopic system models will be the incorporation of results from detailed plasma simulations at the micro-physics level. For example, non-linear transport coefficients developed using small-scale plasma simulations will be used within larger macroscopic system models to predict the photon output. Similarly, source terms for non-thermal or relativistic particles can be developed using plasma simulations, and then applied when the appropriate conditions emerge in a large-scale macroscopic model.

Computational Requirements: Microscopic Plasma Simulations

Most state-of-the-art microscopic plasma simulation codes are currently being run on multiprocessor vector supercomputers, particluarly if more than one spatial dimension is involved.

What does the future hold? Mini-supercomputers will be used increasingly for the less demanding simulations. At present it is not clear whether massively parallel architectures will be well-suited for particle simulations of plasmas except in some special cases, although they may be useful for some types of Vlasov or hybrid algorithms. However, overall there are strong pressures towards moving to next-generation multiprocessor vector supercomputers. The reason for these pressures lies in the need to push beyond the very small volumes that can presently be studied using microscopic plasma simulation methods, and in the need to perform three-dimensional simulations in order to study geometrically complex phenomena such as magnetic field line reconnection. Thus plasma astrophysics has genuine need for supercomputer resources of the class that the NSF and DOE Centers can potentially provide.

Hand-in-hand with the need for supercomputers is the need for advanced graphics and visualization capabilities to interpret the results. Many microscopic plasma simulations follow the evolution of the distribution function of electrons and/or ions in phase space, together with gradients in real space. Thus, present kinetic models are frequently 4- or 5-dimensional (two space dimensions and 2 or 3 velocity dimensions), and future models will add a third space dimension as well. Advanced visualization techniques will be a prerequesite for extracting useful information from simulation models having this high level of complexity.

Computational Requirements: Macroscopic System Models

Simulations which incorporate the results of plasma micro-physics studies into a model of the large-scale astrophysical system have a slightly different computational flavor, although many of the computational requirements are similar to those described in the previous subsection.

In magnetohydrodynamic (MHD) models, only two or possibly three dimensions are involved. Thus it is possible that the memory and speed requirements of these models can be met using the present and next generations of mini-supercomputers, coupled with the type of advanced graphics and visualization tools described above. Massively parallel architectures are also a possibility for future MHD models, although much research remains to be done to optimize performance in this area.

However in the end what is useful for astrophysics is a prediction of the radiation output. Thus some sort of treatment of radiation emission and transport will be a critical element of many macroscopic system models. Once radiation transport is added to a fluid or MHD model, the number of effective dimensions increases, taxing the memory and speed capabilities of (at least today's) mini-supercomputers. Likewise, radiation transport introduces coupling in angle or in frequency which is difficult to treat on massively parallel architectures. Thus the use of state-of-the-art supercomputers will be critical for this type of macroscopic modelling effort.

Case Study C: CCD Optical Images and Image Processing

Large charge-coupled device focal plane imagers in the next- generation instrumentation for very large ground-based telescopes will create massive amounts of data. Real-time automated preprocessing and initial analysis of these data will be required. In the past, image processing in astronomy has generally not been on-line or real-time. The correlators on the VLA radio telescope are a good example of automated data pre-processing, but the image data on that telescope are not automatically processed.

Progress in automated photometry of crowded fields was made with the software packages DAOPHOT and ROMAPHOT. In the radio, the AIPS package contains several semi-automated routines for cleaning. Recently, near real-time preprocessing of digital images has been made possible by standardization of image header information and advances in processor/storage hardware. Data in the image header may be used as process history and keys. That is, stored images (flat field, bias, dark, object exposures) required in the processing may be retrieved automatically by reference to key entries in their headers, including date, time, and filter/spectrograph settings. The IRAF/CCDRED package permits automated pre- processing of large volumes of raw 2-dimensional images, and is gaining popularity in the optical community.

Automated Image Analysis Software

Image databases have been large (several Gb) in optical astronomy for years. The automated detection, classification, and photometry package FOCAS (Faint Object Classification and Analysis System), developed over the last ten years, has enabled statistical studies involving large image databases. FOCAS is a collection of image analysis and automated pattern recognition programs designed for automated reduction and morphological classification of astronomical images. Recent FOCAS releases (now available as part of the IRAF package) include powerful image pre-filtering operators along with interactive color-graphics display programs which allow the user to quickly identify objects with selected properties, such as color or two-dimensional shape.

A 100 million pixel imager

Current silicon CCDs cover about 1% of the quality imaging area in the focal plane of large telescopes. It is now necessary and possible to construct a mosaic of CCDs which cover a larger focal plane area. Let us consider a 5x5 mosaic containing 25 CCD arrays, each 2048x2048 pixels. The peak raw data rate for the camera would be 100 Gb/night. It is nearly possible, by current 1990 technology, to process, reduce, analyze, and archive this imaging database in nearly real-time.

A CCD mosaic imaging survey of a 100 degree patch of the sky would produce vast amounts of data which would have to be managed and processed. The resulting detected object rasters in several wavelength bands, and the FOCAS matched catalog, would comprise 600 Gb. The data rate from these detectors will therefore overwhelm the traditional mini- or micro-computer or workstation. Recording the raw data, without any on- line preprocessing and display, is cost-ineffective: about 100Gb of data would be acquired in each 24hr period.

It is imperative that the design of large CCD imagers make adequate provision for easy and rapid data analysis. The magnitude of the data processing tasks required for this mosaic imager would require a special-purpose system. The characteristics of this system are dictated by the requirement for real-time image correction and automated analysis, but the same hardware would be capable of performing extensive image post analysis.

The design of the data system must emphasize computational power, fast data transfer paths, flexibility, and expandability. Almost as critical as processing the raw data, this would provide the astronomer with a powerful workstation for exploration of the reduced images and catalogs. Due to the need for instant access, image displays, and interactive image analysis, remote supercomputers are not a solution to the computational requirements in optical-IR image processing and analysis. Instead, we look to fast processors based on Digital Signal Processors with GFLOP average speed and multiple wide busses, together with large RAM memory and fast multi-port disks, which are now becoming available.

Consider a multi- wavelength digital sky survey using a 10^8 pixel imager. The final CCD images for each band are passed to the FOCAS automated detector and classifier creating a catalog of properties (isophotal, aperture, and total magnitudes, centroid positions, and several central moments) for each detected object. For this mosaic imaging survey, with a limiting magnitude < 22 magnitude, the sky is sparse and the resulting catalog would be very small (a few Gb) compared with the full processed image databank (8000 Gb). The FOCAS point spread function is automatically determined from the stars in the image. Detection proceeds by convolving the image with a slightly broadened point spread function, and demanding that a real object must have more than say 10 simply connected pixels above 3 times the rms convolved sky noise.

Terabit digital archives

New technologies are emerging which will allow archiving of the resulting large image and catalog databases. Tera-bit optical recorders are now available. If the total imaging survey reduced data (8000 Gb, mostly blank sky) were stored on 2.4 Gb optical disks, it would require 3300 disks, close to $1M. Since access time is not critical for archiving, another technology is very appealing: optical film. Spot density of one per micron and areas up to 1 inch x 2000 feet may be obtained cheaply. Over several hundred Gb of encoded digital data may be archived on such a medium. Hardware for recording and reproducing in this format exists. These recorders can sustain 3 Mb/s data rates. Both the multi-band image rasters and the FOCAS catalogs (less than 600 Gb total) could be archived inexpensively.

It would even be possible to store the entire 8000 Gb in a collection of optical tape reels no larger than a feature length motion picture. It is clear that it is not practical to save the data and analyze it later, an approach which is already causing some problems with the current generation of small CCD cameras: the data simply piles up. To avoid this analysis bottleneck the images must be corrected as they are acquired. The CCD mosaic imager would produce a continuous data stream of 4 Mb / sec. One night's observing would typically produce 100 Gb of data. The final mosaic image is the result of extensive mathematical corrections applied to this data stream. During this correction operation each 1 Gb of data may move from disk to memory and back several times. Mis-alignment of the CCD rows in the mosaic would also be corrected in this processing.

In summary, we will soon have mosaic imager/computer systems capable of pushing the largest existing telescope to its performance limit. High efficiency CCD imagers covering most of the useable focal plane, together with specialized on-line computers using automatic image classification software will radically alter our ability to observe the universe.

Case Study D: A "Typical" Large VLA Data Processing Request

It might be difficult to grasp the enormity of the computing problem for VLA data without an example given in some detail. Most of the computer-limited problems are three-dimensional in nature, usually from spectral-line data. One special problem - imaging low-frequency data, is continuum in nature, but nevertheless requires the spectral line observing mode. Below we describe this computing problem.

The NRAO has recently completed installation of 327 MHz receivers on the VLA. Unique science addressable with this new capability includes steep spectrum objects and objects of large diameter and low surface brightness. Due to the two-dimensional geometry of the VLA, the samples of the visibility function are made throughout a three-dimensional volume, and the conditions under which a two-dimensional Fourier transform can be used to recover the source brightness fail, requiring much more expensive solutions. The simplest of these is a three-dimensional Fourier transform, producing a three-dimensional image 'volume' whose axes are in direction cosines, and within which the desired image is found on a sphere of unit radius. Processing of the image follows the same procedures normally done in two-dimensional processing – for example, deconvolution proceeds in an entirely analogous fashion using the three dimensional image with a three-dimensional beam.

In low-frequency imaging it is necessary to process all the sources within the primary beam. To reach maximum sensitivity, all confusing background sources must be located and removed through deconvolution.

A typical large project at 327 MHz will use all four VLA configurations with perhaps 12 hours observing in each. Because of chromatic aberration, the spectral line correlator must be employed to ensure the bandwidth of each data channel remains small. The result of this is a 16-fold increase in data rate and volume over the continuum case. In this mode, radio frequency interference, can be identified and purged without seriously corrupting adjacent channels. The integration time for each sample must be kept very short to prevent time-averaging smearing. The result is a very large database: Typically 3.5 GBytes, containing over 500 million complex numbers.

Calibration of these basic data is straightforward, and can be accomplished with modest computer resources, providing only that the short-term disk space to contain the data is present. These data must then be written to tape, or other storage medium, perhaps optical disks or high density tapes, with 1 - 2 GBytes.

The imaging needs are exceptionally large. The simple 3-D transform requires making a "dirty" map and beam, each $4096 \times 4096 \times 64$ pixels, which with 4 bytes per pixel requires 8.5 GBytes memory if they are made in the most efficient, straightforward way. Fortunately, another approach is more efficient in memory, although at the cost of I/O and CPU. The image can be built up through a series of a large number of subimages of limited depth. The memory requirement is relatively more modest, about 135 MBytes per facet.

Deconvolution of the image is the next concern. In the simple, single cube approach, the procedure is straightforward, although highly consumptive of memory. No recall of the data is required. In the polyhedron imaging approach, deconvolution can be accomplished with much reduced memory, again at the cost of much increased I/O. A rough estimate is that perhaps 10,000 Fourier transform subtractions of clean components from the data are required, along with 10 re-imagings of the entire field. That is, more than 2500 FFTs, each 1024×1024 will be required. But this does not finish the processing, since the ionospheric corruptions must be removed through self-calibration. Self-calibration and deconvolution are interlinked, the former using the results of the latter to generate a better estimate of the sky brightness. Typically, three loops of self-calibration and deconvolution are required before satisfactory convergence has been achieved. Thus, all the operations described in the previous paragraph must be multiplied by three.

A rough estimate of the time required can be made: Using the Cray Research Inc.'s CRAY-2 supercomputer performance with a more modest case as a benchmark, and multiplying up by the ratio of database sizes and number of fields to process, results in a rough estimate of 250 DAYS for full processing. The time required is dominated by the gridding: Each visibility point must be distributed over about 100 adjacent cells, and averaged with all other visibility values within this cells, resulting in a computation-limited problem. We are confident that useable short-cuts will be found, as detailed studies of solving this computing problem have barely begun. For example, we can probably use a much less expensive gridding algorithm, which could cut processing by a factor of five.

It seems clear that the optimal approach will eventually involve massive parallelism. We could imagine 16 parallel machines, each comparable to, or faster than the CRAY-2 supercomputer. The data are then distributed to each of these machines, each of which is responsible for one sub-field. A central processor will be required to handle the component model subtractions – the model comes from all 16 parallel processors. Factoring in these expected savings, and imagining future, more powerful machines, predicts this particular problem to be soluble with a few hours of computing time.

The essential points of the above example are summarized here: First, there is a need for very fast machines with very high I/O rates and extremely large memory to generate the data cubes. In many cases, parallel processing is clearly advantageous. Second, assuming the NRAO will not be able to obtain such machines, the national supercomputer centers must provide access and support for astronomers requiring this computing. Third, home machines, such as workstations, must be supported to allow proper interaction between astronomer and completed image. And fourth, research and development of computing algorithms must be actively supported, both at the national center(s) and at the NRAO. The latter site is particularly important, for only at the observatory are the problems fully understood, and the vested interest present on a daily and continuing basis – factors which are absolutely essential to ensure progress in imaging science.

The revised VLA computing plan will handle most of the large VLA requests. But perhaps in 10 years, the algorithms and computers will allow the local computing environment to handle even these large requests.

V. NATIONAL HIGH PERFORMANCE NETWORKING: OBSERVATIONAL IMAGES AND THEORETICAL SIMULATIONS

In this section, we describe a major national experiment in networking, just getting underway, with a goal to determine how remote users will be able to interact at high speeds with remote supercomputers, observatories, and digital archives. The Corporation for National Research Initiatives, is organizing a set of five national gigabaud testbeds, which will become an integral part of the High Performance Computing Program. One of the testbeds will be transcontinental in scale and will have as application drivers an astronomical observatory and a distributed dynamical 3D simulation. This testbed, called NRI Blanca Testbed, will provide a first look at how the high performance computing infrastructure of the 1990's will enhance theoretical and observational astronomy.

The NRI Blanca National Gigabaud Testbed

The NRI Blanca National Gigabaud Testbed will create a prototype distributed scientific laboratory involving researchers at a number of universities on a fiber optic network. The plans for the transcontinental testbed network is to start with 45 Mbit/sec rates currently, with a goal of approaching a Gigabit/Sec over the next five years. Supercomputing facilities, large scientific databases, and high- performance visualization workstations will be connected via this Gbit/sec network, with data collection and observatory sites, and with collaborating researchers at each site. Research projects which involve information exchange in the form of data sets or interactive images or both - with volumes that definitely require a network running these speeds, will be supported.

Additional development efforts to be included in this project involve laser disc technology archive systems, image generation algorithm development, and development of a fully distributed, general purpose scientific simulation control and visualization system. The distributed visualization and simulation control system will be of general use, with libraries and client/server processes which can be used by computational scientists regardless of the specific discipline involved.

Simulations and image processing on supercomputers often require access to data bases at remote sites which are too large to be moved and/or which are being collected at high rates. Further, programs running on the supercomputers must be controlled by researchers from remote locations, requiring visualization output at that remote site which is a) of high resolution such as is necessary to determine the accuracy and quality of the run, and b) displayed in real time to allow control of the supercomputer application process.

BIMA–A High Performance Computing Observatory on the Gigabaud Testbed

Future supertelescopes will have as an essential component a very high speed data link between the sensor and a computer. Real-time radio astronomy would revolutionize the field by permitting an observer using a synthesis array to see an image of the radio sky as the observations were being made. Interactive observing could be a reality if the image processing can be done and the images transmitted fast enough. The goal of the BIMA experiment on this national gigabaud testbed is to demonstrate such capabilities and to explore how such capabilities might improve, expand, and extend the power of a telescope system.

The Berkeley-Illinois-Maryland Array (BIMA) is located at the Hat Creek Observatory in northern California and is operated by the University of California at Berkeley, the University of Illinois, and the University of Maryland. The Array is similar in concept to the VLA, but operates at millimeter rather than centimeter wavelengths. By early 1991 BIMA will consist of 6 antennas; there are plans for expansion to as many as 12 antennas. The BIMA system has been chosen for this testbed because the proposed gigabaud network will extend from Berkeley to Urbana, linking the sensor with the supercomputer, and because BIMA will generate data and have computational needs which are a significant fraction of those of the VLA. Although with 6 antennas BIMA has only about 5% of the number of simultaneous interferometers as the VLA, the BIMA spectrometer produces 4 times as many spectral channels and allows observation of up to 8 spectral lines simultaneously; the density of spectral lines in frequency space at millimeter wavelengths means that much of the time this multiplexing capability will be employed usefully. Further, BIMA will be used in spectral-line mode essentially all the time, while the VLA is often used in continuum mode. The BIMA data rate and computational requirements will be about 1/3 those of the VLA. A gigabaud connection between Socorro, New Mexico and one of the supercomputer centers would allow, in principle, similar remote operation of the VLA and the VLBA.

A typical BIMA data set will be in the 100 MB to 1 GB range; such data sets can be transferred from Berkeley to the supercomputer at Illinois at 45 Mbaud (real and sustained) in the period of a 5 minute coffee break. The initial processing of the observed visibility data on the supercomputer will be automatic, under the control of an "expert system" with tunable parameters which may be set in advance by the astronomer. While the observations are in progress, calibration, map making, and an initial deconvolution, self-calibration, and mosaicing (if appropriate) of a partial data set may be carried out and the data cube returned to the astronomer at Berkeley for analysis on a workstation. The astronomer will be able to judge the quality of the data, to see if the signal is strong enough to proceed with the observations, to judge whether the area of sky being mosaiced is correct, and to begin to experiment with processing parameters. Instrumental or atmospheric problems can be detected quickly, and corrections made or re-observations carried out while the telescope is still in the same configuration. Exciting or unexpected results can be pursued immediately. When the project observations are complete, the full data set can be processed interactively on the supercomputer from 2000 miles away. The processing of radio maps is often highly iterative and interactive. The astronomer in Berkeley will be able to examine each step in the deconvolution (CLEAN or MEM) and self-calibration process as it runs on the Cray and fine-tune the algorithm parameters to yield the best possible maps.

Today, such interactive observing is possible only for astronomers in Urbana and only to a limited extent, because of the slow speed of the shared NSFNET. Using the recently developed MIRIAD (Multi-channel Image Reconstruction with Interactive Analysis and Display) software, the sizes of the images which can be processed can reach 4096 x 4096 pixels, images referred to as supermaps. Data are loaded into a supercomputer, which processes it and sends images of the processed data to a frame buffer connected via HPPI (High Performance Parallel Interface) at up to 800 Mbit/sec. This allows the local researcher to observe the calculations in real time; changing parameters and regenerating images interactively. Today, the local researcher can send 2-3 1024 x 1024 x 24 bit images per second (this being the resolution of current display hardware), which allows direct interaction with the image processing of quadrants of supermaps.

During the next 5 years the network capacity will allow 4096 x 4096 x 24 bit images to be transferred in under 0.5 seconds per image, which will enable the same level of interactivity remotely on full "Supermaps" as is available today at Illinois on a single quadrant; MIRIAD can transfer the desired 2-3 images per second and still maintain total interactivity with the image reconstruction.

Combined with applications for multiple simultaneous viewing by separate workstations (multiple collaborators located at multiple remote sites) this will allow for a level of interactive collaboration which is not feasible today. Combined with systems such as the digital archive for astronomical images, this network rate will allow for paging through multiple images. The remote researcher will have the ability to process much of the existing raw data which has not yet been viewed or evaluated over the network.

Remote Control of Fourth Dimension Supercomputers

Tools will be developed in the gigabaud network testbed project to build applications which support real time collaboration among multiple, remote scientists on scientific and computational aspects of a simulation running in real time on a supercomputer. The specific application chosen as a platform with which to demonstrate these tools is the study of storms using a four dimensional numerical model (3 space dimensions and time). The tools developed in this national testbed project should be immediately applicable to similar simulations in theoretical astrophysics.

The distributed interactive execution and analysis of storm simulations is currently limited by disk and network speeds, as the simulation process output is in the range of 32 Mb/s to 320 Mb/s . The critical limitation today, however, is the conversion of data into graphic images, or visualization. Most three dimensional visualization today is done in batch mode using a mini- supercomputer which runs visualization software and can take between several seconds to several minutes to convert raw data into a single animation frame. This delay between simulation and graphic output prevents the researcher from interacting with the model and adjusting algorithms and parameters to yield optimal results. The delay also makes it impractical to collaborate with colleagues during the model verification process, as the scientist must send to the collaborator a finished product (a video) which will arrive several days or weeks after the simulation was done. Further, non-interactive visualization prevents colleagues from collaborating in the area of visualization techniques. Because each model is visualized differently, it is difficult at best to compare the validity of different models. Short term improvements in surface visualization will be obtained by using the supercomputer to do the tessellation component of the process (computing the geometric polygon representation) and to display the images in near-real time using graphic rendering hardware at a scientific workstation on the network. This will allow the researcher to interact with the simulation.

A sample collaborative session between two researchers at different locations could involve several components. Both researchers would have the capability of starting up the simulation or data analysis software from their workstation. Everything that appears on one of the researcher's workstation windows would appear on the other's (this requires screen transfers that can easily exceed 100 Mbits per sec for color). At any time either one of the researchers can take control of the process, or start up visualization from a different dataset for comparison. Surface displays from today's storm models consist of 30,000 to

100,000 polygons and can be coupled with other forms of visualization to qualitatively and quantitatively analyze model information. Animation of these displays being viewed at one site, also appear at the other. A high resolution animation of 8 bit per pixel images can be done with gigabit speeds (1400 frames would take about 15 s to transfer at 1 gigabit/sec). Data sets may have to be moved quickly from one researcher's site to another, depending on how the simulation and data exploration process is distributed and on the capabilities of the local graphics workstations. For collaborative interactive data exploration, this also requires gigabit transfer rates.

Long term improvements, only possible using a gigabit/second wide area network, will allow the display of simulation output , interactive control of the simulation and interactive analysis of the output to take place concurrently at multiple, separate workstations on the network. At this point, real-time collaboration will occur between scientists in the areas of modeling theory as well as visualization techniques. Each scientist will view the simulation variables of most interest to him and in a way which is consistent with the methods he uses to visualize his own model. Thus, the scientists can directly compare the output of two simulation models and begin to determine the strengths and weaknesses of the various modeling techniques.

Further development and increased workstation processing power will conceivably allow these scientists to do the tessellation as well as the rendering on their local workstation using their own custom visualization filters and to jointly analyze the simulation with colleagues across the network. Specific development will include a network software interface similar to the BSD sockets or to the Shared-X-Window system but with mechanisms for specifying experimental network services such as guaranteed minimum throughput, real time services, packet trains, isochronous data stream delivery, maximum tolerable latency, multi-cast, etc. to be implemented by network researchers on the testbed. Further investigation will be made into the transmission over the network of multiple channels to provide voice and image teleconferencing in parallel to simulation output and control.

POLICY OPPORTUNITIES PANEL

RICHARD McCRAY,* University of Colorado, Boulder, *Chair*
JEREMIAH OSTRIKER,* Princeton University Observatory, *Vice-Chair*

LOREN W. ACTON, Lockheed Palo Alto Research Laboratory
NETA A. BAHCALL, Princeton University
ROBERT C. BLESS, University of Wisconsin, Madison
ROBERT A. BROWN,* Space Telescope Science Institute
GEOFFREY BURBIDGE, University of California, San Diego
BERNARD F. BURKE, Massachusetts Institute of Technology
GEORGE W. CLARK, Massachusetts Institute of Technology
FRANCE A. CORDOVA, Pennsylvania State University
HARRIET L. DINERSTEIN,* University of Texas, Austin
ALAN DRESSLER,* Carnegie Observatories
ANDREA K. DUPREE, Harvard-Smithsonian Center for Astrophysics
MOSHE ELITZUR, University of Kentucky
SANDRA FABER,* University of California, Santa Cruz
RICCARDO GIACCONI, Space Telescope Science Institute
DAVID J. HELFAND, Columbia University
NOEL W. HINNERS, Martin Marietta Corporation
STEPHEN S. HOLT,* NASA Goddard Space Flight Center
JEFFREY L. LINSKY,* University of Colorado, Boulder
ROGER F. MALINA, University of California, Berkeley
CLAIRE ELLEN MAX, Lawrence Livermore National Laboratory
GOETZ K. OERTEL,* Association of Universities for Research in Astronomy
BENJAMIN PEERY, Howard University
VERA C. RUBIN, Carnegie Institution of Washington
IRWIN SHAPIRO, Harvard-Smithsonian Center for Astrophysics
PETER ALBERT STRITTMATTER, University of Arizona
SCOTT D. TREMAINE, Canadian Institute for Theoretical Astrophysics
PAUL A. VANDEN BOUT, National Radio Astronomy Observatory
JACQUELINE H. VAN GORKOM, Columbia University
J. CRAIG WHEELER, University of Texas, Austin
SIMON D.M. WHITE, University of Arizona

Policy Opportunities

I. INTRODUCTION

In addition to setting scientific priorities for the discipline, the AASC was charged to assess current policies and practices in the conduct and support of space and ground-based astronomy and to recommend changes that are likely to enhance the productivity of the enterprise. In this Chapter, we consider the relationship and balance between ground-based and space astronomy, strategies for achieving high productivity in both these programs, the support of individual scientists and scientific facilities, international cooperation, the scientific advisory process, and the role of astronomy in education.

In §II we describe the context for our recommendations. In subsequent sections we provide the rationale for our recommendations, the major ones of which are listed below.

The first set of recommendations, discussed in §III, is directed toward the National Science Foundation Astronomy Division. We recommend:

- The NSF should retain primary responsibility for the US ground-based astronomy program, which is a vital component of the nation's overall astronomy effort.
- The budget of the NSF Astronomy Division, in constant dollars, should be doubled during the next 5 years in order to recover ground lost during the past decade, to ensure continued US leadership in ground-based astronomy, and to realize the scientific benefits of the space program.
- NSF-Astronomy should undertake the construction of new facilities *only* in the context of a strategic plan in which the university grants program is adequate to support excellent research programs and the most productive and highest quality observational facilities can be supported adequately. Support for other facilities should be terminated. First call on research funds should be on strengthening the base rather than on building large new facilities.
- The National Optical Astronomy Observatories (NOAO) should provide leadership by building facilities with unique capabilities that will be used by all optical/IR astronomers. In addition, the NOAO should continue to provide access to first-class observing facilities comparable to those at private or university-owned observatories. Whenever possible, the latter mission should be accomplished by cooperation with private and/or state institutions, with emphasis on cost- and technology sharing and the avoidance of duplication of specialized instruments.

The second set of recommendations, discussed in §IV, is directed toward the NASA Space Astrophysics program. We recommend that:

- NASA should carry out the program outlined in the OSSA 5-year Strategic Plan, including full implementation of the Great Observatories program and the deployment of second-generation image-correcting instruments for the Hubble Space Telescope.

- In addition to the 5-year plan, NASA should expand its support of moderate and small programs, implemented by a doubling of the Explorer budget and expansion of the suborbital program.
- NASA should adopt management strategies for the Explorer program with the aim of developing missions faster and at lower cost.
- NASA should support an augmented research and analysis program that is stable and protected against cost overruns of its hardware programs.
- The development of astronomical facilities for the Space Exploration Initiative should follow a logically phased approach. Whenever feasible, the technology development program should include testing through actual astronomical research on the ground, on stratospheric platforms, and/or in earth orbit. The greater expenses of astrophysical observatories on the Moon must be fully justified by the greater scientific return that they are expected to provide.

In §V, we recommend a program for education in astronomy:

- To exploit the unique potential of astronomical and space research to attract young people into scientific and technical careers, astronomers should participate in a broad educational initiative designed to provide more access to the excitement of modern astronomy for students, teachers, and the general public.

Finally, in §VI, we address several other policy issues and recommend that:

- A standing committee of the National Academy of Sciences be established to monitor the overall health of the field and to provide strategic, coordinated advice to all agencies that support research in ground-based and space astronomy.
- Astronomical research will advance most rapidly in a climate of open exchange of information and access to all facilities, foreign and domestic, by the best qualified observers. The agencies should support open access to US facilities and data and should expect other countries to reciprocate.
- We encourage international cooperation on the construction of facilities when each country or entity brings complementary capabilities to the project or when the international nature of the project is uniquely valuable to its performance.
- National and private observatories should formulate policies and make plans for the entry of astronomical data in standard formats into a national archive to enable access by the broad astronomical community. The agencies should encourage these efforts and support their implementation according to scientific merit as determined by peer review.
- NASA and NSF should consider the development of procedures and facilities to enable the simultaneous multi-wavelength observations of variable celestial sources.

II. THE CONTEXT OF THE RECOMMENDATIONS

The NSF and NASA provide primary support for US research in astronomy and astrophysics. During the 1980's, these agencies have responded to the advice of the previous NAS Astronomy Survey Committee (ASC) report.

The National Science Foundation has implemented, fully or partially, many of the ASC recommendations for new ground-based facilities. For example, construction for the Very Long Baseline Radio Telescope Array (VLBA) is now well-advanced. NSF provided partial support to build new 2 - 4 meter-class telescopes at universities, one of which is already operational, and the NOAO has formed partnerships with private universities to build and operate such telescopes at KPNO and CTIO. NSF financed the construction of a submillimeter telescope on Mauna Kea. The Agency supported several outstanding young astrophysicists through the Presidential Young Investigator program. It provides advanced computing power to astrophysicists through the national supercomputing centers and the national research network.

In addition, the NSF Astronomy Division initiated the solar Global Oscillation Network project and responded to the collapse of the Green Bank radio telescope by funding the construction of a modern fully-steerable 100-m radio telescope. The NSF Division of Polar Programs supported the development of astrophysics programs to exploit the unique advantages of the South Pole.

It is also encouraging that the NSF has improved its budgeting for major projects by setting up a separate project line item within the Division of Mathematical and Physical Sciences.

After the Challenger accident, NASA recognized a crisis in space science and responded by a commitment to do space science for its own sake, independently of the manned program. The Office of Space Science and Applications (OSSA) developed a balanced Five Year Strategic Plan that includes an exciting program for astrophysics research. In this plan, NASA recognizes that: (1) the OSSA Budget should be a stable fraction (20%) of the total NASA budget; (2) much of space science can be done best with expendable launch vehicles; and (3) NASA must provide stable support for a healthy infrastructure of theory, data analysis, and instrument development at universities in order to realize the benefits of its science missions and to ensure the future vitality of space science.

The rich scientific yield of the COBE satellite proves NASA's wisdom in re-structuring that mission for launch with an expendable rocket rather than the Space Shuttle. NASA should avoid using the Shuttle to launch free-flying satellites for astrophysics unless its unique capabilities are *required* for the mission.

NASA has steadfastly supported the Great Observatories Strategy recommended by the ASC Report. The first of these, the Hubble Space Telescope (HST), has been launched, the Gamma Ray Observatory (GRO) will launched in early 1991, the Advanced X-Ray Astronomy Observatory (AXAF) has been started, and the Space Infrared Telescope Facility (SIRTF) has a high priority in NASA's five year strategic plan. The budget for Explorer satellites was increased, and the first-ranked intermediate program, the Far Ultraviolet Spectroscopic Explorer (FUSE), is under development.

Beyond that, the NASA Astrophysics Division launched two astronomical Explorer missions (IRAS and COBE) and sustained a vigorous suborbital program on rockets, balloons, and aircraft. The scientific value of the suborbital program was illustrated by NASA's quick and decisive response to the extraordinary scientific opportunity presented by Supernova 1987A, which yielded irreplaceable data on the infrared and gamma ray emission from this unique event. OSSA initiated a new program for Small Explorers (SMEX). The NASA Astrophysics Division also developed a creative program of international cooperation involving US participation in several European, Japanese, and Soviet astrophysics missions as well as participation by several other countries in NASA missions.

NASA's Astrophysics Division has made plans to strengthen the infrastructure of university science in support of its missions. These include support for individual scientists through the Astrophysics Data Program and data analysis programs of individual missions, and a commitment to make space data available to all qualified scientists. Recognizing that scientific insight often comes from integrating data obtained at different wavelengths, NASA-Astrophysics provides science-oriented funding in addition to mission-oriented funding for data analysis.

It makes sense for NSF and NASA to work together toward the common goal of understanding of astronomical phenomena. NASA's Planetary Division supports the Infrared Telescope Facility on Mauna Kea. NASA has also cooperated with the NSF to upgrade the Arecibo radio telescope of the National Astronomy and Ionosphere Center (NAIC) and the VLA in order to enhance their capabilities for deep space tracking, planetary radar, and astronomical research. NASA and NSF have worked together to distribute advanced detectors, originally developed for space missions, to ground-based observatories, greatly enhancing the power of their optical telescopes.

These activities have been carried out in the face of major obstacles. The NSF Astronomy Division has received insufficient funding to respond fully to the ASC recommendations. NASA's space science programs for the 1980's were set back severely by science funding shortfalls, delays in the Shuttle Program, and the Challenger accident. The success of NSF and NASA in implementing part of the recommended program despite these handicaps is a tribute to the hard work by dedicated people at NSF and NASA. We appreciate their service. We commend both NASA and NSF for relying on the peer review process in making decisions for funding.

We endorse the agencies' practice of using "rotators," research scientists on temporary leave from universities, on agency staff. We encourage universities to recognize and reward such service and call on the AAS Council to encourage members to participate.

The Departments of Defense and Energy have also contributed substantially to astrophysics research and technology development, and we expect this synergism to continue. Between the end of World War II and 1960, DoD support provided the foundation for the great expansion of astronomy that has occurred

ASTRONOMY AND ASTROPHYSICS PANEL REPORTS

since then and for our present preeminence. DoD continues to support research in astrometry and optical interferometry, and technology developments in cryogenics, infrared detectors, adaptive optics, fiber optics, and parallel computer architectures. DoE scientists have been leaders in calculations of gravitational collapse, supernova explosions, nucleosynthesis, and stellar opacity, and in observations of cosmic X-ray and gamma ray sources.

The Space Exploration Initiative has the potential to provide new astronomical observing capabilities that can qualitatively improve astronomical resolution and sensitivity. A Presidential Decision of February 16, 1990 enlists the help of the Departments of Defense and Energy in this initiative.

The Smithsonian Institution, through its Astrophysical Observatory, has supported research in astrophysics for a century. The Observatory is now engaged in a broad range of research efforts, including studies of large-scale structure of the universe, high energy phenomena, atomic and molecular interactions, radiative transfer, stellar atmospheres, cosmic masers, molecular clouds, star formation, and the solar system. Facilities include a major optical observatory, available in part to visitors.

The National Institute of Standards and Technology supported measurement and theoretical calculations of atomic and molecular processes that are fundamental to the understanding of many astronomical observations and phenomena.

Although the policies of the 1980's have yielded many successes, our panel has identified several opportunities where the implementation of new policy recommendations would improve the productivity of the enterprise:

- Astrophysics suffers because the ground-based astronomy program is too small to yield full scientific value from the nation's investment in space. There has been a serious decline in the infrastructure of ground-based astronomy, including both the support of existing facilities and of individual researchers. We make recommendations to correct this problem in §III.

- NASA can improve productivity, stimulate inventiveness, and train a new generation of space scientists and managers by devoting more resources to missions of reduced complexity that can be developed and launched within about three years. These and other issues are discussed in §IV.

- The national astrophysics program can attract more talented people into scientific careers. It also can contribute substantially to improving scientific and technical literacy. In §V we give our rationale and present recommendations to accomplish these goals.

- We provide some guidelines for the scientific advisory process, international cooperation, archiving of astronomical data, and multi-wavelength observations, in §VI.

After this Chapter was nearly completed, we learned that the Hubble Space Telescope was launched with a defective mirror that seriously compromises the present ability of the telescope to do the frontier science for which it was designed. Our Panel cannot assess the specific causes for this problem; that task has been assigned to other Committees specifically charged to do so. There are, however, important lessons to be learned from our experience with the HST that are independent of the specific causes of the mirror problem; we discuss them in §IV(g).

III. REVIVING THE NATION'S GROUND-BASED ASTRONOMY PROGRAM

Once a star performer among U.S. science programs, ground-based observational and theoretical astronomy is now imperiled by continuing budget cuts, with consequent decay of major facilities and loss of key staff personnel at national observatories. The cause of this decline is twofold – the NSF basic research budget did not keep pace with the scientific needs of the nation and the relative priority for astronomy within the Foundation declined.

Ground-based observational astronomy and associated theory are the essential core of astronomy. Without adequate support, the U.S. risks losing the fruits of its entire astronomy program, including the space effort. A modest augmentation of the ground-based effort can strongly enhance the total yield of the space program. Astronomy drives technology and science education; it is appreciated and admired by the general public and provides, for many, their only glimpse into what science is all about. Thus, the current crisis of support for ground-based astronomy is a national problem – for astronomy, for scientific efficiency, for science education, and for scientific prestige.

a) Why Ground-based Astronomy?

The number of U.S. ground-oriented observational and theoretical astronomers has doubled since 1970, reflecting the excitement of the subject. Of all recent astronomical papers that refer to observational data, 72% relied mainly on ground-based data, while an overwhelming 83% contained at least some ground-based data. Space observations stimulate ground-based activities rather than replace them. Of the space-oriented papers studied, 39% also utilized ground-based data, and most of these reported new ground-based data acquired specifically to follow up and support the space discoveries. Although these statistics are probably influenced by the lack of launch opportunities for astronomical spacecraft during the 1980's, it is clear that ground-based observations are fundamental to astronomical research. The case for an excellent ground-based observational program is even more compelling in view of the relatively modest cost of ground-based facilities. Clearly, whatever can be done on the ground, should be done on the ground.

Without adequate ground-based follow-up to space observations, America risks losing much of the cream of its space science program. Astronomers in other countries have invested heavily in ground equipment. They can easily obtain data from our open space archives and follow up with superior ground-based facilities. The solution is to augment our own ground-based capabilities with comparatively modest expenditures, to a level appropriate to realize the full potential of the space effort. To fail to do so would be a serious mis-allocation of national resources.

b) The Unique Role of the National Science Foundation in American Astronomy: The Importance of a Strong NSF Program

NSF, the only Federal agency with a mandate to support basic research, has unique responsibilities and abilities. NSF funds many grants and small projects with short lead times and great flexibility. This is especially important for theory, which would become seriously distorted if it were tied too closely to specific missions. NSF should therefore remain the principal custodian of funding for basic theoretical science in the U.S. Averaged over the past decade, the Astronomy Division of the NSF devoted only nine percent of its University budget to theory, a relatively low percentage compared to other Federal offices that support theory.

The NSF peer review process is generally perceived to be fair. The Astronomy Program Directors have a good track record of embracing the priorities in the Astronomy Decade Reviews. However, they lack the resources to do the job properly.

Astronomical research has historically been centered in university departments, the main source of advanced education and professional training. Combining the roles of basic research, training, and education at public and private universities is the unique property of the US scientific research system. Maintenance of this system, which is the envy of other scientifically advanced nations, is a primary responsibility of the NSF. Finally, astronomy stimulates extraordinary popular interest, and is a major asset to the NSF in its mission to cultivate public awareness of science.

c) The Decline in U.S. Ground-based Astronomy

The current crisis follows directly from the long-term funding history. New facilities have been constructed and the number of ground-oriented observational astronomers has doubled since 1970, but the budget (in Consumer Price Index-adjusted dollars) to operate facilities and conduct basic research has not increased in the past twenty years, despite a substantial expansion in the scope of the science. NOAO opened two new 4-m optical telescopes and absorbed the operation of Sacramento Peak Solar Observatory within its declining budget. NRAO opened the VLA, staffed a new site in Socorro, NM, and began to operate the VLBA. At the same time, improvements in astronomical instrumentation and data analysis algorithms have increased the power of optical and radio telescopes greatly. The cost of operating the NAIC was transferred from DARPA to the NSF. The number of visiting observers at all national observatories has tripled in the past twenty years. At the same time, their staffs have declined. To expect efficient operation of a larger program with less staff and smaller budgets is unrealistic, and it shows.

The decline of the budget is illustrated in Table 1. *Corrected for the Consumer Price Index (CPI), the budget has been flat for twenty years. Corrected for the actual cost escalation in technical environments of*

8.5% per year during the decade 1980-1989, the NSF base budget declined in real spending power to only 71%, and the spending per astronomer to 36%, of what they were in 1970.

Construction of new facilities and advanced technology development at national observatories was achieved only through a diversion of resources from other critical areas, resulting in deferred maintenance, and deferred purchase of new equipment. Long-term deleterious consequences of these policies, now apparent, are detailed below.

The specific effects for the University Grants Program and the National Observatories illustrate the impact (all budgetary figures are expressed in real dollars corrected using NSAC [see Table 1] inflation):

University Grants Program (36% of base budget). The NSF is the source of almost all grants to support ground-based observational work. Most astronomers have no alternative funding sources for this work.

- The purchasing power of an average grant fell by more than a factor of two since 1980.
- Available grant funds per U.S. astronomer have been reduced by 1/2 since 1980.
- The number of funded postdoctoral fellows fell by 20% since 1980.
- The success rate of new proposal applicants, mostly young investigators, fell to 10%. Most new applicants have no alternative sources for funding their research programs.
- Many leading researchers had grants delayed, slashed, or cancelled.

National Optical Astronomy Observatories (NOAO) (30% of base budget):

- Staff level was cut by 15% since 1984, 25% since 1979.
- Budget was cut by 21% since 1984.
- One heavily utilized telescope was permanently closed, more closures are under review.
- Two-thirds of observing requests are now rejected for lack of facilities.
- Travel support has been suspended for visiting observers. Many observers now pay travel out of their own pockets for lack of NSF grants (even to CTIO in Chile).
- The Advanced Projects Group at NOAO was closed for lack of funds. Group leader was hired away by European Southern Observatory to build world's largest telescope (ESO VLT project). NOAO lost its leadership in optical telescope construction and advanced optics.
- Other countries now dominate large-telescope construction. Seven large telescopes were built abroad since 1975. In the same period NSF funds built 1/2 of one telescope in the U.S. Relative decline in forefront optical and electronic technology is comparable.

National Radio Astronomy Observatory (NRAO) (27% of the current base budget):

- Operations staff was cut by 15% in last 5 years.
- Operations budget was cut by 30% over last 5 years.
- Leadership in millimeter astronomy – developed by the US – was lost to Europeans and Japanese. No major new telescopes were built, despite an elegant proposal.
- Deferred maintenance, such as the VLA track system, requires one-time funding of several million dollars.
- Cannot operate VLA at full capability or exploit new image processing techniques for lack of modern receivers and adequate computers – despite excellent peer reviews.
- Operations funds for the VLBA are ramping up at only half the rate required to put antennae into service.
- The world-famous NRAO technical group is threatened due to low salaries and low morale. Director says, "If the core technical team disbands, the Observatory has no future."

National Astronomy and Ionosphere Center (NAIC) (7% of base budget):

- Staff was cut by 10% in last five years, 24% since 1979.
- Budget was cut by 35% since 1984.
- Decaying scientific equipment, some of it 25 years old.

Table 1. Long-Term History of the NSF Astronomy Base Budget[1]

Year	1970	1980	1989
Actual-year dollars	$23.8M	$52.2M	$77.4M
CPI adjusted[2]	1.00[3]	1.03	1.06
NSAC inflation adjusted[4]	1.00[3]	1.03	0.71
Per U.S. astronomer (CPI)[5]	1.00[3]	0.76	0.55
Per U.S. astronomer (NSAC)[6]	1.00[3]	0.76	0.36
Rel. to total NSF budget[7]	1.00[3]	0.96	0.80
Rel. to MPS budget[8]	—	1.00[8]	0.78
Rel. to U.S. GNP[9]	1.00[3]	0.81	0.63

[1] The NSF base budget includes the university grants program and funds for the operation and maintenance of the National Observatories. New construction is omitted.

[2] Adjusted for increases in the Consumer Price Index.

[3] Set to 1.00 in 1970.

[4] Adjusted for "technology inflation", as estimated by the Nuclear Science Advisory Committee subcommittee on inflation. We have applied a correction of 4.6% per year to this figure to the period 1980-1989. No adjustment has been applied to the period 1970-1980, for which no comparable data are available.

[5] Adjusted using the increase in total AAS membership and CPI inflation.

[6] Same as the previous row, but with NSAC inflation assumed.

[7] Adjusted for increase in the total NSF budget.

[8] Adjusted for increase in the Mathematical and Physical Sciences budget (set to 1.00 in 1980, since MPS did not exist in 1970).

[9] Adjusted for increase in U.S. national GNP. Shows NSF astronomy as a fraction of national effort.

d) Augmenting the NSF Astronomy Budget

Funding for ground-based astronomy has fallen so far below par that a concerted effort of restoration is required. The figures in Table 1 mandate, at minimum, a doubling of the NSF base budget for astronomy during the next five years. Astronomy has not shared in the growth of the NSF budget for many years. Given the situation described above and the spectacular continuing advances possible in this field, a period of above average increases is warranted.

e) The Role of NOAO in Ground-based Night-time Astronomical Research

Astronomers hold conflicting views of the role of NOAO in ground-based, nighttime optical astronomy. On the one hand, the observatories were established and continue to be used to provide night-time facilities that are comparable in aperture to those run by universities or private institutions to which access is usually restricted by affiliation. On the other hand, many astronomers believe that NOAO should focus its efforts on providing unique facilities, such as the European VLT (Very Large Telescope).

About 20 percent of NOAO users come from groups planning to construct their own 8-m and 10-m telescopes. The remaining 80 percent typically do not have access to such facilities, and most do not have access even to 1-4 meter class telescopes. Given this situation, there is substantial resistance among the latter group to expenditure of NOAO's budget for building more ambitious, state-of-the-art telescopes if doing so precludes the continued operation and enhancement of the existing telescopes.

For long-lived facilities like telescopes, operation costs over their lifetimes exceed initial capital costs. As larger and more modern facilities become more common, smaller and aging telescopes appear less attractive and cost-effective. Changes in scientific emphasis also favor larger aperture.

As guidelines towards definition of an achievable mission that will serve a wider segment of the U.S. astronomical community, we recommend:

- That NOAO further reduce its support of those telescopes with relatively low oversubscription rates, seeking arrangements, whenever possible, to transfer them to private institutions.
- The NOAO should endeavor to increase the time it can make available on 4-m class telescopes by seeking more partnership arrangements like the several efficient cost-sharing arrangements with university consortia that it has recently undertaken.
- NSF-astronomy and NOAO should establish provisions for trading or purchasing telescope time from private and university groups operating 8-m and 10-m class telescopes and smaller special-purpose telescopes in order to ensure that the full range of observing facilities is available to the whole community. Astronomy in the U.S. would be well served by a cooperative interdependency of the private and public sector.
- By adding 8-m telescopes, NOAO will continue to fulfill its mission of providing access to front-line facilities to all astronomers independent of institutional affiliation. The construction and operation of 8-m telescopes and a 4-m telescope dedicated to fiber optics spectroscopy, and technology development for optical and infrared interferometry, are also important steps for development of a technical base upon which even more ambitious, unique facilities will be built.

The development of *unique* facilities, which are distinguished either by scale or function, is a crucial step by which NOAO can broaden its support of the entire astronomical community. To maintain leadership in optical astronomy, NOAO needs the *active* involvement of the nation's leading astronomers, including those with access to private 4-m, 8-m and 10-m telescopes. They must be involved in the definition of NOAO projects and their implementation, and most importantly, they must *use* these facilities. The presently planned 8-m telescopes for NOAO are an important step along the way. A larger world-class, unique facility for optical astronomy will be needed to ensure scientific leadership by NOAO and to exploit fully the U.S. advantage of combining powerful resources in both the private and public sector.

- NOAO cannot be the site for all of the necessary technological innovation, but it can play a vital role as a clearinghouse for such technology. To this end, NOAO should encourage outside partnerships in detector and instrument development.

The health and success of U.S. optical astronomy has been based on a combination of a strong *national* observatory together with *non-federal* funding for private and university facilities that is unique in the world. A partnership that enhances the strengths of these two elements will ensure continued U.S. leadership in optical astronomy.

IV. A VIGOROUS PROGRAM OF SPACE ASTROPHYSICS

NASA's agenda of unfinished astrophysics missions remains substantial. For example, the HST needs repair, SIRTF has not yet been started, and AXAF and most of the Explorer missions approved during the 1970's and 80's will not be launched until the late 1990's.

OSSA's five year strategic plan incorporates this unfinished agenda. We endorse this plan without reservation. In doing so, we recognize that its completion establishes NASA's strategy for most astrophysics missions to be launched until the late 1990's. Therefore, our recommendations cannot affect this plan in a major way. They should, however, affect the process by which NASA will select and implement astrophysics missions to be started during this decade and beyond.

a) An Enhanced Explorer Program

We recommend that NASA develop a more vigorous program of missions with reduced complexity and shorter times from inception to completion. At present, the funding is weighted toward large missions costing more than $300 M, such as HST, GRO, and AXAF. Averaged over 1984 - 1989, the fraction of NASA astrophysics project funding devoted to large missions was 73%, compared to 12% for moderate ($100 - 300 M) missions and 15% for small (< $100 M) missions, including rockets, balloons, and aircraft.

Large missions such as the Great Observatories have revolutionary capabilities that cannot be matched by moderate and small missions. The latter, however, can add a dimension to NASA's space science program that is vital and cannot be provided by the large missions: the ability to deploy new instrumental technology into space on a timescale of a few years. The prospect of rapid access to space is a strong driver of innovation. This opportunity attracts and permits the training of talented young instrumentalists, engineers, and project managers who are essential, not only for the health and future of NASA's space science programs, but also for the nation's future technical competitiveness. Yet, only two Explorer missions devoted to astronomy were launched during the period 1980-89.

There are many good ideas for scientific payloads for small and moderate missions to make critical scientific observations that cannot be done with any other planned mission. For example, a 1988 NASA solicitation yielded 27 proposals for Delta-class Explorer missions for astrophysics, 7 of which were ranked with highest ("category 1") scientific priority, and a 1989 solicitation yielded 17 proposals for Scout Class Small Explorers (SMEX) for astrophysics, 3 of which were ranked category 1. However, it was only possible for NASA to select one mission for development from each of these competitions owing to the constraints of the Explorer budget. Moreover, some of the most innovative instruments developed by U.S. space scientists are now being flown first on foreign spacecraft for lack of NASA launch opportunities.

The present level of the Explorer budget is approximately \$60 M/yr for Delta-class missions and \$30 M/yr for SMEX. Assuming optimistically that Delta-class missions will cost \$120 M each and SMEX \$30 M each, the Explorer budget will then permit one Delta-class mission every two years and one SMEX per year. That is approximately the necessary rate for a robust Explorer program for astrophysics alone, but the current Explorer budget must also support missions for several other disciplines of space science. Thus, a doubling of the Explorer budget is the minimum needed to maintain a vigorous program of astrophysics Explorer missions assuming that half the budget will be devoted to astrophysics missions.

Presently, the scientific opportunities for small and moderate Explorer missions are constrained by NASA's lack of an expendable rocket with payload intermediate between those of the Scouts and Deltas. Important new opportunities for powerful but relatively inexpensive astrophysics missions will appear when OSSA procures such a vehicle.

b) Costs and Management of Small and Moderate Missions

Even with a doubling of the Explorer budget, we will be able to achieve rapid and steady access to space only by holding mission development to cost and schedule. The productivity of the Explorer program will be maximized by having more frequent cost- and schedule-constrained missions rather than by maximizing the scientific performance of each individual mission. We have seen examples, such as the Japanese *ASTRO* program, where this strategy has enabled a robust program of X-ray astronomy missions launched at regular intervals.

Cost and schedule control begins with the Explorer selection process. NASA has begun to introduce incentives by supporting a greater number of missions for the definition phase (Phase A) and then conducting a second competition to select missions for development (Phases B,C,D). To ensure that this strategy is successful, NASA should also: (1) include mission costs and their impact on the Explorer program as criteria of the peer review process in both the Phase A and Phase B competitions; (2) hold the management teams to their budgets, *even if it becomes necessary to scale down performance specifications.*

It is vital for cost containment that missions have *all* critical technologies under control before they are selected for development. To meet this requirement, NASA should invest adequately in technology development in its Research and Analysis program and in Phase A. If so, we see no reason that mission development should require more than three years from the beginning of Phase B to launch.

To achieve an optimum result within budget and schedule, the project management team must be able to trade off scientific performance and cost, and take risks if necessary. In order to enable this process, NASA should: (1) vest full authority, including control over budget, staff, and procurement in a project management team consisting of the Project Manager and the Principal Scientific Investigator; and (2) provide full funding as planned to support the master schedule.

There are necessary risks to such a strategy. If the project management team fails to meet milestones or exceeds costs, NASA must decide whether to stretch the schedule and augment the funding or to cancel the mission. Such decisions should be based on a careful assessment of the options and their impact on the overall Explorer program, with advice from the scientific community. The gains – in technical development, management experience and program discipline – may outweigh the losses if an occasional mission is cancelled.

c) A Renewed Partnership with Universities and Industry

If moderate and small missions can be launched at a healthy rate, we think that NASA can achieve a more productive overall small/moderate mission program by involving project management and systems engineering at universities working with private industry, and/or NASA centers. Universities have unique advantages for attracting and training people for careers in engineering and management as well as in basic science. By providing opportunities to work on all aspects of space missions at universities, NASA can help provide a healthy supply of technically proficient and talented people, not only for its own needs but for the nation in general.

The time is ripe for a more vigorous partnership between NASA, universities, and the aerospace industry in space science projects. The aerospace industry has a rich reservoir of management and technical expertise for the building of space hardware, and space science could benefit greatly if more of this capacity became available as a result of decreased demand for defense systems.

We therefore recommend that NASA carry out its Explorer program in the context of a "mixed economy," in which some missions are developed by NASA centers and others are developed by management teams from universities working with private industry. Within such an economy, NASA should compare cost and productivity and seek an optimum mix.

d) Astrophysics within the Space Exploration Initiative

NASA's Space Exploration initiative presents exciting prospects for astronomical observations on the lunar surface. There are, however, great uncertainties about the technical and logistical infrastructure to support such facilities, the timescale for their development, and the cost. It is prudent for astrophysicists to work with NASA to understand better the opportunities and problems of doing astronomy on the Moon. NASA should develop the required technology in logical phases.

To ensure that the required technology is effective, NASA should, whenever feasible, test it on the ground, on suborbital platforms, and/or in Earth orbit. The requirement to produce actual scientific results in these tests introduces a technical rigor to the program that paper studies cannot provide. Further, the investment required to test the scientific and technical systems on the ground or in Earth orbit is relatively small. It is the best way to ensure maximum return from the much greater investment that will be required to install and operate an observatory on the Moon. Since the development phase of this initiative will be long, the opportunity to do scientific observations during this phase would help to attract and train the highly talented scientists whose energies and skills would be essential to the success of this initiative.

e) A Vigorous Program of Suborbital and Airborne Research

Our recommendation that a greater fraction of NASA's resources be allocated to Explorers is motivated by several important goals: (1) the training of young astronomers and instrumentalists; (2) fast turn-around and frequent opportunities for testing and developing new instrumentation and techniques; and (3) improved cost-risk-benefit ratios to foster innovation. All of these desiderata are met extremely effectively by NASA's suborbital programs of rocket, balloon, and airborne astronomical research.

These suborbital platforms play a unique and critical role as test-beds for new instruments. Because of the nature and operating procedures of these programs, astronomers have excellent access for adjustments and modifications to their instruments. In the airborne program, as epitomized by the highly successful Kuiper Airborne Observatory (KAO), most groups have continuous access to their instruments during operation and can make minor adjustments even *during* a research flight. More significant adjustments and modifications can be made between flights on the KAO, or between launch opportunities for rockets and balloons.

This "hands-on" mode of operation also provides a special opportunity for the training of young instrumentalists. The pay-off provided by the opportunity to participate directly in instrument development is apparent from the established track records of suborbital programs. Examples include: the explosive growth in our understanding of the interstellar medium due to the development of ultraviolet spectroscopy, initiated through the rocket program; the development of powerful new gamma ray telescopes through the balloon program; and the invaluable role of the KAO in the professional development of most currently active and prominent researchers in infrared and submillimeter astronomy. A recent survey of participants in the KAO program, which provides flight opportunities for about two dozen research groups per year, has shown that the KAO has supported the Ph.D. research of about 40 scientists who are currently active in these fields.

The ready access to the instruments in the suborbital program leads to a balance of costs, risks, and benefits that is very different from that for instruments on spacecraft. The high degree of reliability required for instruments on spacecraft drives up their cost, precluding frequent launch opportunities. As a result, high-risk but innovative instrumental development is inhibited. In contrast, suborbital platforms are ideal for trying out truly innovative but risky ideas, since the price of a single failure is relatively minor. Thus the suborbital programs are an essential component of a well-balanced strategy for instrumental and scientific progress. Furthermore, the intrinsic flexibility in scheduling the suborbital platforms allows short-term redirection of these facilities to take advantage of targets of opportunity, such as bright comets or SN 1987A.

For all of these reasons, we strongly endorse NASA's suborbital research program, and support vigorous expansion of this program in areas where the scientific and instrumental opportunities are well-defined and offer the potential of high rewards.

f) The NASA Research and Analysis Programs

NASA's research and analysis program supports activities that are essential to its space science program, including:
 – research and development of new instruments for its flight programs
 – data analysis for specific flight programs and panchromatic analysis of archival data sets.
 – theoretical investigations needed to interpret data from space missions, to enable better planning of future missions, and to gain a deeper understanding of the universe.

NASA has recognized that the support of these activities is inadequate and plans to increase its support of these activities through the Astrophysics Data Program, the Astrophysical Theory Program, the Hubble Fellowship Program, the Long-Term Space Analysis Program, and individual research grants. NASA must sustain this commitment and protect the planned growth of these programs in order to ensure a stable and strong infrastructure for its space science programs.

These programs are also NASA's primary mechanism for support of space science research at universities. Funding through grants to individuals and small groups at universities has proved to be a highly productive way to support scientific research and is NASA's *only* way to train the next generation of space scientists. Larger research centers also have essential roles to play in supporting instrument development, data analysis, and theory.

NASA's research and analysis programs gain strength through a healthy competition among universities and centers for limited resources. In order to ensure that the programs have maximum productivity and to maintain a healthy balance between funding of individual scientists and larger centers, it is important that all programs for support of research and analysis be reviewed periodically, with community input.

g) The Problems of the Hubble Space Telescope

Although this Panel cannot assess the causes of the defective mirror on the HST, we wish to make a few comments here that are independent of the findings of those committees charged to investigate that problem.

First, the HST should eventually be able to fulfill its role as the centerpiece of NASA's Space Astrophysics research program for the 1990s. Except for the mirror, the telescope and all its instruments are working well. It can be used now to do frontier science, and it can be restored to nearly full capability provided that its focal plane instruments can be replaced by new ones containing correcting optics. NASA should make these replacements as soon as it can do so according to priorities established by its scientific advisory committees.

Second, the HST mirror problem does not lead us to modify the major recommendations of this Chapter; indeed, the problems of the HST only reinforce some of the conclusions that we had already reached.

Third, the HST mirror problem reminds us that no space science program can be immune to failure. When the fraction of resources concentrated in missions of large scale and long development times becomes too large, the space science program becomes brittle and unforgiving of failure and cost-overruns. Failure can be tolerated least in the most expensive missions; that fact drives their development costs higher. NASA should have a space science program that is more resilient to the occasional failures that will occur. That is one of several reasons why our Panel has urged NASA to devote a greater fraction of its resources to missions of small and moderate size.

Fourth, the most serious problems of the HST project (not only the mirror problem) can be traced to errors that were made in its early stages. At that time there was inadequate involvement of scientists in technical oversight and decision-making. NASA should require the scientists who have most at stake in a space mission to assume a greater share of the responsibility and accountability for technical decisions. Participation in oversight committees is not enough; the scientists must be involved at the working level. In particular, scientists with personal commitments to the ultimate uses of a mission must participate in the design and implementation of systems tests that verify the correct performance of all aspects of the mission before launch, and NASA must support such tests. In order to ensure this involvement, NASA must provide scientists with adequate resources for technical support, especially during the critical early development stages.

Fifth, the development of the HST was troubled by diffusion of authority and responsibility among different centers and the lack of a prime contractor. Moreover, the long delays in development contributed to a lack of continuity in key management, scientific, and technical personnel. To minimize risk in its large space projects, NASA should vest authority and accountability in a tight management team including the principal scientists led by a single individual of renown and ability who has direct oversight and responsibility for the scientific success of the mission.

Finally, we wish to emphasize that NASA has learned many of these lessons and is already applying them to the management of future Great Observatories such as AXAF and SIRTF. If NASA continues in this path, these missions should not be susceptible to the most serious problems that have troubled the HST project.

V. AN EDUCATION INITIATIVE IN ASTRONOMY

In his 1990 State of the Union Address, President Bush declared, "By the year 2000, U.S. students must be first in the world in math and science achievement." Astronomers have much to contribute to this high national goal. Here we recommend a strategy for astronomers to increase their current contributions to science education and national scientific literacy.

Education in astronomy has several distinct but overlapping goals. One is the training of astrophysicists. Modern astrophysics research requires people of high talent trained in diverse specializations. The supply of newly-trained astrophysicists that existed in the 1970's and early 80's is shrinking rapidly, while at the same time the research opportunities and demands for trained personnel are growing. Therefore, we must attract and train people in the field with renewed vigor. This is especially true for women and minorities, who remain severely underrepresented in the profession.

A second goal is to contribute to the broader national pool of professional scientists and engineers. Many physicists, chemists, engineers, and computer scientists are drawn into their careers through an initial interest in astronomy, and many people trained as astrophysicists pursue careers in other areas of science and technology.

A third goal is to help raise scientific literacy in the nation. For this goal, astronomy has a number of advantages, such as: the intrinsic fascination of the cosmos; astronomy's central role in the history of physical science; the accessibility, diversity and universality of astronomical concepts and techniques; the glamour of the space program; and the vitality of amateur astronomy.

The nation's colleges and universities are doing a good job of training scientists. However, too few students enter college with adequate scientific literacy or the intention of pursuing scientific careers. A serious attempt to address the three goals stated above must emphasize the pre-college (K - 12) years, and that is the focus of this section.

Astronomers are involved in many aspects of science education. In colleges and universities throughout the nation, introductory astronomy courses have high enrollments and provide the main exposure to physical science for many undergraduates. Teaching undergraduate and graduate courses in astronomy and astrophysics is the primary educational commitment for many of the nation's research astronomers. Astronomers also make contributions beyond the college classroom. Most college and university departments welcome the general public to their telescopes and research facilities in public "open nights." Astronomers write articles for popular science journals, they contribute to educational television programs, and they speak at local schools, museums and planetariums, amateur astronomy clubs, and other civic groups.

Astronomers also support more organized efforts for science education beyond the university. The American Astronomical Society (AAS) conducts workshops for local schoolteachers and public lectures at its semiannual meetings and supports expenses for astronomers to visit colleges and universities that do not have astronomers on their faculties to give lectures in classrooms and public lectures for the local communities. The Astronomical Society of the Pacific (ASP) conducts meetings for amateur astronomers and publishes "Mercury," a non-technical journal of astronomy that is written for the public, educators, and amateur astronomers as well as "The Universe in the Classroom," a newsletter of wide circulation. The Harvard-Smithsonian Center for Astrophysics, with NSF support, is developing a year-long "hands-on" high school course that uses examples from astronomy to teach fundamental principles of physics and mathematics. Astronomers, sponsored by the NSF Education Division, have conducted summer workshops for schoolteachers at national observatories and universities. The International Astronomical Union's Teaching Commission provides liaison with astronomy educators around the world.

The enthusiastic response to these activities reveals a great unmet public demand for more exposure to the fruits of modern astronomical research. How can the nation's astronomers meet this demand? There are millions of schoolchildren who need more exposure to astronomy, but only a few thousand professional astronomers, most of whom are already heavily committed to college teaching and research. In fashioning a response to this challenge, we must identify programs that will give high leverage to the limited human resources that we can provide.

A coordinated educational initiative in astronomy should be funded by the educational directorates of the respective agencies, at a level commensurate with the national priority of education, in order to establish a stable interface whereby researchers can provide their resources to the nation's professional educators. This initiative should be guided by three principles: (1) engage the nation's teachers to multiply the astronomers' efforts; (2) make the program highly visible in order to attract the most talented teachers and students; and (3) stress affirmative action.

Several specific strategies consistent with these guidelines have been described in the document "An Educational Initiative in Astronomy" [R. Brown, ed., published by the Space Telescope Science Institute, 1990]. They include, among other ideas:

- Expand summer programs and workshops at universities and national research centers for in-service training of science teachers. These workshops provide the best opportunities for science teachers to gain direct experience with modern astronomical research and to make ongoing contacts with astronomers who are committed to improving science education. The workshops will be particularly effective if they can attract those master teachers who are involved in the development of curriculum materials and the training of other teachers.

- Expand the educational programs of the AAS and ASP. The agencies responsible for funding public education should be responsive to proposals for in-service training of teachers and the development of curriculum materials.

- Increase astronomer participation in textbook and curriculum development at the K-12 levels (astronomers are already heavily involved in the writing of college textbooks). Involvement in commercial endeavors may also generate additional resources and ensure wide dissemination of the products. Professional astronomers should also participate in school or state evaluation and adoption procedures.

- Adapt innovative technologies for use in schools. For example, CD-ROM data bases and image processing programs for the analysis of astronomical data on personal computers can give students the opportunity to become familiar with modern computer technology while they explore the sky in all its wavelength bands.

- Develop special programs for gifted students. High-visibility prestige programs not only help to identify and recruit the future leaders of astronomy, but also send a message to young people that a career in science can be as feasible and rewarding as a career in, say, law or professional sports. We recommend that each state identify one or two outstanding high school students as State Fellows in Astronomy. The students will serve as paid interns at one of the major national or private observatories and will participate as assistants in the active research of the professional staff.

- Recognize and reward astronomers for their contributions to education. The community can encourage astronomers to contribute to an educational initiative by increasing the reward structure. For example, the AAS can establish awards for contributions to education, science writing, and public service, and universities can reward their faculty for improving public education as well as for their teaching and research.

We have two recommendations for these strategies to be realized in a coherent educational initiative. First, in order to bring the excitement of NASA's space science programs into the nation's classrooms, NASA should provide support to the education initiative that is complementary to the support provided by the NSF education directorate. Second, offices for implementing a major part of the educational initiative should be established at centers operating national research facilities. Such offices should act as clearinghouses for the dissemination of information and curriculum materials, sponsor workshops, and coordinate the State Fellowship program. We recommend that NSF and NASA each establish such an office at one of their research centers and consider establishing additional offices, depending on experience and demand.

VI. OTHER POLICY ISSUES

a) Science Advice to the Government Regarding Astronomy and Astrophysics

NASA and NSF, the agencies providing primary support for research in astronomy and astrophysics, already have in place advisory committees of leading scientists. The National Science Foundation has an Advisory Committee for Astronomical Sciences (ACAST) that reports to NSF management, and NASA has a number of Management and Operations Working Groups (MOWG's) and Astrophysics and Solar System Exploration Subcommittees that report to the Astrophysics Division and Solar System Exploration Division managers and their Directors, respectively. In addition, NASA benefits from advice from the NAS Space Studies Board and its subcommittees for astrophysics (CSAA), planetary and lunar exploration (COMPLEX), and solar-terrestrial physics (CSSP/CSTR). These committees are appropriate entities for advising the agencies on scientific priorities.

However, we believe that astronomy would benefit from an independent standing broad-based Astrophysics Strategic Advisory Committee of the NAS (here called ASAC), constituted of leading scientists from every major sector of astronomical and astrophysical research, including those supported by NSF, NASA, other agencies, private observatories, and industrial research labs. The ASAC should monitor the overall health of the field and provide strategic advice regarding how to maintain a balanced program of astronomy and astrophysics within the guidelines of this Report. The ASAC should not provide continuing tactical advice to the agencies regarding scientific priorities within their purview; that role properly belongs to the ACAST and the MOWG's. Nor should it review or revise the plan presented in the AASC report, which is the result of an extensive effort by a large fraction of the astronomical community.

There are important roles for the ASAC that no other advisory committee can play. One is to provide advice on the global issues in astrophysics research to all agencies to maximize the scientific and educational benefits by working together and with the private sector. Another role is to advise the agencies, the Congress, and the Executive Branch how to achieve the scientific objectives of the decade plan when unforeseen circumstances arise. Examples of such circumstances might be new technical opportunities, failure of a major facility, opportunities for international cooperation, or changes in the budgetary picture beyond the assumptions of the AASC plan. Finally, to have one broad-based standing committee of distinguished astrophysicists may prove especially useful when consideration is given to programs involving more than one agency.

b) International Cooperation and Competition

Research in astronomy and astrophysics has always been an international enterprise. Important astronomical research programs are now being carried out by many nations. Several nations and groups of nations have built and are planning to build research facilities with capabilities that are unmatched in the U.S. The vitality of research in the field, therefore, depends on healthy international cooperation and competition.

The most productive kind of international cooperation occurs when individual scientists exchange knowledge and share facilities to solve specific scientific problems. The scientific programs of all nations flourish when scientists can freely exchange data and technical knowledge and when qualified scientists of any nationality have access to unique facilities throughout the world. Such exchange is essential for effective international cooperation in the construction of instruments and facilities. The United States has and should continue to set a standard for open scientific research that is exemplary to the world. We commend NASA and NSF for opening the competition for observing time on the Hubble Space Telescope and on the National Radio Observatories to qualified observers from all nations. In turn, the United States should expect that other countries will foster an equally open scientific atmosphere.

International cooperation is also fostered by a vigorous program of scientific exchange through workshops and visiting scientist programs, and by opening educational and employment opportunities to highly qualified foreigners. This is another area in which the U.S. has been world leader, and in which it has benefitted greatly by attracting outstanding scientists from all nations, many of whom choose to remain in the U.S. and become citizens. We recommend that this be enhanced by removing the restrictions to US nationals from fellowship and other employment opportunities. The Immigration and Naturalization Service can also help by removing obstacles to permanent employment for highly talented foreign scientists.

International cooperation in building major facilities is appropriate when the nature of the project is inherently international, when the project combines complementary capabilities that exist in different nations, or when the project is too complex or costly for individual nations. Examples which meet these criteria are the Global Oscillation Network for studying Solar seismology, intercontinental radio interferometry, several space missions, and, possibly, a permanent manned observatory on the moon.

However, international cooperation can become complicated when more than one country is involved in the construction and operation of major facilities. In those cases guidelines are needed to manage the greater administrative complexity, costs, and delays that can arise from the need to coordinate technical interfaces and independent national bureaucracies. Before undertaking such projects, the agencies, together with their scientific advisory committees, should scrutinize them carefully to determine whether the benefits of the cooperation outweigh the costs. For some projects, independent supra-national entities like CERN or ESO may be appropriate. In many cases, science will advance more rapidly if nations elect to build unique specialized facilities on their own rather than collaborating with other nations. Such a strategy becomes most attractive if nations agree to make their unique facilities available to qualified scientists from any nation.

c) Archiving and Distribution of Astronomical Data

Modern astronomical instruments produce multidimensional data of enormous size and complexity, which often do not reveal all of their secrets and subtleties upon initial examination. Upon further examination, data archives can yield answers to questions that the original observers may never have considered. Archival research can be especially productive when data taken at different times or wavelength bands are compared. Investigators can also use archival data to study large volume-limited or flux-limited samples of objects and to assess the feasibility and merit of proposed observing programs. Rapid advances in technology now allow us to store very large astronomical databases and to make them readily available to qualified researchers. We expect archival research to become an increasingly important part of astrophysics, as more investigators gain access to and become familiar with these data sets and the powerful workstations and software for analyzing them.

Public access after an appropriate proprietary period can maximize the scientific return on data obtained at great effort and cost. It also encourages the timely publication of the results of observing programs. We commend the steps that have already been taken. All data obtained from NASA space programs are available to the public after a proprietary period of typically one year. NASA has made a major commitment to and financial investment in preparing calibrated archives of most data obtained with its space-borne instruments. The Agency has also developed an on-line central directory to identify and locate the data, and it has supported the dissemination and analysis of these data through the peer-reviewed Astrophysics Data Program. The Space Telescope Science Institute has developed an archive system to make data from the HST available to all qualified scientists. The NRAO has created a raw data archive of all VLA data with public access after 18 months. The NOAO and NRAO have established the standard

FITS formats for data transfer and have provided the IRAF and AIPS data analysis tools to the community. The IAU has adopted and is working to extend the FITS standards, and NASA has established a FITS Support Office.

The development and preparation of archives is often expensive, especially when efforts are undertaken to produce data products for distribution, i.e., data that have been processed to remove instrumental signatures, calibrated, and reduced to simplify access by users. Techniques and algorithms for processing archival data sets continue to evolve as scientists gain a better understanding of the instrument and the data set. Therefore, it may be better to defer processing the entire data set.

It may soon become practical and worthwhile to incorporate many kinds of astronomical data, including ground-based data, into an international network that is accessible to all qualified researchers. To make this happen the NSF and NASA should work together and with scientific agencies of other nations to provide the resources and standards, with the active involvement of the community. Because the choices for storage media and means of handling and accessing data are evolving rapidly, and costs are still unclear, it is prudent to adopt a measured approach toward this goal.

As a first step, NOAO and other ground-based observatories with Federal support should formulate and implement policies for archiving newly generated data, building on the experience of efforts that are already underway. The policies should address questions such as: when does an instrument become sufficiently standard for the data to be archived routinely; what formats should be used; what documentation is required for its calibration and interpretation; what are the most cost-effective means for storing and distributing the data; and for what period should the data remain proprietary.

We also urge private observatories to formulate and implement their own policies for archiving and distributing their data. The NSF should respond to requests from private observatories for capital expenses associated with the establishment of archives with the advice of peer review. Since the production of archived data can add scientific value to observing projects, observatories and individual investigators should be requested to describe their plans and performance in archiving data in proposals for research support.

NASA and NSF should work together to create a unified on-line directory to provide access to all archived ground-based and space data. NASA should provide information, assistance, and coordination for software and connectivity to the ground-based National and private Observatories. The National Observatories should provide catalogs of archived data for the NASA Astrophysics Directory Service.

Subsequent steps toward an international archive of astronomical data may include facilities to make archived data themselves accessible through electronic networks, the incorporation of previously archived data into modern storage media, the production of archives of processed data products, and the critical evaluation of astronomical data bases.

The agencies should support the steps described above on a case-by-case basis when the scientific benefits of the project justify the cost. Such decisions should always be based on a peer-reviewed assessment of the scientific merit compared to other activities, such as instrument development and observing, that compete for the same scarce resources.

The ultimate form of processed data is its publication in a refereed professional journal. It may become feasible and desirable to include such data in an international archive base. The journals should adopt policies to enable retrieval of archival data described in submitted papers, as is done in some other research fields.

d) Multi-wavelength Observations of Variable Sources

Throughout the history of astronomy, observations of the time variation of celestial objects have yielded exciting new discoveries and fundamental insights into the nature of cosmic systems. Today, the ability to observe variable celestial sources in all wavelength bands of the electromagnetic spectrum has revealed entirely new phenomena, such as active galactic nuclei, X-ray binaries, and gamma ray burst sources, the interpretation of which continue to challenge astrophysicists. Analyses of the time variability of such systems at various wavelength bands have provided the most basic clues to their nature.

The revolution in modern astronomy has been driven largely by the ability to compare observations of celestial sources in different wavelength bands, and this certainly must be true for variable sources as well as for steady ones. Indeed, simultaneous observations at different wavelengths can yield fundamental information about variable sources that can be obtained in no other way. However, the task of coordinating such observations with telescopes on the ground and in space presents logistical problems that are always difficult and often technically impossible with current organizational procedures and available instrumentation.

For these reasons, we recommend that NASA and NSF pay particular attention to finding ways to seize the scientific opportunities of multi-wavelength observations of variable sources. NASA should consider the development of spacecraft dedicated to multi-wavelength observations of variable sources, and NASA and NSF should work together to establish a dedicated capability for providing simultaneous ground-based optical, infrared, and radio observations of sources that are observed by NASA spacecraft.

BENEFITS TO THE NATION FROM ASTRONOMY AND ASTROPHYSICS PANEL

VIRGINIA TRIMBLE, University of California, Irvine; University of Maryland, *Chairman*
JOHN N. BAHCALL, Institute for Advanced Study, *Vice-Chair*

ERIC CHAISSON, Space Telescope Science Institute
ARTHUR CODE, University of Wisconsin, Madison
EDWARD K. CONKLIN, FORTH, Inc.
JOHN COWAN, University of Oklahoma
ALEXANDER DALGARNO, Harvard-Smithsonian Center for Astrophysics
FRANK DRAKE, University of California, Santa Cruz
REBECCA ELSON, Harvard-Smithsonian Center for Astrophysics
GEORGE FIELD, Harvard University
ANDREW FRAKNOI, Astronomical Society of the Pacific
HERBERT FRIEDMAN, Naval Research Laboratory
JOHN S. GALLAGHER, AURA
JESSE GREENSTEIN, California Institute of Technology
HERBERT GURSKY, Naval Research Laboratory
STEPHEN MARAN, NASA Goddard Space Flight Center
PHILIP MORRISON, Massachusetts Institute of Technology
CATHERINE A. PILACHOWSKI, Kitt Peak National Observatory
PHILIP SADLER, Harvard-Smithsonian Center for Astrophysics
CARL SAGAN, Cornell University
PHILIP SCHWARTZ, Naval Research Laboratory
STEVEN SHORE, NASA Goddard Space Flight Center
ALEXANDER G. SMITH, University of Florida
HARLAN SMITH, University of Texas, Austin
MICHAEL S. TURNER, Fermi National Accelerator Laboratory
PAUL A. VANDEN BOUT, National Radio Astronomy Observatory
GART WESTERHOUT, U.S. Naval Observatory
JAMES WESTPHAL, California Institute of Technology
R. STEPHEN WHITE, University of California, Riverside
ALBERT WHITFORD, University of California, Santa Cruz

Benefits to the Nation
from Astronomy and Astrophysics

KEY POINTS

- Astronomy makes unexpectedly large contributions to formal and informal science education, given the small number of research astronomers.
- Technology transfer and spin-offs from astronomy have important applications in medicine, industry, defense, environmental monitoring, and consumer products.
- Mankind's view of its place in the world as a whole is strongly influenced by the results of astronomical research.
- Astronomy provides unusually promising opportunities for international cooperation.
- Other sciences benefit from synergistic interactions with astronomy.

I. INTRODUCTION

Astronomy and astrophysics could not exist in their present form in this country without firm public support, expressed through the funding of research by federal and other agencies, public and private. The providers of this support can quite reasonably ask what they are getting in return for their money. The primary answer is, of course, scientific knowledge and all that it implies. (Identifying how that knowledge can best be extended in the future was the principle task of the Astronomy and Astrophysics Survey Committee (AASC)). But there are other, less obvious, returns, and the Panel on Astronomy and Astrophysics as National Assets was charged with identifying and documenting these.

It is not the intention to claim that these educational, cultural, and technological spin-offs are the sole, or even the major, justification for astronomical research, but only that they are a real part of the total picture of how science interacts with the rest of society. In addition, because astronomical objects and ideas are relatively appealing to non- scientists, it seems plausible that the subject may be able to play a significant role in the essential task of revitalizing American leadership in science and technology, both by encouraging young people to consider careers in these areas and by promoting scientific awareness among the general public.

The chapter on National Assets in Volume I of the AASC report presents an overview of the synergistic, educational, and cultural contributions of astronomy and astrophysics. This panel report includes a number of additional examples and technical details of a few outstanding ones. Space did not permit including all of the items collected by the panel or complete crediting of the information to the colleagues who provided it, though contributors other than panel members are listed at the end of the chapter.

II. SCIENCE EDUCATION AND LITERACY

The need for a scientifically sophisticated electorate and how far we are from achieving this have received enough publicity in recent years to require no further explication here. But, of the little science that most people are exposed to–and that they choose to expose themselves to–astronomy forms a surprisingly large part. People trained in astronomy also form part of the general technologically-educated manpower pool.

A. Formal Education

1. College-level courses

Formal astronomy classes have their largest impact at the non-major undergraduate level. The college and universities with astronomy (or physics and astronomy) departments had 1.2 million undergraduates in 1988; 103,300 of them were taking introductory astronomy (Ellis 1988). This means that (integrated over a 4.5- year average curriculum) 35-40 percent of the graduates of these institutions fulfill their science breadth requirements with astronomy, generally as their only exposure to physical science. Astronomers typically make up 5-10 percent of the physical science faculties at these institutions.

There is also considerable demand for astronomy at colleges with no separate department of the subject. Each year the American Astronomical Society receives more than 100 requests for visits by research astronomers to these institutions through its Shapley Program. About 90 requests can be filled. The primary purpose is to talk with classes and student groups, but most visits include a public lecture and meetings with administrators well (C.R. Tolbert, University of Virginia, personal communication 1990). Text book sales indicate that a total of 200-250,000 students per year enroll in an astronomy course (M. Zeilik, University of New Mexico, personal communication 1990).

While taking these classes, students both increase their knowledge of the specific subject and change their attitudes toward science in general. A standardized test, administered as part of the planning for Project STAR (Section II.A.2), shows that those who complete an introductory class know about as much astronomy as the average secondary school teacher. Those just starting the class do considerably less well and score at about the same level as elementary school teachers.

Attitudes toward science were probed with an anonymous questionnaire given to undergraduates at Cornell University, University of Maryland, and University of Wisconsin at the end of one-semester courses. Table 1 shows the results. More than 70 percent of the 1260 students polled reported that they thought understanding science was more important than they had at the beginning of the semester. The majority also said that they were more likely to read about science and to vote for pro-science candidates for political office. All but a few percent of the rest reported their views as unchanged (some explicitly volunteering the information that they had been fairly pro- science to begin with).

Most university departments also offer adult education and extension courses in astronomy and report (e.g., from UCLA and Harvard) that these are among the most popular and successful of their offerings.

2. Pre-College Education and Teacher Training

After prolonged near-absence, astronomy is beginning to reappear in elementary and high school curricula. The 1989 National Science Foundation's (NSF) grants for astronomy education included two high school student summer programs; one each for teachers in high schools, two-year colleges, and elementary and middle schools; and three projects to develop teaching materials for middle and high schools. Many other programs are supported by schools, colleges, and research organizations. A representative sampling follows.

The Astronomical Society of the Pacific "Universe in the Classroom" one-week summer workshop for grade 3-12 educators has had 2500 alumni over the past 12 years. The Society also provides a catalog of educational materials to about 250,000 people world-wide; and a newsletter "Universe in the Classroom" goes to 22,000 teachers, with further reproduction by school districts and planetariums and translation into five foreign languages.

The Space Telescope Science Institute (StScI) currently supplies speakers on request to school classes in its area at a rate of about one per day. Astronomers at nearly every university, lab, and observatory talk to grade and high school classes and clubs on a regular basis.

TABLE 1. CHANGES IN STUDENT ATTITUDES TOWARD SCIENCE DURING
ONE-SEMESTER INTRODUCTORY ASTRONOMY COURSES

Understanding science is:		Understanding science is:	
Much easier*	131 (10.4%)	Much more important*	385 (30.7%)
Somewhat easier	495 (39.4%)	Somewhat more important	513 (40.8%)
About the same	533 (42.4%)	About the same	541 (27.1%)
Somewhat harder	80 (6.4%)	Somewhat less important	14 (1.1%)
Much harder	22 (1.8%)	Much less important	3 (0.2%)
Probability of reading about science in the future		Probability of voting for candidates favoring support for scientific research	
Much higher*	258 (20.5%)	Much higher*	303 (24.1%)
Somewhat higher	593 (47.2%)	Somewhat higher	478 (38.0%)
About the same	375 (29.8%)	About the same	445 (35.4%)
Somewhat lower	17 (1.4%)	Somewhat lower	19 (1.5%)
Much lower	14 (1.1%)	Much lower	12 (1.0%)

* Number of students expressing this opinion after a one semester
introductory astronomy course.

StScI also participates in (1) a summer workshop for science teachers that is expected to have about 300 participants from across the nation in 1990, (2) enrichment programs for scientifically-interested high school students from under-represented minorities, and (3) production of a 32-part instructional television series for middle schools, with broadcast in Maryland and elsewhere to begin in fall 1990.

SPICA at the Center for Astrophysics is an unusually highly- leveraged project whose participants, secondary school teachers, in turn present workshops for elementary and junior high teachers in their home districts.

The National Radio Astronomy Observatory (NRAO) in cooperation with West Virginia University operates a summer workshop for high school teachers, whose funding for 1990 is being taken over by the Claude Worthington Benedum Foundation from NSF.

The Naval Research Laboratory (NRL) and six other Washington-area research institutions provide opportunities for about 100 high school students a year to get involved in astronomical research. Most go on to careers in science and engineering.

More than 500 Starlab portable planetariums (16-foot inflatable domes from Learning Technologies, Cambridge, Massachusetts) have reached some five million school children (mostly in the earlier grades, and including many inner city and disadvantaged kids)

At the Thacher School Summer Science Program, about 1000 students over the past 30 years have worked on an astronomical research project (determining asteroid orbits from photographs and mastering the necessary associated math and physics). All participants go on to college. About 37 percent of the pre-1985 graduates are now working in science and medicine, and 34 percent in engineering, mathematics, and computer science (including the founder of Lotus Development Corporation).

Haystack Observatory has a similarly-successful summer internship for middle school students and the University of Illinois has one for high school students.

Six inner-city San Antonio schools are pioneering a junior- level year of high school science consisting of astronomy and marine biology as part of Project 2061. The real surprise is that most of the students have chosen to take another year of science as an elective in their senior years.

Project STAR (Science Through its Astronomical Roots), one of the most extensive NSF-funded programs, is being developed at the Center for Astrophysics as a serious, quantitative alternative to high school chemistry and physics.

Many of these projects were initiated within the astronomical community, and all have had some input from researchers. But this is an area where more can and should be done. Specific initiatives are proposed by other panels. It is at the high school level and earlier that science must be made attractive to students, before they decide not to take the necessary mathematics.

B. Informal Education and Scientific Literacy

The activities discussed here have three connections with astronomical and other scientific research. First, virtually all of them have either been initiated by or had significant input from research-oriented astronomers. Second, many lines of anecdotal evidence indicate that informal exposure to astronomy motivates people to take a serious interest in science and technology as potential careers. And, finally, in order for books, television programs, planetarium shows, and other presentations about astronomy to remain as popular as they are, there has to be a continuing stream of exciting new results to present.

1. Television

Cosmos is the most successful public television series in history, seen by about 400 million people in 60 countries. The book version is the best-selling English-language science book ever, and the home video version had 100,000 orders placed for the full 13 episodes before release, an unprecedented number for any kind of videotape. Other astronomical television items include:

Project Universe, a series of 30 half-hour programs which reached its 100th showing in 1989, most broadcasts being on local stations in cooperation with nearby colleges offering credit for the series as a course.

Extensive, widely-watched coverage of the Voyager Neptune encounter, whose audiences included millions of young people in and out of school, and a large number of Pasadena residents and visitors (even some European amateur astronomers who flew in for the occasion), who watched in real time at an auditorium near the Jet Propulsion Laboratory (JPL).

A Galactic Odyssey (funded and produced by Japanese National Television) and The Astronomers' Universe (funded by the Keck Foundation and produced by KCET), which are 6-8 hour series focusing on astronomy and the people who do it, scheduled for 1990-91 broadcasts. One of the stated purposes of the Keck- sponsored series is to motivate pre-college students to consider careers in science and technology.

2. Astronomy in Print

Astronomy is one of the few sciences with its own (profit-making) book club. A volume featured by book clubs will sell in the range of 40,000 copies (e.g., Herbert Friedman's *Sun and Earth*), while one of the all-time winners, Stephen W. Hawking's *A Brief History of Time*, has reached the one-million mark and spent two years on the New York Times best seller list. The 1988 New York Times list of ten best non-fiction books included three on astronomy, and the subject is similarly over-represented among the winners of the American Institute of Physics science writing award.

Sales of magazines in 1988 reveal 632,500 regular readers of Scientific American, 95,000 of Sky and Telescope, and 165,000 of Astronomy, indicating that 20-25 percent of the audience for science at this level is specifically an audience for astronomy. Within the broader-based magazines (Discover, Science Digest, Scientific American, and Science 80-86) about 7 percent of the articles over the past decade have dealt with astronomy. In contrast, professional astronomical journals make up 0.5 percent of the 3300 covered by Science Citation Index, and astronomy Ph.D.'s make up about 0.7 percent of the 18,000 awarded each year in physical biological, social, health, engineering, and computer sciences (Kidd 1989).

While few papers cover astronomy as regularly as astrology, the subject is over-represented relative to other sciences in newspapers as well as magazines. For instance, 10 years of articles in the New York Times, Wall Street Journal, Washington Post, and LA Times include 325 items on astronomy and space sciences, 360 on physics, and 280 on biology (excluding medicine; J. Cornell, Center for Astrophysics, personal communication 1990).

3. Observatories, Planetariums, and Museums

Of the 9.4 million visitors to the National Air and Space Museum in 1988, 36.2 percent (based on random sampling of departing visitors) found the astronomy and space exhibits more interesting than the aviation ones. About a third of a million visitors saw the planetarium show. On smaller scales:

All but two of the fifty states have observatories or planetariums regularly open to the public.

McDonald, Palomar, and Kitt Peak Observatories report that about 100,000 people per year travel the relatively large distances necessary to visit each of them. McDonald Observatory has been featured in the monthly Texas hotel magazine for tourists.

Griffith Observatory, near Los Angeles, more accessible than the research observatories, hosted 1.7 million people in 1989, as many as the Los Angeles Museum of Art and the John Paul Getty Museum together (879,000 and 338,800 respectively). The Adler Planetarium in Chicago records about 700,000 visitors per year. The total number of planetariums in the U.S. is about 1000.

4. Radio and Telephone Hot Lines

Stardate–daily five-minute programs produced by the University of Texas–is carried (and paid for) by about 200 radio stations, including some large ones like KNX in Los Angeles and KCBS in San Francisco. It has received a Corporation for Public Broadcasting award for excellence and attracted half a million letters from listeners over the past decade. Its spin-offs include a Spanish-language version, one minute TV news spots, and part of a CD-ROM computer commercial demonstration disk.

At least 20 astronomical telephone hot lines operate in the US. Most change about weekly and feature a mix of local observing information (moon phases, planets, and so on) and research news. A typical one, Starwatch at University of Minnesota, receives about 30 calls a day (more during Voyager encounters, Halley perihelion, etc.) and portions of its content are carried by a dozen local newspapers. Incoming students at the University sometimes mention that Starwatch was a factor in their choosing the institution and a science major.

5. Amateur Astronomy

Every state in the union has at least one active astronomy club. More than 240 dealers and manufacturers are engaged in the business of providing telescopes, accessories, and software for observers who are not professional astronomers. Some highlight activities are the following:

Telescope and magazine sales suggest that roughly 200,000 people take some interest in amateur astronomy. Of these, more than 14,000 belong to the main umbrella groups, the Astronomical League and the Western Amateur Astronomers.

The Planetary Society, whose members contribute $25 per year toward the cause of exploration of the planets and the search for extra-terrestrial intelligence, has about 130,000 members. Their publication, The Planetary Report, recently reported results of a random survey by the Public Opinion Laboratory at Northern Illinois University indicating that half or more of adult Americans support the Society's goals.

The American Association of Variable Star Observers provides a bridge between the amateur and professional communities. Of its 1100 members, about half each year provide about 250,000 observations of 3500 stars to a central data depository. Amateur observers of variable stars thus outnumber the professionals (about 100 of whom per year make use of AAVSO data). The total number of contributors over the history of the society is nearly 5000, equal to the current membership of the AAS.

Another important AAVSO contribution is providing data to educators for astronomically-based labs and science projects. Association membership data indicate that amateur astronomy participation among young people serves to recruit both future astronomers and scientists, engineers, and programmers in other disciplines.

Amateur astronomers frequently share their interests and expertise with Scout troops, school classes, and other groups of young people.

B. Contributions to the Pool of Scientifically Trained Personnel

About 70 American colleges and Universities currently offer degrees in astronomy or closely related fields, awarding about 100 Ph.D.'s per year, 160 B.A.'s and B.S.'s, and 40 terminal M.A.'s. Most of the recipients who do not pursue long-term careers in astronomy do remain part of the manpower pool in science- and technology-intensive fields.

1. Ph.D. Recipients

As indicated in the report of the Panel on Status of the Profession, about half as many astronomers leave the field each year as receive new Ph.D. degrees. Complete samples of 106 doctoral recipients (1952-88) from the California Institute of Technology and 94 (166-88) from the University of Maryland confirm this. About half are primarily engaged in astronomical research; 20 percent are employed in other sciences and in industry; 7 percent hold teaching or science administration positions; and most of the rest work on hardware or software in support of astronomical or related research.

2. B.A./B.S. recipients

Of recent astronomy bachelors, a little more than half go directly on to graduate school (a third of them in astronomy) and the others enter the work force directly (Ellis and Mulvey 1988). Complete samples from a few institutions over a longer time period confirm this pattern. The samples include Swarthmore College (28 B.A.'s, 1940-85), California Inst. of Technology (140 B.S.'s 1956-88). and Williams College (26 B.A.'s 1974- 89). 35 percent are in astronomy (research, supporting activities, or graduate school); 39 percent are engaged in other sciences or are employed in technologically intensive industries; 11 percent are teaching; and 15 percent are in non-science occupations (including law, photography, writing, and many others).

Undergraduates in astronomy are much more likely than those in most other sciences to engage in significant, publishable research; and this may contribute to the high retention rate. If so, there might be a useful example to be followed by other sciences where the ratio of Ph.D.'s to B.A./B.S. degrees is much lower, averaging about 5 percent over all the natural sciences.

III. TECHNOLOGY TRANSFER, SPIN-OFFS, AND THE PRIVATE SECTOR

Astronomy has benefited from technological advances made in many fields in science and engineering, but astronomy also contributes to technological advances in two ways. First, the demands of researchers for devices at the very edge of what is possible have sometimes been the drivers for industrial development whose products were then useful elsewhere. Photographic emulsions are a classic example. Second, ideas, algorithms, devices, processes, materials, and so forth invented within the astronomical community are from time to time modified for use in other areas: the radio astronomy technique of aperture synthesis is such a case. The first four subsections categorize items by the fields in which they are applied rather than the part of astronomy within which they originated. The concluding subsections briefly address some potential areas for future technology transfer and the support of astronomy by the private sector.

A. Medicine

The single largest problem shared by medicine and astronomy is that of imaging things you cannot get to and of reconstructing two or three dimensional structures from a number of one or two dimensional scans. Astronomers, especially radio astronomers, led the way in solving this problem. Martin Ryle's Nobel Prize cited his development of aperture synthesis, and the solution to image reconstruction pioneered by Bracewell and Riddle (1967) is now used in CAT scanners, magnetic resonance imaging, positron emission tomography, and other medical imaging methods.

Specific computer languages and ways of handling large data arrays have also proven transferable from astronomy to medicine. IDL (Interactive Data Language) and IRAF (a very flexible image processing system) are products of optical astronomy. Their medical applications include
- Study of activity and chemistry of neutron in the brain (University of Southern California).
- Cardiac angiography and PET scans (University of Michigan).
- Magnetic resonance imaging (National Inst. of Health).
- Medical imaging and product development (Mallinkrodt Institute of Radiology and Siemens Gamma-sonics).
- X-ray computer tomography (PDA Engineering).

The need for clean environments is another problem common to medicine and astronomy. A version of the positive pressure clean room designed at the University of Wisconsin for work on the OAO- 1 satellite is now in many hospitals. NASA's needs for contamination-free environments led to data bases, handbooks, and courses for clean room personnel, as well as air handlers and "bunny suits" whose commercial versions appear in hospitals and pharmaceutical labs.

A U.S. drug company has teamed up with the Cambridge (U.K.) Automatic Plate Measuring facility to use its expertise in scanning and interpreting images to analyze blood samples from leukemia patients. This permits much more rapid detection of responses to changes in medication and other pharmacologlcal effects than would otherwise be possible.

Radio astronomers have adapted their methods of measuring microwave temperature for non-invasive detection of tumors and other regions of vascular insufficiency. Microwaves have poorer angular resolution than infrared but are more sensitive to deep tissue temperatures. The combination of microwave and infrared thermographic data provides a true-positive detection rate of 96 percent, better than either alone, for breast cancer (Barrett *et al.* 1978)

Tiny paste-on thermal sensors first designed to keep ultraviolet detectors within their narrow operating temperature range have been adapted for controlling heat lamps in neonatology units.

Finally, the X-raying of people shares with X-ray astronomy the problem of having fewer photons than you would like to work with. Thus the Lixiscope (low intensity x-ray imaging scope), a portable, low-energy X-ray scanner to which NASA holds the patents, is widely used in neonatology, out-patient surgery, diagnosis of sports injuries, and third world clinics. The FDA even used it to search for poisoned capsules during the Tylenol scare a few years ago. A second generation spin-off, the Fluoroscan imaging system, has a variable power X-ray tube source among other improvements and a wider range of applications, including catheter placement.

B. Industry

The two kinds of spin-off (driving development and originating ideas) are illustrated by astronomical interactions with photography and the communications industry.

As early as 1912, C.E.K. Mees (the first research director at Eastman Kodak) initiated research leading to special series of spectroscopic plates to meet astronomical needs. The sensitizing dyes and emulsion-making techniques resulting from this work led to products of wide utility. One example is gold sensitization, which made possible Tri-X and a number of other 400-speed films from Kodak and other manufacturers. These have dominated the professional and amateur high speed film market for a number of years.

Kodak Technical Pan film, whose sharp resolution and fine grain permit enormous enlargements, is used by medical and industrial spectroscopists, industrial photographers, and serious fine-art photographers. It was first developed for solar astronomers interested in recording changes in fine scale surface structure.

Red and infrared-sensitive emulsions, evolved for spectroscopic plates, now penetrate military camouflage, and detect diseased crops and forests. Other applications include dentistry, medical diagnosis, and probing below the surface of paintings for evidence of forgery or pentimentos. Hypersensitization techniques, developed by astronomers during the 1970s, show promise in medical and industrial microscopy and in autoradiography.

Radio astronomy has been a copious source of transferable technology, algorithms, and people interested in applying them, especially in communications. Millitech, whose founders came from the University of Massachusetts radio astronomy group, now builds millimeter wavelength components based on devices used in radio astronomy, for the communications industry. Their products include varactor multipliers, voltage-tunable Gunn oscillators, and cooled GaAs Schottky mixers (Weinreb and Kerr 1983).

Radio astronomers also founded Interferometrics (Vienna, Virginia) which tests and evaluates antennas using holographic methods first reduced to practice by British radio astronomers. A holographic map of a dish surface takes a few hours (versus several days for a mechanical survey) to reveal high and low spots that must be corrected before (for instance) sidebands are low enough to meet FCC standards for satellite communication links.

High density recording techniques have come from both NRAO and Haystack Observatory. Digi Data of Maryland is marketing several versions of the NRAO version (which achieves 2.5 Gbyte capacity and 120 kbyte/sec data rate by storing digital data in analog form) for archiving of business data, disk backup, and other applications. The Haystack technique uses a 36 channel, high accuracy narrow-track headstack that can be moved precisely across tape to increase the density of the recorded information by a factor of more than 12, so that a single reel accommodates nearly 6 terabits and can record at a rate in excess of 1 Gbit/sec. Honeywell of Denver is now producing these high- density headstacks as a standard component.

Radio astronomers have been both drivers and developers of low noise amplifiers, including cryogenically-cooled gallium- arsenide field effect transistors (now marketed by Berkshire Technologies, also founded by radio astronomers) and high electron mobility transistors, which may replace masers in some communications amplifiers.

The computer control language FORTH was invented by a professional programmer with a strong interest in astronomy and first applied by him to coordinate telescope operation, data acquisition, and initial reduction for the NRAO 36-foot dish at Kitt Peak. It has grown into a profitable company (Forth, Inc., Manhattan Beach, California) and been modified for a wide range of purposes in manufacturing and service industries. About 20 vendors supply Forth systems for hardware from handheld computers to VAX mainframes. Some computers (most recently the Harris RTX 2000 microprocessor) execute Forth directly. The system is currently used in a rule based ("expert system") automobile engine analyzer at over 20,000 service stations world wide and in a high-accuracy densitometer used by Kodak for quality control in film manufacture. The initial support from NRAO and wide diffusion of Forth through the astronomical community were instrumental in its development into a broadly-applicable system.

Other examples of fruitful technology transfer from astronomy include:

- Use of AIPS (a set of image processing programs from radio astronomy) by Boeing to test computer hardware (several vendors, including Convex and International Imaging Systems advertise that their systems support AIPS).
- General Motors' application of IDL to analyzing data on car crashes.
- Acquisition of the patents for the first gravitational radiation detectors by Hughes Research Laboratory for use in modified form to sense gravity anomalies associated with underground oil pools.
- Use of the IRAF image processing program at AT&T for solid state physics graphics and computer systems analysis.
- Cold spot welding techniques that do not distort the underlying metal, developed at University of Wisconsin during construction of OAO-l.

C. Defense

The common technological needs of astronomical observations and of certain defense programs have often resulted in one research community developing techniques or making observations useful to the other. For example, satellite and aerial surveillance have replaced many ground-based intelligence activities. The resulting increased certainty (on both sides) that accurate information will be available has contributed to recent progress in arms reduction. Surveillance requires telescopes with large accurate mirrors, precision optics, and the ability to process numerous imperfect images and extract the maximum possible amount of information. The necessary large mirror technology, adaptive optics, and processing algorithms have all had significant input from techniques developed within astronomy and by people trained as astronomers, from the time of the U2 cameras to the present. Some specific examples which extend across the electromagnetic spectrum:

- A recent investigation at Grumman on recognizing rocket plumes for strategic warning purposes made use both of observations of stars and of model stellar atmospheres to discriminate plumes from cosmic objects.
- Aperture synthesis radar is the remote descendent of the radio astronomy technique for which Martin Ryle won the Nobel Prize.
- Development of the channeltron was supported originally for ultraviolet astronomy, but it has since found its way into various uv military cameras.
- Expertise developed in conjunction with the Kuiper Airborne Observatory has provided direct support to several Navy and Air Force airborne infrared sensor development programs.
- Star counts and models of stellar spatial distribution are used to assess data rates for spaceborne signal processors and sensors as well as for satellite pointing and calibration.

- Astronomers who had been working on X- and gamma-ray detectors at Los Alamos helped build the instruments for the Vela satellite monitors.
- Solar blind photon counters were invented for uv astronomy and later adapted to sensing the uv corona round supersonic objects in daylight and for toxic gas detection.
- The Air Force Weapons Laboratory at Albuquerque has issued a number of contracts to astronomers to investigate topics like optical imaging of satellites in geosynchronous orbits using 10-30 meter baseline optical interferometry.
- The infrared maps of the sky obtained by IRAS met DoD needs for information that the Air Force Geophysics Lab rocket program had been unable to provide.
- The techniques being developed by Itek, LBL, and others for stress polishing of off-axis mirror segments for the Keck telescope have potential defense uses.

The early development of thermonuclear weapons made extensive use of astrophysical knowledge of radiative transfer and temperature/density diagnostics. At the present time, 69 American Astronomical Society (AAS) members are employed at Los Alamos and Sandia National Laboratories and another 32 at Lawrence Livermore National Laboratory, most of them at least partly on programmatic work. A background in astrophysics appears to provide flexibility and skills in carrying out approximate calculations based on integrating information from a variety of sources that are a good match to defense laboratory needs.

The presence of Soviet reactors in space has apparently been known to DoD for some time, but astronomical gamma-ray detectors on the Solar Maximum Mission and on a University of California, Riverside balloon-borne experiment made independent discoveries of the phenomenon (Rieger *et al.* 1989; O'Neill *et al.* 1989).

Looking ahead, the Navy is supporting neutrino astronomy for its long-term potential for communicating through the earth and for long distances under water. Solving the engineering problems associated with DUMAND (Deep Underwater Muon And Neutrino Detector) should lead to valuable new oceanographic technology as well. Grazing-incidence X-ray optical devices, which have been reduced to practice for solar astronomy, are likely to find future applications in laser weapons.

Another area where astronomical and defense interests overlap is in the need for precise coordinate systems, times, and time intervals, for use in navigation, clock synchronization, guidance, and secure communications as well as in astrophysics. The fundamental time standards are now atomic clocks, not the earth's rotation, but the determination and dissemination of time data for the U.S. are still the responsibility of the U.S. Naval Observatory (USNO). Accurate measurements of the earth's rotation rate are needed to keep civil time in step with astronomical time. This must be done for navigational and other purposes and is accomplished by a network of radio and optical observing stations, maintained by USNO and observatories of many other nations. Very Long Baseline Interferometry between widely separated radio telescopes was the original driver to turn hydrogen maser clocks into rugged, off-the-shelf items, whose main users are now space communications and DoD. In addition, VLBI methods are currently used to synchronize widely separated clocks at the nanosecond level.

The fundamental celestial coordinate system used for navigation is now a radio based one. The locations of the artificial satellites which make up the Global Positioning System and which transmit their own radio signals are in the process of being tied to the positions of quasars and other distant sources. Inertial guidance systems (for missiles and other purposes) require this accurate astronomical coordinate system for their calibration. Accurate optical star positions are used in surveying and in automated star-tracker guidance systems. The tying together of accurate radio and optical coordinate systems is a topic of current intense study. Finally, because satellite orbits are blind to the assorted wobbles of the earth beneath, correct location of terrestrial targets (for environmental and surveillance imaging as well as bombing) requires accurate forecasts of earth orientation. USNO is also responsible for providing and disseminating this information, which comes largely from VLBI observations of quasars, in the U.S.

D. Energy and the Environment

The search for fossil fuels and alternative energy sources has benefited from astronomical spin-offs in several contexts. For instance,

- Texaco, Inc. and BP America both use the image processing program IDL for analysis of drilling core samples and other aspects of petroleum research.
- SAIC (San Diego) has built solar radiation collectors up to 16 meters in diameter using graphite composite materials first developed in design studies for a proposed orbiting telescope called the LDR (Large Deployable Array).
- Grazing incidence X-ray optics was reduced to practice for solar astronomy and now finds application in plasma diagnostics for magnetically confined plasma fusion. Detailed knowledge of atomic spectra at high temperatures, gained from study of the solar corona, is also important in this context.
- Plasma and magneto-hydrodynamic phenomena, including magnetic reconnection and radiation-driven thermal instabilities, were first explored in solar and space physics environments. They also occur in fusion plasmas (and are deleterious there).

Remote sensing from orbiting satellites is now the method of choice for keeping track of an enormous range of ecologically important factors—the extent of the Arctic ice pack; the moisture content of soil in the Sahel; upper atmosphere profiles of temperature, density, and trace constituents; sea surface temperatures; and many others. Astronomically-derived image processing algorithms are widely used in these applications. Several of these are mentioned elsewhere. Another with many remote sensing and oceanographic uses is a digital correlation technique for spectral analysis of broadband signals which came out of radio astronomy (Weinreb 1963; Cooper 1976)

Specific radio, microwave, and infrared spectroscopic methods from astronomy have also proven useful in environmental applications from space and ground. Downward looking millimeter wave sounding traces back to work on the atmospheres of Venus and Mars and was validated for the earth by radio astronomers using balloon borne telescopes. The technique is operational on the current Defense Meteorological Satellite Program (DMSP) and will be the primary temperature sensor on the next generation of NOAA satellites in the 1990s.

Millimeter wave technology in space (e.g., Staelin 1981) is sensitive to composition as well as temperature of the atmosphere, including greenhouse gases in low concentrations. Microwave sounders, scheduled for the ATLAS series of spacelab experiments and for the Earth Observation Satellites, were developed by a consortium of American and European radio astronomers and atmospheric scientists.

A particularly timely application of microwave astronomy techniques from the ground is study of chlorine chemistry (relevant to ozone depletion) in the Antarctic. In September 1986, instruments developed by radio astronomers at SUNY, Stony Brook found a hundred times the normal concentration of chlorine oxide at an altitude of 15-20 km in the Antarctic ozone hole. The excess disappeared in October, verifying the role of manmade chlorine compounds in ozone depletion. The detailed chemistry had earlier been tested by the group's measurements of the diurnal variation of chlorine oxide in the middle stratosphere above Mauna Kea. The Antarctic spring cycle of chlorine oxide rise and fall was followed through the 1987 season with better instrumentation yielding the full concentration profile from 16 to 40 km. Monitoring at about five sites around the world over the next 10-20 years is planned as part of the NASA-sponsored Network for the Detection of Stratospheric Change.

E. Everyday Life

Many of us benefit regularly from the machinery used to X-ray luggage in airports, whose design descends from that of the earth rocket and satellite borne X-ray telescopes. Airport surveillance for drugs and explosives makes use of a particular gas chromatograph design supported by NASA for use on Mars. Some other mundane spin-offs from ground and space-based astronomy include:

- A hand-held COD photometer developed by astronomers at University of Hawaii for use by policemen checking the transparency of automobile windshields
- A non-invasive probe for contaminants likely to cause structural weakening in historic buildings; it has a neutron source and gamma-ray spectrometer, was first used to analyze lunar soil, and has been tried by astronomers at GSFC in a Colonial Williamsburg smoke house and at St. Mark's Basilica in Venice to look inside the walls behind fragile mosaics.
- Software to process two-dimensional images on a personal computer, developed by Michael Norman at the National Center for Supercomputing (Illinois) for his own astronomical purposes and modified for public consumption; about 10,000 copies have been sold.

- Use of Forth in the hand-held computers carried by the 40,000 delivery agents of one of the major express mail firms.
- Application to industrial and amateur photography of enhancement techniques developed by David Malin for handling astronomical images from large telescopes (Malin 1982, 1990).

F. Looking Ahead

Technology transfer is an ongoing process. For instance, observers are currently driving COD technology (as they did photography earlier) in the direction of thinning the chips to broaden the range of wavelengths over which they are sensitive. And astronomers are pushing for cryogenic infrared array detectors with very low backgrounds and long integration times, so that they can be used at low light levels. These technologies are likely to prove useful for non-astronomical purposes.

X-ray astronomers have been responsible for the development of bolometers and superconducting devices as non-dispersive spectrometers. The entire energy of the of the absorbed X-ray is transformed into an electrical signal via phonons, producing a much larger response for a given X-ray energy than in photoelectric detectors. These have potential applications in non-destructive testing and in medicine, where getting the largest possible signal out of the fewest possible X-rays is also important.

Many radio astronomy observatories with millimeter-wave antennas are currently developing SIS (superconducting- insulating-superconducting) mixers for low noise receivers. NRAO is among these and has begun technology exchange with several commercial and government organizations (Hypress, NRL, the National Security Agency, etc.) who are interested in non-astronomical applications. Millimeter- wave astronomers are also working on error- correcting secondary mirrors and lenses. Such error-correcting optics is likely to be part of high-performance communication, surveillance, and other non-astronomical antennas and telescopes of the future.

G. Astronomy and the Private Sector

No other branch of science, except medicine, has had as much support as astronomy from private individuals, industrial firms, and foundations. Two of our great observatories, Lick and McDonald, bear the names of the men whose bequests founded them. Both are now largely maintained by state and local, not federal, funding. Contributions from Rockefeller and Carnegie and the foundations they established have built and helped maintain the Yerkes, Mt. Palomar, Mt. Wilson, and Las Companas Observatories. More recently, Oscar Meyer provided some much-needed new buildings for Palomar. And money from the Keck Foundation is even now being transformed into a ten-meter telescope that will be the largest American optical observing facility for the next generation.

The motive for this generosity (apart from tax laws) appears to have been the breadth of vision needed to span a nation with railroads or to build up a steel industry appreciating the breadth of vision needed to span the Universe and build an understanding of it. Other interactions have been of more obvious mutual benefit. Kodak has donated the several thousand 14"x14" photographic plates needed for the second Palomar Observatory Sky Survey because this use with long exposure times and low light levels provides a critical test of their emulsions. A recent document from the American Institute of Aeronautics and Astronautics (1989) encourages federal support of the Hubble Space Telescope and similar projects because "such cutting edge technology programs stimulate commercial spin-offs of potentially great value to industry and to the nation's economy."

The process of compiling this report revealed that people whose livelihoods in no way depend upon astronomy can nevertheless feel that it is an essential activity. Whenever the AASC received a bit of publicity, they wrote, phoned, and sent photocopies emphasizing that astronomy is needed to attract students into science and technology, to inspire long-range advances (e.g., neutrino communication), and to form part of the human intellectual adventure–much the same points the Panel has identified.

IV. MAN'S PLACE IN THE UNIVERSE

As far back as history records, peoples have attempted to understand how the world got to be the way it is, what the big picture is, and how we fit into it. Anthropologists call the answers (even answers they believe) creation myths. Our modern Western myths have a long history, with input from Greek philosophy, from Judeo-Christian religious ideas, and, at several critical points, from astronomical research.

The Copernican revolution was the most obvious and far reaching of these. The earth ceased to be the unique center of everything and declined to merely one of several planets orbiting the sun. With further celestial study, our sun, in turn, metamorphosed into a typical, undistinguished star, not even at the center of the Galaxy. A third of the way into the 20th century, our Milky Way Galaxy itself had shrunk to a status neither special nor central to anything. In fact, cosmic models incorporating general relativity show that all places in the Universe are equivalent, there being neither any center nor any edges. And in just the last few years, a picture of the very early Universe motivated by theory on the frontier between cosmology and particle physics has made it seem plausible that the Universe–the entire four-dimensional space-time with which we might ever communicate–is only one of many universes, dictionary definitions notwithstanding.

Curiously, other recent astronomical research has pushed our thinking back a little bit in the other direction. The life- bearing earth really is very different from the other nearby planets. Looking down at it from space, we can see our home as a single, small, fragile entity, whose residents all have a common, profound interest in its well being.

Other 19th and 20th century discoveries clarify other aspects of our relationship to the rest of the Universe. The spectra of the sun and stars show absorption and emission lines in just the same patterns that are radiated by common chemical elements when you heat them in the laboratory. Thus celestial objects do not consist of some "quintessence" or substance unique to them. They are made of the same stuff that we are, and even in more or less the same proportions. Apart from helium (which forms no stable compounds), the commonest elements in the stars are the hydrogen, oxygen, and carbon that make up most of our bodies. Close study of spectra of distant galaxies and quasars reveals not only this commonalty of composition but also that the constants and laws of physics are the same at distant times and places as they are here and now on earth.

The totality of modern astronomy makes up a major part of our Western creation myth, answering many of the traditional questions about how big, how old, and what came before. The world, or Universe, is large. It is the same in all directions (on large enough scales). It expands and is only three or four times older than the earth itself (but the earth is some 100 million times older than the span of an average human life). We are made of starstuff–chemical elements built up from hydrogen atoms by nuclear reactions in massive stars. And chemical reactions in interstellar gas and in the material that formed the meteorites and comets have produced the same molecules that are the building blocks of living creatures on earth.

The task of clarifying our relationship to the rest of the Universe is an on-going one, with many important questions still incompletely answered. It is, for instance, just becoming meaningful to ask whether the Universe could have been very different from what it is (in size, age, laws of physics, kinds of particles, and so forth) and whether such a different Universe could have life arise in it. On smaller scales, detailed studies of Mars and Venus will play an important part in understanding the early evolution of the earth's atmosphere, oceans, and biosphere, and in determining just how delicate the present state of terrestrial habitability is likely to be.

The "where do we belong" aspect of astronomy seems to be responsible for most of the popular interest in the subject. The potential for technology transfer and for attracting students into the sciences may be good (though not central) reasons for funding astronomical research. But they are not the reasons that people watch Cosmos, buy and build small telescopes, or read books and articles about astronomy. Rather, these people are seeking new answers to the old human questions about the world and our place in it.

An important property of the modern creation myth is that its answers are neither static nor given by fiat. Everything (or nearly everything) within the sciences is subject to change without notice. Our picture of the Universe expands and evolves as our knowledge expands and evolves. A vigorous continuation of this process can help to keep human minds flexible enough to deal with immediate practical problems that now also change on timescales much less than a human lifespan. Practicing astronomers feel a great deal of certainty that, although any given piece of information may turn out to be wrong, the basic process of inquiry is sound and leads to continuously better understanding of the world around us. Confidence that the Universe is neither incomprehensible nor intrinsically hostile is perhaps the most important return astronomers can offer to their fellow citizens.

Astronomy as part of our world view has a less serious side as well. For instance, the phrase "black hole" seems to have entered the everyday vocabularies of Rep. Bill Frenzel (R-Minn.), New York Yankee Don Mattingly, and Supreme Court Justice Sandra Day O'Connor in various contexts (Montgomery Journal, April 26, 1990, p.2). "Astronomical" distances and amounts of money and "zeniths" and "nadirs" of achievement and despair are also common phrases.

Modern astrophysics has not inspired any artistic works comparable with Dante's treatment of medieval cosmology's circles of heaven and hell, but Van Gogh's "Starry Night" reveals a mind not unmoved by the Universe as it is now understood. And each new astronomical–discovery novae, supernovae, neutron stars, black holes, multiple Universes, and many others–has inspired science fiction films, stories, and novels tying these discoveries to possible individual lives.

V. INTERNATIONAL COMPETITION AND COOPERATION IN ASTRONOMY

Science and technology are normally perceived as factors in international competition, both military and economic. That aspect is by no means absent in astronomy. Most of us are really rather proud of the long list of American "firsts" and "bests" (the Apollo Program and Viking Landers; large optical telescopes; Uhuru and COBE; the VLA and the VLBA; and so forth) and pleased by the leading role that the U.S. has played in the International Ultraviolet Explorer (IUE), the Infrared Astronomy Satellite (IRAS), the Hubble Space Telescope, and the establishment of intercontinental networks of VLBI stations.

We believe that it is important for the U.S. to continue to take, and be seen to take, a position of leadership in astronomy, astrophysics, and space exploration. Benefits of remaining at the forefront in research include a strong positive image in the eyes of the nations we interact with, the potential for future spin-offs, and opportunities for fruitful international collaborations. The ability of the U.S. to continue to attract outstanding students and young researchers from abroad is also vital for the continued health of science and engineering here. About one-quarter of the astronomical research community in both senior and entry-level positions is foreign born (Trimble 1988). Among graduate students in engineering and physics, the proportion is roughly one half. Because the results of astronomical research receive a good deal of media attention, leadership in this field can contribute disproportionately to a positive American image abroad.

The world is, however, entering an era in which international cooperation will replace competition, at least so we all devoutly hope and possibly even rationally expect. Among the sciences, astronomy has an unusually long and rich history of internationalism dating back to before the 19th century. One important driver for collaborations is the absolute necessity of observatories spaced over the surface of the earth necessary to see the whole sky. In addition, the perceived impracticality of astrophysical research has probably helped it to be seen as a safe area for contact when there were not many others. Beginning in 1887, two American observatories were among the 18 world wide that banded together to produce a map of the entire sky (Carte du Ciel). The International Astronomical Union was the first of the modern international scientific unions organized, in 1920, under the Treaty of Versailles, with the U.S. as one of the founders.

A large fraction of new observing facilities, spacecraft, and research programs are international collaborations. The European Southern Observatory and the Canada-France-Hawaii telescope are self- explanatory. The latter peacefully shares the top of Mauna Kea with a British infrared telescope and several American projects. Construction of a Japanese observatory there is expected to begin in the next decade. Other recent success stories include the following:

- The sharing of the International Ultraviolet Explorer observing time between the European Space Agency and NASA (in ratio 1:2) over the past 13 years.
- An American (University of California–Berkeley) spectrometer launched by a Japanese rocket to study the cosmic microwave background.
- European-built instruments on IRAS and HST, with proportionate sharing of the observing time.
- An American instrument on the Soviet Vega 1 and 2 spacecraft that flew past comet Halley. Soviet scientists participated in the Voyager 2 encounter with Neptune.
- The 20 percent of the papers published in the Astrophysical Journal in 1990 that had authors headquartered in the US and at least one other country (Abt 1990).

Scientists and administrators in the United States and the Soviet Union have even begun exploring the idea that the first manned mission to Mars should be a joint one.

On a more personal level, most American astrophysicists count at least a few foreign colleagues, often including some from ideologically very diverse countries, among their closest friends, though they may not meet more often than the triennial General Assemblies of the International Astronomical Union. Perhaps contemplating very large scale phenomena makes small scale differences a little easier to overlook. Because astronomy and the defense industry already share some hardware, techniques, and personnel, gradual conversion of additional resources from competition in the latter to cooperation in the former may not be an unrealistic goal.

VI. SYNERGISM WITH OTHER SCIENCES

A. Historical Examples

Back when science was called natural philosophy, its practitioners were supposed to know a good deal about all of it. Thus it is no surprise that observations of planetary motions played a central role, through Newton's work, both in the elucidation of the properties of gravitation and mechanics and in the formulation of the calculus. Astronomy and mathematics continued to be closely linked into the 19th century. Gauss, for instance, invented the least squares method while struggling to describe discordant observations of the first asteroid (Ceres, found in 1801) by a single best orbit. This led to his subsequent investigation of the normal error distribution, crucial to a variety of statistical studies.

Later in the 19th century, astronomy came to be more closely linked to chemistry, through spectroscopy. Helium, the second most abundant element, first revealed itself as the source of an emission line in the solar spectrum recorded by Janssen at the eclipse of 1868. Later on, lines from high ionization and excitation states of elements like europium were often seen in stars before they had been studied in the lab.

In the 20th century, the closest links have been with physics. Cosmic rays were the main source of very high energy particles until about 1950, and the positron, muon, and pi meson were first discovered among cosmic ray secondaries. Eddington, contemplating the problem of stellar longevity in the 1920s, was the first to appreciate nuclear fusion as a potential source of enormous quantities of energy. The correct reaction chains describing these stellar processes were identified in 1939, just before nuclear physics disappeared into weapons laboratories all over the world. Hans Bethe and others who made fundamental contributions toward nuclear and thermonuclear weapon theory had cut their scientific teeth on the stellar problem.

In our own time, science has expanded far beyond any one person's powers of comprehension, and most scientists are, at best, intelligent laymen anywhere outside their own subdisciplines. Surprisingly, fruitful interactions continue to occur among sciences, often to the pleased surprise of those involved when they recognize, for instance, that statistical techniques invented to describe the survival of medical patients can be useful in interpreting the celestial X-ray background. The following sections present cases where astronomical data, concepts, and methods have been applied in other branches of science, including physics, chemistry, and earth and environmental sciences.

Many of these symbioses occur because the Universe is a laboratory far grander than any we can build. It accelerates particles to energies beyond those projected for the Superconducting Supercollider (though unfortunately not very many in any one place at the present time). Cosmic objects demonstrate the effects of stronger magnetic fields, emptier vacuums, and louder sonic booms (shock waves) that are beyond the reach of terrestrial phenomena. Thus a critical test of our understanding of a local phenomenon is sometimes whether the same model works when applied to the more extreme astronomical case.

B. Nuclear and Particle Physics

In recent years, astrophysics has pushed experimental and theoretical nuclear physics to their limits. Modeling nucleosynthesis, supernovae, neutron stars and the rest requires (1) masses, lifetimes, and cross-sections for highly unstable nuclei, (2) the equation of state of very dense matter at both zero and finite temperatures (up to densities where pions, hyperons, or strange matter may replace neutrons as the dominant particles), (3) cross-sections for common reactions that, for example, convert carbon into oxygen at such low energies that the laboratory reaction rate defies measurement, and (4) the lifetime of the neutron to unprecedented accuracy.

Topics in particle physics to which astrophysics has recently made contributions include the number of groups of fundamental particles that exist, properties of the neutrino, and the existence and nature of dark matter. Laboratory experiments between 1950 and 1980 had gradually demonstrated the existence of three groups of particles (associated with electrons, muons, and tau leptons) and seemed likely to continue to reveal additional groups. But the calculated abundance of helium produced in the early Universe (in comparison with the amount observed in relatively unprocessed astrophysical plasmas) was only marginally consistent with even a fourth group and clearly ruled out a fifth (Steigman 1989). Measurements from terrestrial particle accelerators (Abrams *et al.* 1989) now strongly suggest that there are only three groups.

A table of neutrino properties consists largely of upper limits. The tightest limits on charge and magnetic moment now come from analysis of the neutrino burst from supernova 1987A. The supernova limits on electron neutrino rest mass and on oscillations among neutrino types are comparable with the laboratory limits but are likely to be easier to improve.

Non-luminous material–dark matter–makes up 90 percent or more of the gravitating mass of the Universe as revealed by rotation of individual galaxies and motions of galaxies within clusters. The fraction rises to 99 percent if the Universe has, as many physicists suspect, precisely the critical density that separates ever- expanding from recontracting future dynamics. Particle physics beyond the standard model has provided a wealth of potential candidates for dark matter in the form of weakly interacting massive particles, topological defects, and many other strange entities.

None of the candidates has yet been unequivocally seen in the lab (though searches are underway) and the best constraints on their properties come from a variety of astrophysical processes: (a) they must not disturb energy transport in stars or the magnetic fields of pulsars and galaxies; (b) they must not distort nuclear reactions in the early Universe so as to produce the wrong proportions of hydrogen, helium, and lithium; (c) they must not have interfered with neutrinos and light from SN 1987A reaching us a few hours apart and in the expected strength; (d) they must not produce more gravitational lensing or more gravitational radiation than observations of quasars and pulsars tell us can exist; and (e) their behavior during the epoch of galaxy formation must promote, or at least not prevent, the formation of the structures we see.

This is currently a very rapidly evolving topic with many controversies, but if the dominant constituent in the Universe is not ordinary baryonic material (protons, neutrons, and the atoms made from them), then the astronomical observations probably have at least as good a chance as the laboratory tests of telling us what the stuff is.

The solar neutrino problem, on the other hand, has been with us in nearly its present form for twenty years. The three-fold deficiency of observed solar electron neutrinos compared to the best theoretical predictions may turn out to have been the first hint we had of fundamental new physics linking the three particle groups of the standard model. The right answer could also turn out to be some even more exotic new physics not yet thought of, or some fairly uninteresting erroneous detail in stellar structure modeling.

C. Physics of Fluids and Plasmas

Astronomical objects present a much wider range of scales and properties than can be achieved in the laboratory. Pulsar magnetic fields are a million times stronger, solar convection Reynolds numbers a million times larger, and intergalactic vacuums more than a million times emptier than the most extreme terrestrial conditions. But the underlying physics is the same. Hence astronomical phenomena can provide valuable constraints on our understanding of terrestrial physics, for instance,

- Pulsar radiation mechanisms for microwave generating devices
- Solar flares for magnetic reconnection in Tokomak plasmas
- Astronomical cooling functions, Stark coefficients, and oscillator strengths for controlled and uncontrolled thermonuclear fusion studies.
- Instabilities in quasar jets for confinement of laboratory plasmas.

Some theoretical methods have also proven transferable. Smoothed particle hydrodynamics, now the most widely used method of modeling complex three dimensional fluid flows, was developed by a Columbia University astrophysicist studying binary stars.

D. Chemistry, Spectroscopy, and Atomic Physics

The first comprehensive table of spectral lines reveals its origins in its title, A Table of Multiplets of Astrophysical Interest (by Charlotte E. Moore) and its initial publication as Contribution No. 20 of the Princeton University Observatory. The current edition retains the astrophysical title, but is used in industrial analytic spectroscopy and a variety of other non- astronomical contexts. The standard table of ultraviolet lines (Kelley 1987) similarly incorporates astrophysical data.

Microwave and infrared observations of astronomical objects have revealed molecules not previously known on earth, including HC_nN (with n up to at least 11), C_3H_2 (whose structure was inferred from the astronomical data, permitting later laboratory study), and HCO^+ in interstellar gas; C_3NH_2, and H_2O^+ in comets; and SiC and SiC_2 in stellar shells. Some of these are actually not possible in lab settings, where densities are always high enough for frequent collisions. HCO^+ was the first molecular ion identified anywhere. Study of these is now a large, productive branch of chemistry.

Meteorites also contain some familiar substances, including tiny grains of diamond and silicon carbide. These cannot have been made by the high pressure processes operating in Kimberlite deep underground. They must have arisen from something analogous to vapor deposition. This is now being studied by industrial materials scientists in hopes of greatly reducing production costs of these useful substances.

On the theoretical front, the attempt to explain intensities of spectral lines from interstellar molecules first made clear the need to consider departures from thermal equilibrium and the populations of individual energy levels when calculating reaction rates. And the process of dielectronic recombination, now known to be important in setting the ionization structure in Tokomak plasmas, was first recognized as occurring in the solar corona.

E. Geophysics

Very Long Baseline Interferometry, as well as establishing precise coordinates of the radio sources observed, tells you the distances between the antenna pairs with a precision of about a centimeter for baselines up to several thousand kilometers. These are of geophysical interest where baselines span tectonic plate boundaries and known faults. Measurements across boundaries have confirmed that continental plates are currently moving at the average rates implied for the past few million years by seafloor data (typically 1-5 cm/yr).

Of more immediate importance to society are the measurements around active fault zones. These confirm slippage and deformation along the San Andreas and other active faults, again typically a few cm/yr when and where plate motion is occurring smoothly, and little or no motion when it is not. The resulting built-up stresses are released as earthquakes. For instance, measurements made before and after the November 1989 Loma Prieta quake showed that about 5 cm slippage had occurred during the quake in the Santa Cruz area, but very little around San Francisco, where there had been smooth motion at about 5 cm/yr before the quake.

Post-quake measurements at Loma Prieta and after two large events in the Gulf of Alaska show that significant sudden displacements occur as far as 40-200 km from the epicenters. These measurements are crucial to understanding the nature of the energy release in earthquakes, which in turn may provide clues that will allow us to predict them.

F. Environmental Sciences

If one or more of the major episodes of extinction of terrestrial species resulted from asteroids or comets hitting the earth, this is surely the most spectacular of the synergisms. Astronomical aspects of testing this model for the extinction of the dinosaurs (etc.) include (1) inventorying the numbers and orbits of potentially earth-impacting asteroids, (2) comparing the history of impact cratering on the earth with that on other planets, and (3) looking for things that might cause sudden drastic changes in orbits of objects in the outer solar system and thereby increase the numbers of earth-crossing comets and asteroids. A distant companion star and periodic crossings of the galactic plane have both been suggested.

The hypothetical phenomenon called nuclear winter (Turco *et al.* 1983) is closely related to this interpretation of extinction episodes in the Earth's past. The idea is that all-out nuclear war would place so much dust and smoke in the earth's atmosphere that the resultant cooling and inhibition of photosynthesis might actually be the most serious effects of the nuclear explosions. The best way of deciding whether extensive nuclear explosions will cool the earth, heat it, or do something entirely different is undoubtedly through modeling, tested by its ability to explain the historical record on earth and the other terrestrial planets.

Another vital area of interaction is solar-terrestrial relations. Two rather different kinds of phenomena are involved. First are well-established, short-term effects (mostly unpleasant) of sudden increases in the ultraviolet, x-ray, and particle radiation coming from the sun. The primary goal here is to understand how and why these increases occur well enough to predict them. Second are tantalizing hints of longer-term, more subtle effects on terrestrial weather and climate associated with small changes in the visible light output of the sun and other solar cycle effects.

On the short term, the 11 year cycle of solar activity (spots, flares, and other surface manifestations of changing interior magnetic fields) is just now coming down from a peak as high as any reached in the past 300 years. 1989 saw three flares of strengths that generally happen only about once a decade. The high energy radiation from such flares hits the earth's upper atmosphere, heating and ionizing it, and also modifying terrestrial electric and magnetic field structures right down to the ground.

Documented effects of the 1989 March and September flares include (1) a nine-hour power outage affecting six million customers in Quebec, Canada, resulting from currents induced in power lines and transmitters by the magnetic changes, (2) a billowing up of the heated atmosphere high enough that drag on assorted satellites and space debris changed so many orbits that the Air Force tracking system responsible for them suffered an overload, (3) changes in atmospheric ionization that disrupted radio communication at most frequencies, and caused a U.S. military alert when over-the-horizon radar was lost, (4) deviations of magnetic compasses of up to 10x from normal magnetic north, and (5) an increase in the amount of energetic radiation reaching aircraft altitudes to the point where a Concorde pilot would have exceeded his yearly permitted dose in 35 hours of flying; women in weeks 8-15 of pregnancy may also exceed radiation guidelines when flying under peak flare conditions (New York Times, Wednesday, 14 February 1990). Even short term warning of flare activity are useful against some of these effects.

Adverse effects of solar activity are likely to become more serious as power grids become more extended; satellites become increasingly vital for communication, navigation, and weather forecasting; and humans venture further outside the earth's protective atmosphere and magnetosphere. Clearly the main astronomical problem to be solved is that of accurate predictions of solar events. Already, seeing the beginning of a flare through optical monitoring permits a few hours' warning of the main particle impact, allowing astronauts and power companies to hunker down. Over days to weeks, we can forecast at least the likelihood of terrestrially significant activity, because active areas on the solar surface live for several solar rotation periods and will point at us once per period. Prediction through and between 11 year cycles requires a level of understanding we do not yet begin to have. A promising avenue for acquiring it seems to be a comparative study of the sun with the many other stars known to have 5-20 year cycles, correlating the patterns of activity to rotation periods, magnetic field strengths, and atmospheric structures of the stars. Astronomers are actively working on this.

On longer time scales. there are tantalizing hints of sun- driven climate changes. Measurements during the most recent solar cycles (for instance from the Solar Maximum Mission satellite) show that the total brightness of the sun changes about 0.1 percent between maximum and minimum activity. Most models say that this should cause a global temperature change of about 0.1 ° C. It may or may not be significant (1) that the integrated solar activity, smoothed over several cycles, has increased roughly in phase with an apparent 100-year warming trend, and (2) that the Little Ice Age in the 1600s coincided with a time of exceptionally low solar activity. In addition to temperature changes cause by solar brightness changes, the climate might, in principle, be responding to changes in solar ultraviolet production of ozone, to stratospheric heating and ionization by solar energetic particles and radiation, and to extra-solar cosmic rays, whose penetration to the earth varies inversely with solar activity. We need to know more about both activity cycles and associated changes in solar brightness to decide how important these effects are. Again, detailed comparison of the sun with other cyclicly active stars is a promising line of attack.

G. Looking Ahead

Man-made changes in the earth's atmosphere may be coming perilously close to turning some aspects of solar-terrestrial relations into an experimental subject. Fortunately, nature has provided informative laboratories for studying the results of higher and lower solar illumination in the forms of Venus and Mars.

Some of the models now used at Goddard Institute for Space Studies to study large scale weather patterns on earth had their roots in work by James Hansen on the atmosphere of Venus. That planet, owing to its high albedo, actually absorbs less solar energy per unit area now than does the earth. Its high temperature is, therefore, due entirely to greenhouse warming.

The "runaway greenhouse" and "runaway ice age" characteristics of Venus and Mars set limits to the tolerance levels for habitable planets. The implications of current studies is that these tolerances are rather narrow—if not in comparison with the changes we are likely to induce on the earth soon, at any rate in comparison with the effects of the brightening of the sun that will occur over the next few billion years.

Astronomy will soon be in a position to provide auxiliary evidence of global warming, whether naturally or artificially induced, if it occurs. The ability to measure vertical motions and height changes of VLBI sites is still improving. These vertical measurements can be used to studying small changes in sea level which occur with warming (due both to melting of ice and to expansion of warming sea water). Sea level changes as small as a few millimeters are of interest and will soon be measurable.

Despite the slightly pessimistic implications of Mars and Venus, astronomers have been in the vanguard urging serious searches for life elsewhere in the Universe and attempts to communicate with it. Only too clearly, no one knows for sure how to do this or even whether there is anything to look for. But the most promising methods so far devised for SETI (Search for Extra-Terrestrial Intelligence) use the telescopes, receivers, amplifiers, and correlators of radio astronomy. The discovery of life elsewhere in the Universe could well be the most exciting scientific event of the millennium in which it occurs; and it has a good chance of being an astronomical discovery.

VII. ACKNOWLEDGEMENTS

The following colleagues, who are not members of the panel, have generously shared their ideas and expertise with us: Thomas F. Adams, Laura P. Bautz, Roger Bell, Gregory Benford, Leo Blitz, Matt Bobrowsky, Ralph Bohlin, Phil Bowers, Peter Boyce, Robert A. Brown, Robert Brownlee, Claude Canizares, Stirling Colgate, Edward K. Conklin, James Cornell, Tim Cornwell, Art Cox, Carol Crannell, David Cudaback, Grace Deming, Helene Dickel, Ann Druyan, Harland Epps, Eric Feigelson, Paul Feldman, John Gaustad, Mark Gordon, Martin Harwit, Robert C. Haymes, David Helfand, Ernest Hildner, James Houck, John Huchra, William Irvine, Tom Jones, Judith Karpen, Ken Kellermann, Jerry Krassner, Ed Krupp, Harold P. Larson, Barry Lasker, John Leibacher, Frank Low, Craig MacKay, John Mathis, Lucy McFadden, Christopher McKee, Adrian Melott, Michael Molnar, John Mulchaey, Richard Muller, Robert Noyes, R. Bruce Partridge, Jay Pasachoff, Peter Pesch, David Pierce, Caty Pilachowski, William Press, Ralph Pudritz, Kumar Ranohalli, John Rather, Steve Ridgway, Larry Rudnick, Joseph E. Salah, Blair Savage, David Schramm, Philip R. Schwartz, James S. Schweizer, Frederick Seward, David Shaffer, Steven Shore, Alex G. Smith, Peter L. Smith, Phil Solomon, Linda Sparke, Laird Thompson, Rodger Thompson, Harley Thronson, Charles Tolbert, Hugh Van Horn, Gerrit Verschuur, Thomas Weldon, Gart Westerhout, George Withbroe, Sidney Wolff, Andrew Young, Michael Zeilik.

VIII. REFERENCES

Abrams, G.S. et al. 1989. Phys. Rev. Lett. 63, 2123

Abt, H.A. 1990. Publ. Astron. Soc. Pacific 102, 368 Barrett, A.H.

P.C. Myers, & N.L. Sadowsky 1978. "Microwave Thermography of Normal and Cancerous Breast Tissue" Proc. 3rd Intern. Symp. on Detection and Prevention of Cancer (Decker: New York)

Bracewell, R.N. & A.C. Riddle 1967. Astrophys. J. 150, 427

Cooper, B.F.C. 1976. "Autocorrelation Spectrometers" Meth. of Exp. Phys. 12B, 280

Ellis, S.D. 1988. Enrollments and Degrees, Publ. No. R-151,25, Amer. Inst. Phys. NY Ellis, S.D. & P.J.M.

Mulvey 1989. 1987-88 Survey of Physics and Astronomy Bachelor's Degree Recipients. Publ. No. R-211.20, Amer. Inst. of Physics, New York

Kelley, R.L. 1987. "Atomic and Ionic Spectral Lines Below 2000 Angstroms" J. Phys. Chem. Ref. Data 16, Suppl. No. 1

Kidd, J.S. 1989. Scientometrics 15, 241

Malin, D.F. 1982. "Photographic Image Enhancement and Reduction: A Unified Optical Approach" J. Phot. Sci. 30, 87

Malin, D.F. 1990. Kodak Techbits, 1990 Issue No. 1, 1-10

O'Neill, T.J. et al. 1989. Science 244, 451

Reiger, E. et al. 1989. Science 244, 441

Staelin, D. 1980. "Passive Microwave Techniques for Geophysical Sensing of the Earth from Satellites" IEEE Trans. Ant. & Prob. AP-29, 683

Steigman, G. 1989. Ann. NY Acad. Sci. 571, 1 (14th Texas Symposium)

Trimble, V. 1988. Publ. Astron. Soc. Pacific 100, 646

Turco, R.P. et al. 1983. Science 222, 1283

Weinreb, S. "Digital Spectral Analysis Technique and its Application to Radio Astronomy" MIT Research Lab of Electronics, Tech. Report 412, 1963

Weinreb, S. & A.R. Kerr 1983. "Cryogenic Cooling of Mixers for Millimeter and Centimeter Wavelengths" IEEE Journal of Solid State Circuits SC-8

STATUS OF THE PROFESSION PANEL

PETER B. BOYCE,* American Astronomical Society, *Chair*
CHARLES A. BEICHMAN,* Institute for Advanced Study, *Vice-Chair*

HELMUT A. ABT,* National Optical Astronomy Observatories
WENDY HAGEN BAUER,* Wellesley College
GEOFFREY BURBIDGE,* University of California, San Diego
ANITA L. COCHRAN,* University of Texas, Austin
ROBERT DORFMAN, University of Maryland
HUGH HARRIS,* U.S. Naval Observatory
ROBERT HAVLEN,* National Radio Astronomy Observatory
CHRISTINE JONES, Smithsonian Astrophysical Observatory
JEFFREY L. LINSKY, University of Colorado, Boulder
JULIE LUTZ,* Washington State University
LEE G. MUNDY,* University of Maryland
COLIN A. NORMAN, Space Telescope Science Institute
PATRICK S. OSMER, National Optical Astronomy Observatories
JAY M. PASACHOFF, Williams College
BENJAMIN PEERY, Howard University
R. MARCUS PRICE, University of New Mexico
HARRY SHIPMAN,* University of Delaware
JILL C. TARTER,* NASA Ames Research Center
HARLEY THRONSON,* University of Wyoming

Status of the Profession

KEY POINTS ON THE STATUS OF THE PROFESSION

- The number of astronomers has grown by about 40% over the past decade. While the number of women entering the field has increased substantially, minorities are under-represented.
- The number of astronomers with jobs in industry, or with long term, non-tenured, jobs has increased dramatically compared with traditional faculty positions.
- The increase in the number of astronomers and the declining share of the NSF budget going to astronomy has led to extreme difficulties in the NSF grants program and in the support of the National Observatories.
- In 1989, direct NASA support of astronomers through the grants program exceeds that of NSF, although the total of the NSF grants program over the decade far exceeds that of NASA.
- Access to major new telescopes will be an important issue for the 1990s. US astronomers, who once had a monopoly on telescopes larger than 3 meters, will, by the year 2000, have access to just half of the world's optical telescope area.

INTRODUCTION

Astronomy in the 1990s is on the threshold of a great revolution with the advent of NASA's Great Observatories and with the promise of several ground-based optical telescopes that will exceed the light gathering power of the Palomar 5 m telescope. Dramatic progress will be made at other wavelengths as well, with the operation of the VLBA radio array scheduled for 1992 and the flights of the gamma ray, X-ray and infrared Great Observatories planned for the coming decade. The excitement generated by astronomy has resulted in a growth of the number of astronomers by 42 percent over the 1980s. However, the growth of the number of astronomers, the failure to keep pace with funding levels attained by other branches of science and the expense of new instruments has caused significant difficulties for several parts of astronomy. Problem areas include:

- Funding from the National Science Foundation, traditionally the patron of ground-based astronomy, has failed to keep up with the needs for grants to individual researchers, for maintenance of existing observatories, and for development of innovative techniques such as optical and infrared interferometry.
- The US is danger of losing its pre-eminence in astronomy as other countries undertake major new initiatives to build major new ground based and orbiting observatories.

- NASA's focus on large projects that take 10 to 20 years to complete has eliminated many more modest efforts that span the career of graduate students or postdoctoral fellows.

This chapter raises a number of issues without recommending any solutions, for this is the purview of the report of the Policy Panel. Topics discussed below include the demographics of the profession, the impact of the previous survey report, trends in the funding of research by NASA and the NSF, the access to major research facilities, a snapshot of the employment record of astronomers, the productivity of research astronomers, and the status of women and minorities in the field.

THE DEMOGRAPHICS OF ASTRONOMY

How Many Astronomers Are There?

The number of astronomers is important when we consider whether there will be enough astronomers available in the next decade to replace the number of expected retirements from astronomical faculties and to analyze the data from new instruments. A significant number of scientists trained in other fields, particularly physics, are attracted to the science of astronomy. Therefore our definition of who is an astronomer must include people who have migrated into the field and are doing astronomy as well as those who have been specifically trained as astronomers. To maintain consistency with the Field Committee Report, we use an operational definition of an astronomer as a scientist with a PhD, or equivalent, who is either teaching or performing research in astronomy. Although a broad definition, this set represents the pool of qualified astronomers in the U.S. capable of using astronomical data.

One of the best ways of estimating the number of astronomers, and of establishing the characteristics of the astronomical community, is to use the membership of the American Astronomical Society. The American Institute of Physics periodically surveys a sample of the AAS Membership, so their characteristics as to age, employer, primary work activity, etc., are known.

Before examining the results of those surveys, we must first confirm that the AAS membership fairly represents the astronomical community. At the time of the Field Committee report, it was found that about 80 percent of working astronomers were AAS members. Recent estimates by AIP and a spot check in the Washington, DC area indicate this number is the same today. We assume that the AAS membership is representative of the community and that information about the AAS membership is valid for the community as a whole.

Figure 1 shows that the membership of the AAS has grown steadily over the last decade. The size of the pool of astronomers can be estimated from the AAS membership by applying various correction factors to account, for example, for the number of foreign members, incompleteness of the AAS membership etc. An estimate of these factors by the Field committee led to an estimate of the nation's pool of astronomers to be 0.82 times the membership of the AAS. Now, ten years later, a similar factor still seems to be appropriate, and we estimate that the U.S. has a pool of nearly 4200 astronomers, up by 42 percent since 1980.

The pool of astronomers is not, however, the same as the number of active research astronomers. If the latter is defined as someone who at least co-authors one paper year, then a survey by Abt (1990a) indicates there are about 2800 active researchers. However, research activity is limited both by available funding and time, particularly for faculty who work at institutions which are oriented primarily toward teaching. Given modest support, astronomers at smaller institutions oriented toward teaching can, and do, make significant contributions to astronomical research. These astronomers can readily involve students in their work and thus make even greater contributions to the educational process. With additional resources the pool of research astronomers could increase significantly.

The Solar and Planetary divisions of the AAS provide an initial indication of the numbers of solar system astronomers. The Solar Physics division numbers about 350 U.S. members, or approximately 6 percent of the AAS membership. This value is considerably lower than similar ratios reported for other countries, e.g. France with 15 percent, West Germany with 8 percent and Japan with 12 percent. The Division of Planetary Sciences has 700 members, about 13 percent of the total AAS membership.

How Active Are the Astronomers?

The number of papers published in astronomy is a measure of research activity. Abt (1989a,b) has

AAS Membership
and Number of U.S. Research Papers

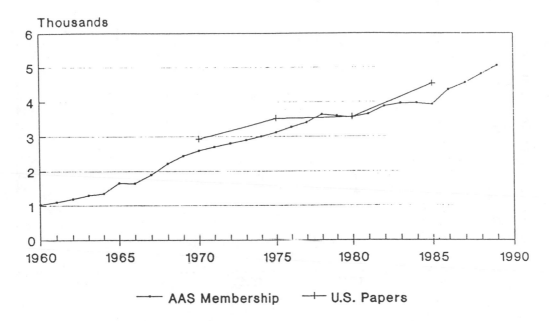

FIGURE 1   The membership of the AAS has increased by 42 percent in the last decade. The number of scientific papers by US authors has increased by roughly the same proportion.

studied the publication practices of astronomers. Over the past two decades the number of research papers produced by U.S. astronomers has grown significantly, just about as rapidly as the AAS membership. U.S. astronomers published 4527 papers in 1985, a growth of 28 percent in ten years. Over the same time, the AAS membership grew by 26 percent (Figure 1). However, the growth of U.S. papers has lagged behind the rest of the world. The number of non-U.S. papers grew by 57 percent over those ten years, and now they account for a dominant 68 percent of all astronomy papers.

Abt also finds that the average full member of the AAS publishes 3.3 items per year, of which about half are refereed papers describing original research. Correcting for multiple authors, Abt concludes that the average AAS member produces the equivalent of one-half of an original research paper per year. However, about a quarter of the AAS members are more than twice as productive as the average, producing the equivalent of one single-author research paper per year.

Astronomy is rapidly becoming more international. Nowhere is this more evident than in the growth in the number of papers from foreign authors in astronomical journals around the world. Abt (1989b) has examined the statistics of foreign authorship in astronomy by defining four categories of papers: 1) solely American authorship; 2) lead US authorship but significant foreign participation; 3) lead foreign author but significant American participation; and 4) solely foreign authorship. If foreign papers are defined as half of those in categories 2) and 3), plus all of those in category 4), then Figure 2 shows a dramatic increase in foreign papers since 1970 in the *Astrophysical Journal*; similar numbers apply to the *the Astronomical Journal* as well.

The number of papers with international collaboration (categories 2 and 3) are becoming much more common than they were before the 1960s when they represented less than 5 percent of the total. Starting in the early 1970s, the *Astrophysical Journal, Astronomical Journal, the Monthly Notices of the Royal*

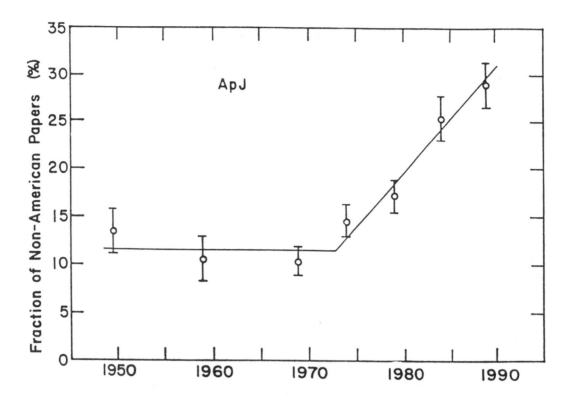

FIGURE 2   The number of papers with significant foreign participation appearing in the Astrophysical Journal has increased dramatically in the last 20 years (1990b). Courtesy of the H. Abt and the Astronomical Society of the Pacific

Astronomical Society and *Astronomy and Astrophysics*, all showed a shift from purely national papers to the current percentage of 25 percent multi-national papers. International cooperation is quickly becoming the standard way of doing research, just as before 1970, papers by a sole author were the norm, but thereafter papers by multiple authors became usual.

How Can We Characterize the Astronomical Community?

Astronomy is being carried out by a young and growing community. Traditionally concentrated in academic institutions, astronomers are now employed by an increasingly diversified group of institutions. One-half of U.S. astronomers now work outside of academe. The number of astronomers working with industrial or commercial companies is showing the most rapid growth, a factor of three over the past decade.

In 1987, the median age of U.S. astronomers was 42, the youngest for all the societies sampled by the American Institute of Physics (Figure 3). Half the U.S. astronomers were between the ages of 35 and 50. In contrast, in 1973 astronomers were, on the average, eight years younger, with a median age of 34. In those days, half the astronomers were concentrated in the ten year age group of 27 to 37. Given a continuation of present trends, we would expect the median age to move up by another five years in the coming decade. The profession will not suffer any serious effects due to retirement in the coming decade.

The percentage of women in astronomy has grown by almost fifty percent during the last decade; from eight percent to twelve percent. Examination of the graduate student population shows signs of a slow, continuing growth in the number of women in astronomy.

Recent surveys show that twelve percent of the doctoral degrees awarded by astronomy departments in 1987 went to women, and that twenty percent of the total astronomy graduate population is female. These

Age Distribution
AAS Membership and All Physicists

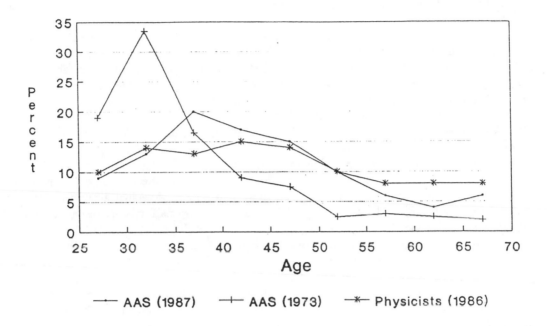

FIGURE 3   The average age of astronomers and physicists as determined by membership surveys by the AIP. In general, a astronomers are younger than physicists.

fractions are similar to or higher than the female membership of the AAS and lead us to expect that the percentage of women in astronomy will continue to rise slowly over the next decade.

Ethnic minorities are significantly under-represented in astronomy. Preliminary results from a survey of the AAS membership show that 93 percent of the AAS members classify themselves as white, 4 percent as Asian and 1 percent as Hispanic. Afro-Americans comprise less than 0.5 percent of the AAS membership.

Results of the Field Committee Report on Personnel Issues

The previous Astronomy Survey Committee, chaired by George Field, emphasized that successful basic research is dependent upon an active and vigorous community which can only be maintained through adequate funding of the research infrastructure. The Field committee made two important recommendations with regard to providing long-term, federal funding in astronomy. First, they recommended an increase in the money for theory and data analysis. Second, anticipating increased retirements in the 1990s, the Committee also recommended that ten to twenty competitive five-year positions, funded by NSF, be established for young astronomers.

Partly in response to these recommendations, NSF established the Presidential Young Investigator awards designed to encourage young scientists, and NASA initiated a specific program for supporting astrophysics theory groups. NASA also recognized the need for longer term funding for young astronomers and has recently converted 10 percent of their growing data analysis funding into a program of five-year awards.

THE FUNDING OF ASTRONOMICAL RESEARCH

Although astronomy has a long tradition of private and state support, federal funding continues to provide the bulk of the funding for astronomical research. The NSF provides the major share of funding for ground-based astronomy, while NASA provides nearly all the funding for space astronomy. Historically, ground-based optical astronomy has attracted additional private and state support; radio astronomy has depended heavily upon federal funding; and space astronomy has been entirely federally supported.

Observational astronomy depends heavily upon modern, effective instrumentation. The faintness of the objects being studied requires massive and expensive telescopes which, in turn, limits the number of research instruments that can be built. Since their establishment nearly three decades ago, the national observatories, funded by NSF, have provided unrestricted access to first class research telescopes and facilities on the basis of merit. NASA, likewise, has adopted the general policy that observing time should be accessible to anyone in the astronomical community and that use of the space facilities should be awarded on the basis of scientific proposals which are reviewed by a panel of scientific peers.

But it takes more than instruments to make a strong and vital scientific program: it takes people to do the research. Programs of competitive research grants for individual astronomers have been one of the most important components of U.S. astronomy. In combination with telescopes and instrumentation available to the entire community, these grant programs make possible the broad geographical distribution which characterizes U.S. astronomical research. Both NSF and NASA have such programs. In addition, the Departments of Energy (DoE) and Defense (DoD) have grant programs for basic research, and both award some grants that are astronomical in nature.

The impact of private and state funding upon the total competitive research grant picture is small. There is a substantial amount of private and state money in various institutions that supports astronomical research by the staff and faculty of the institution. However, the NSF and NASA grant programs are of paramount importance to the health of the field. Thronson has tabulated the credits for funding which appear in articles by U.S. researchers in the *Astrophysical Journal* in 1989. He finds that NASA and NSF are each cited about 40 percent of the time and other federal agencies (especially DoD and DoE) are credited in about 10 percent of the papers.

We will, therefore, analyze the funding trends in the two largest grant programs in which the entire program is restricted to astronomy projects, the NSF program of astronomy grants and the various grant programs in NASA's Astrophysics Division (Figure 4). Measuring trends in the two major programs will suffice to give an indication of the health of astronomy funding in the two agencies. All references to budgets and amount of support have been converted to 1989 dollars using the OMB correction factors for basic non-military research, which are very close to the standard cost-of-living index.

Support from the National Science Foundation

The Grants Program Funding from the NSF Division of Astronomical Sciences, which supports most astronomy from the ground, goes into two major programs: the national observatories and the program of individual research grants. This latter category includes both grants to individual astronomers and grants to universities to support various facilities.

The NSF grants program funds projects that range widely in size, from small grants for various supplements, publication of meeting proceedings, etc. up to grants of over a million dollars for the operation of entire university observatories (Figure 6a). The bulk of the NSF grants falls between these extremes and supports individual research projects. These grants are critically important to the individual researchers, particularly in the area of theory. Figure 6b shows a steady erosion in the grants to individuals and which account for about 2/3 of the overall grants program; excluded are major grants to university observatories. Another measure of the erosion in support to individual astronomers is the decrease in the average size of grants in the $25-250k range. With typical costs of summer salary, publications and a graduate student salary plus overhead now exceeding $70,000 per year, the grants program has dropped below a critical threshold. While an individual can, and often must, have more than one grant, such a system of support suffers from inefficiencies and often produces more proposals than papers.

Overall, the picture is rather bleak within the NSF grants budget, except for a 24 percent increase in

Astronomy Grant Funding
in 1989 Dollars ($M)

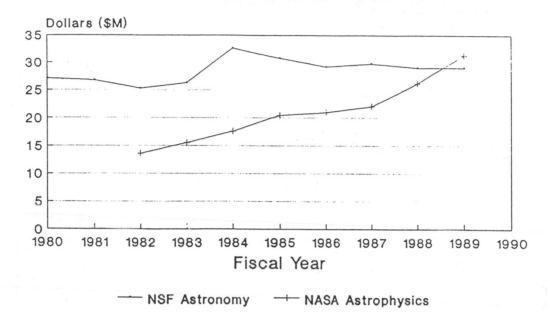

— NSF Astronomy   —+— NASA Astrophysics

FIGURE 4   Overall funding of astronomy grants by NASA and the NSF in constant collars. In the case of NSF, funding of university observatory facilities is included.

1984 that was eaten away in subsequent years. While all the indicators of activity within astronomy were rising with the number of publications up by 27 percent in the first five years of the decade and the number of astronomers increasing by 40 percent since 1980, the 1989 grants budget is only 11 percent above the 1980 level in constant dollars. The number of grants awarded has risen from about 250 in 1980 to 372 in 1989, but the average size of grants to individuals is down by almost one-half in 1989 dollars, from just over $90k in 1981 to $55k.

The increase in 1984 appears to have resulted from the publication of the Field Committee Report, which noted the sad state of the infrastructure that enables astronomical research to proceed, and recommended an augmentation. Unfortunately, after 1984, the research base at NSF has been eroded year by year. In fact, Figure 7 shows that the construction funds for building the VLBA have come from the research base, a move that leaves the research infrastructure as poorly funded as it was a decade ago.

Support for Facilities   Support for national facilities has deteriorated during the last decade. Data for NRAO are representative of the problem (Figure 7). The integrated shortfall in the operations budget between 1984-1989 is almost $15 million, with drastic, adverse effects in staff compensation and morale, telescope and receiver maintenance, development of new receivers, computing capabilities and cutbacks in critical educational and postdoctoral programs.

Support from NASA

In marked contrast to NSF, the NASA grant programs in astrophysics over the decade have more closely tracked the growth in the number of astronomers and the publication activity. The NASA programs were reorganized several times during the decade, and it is not possible to derive comparable numbers for years before 1982. However, from 1982 to 1989, the total of the various NASA grant programs has grown from $17.5M to $31.3M, or 79 percent. The average grant size has remained stable at about $40,000 throughout this time. The number of grants awarded per year rose from 336 in 1981 to 574 in 1989. The small grant size is caused by the fact that NASA gives grants to pay for the expenses associated with observing with IUE.

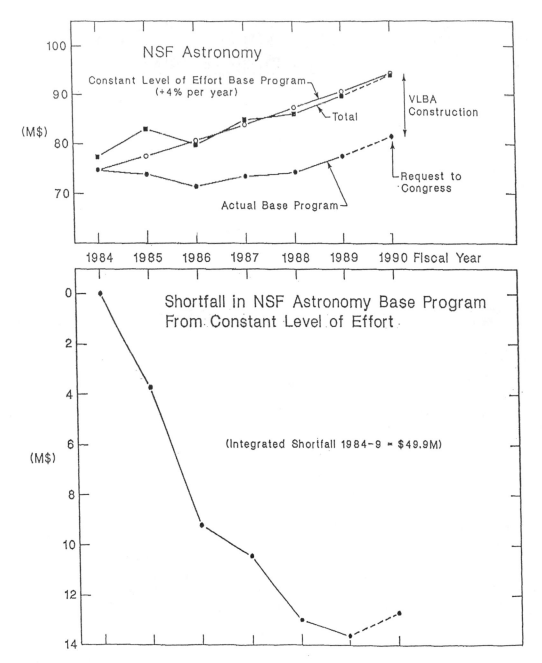

FIGURE 5   The NSF astronomy program has not received adequate funding to maintain a constant level effort of program over the last five years. Construction of the VLBA has had a marked impact on the support available for research grants.

It should be pointed out, however, that the average grant is too small to support a post-doctoral fellow or, in some cases, even a graduate student. Many IUE researchers have to write multiple proposals to support themselves and their postdocs and graduate students. In the rest of the Astrophysics Data Program the grant sizes are significantly larger, but still often low enough that multiple grants are required to support a researcher.

Behind the rise in the number of grants by NASA is a specific new policy to provide sufficient resources to analyze adequately the data from their space science missions. This policy includes setting up multi-wavelength data archives, provision of data analysis funding long after missions have ceased operation, and the establishment of senior fellowships. Nor has theory been neglected. There is now a strong program that

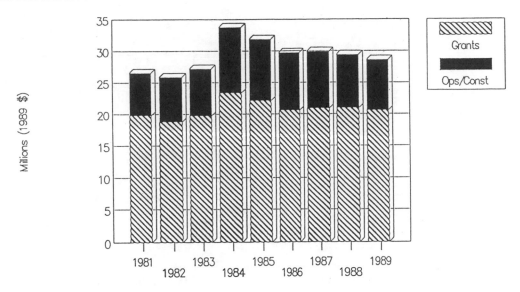

FIGURE 6a The division between funding of research grants and the operation and construction of university observatories over the last decade.

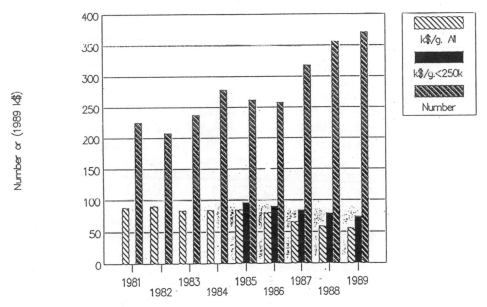

FIGURE 6b The numbers and average size of NSF research grants, corrected for the portion of the research budget devoted to operation and construction of facilities. The middle bar shows the decreasing size of grants in the $ 25-250 thousand range.

supports work on theoretical astrophysics, a vital component needed for the interpretation of astronomical data.

One important trend to emerge from this analysis of the grants program is the emergence of NASA as the dominant agency in astronomy grant funding. In 1982, NSF provided almost 60 percent of the grant money, but in 1989 NASA, for the first time, provided more money for astronomy grants than did the NSF. Since several large, long-lived missions will be launched in the early 1990s, NASA's future spending plans call for a major increase in the amount of data analysis money which will go the community in the form of grants. If present plans come to pass, NASA will be providing two-thirds of all the astronomy grant money by 1995.

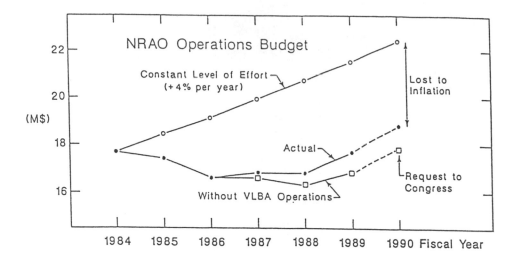

FIGURE 7 The funding of the National Radio Astronomy Observatory shows that funding has fallen significantly short of maintaining a constant base program with concomitant impact on all aspects of observatory operation.

ACCESS TO MAJOR TELESCOPES

Astronomy is, more than almost any other physical science, an observational science. Two aspects are of critical importance for maintaining American leadership in the field. First, we must build the best telescopes and equip them with the best instruments. Second, we must guarantee broad access by US astronomers to those instruments. The US faces serious challenges to its long-standing pre-eminence in optical astronomy. Since the 1950s, the US had almost all of the collecting area in telescopes larger than 3 m (Figure 8 and Table 1). By the end of the century the US fraction will drop to about 50 percent with Japan and a united Europe making major national commitments to large, 7-8 m, optical telescopes. Without the *privately* funded instruments like Keck, Columbus and Magellan, and without the proposed (shared) NOAO 8 m telescopes, the U.S. fraction would drop to about 35 percent of the world's collecting area.

Different branches of astronomy have similar problems:

- In the fifteen years between Einstein and AXAF, American X-ray astronomers will have relied solely on foreign telescopes to make advances in a field they invented.
- In centimeter wavelength radio astronomy, the aging VLA is still the supreme instrument, but in the rapidly growing field of millimeter radio astronomy, the IRAM 30 m telescope and the new millimeter arrays in Europe and Japan challenge our leadership in another field we developed.
- The US led the development of infrared astronomy, based in large part on the strength of our detector technology, and opened space infrared astronomy with the IRAS satellite. Leadership in infrared astronomy will pass to the Europeans when they launch the ISO mission and we wait until 2000 or later for the flight of NASA's infrared Great Observatory, SIRTF.
- The US pioneered neutrino astronomy, but the major facilities in particle astronomy are now in Europe, Japan and the Soviet Union with a major new Canadian observatory expected soon.

The competition for observing time on major new telescopes will be severe. For instance, the Hubble Space Telescope (HST) made initial time assignments in response to proposals for the first year of operation. Nine times as much observing time was requested as was available and in the first year only one proposal in five was assigned time. This is a remarkably high demand for an instrument whose launch date was two years in the future at the time of proposal submission and whose launch had frequently been postponed. Even with reduced capabilities due to the flaw in its primary mirror, HST will be oversubscribed by a factor of about five to one for ultraviolet imaging and spectroscopy where its powers are unequaled by other facilities. Another less extreme case is the four meter telescopes at the National Observatories, which are

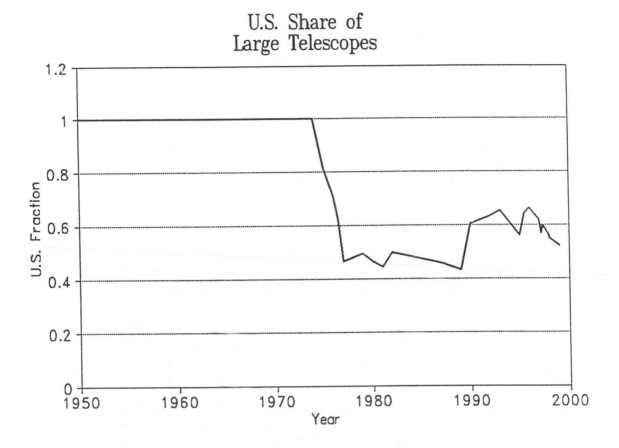

FIGURE 8 The fraction of the world's telescope area accessible to US astronomers, for optical telescopes larger than 3m. The figure assumes all the telescopes described in Table 1 are built.

more than ten years old, but where the demand typically runs four times more than time available. Requests for observing time on other U.S. telescopes typically exceed the number of available hours by factors of two or more.

ASTRONOMY AS A PROFESSION

The Growth of "Firm Money" Positions

The types of jobs that astronomers now occupy have changed over the last decade. Traditionally, an astronomer achieved a permanent job on the tenured faculty at a university or college. Industrial and academic research jobs, with their perceived impermanence and lack of academic freedom, were considered less desirable. However, as predicted in the Field Committee Report, the rapid growth of astronomy in the late 1960s and the concomitant youthful age of the astronomical faculty has meant that faculty retirements have been few and far between. This has, in turn, severely limited the number of tenured faculty jobs that opened up during the last ten years. Industrial and national research facilities have taken up much of the slack by employing increasing number of astronomers who still carry out astronomical research.

Academic institutions still provide jobs for half of the AAS members (Figure 9a), but other types of employment have become more important. Nearly 75 percent of the employed in academia hold permanent, tenured jobs (Figure 9b). Exclusive of astronomers holding postdoctoral appointments, 17 percent of the academic astronomers have positions as research associates.

Table 1. U.S. Share of Large (> 3m) Telescopes

Telescope Name	US Share	Mirror Diam (m)	# Mirrors	Date	World Telescope Area (m^2)	US Fraction
Palomar	1	5	1	1950	19.6	1.00
Lick	1	3	1	1959	26.7	1.00
NOAO-KPNO	1	4	1	1974	39.3	1.00
CTIO	1	4	1	1975	51.8	1.00
AAT	0	3.9	1	1975	63.8	0.81
Calar Alto	0	3.5	1	1976	73.4	0.71
ESO	0	3.6	1	1976	83.6	0.62
USSR	0	6	1	1976	111.9	0.46
IRTF	1	3	1	1979	118.9	0.50
UKIRT	0.15	3.8	1	1979	130.3	0.47
MMT	1	1.8	6	1980	145.5	0.52
CFHT	0.15	3.6	1	1980	155.7	0.50
Herschel	0	4.2	1	1987	169.6	0.46
ESO NTT	0	3.6	1	1989	179.2	0.43
Keck-1	1	10	1	1991	257.7	0.61
ARC	1	3.5	1	1991	267.3	0.62
WIYN	1	3.5	1	1992	277.0	0.63
MMT-upgrade	1	6.5	1	1994	294.9	0.65
VLT-1	0	8	1	1995	345.2	0.56
Keck-2	1	10	1	1996	423.7	0.64
Columbus	0.75	8	2	1996	524.2	0.66
Japan	0.1	7.5	1	1997	568.4	0.62
VLT-2	0	8	1	1997	618.7	0.57
Magellan	1	8	1	1997	668.9	0.60
VLT-3	0	8	1	1998	719.2	0.56
NOAO	0.5	8	2	1997	819.7	0.55
VLT-4	0	8	1	1999	870.0	0.52

In contrast to postdoctoral appointments, which are normally awarded for a short, fixed term, usually two or three years, the research associate position, while still dependent upon outside funding, is available for a longer term. In some cases it may be continued as long as money is available. A significant number of universities are showing a willingness to guarantee some degree of permanence for their research associates, typically by guaranteeing a year or two of support to weather the loss of a grant or contract until new support can be found. The emerging importance of this type of job as a long-term career choice for astronomers deserves highlighting. We refer to such jobs, that carry a degree of permanence which exceeds the normal three year postdoc position, as "firm money" positions.

Industrial jobs, though often dependent upon a company's ability to obtain outside funding, also have a degree of permanence. Although companies may shift employees from one contract to another, the larger salaries and the challenges of the corporate world are attracting more astronomers than ever before. The percentage of U.S. astronomers who work in a corporate setting has nearly doubled over the last decade. In 1987, the number of astronomers who work for industry or are self-employed exceeded the number who work in Federally Funded Research and Development Centers, such as the national observatories. Corporate positions, which have no guarantees but which do have a reasonable degree of permanence, would also be

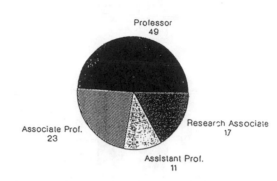

Where Astronomers Work
AIP Survey - 1987

University 45
College/Junior 5
Non-Profit/Other 4
Government 18
FFR&DC 13
Industry/Self 14

Academic Positions Of AAS Members
AIP Survey - 1987
Postdoctoral Positions Excluded

Professor 49
Associate Prof. 23
Research Associate 17
Assistant Prof. 11

FIGURE 9 a) The fraction of astronomers in various types of jobs. FFR&DC refers to federally funded research and development centers; b) The fraction of astronomers holding different ranks within universities and colleges.

classified as "firm money" positions. Interestingly, it is not exclusively the young astronomers who are choosing careers outside of academia. The median ages of academic and non-academic astronomers are identical.

Trends in employment become particularly apparent when a tabulation is made of the type of position astronomers hold ten years after receiving their Ph.D. A survey was made of the 117 astronomers who received their Ph.D. in 1980. In comparison with the class of 1970 surveyed ten years later two conclusion can be drawn. First, there has been a significant decrease in the number of graduates who are working in universities, particularly true for tenure-track faculty positions. Only 32 percent of the class of 1980 are now in universities, 22 percent are in tenured or tenure track positions. The numbers for 1970 are 42 percent and 30 percent respectively. Second, there has also been a substantial growth in the number of people who are working in industry and some growth in the percentage working in Federally Funded R&D (FFR&D) Centers. 12 percent of the class of 1980 work in industry, up from 1 percent for the class of 1970. 15 percent now work in FFR&D Centers, up from 11 percent. Finally, the number of people who have left the field is a roughly constant 30 percent for both the 1970 and 1980 classes.

Research is an important part of the life of the average U.S. astronomer. Not only is this fact manifested through the large number of papers produced every year, more than one for each member of the astronomical community, but it also emerges from direct surveys of the community. Over half of all AAS members surveyed claim to work primarily on basic research and only 9 percent cite teaching as their primary activity, but another 25 spend a substantial amount of their time in teaching. The other three major activities at which astronomers work are administration, applied research and design, development or engineering.

The records of the Solar Physics Division of the AAS show that solar astronomers are not well represented on university faculties. Approximately 45 out of the 350 members of the solar physics division are found in academic environments, or roughly 10-20 percent. Most of the remainder of the solar astronomers are found in industry of national facilities. This value contrasts with the roughly 50 percent of astronomers who hold university jobs. In contrast to solar astronomers, the distribution of planetary astronomers mirrors that for U.S. astronomers as a whole. Fifty-one percent of the members of the AAS Division of Planetary Sciences are employed by universities. Thirty percent of the planetary astronomers say they work in government labs, comparable to the percentage of all astronomers which work in government labs and federally funded research centers.

What Can We Say About the "Typical" Astronomer's Career?

In order to understand the workforce in astronomy we have to know where people who work as

Number of Astronomy Degrees

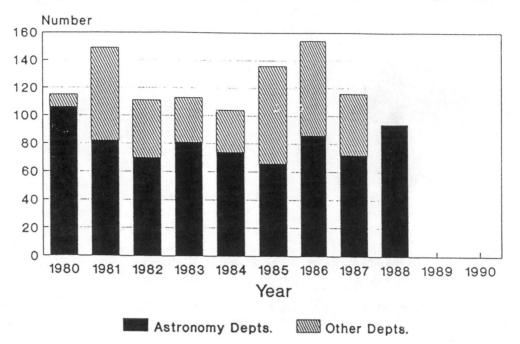

FIGURE 10 The number of PhDs granted in astronomy has been relatively steady over the decade. The number of scientists entering astronomy from other fields has increased since 1980.

astronomers are trained, what the progression of jobs is in an "average" career, and how and when they leave the field.

The rate of production of astronomers has remained virtually constant over the past two decades. The number of doctoral degrees awarded for dissertations in astronomical topics over the last decade has averaged 125 degrees per year (Figure 10). Of these, 81 degrees come from the 69 institutions that grant astronomy degrees. The other 44 are degrees in physics, or some other science, where the thesis topic is clearly identified as astronomy or astrophysics.

What undergraduate institutions do astronomy graduate students come from? AIP surveys show that 57 percent come from PhD granting institutions, whereas 15 percent are produced by the 4-year colleges and foreign institutions account for 24 percent. What did astronomy graduate students major in as undergraduates? Generally it was physics (52 percent), followed by astronomy (28 percent), mathematics and other (16 percent), and engineering (4 percent).

The flow of new people into the AAS has been twice the rate of degree production, averaging 250 per year. Thus, half the people entering astronomy in the past decade were trained in a science other than astronomy and migrate into the field later in their career. In the recent survey of the AAS membership, 1/3 of the respondents held their degree in a field other than astronomy. Such a high fraction of scientists coming in from outside the field is quite unusual and attests to the intellectual excitement of astronomy.

The other critical factor regulating the size of the pool of astronomers is the flow out of the field. Many people who were trained in astronomy or who once worked as astronomers and have, for personal or economic reasons, moved into another field typically tend to continue their membership in the AAS for some time after they leave the field. Under our definition such people continue to be counted as astronomers for a few additional years, only disappearing from our counts when they drop their AAS membership. Over time, this effect will average out and should not introduce a significant error into our estimates.

Specific studies done for the Field Committee ten years ago found that the attrition rate was greatest immediately after the postdoctoral period. Similar studies have not been carried out recently. However, the number of people dropping their AAS membership each year has remained unchanged during the past decade, about 50 persons per year.

What is the Job Situation for Astronomers?

One critical question is how closely the number of astronomers matches the number of available jobs. Since astronomy is a small and relatively well defined field, it is possible to get fairly comprehensive information about the employment situation. The American Astronomical Society publishes a monthly listing of jobs of interest to people trained in astronomy. The AAS Job Register was started in 1977, and in the early eighties a successful effort was made to ensure that it provided a complete listing of astronomy jobs. Since then, the AAS Job Register has provided an excellent indication of the astronomy job market, listing most astronomy-related jobs likely to be filled by Ph.D. astronomers. Of course, it also includes some jobs for people with Bachelors and Masters degrees, as well as jobs that involve primarily engineering, computer work, administration, and other non-astronomy work that may be of interest to astronomers. However the great majority of listed jobs are for Ph.D. level scientists.

The total number of jobs advertised each year in the AAS Job Register has risen steadily from fewer than 200 per year in the early 1980s to about 280 per year in the late 1980s, a net increase of about 30 percent over the decade. While these are not all new jobs, they do indicate an increased opportunity to find or switch jobs. The largest single increase came at the time the Space Telescope Science Institute was established and clearly represents a growth in the number of jobs in astronomy.

Two-thirds of the listed jobs are non-permanent, either post-doctoral positions or work on contracts expected to be of limited duration. These jobs serve an important role in providing transitional positions between graduate school and permanent work, but they do not contribute to the permanent job market. Instead permanent and semi-permanent jobs provide the long-term location and job security sought by most astronomers as their careers advance. These include academic and observatory jobs that are described as tenured or tenure-track, civil service positions at government labs, and jobs that have no definite duration and have a high expectation of continuing for several years.

Excluding tenured or senior jobs that are not open to young astronomers, the number of new, permanent and semi-permanent jobs has risen only slightly from about 60 per year in 1980-1982 to just over 70 per year in 1986-1988. Because this number is only about one-half the number of new Ph.D. astronomers, the competition for permanent jobs in astronomy has remained high.

As confirmed by the growth in the number of astronomers working in the commercial sector, it is apparent that substantial numbers of Ph.D. astronomers have joined a pool of people not employed in traditional permanent astronomy jobs. While some people have left astronomy altogether, most have maintained employment in astronomy-related jobs of various types. They are often doing applied research, hardware design or construction, and computer system or software development, often on contracts for particular projects. Many of these jobs are funded by federal sources, either directly by NASA or NSF grants, or less directly by contracts for NASA projects.

Projections of the job situation into the next decade are clouded by uncertainty in several areas. First, the recent lifting of mandatory retirement ages may change the normal retirement patterns, particularly for academic astronomers. Second, the trend toward longer term, more stable research funding by NASA coupled with the growing acceptance of "firm money" jobs by astronomers makes the job situation particularly sensitive to federal research funding. Third, when half the practicing astronomers come from physics, which has ten times the number of people, the field is particularly susceptible to changes in the migration pattern. These changes might be small in terms of the other field, but could severely disrupt the projections of growth in astronomy.

EDUCATION AND HUMAN RESOURCES

What Can Astronomers Do About Improving General Science Education?

According to all the experts, including the Presidential Commission that wrote the report entitled *A Nation at Risk*, the U.S. is facing a technical manpower shortage. A major effort is under way to improve the technical competence of the U.S. workforce before Halley's Comet returns in 2061. What part can astronomy play in this effort? With its broad public appeal, astronomy often attracts students at all levels into science courses, and it is only these students who are able to go into careers in science and engineering. Astronomers can work with local schools and with science museums and planetariums to provide up-to-date, interesting, and scientifically accurate materials. They can cooperate with local reporters to see that exciting advances in astronomy are brought to the attention of the public.

The astronomical community has a responsibility to help at all levels, from working with local schools to encouraging the federal government to take action. The supply of new scientific talent, both for the benefit of our country and for maintaining excellence in U.S. astronomy, will depend, in part, on our efforts in this important task. The diverse issues of astronomy and education are discussed in the report of the Policy Panel, in Chapter 7 of the Volume I of this report, and in *An Educational Initiative in Astronomy*, edited by R. A. Brown.

What Can Astronomers Do About Improving Pre-College Science Education?

Though almost all students take science courses in the junior high-school years, few students choose to take more advanced high-school science courses. Contributing to this drastic sieve is the content of high-school physics courses and the reputation among high-school students of the courses and their teachers. If more high-school students took courses in the physical sciences, then the colleges and universities would have more chance to train students technically in astronomy, physics, mathematics, and geology. Astronomers could help high schools include contemporary astronomical topics in their courses, as is done in biology and other science courses, rather than limiting astronomical discussion largely to the important historical origins of astronomical thought in centuries past.

On the elementary-school and junior-high levels, science courses are usually taught by teachers with little science training. Astronomers can provide up-to-date information about current astronomy projects to local teachers and schools, and offer to assist in presenting current and exciting science topics to young students. Astronomers can encourage local schools to take their students to local facilities such as science museums and planetariums, or to visit telescopes at the astronomers' institutions.

On all secondary levels, astronomers can offer to cooperate with local school boards or state adoption committees, as appropriate, to see that textbooks and other materials have extensive and accurate information about astronomy. Astronomers can work with publishers and other providers of educational material to see that astronomy is well represented in the materials available.

What Can Astronomers Do About Improving College Education?

Astronomers can take active roles in their own institutions' teaching of both non-science-major and science-major undergraduates. They can see that personnel and equipment are devoted to providing the best possible environment for undergraduate courses in astronomy. They can make available courses for all academic levels of undergraduates, with particular attention to providing courses beyond the beginning survey not only for astronomy or physics majors but also for majors in other sciences and in non-science fields.

Astronomers can improve the status of astronomy teaching by making certain that the best professors take their turns in teaching lower-level courses. They can provide rewards for good teaching among graduate students. At present, most first-year students (60 percent) are supported on teaching assistantships, while research assistantships are provided later. Astronomers could try, instead, to provide structure in which the most qualified and experienced graduate students work as teaching assistants for part of their time. This would not only assist the undergraduate students in obtaining an interesting introduction to astronomy but also provide better training in teaching for the future astronomy professors. Providing the best education to liberal-arts college majors is a way of influencing the country's future lawyers, businessmen, and politicians, who can eventually play important roles in governmental and private funding of science.

The national agencies, including NSF and NASA, can extend and improve their programs that involve in research professors who have heavy time commitments involving teaching. Such programs as NSF's *Research in Undergraduate Institutions* can be expanded to increase the exposure of undergraduate students to contemporary astronomical research. During fiscal 1988, seven new awards were made under this program for a total of $369,000 plus three supplemental grants for $170,000. From regular research awards, an additional 6 grants were awarded to undergraduate institutions for $287,000 plus 8 continuing awards for $185,000 totaling $472,000.

Programs of liaison between research universities and surrounding colleges and universities can be extended. Regional informal meetings of astronomers are valuable ways of keeping members of the astronomical community, and their students, in touch with modern astronomical research.

Programs at observatories and research institutes to involve undergraduates in summer or term-time programs can be enlarged. Such programs tend to be prominent candidates for cutting at times of fiscal stress.

Astronomers can be active in incorporating recent research, including not only optical astronomy but also non-optical astronomy, into laboratory work for students at all levels.

What education issues face the astronomical community?

Training of instrumentalists. Based on anecdotal opinions there seems to be a falloff in the number of astronomers being trained in the design and construction of astronomical instruments. A questionnaire to all astronomy department chairs as part of this survey revealed that half of them them feel that their departments do not have enough faculty and staff who specialize in the development of astronomical instrumentation. More than half also foresee the need for an increase in the number of faculty who specialize in instrumentation during the next five years.

Problems of couples in astronomy. As the number of women in astronomy grows, the hiring of qualified applicants is more and more dependent upon finding employment for the applicant's spouse. Since women astronomers are more likely to be married to other astronomers than are their male colleagues, this problem is much more severe for the female members of the profession.

The survey of department chairs shows that 60 percent of the departments have tried to hire an academic couple during the past decade. None of the departments had anti-nepotism policies which would prevent them from hiring both persons, and there were a fair number of successful arrangements where both parties were accommodated. However, there were a larger number of cases where substantial difficulties were encountered. In the majority of cases all parties made a good faith effort to make things work out, but, in a few cases, the department or the central administration was simply unwilling to bend.

The most common difficulty was that making a spousal arrangement took time. Some spouses had to accept distinctly second-class positions such as adjunct faculty appointments. The number of cases reported indicate that this is a substantial problem which requires sensitivity, good faith and time to work out. With the rising number of women in astronomy it is important that all astronomy departments and university administrations increase their efforts to find adequate solutions to the problem of hiring astronomy couples.

Women and minorities in astronomy. Issues of equality, under-representation, role models, etc. are important. The growing percentage of women in astronomy will bring these issues into sharper focus in the next decade. The AAS has surveyed its membership on the question of discrimination. On the basis of answers to the survey, which had a response rate in excess of forty percent, the incidence of discrimination is not large. 10 percent of the respondents report having seen or experienced discrimination against women in the areas of hiring practices, pay and promotions. 16 percent report having seen discrimination in the general social treatment of women.

It is important to follow up on these data with a more detailed investigation. In 1979 the AAS Committee on the Status of Women made a detailed study of the perceptions of female members. This study should be repeated so a comparison can be made with the attitudes and perceptions of the previous decade. It is important to pinpoint areas in which progress has been achieved as well as areas in which more needs to be done. We owe it to our profession to see that the study of astronomy is equally accessible both males and females. We owe it to our colleagues to ensure that jobs at all levels are, likewise, equally accessible to everyone.

REFERENCES

Abt, H. 1990a, *Publ. Astron. Soc. Pacific*, in press.

Abt, H. 1990b, *Publ. Astron. Soc. Pacific*, in press.

AIP Membership Profile: Employment Mobility and Career Change, 1989, Porter, B. F. and Kellman, D., (American Institute of Physics Report R-306.2)

The American Astronomical Society 1990 Membership Survey Preliminary Report, AAS internal report.

An Education Initiative In Astronomy, Report from a Workshop held in Washington, D. C., February 1990, ed. by R. A. Brown

A Nation at Risk: The Imperative for Educational Reform, 1983 (Department of Education: Washington, D.C.)

SCIENCE OPPORTUNITIES PANEL

ALAN LIGHTMAN, Massachusetts Institute of Technology, *Chair*
JOHN N. BAHCALL, Institute for Advanced Study, *Vice-Chair*

SALLIE L. BALIUNAS, Harvard-Smithsonian Center for Astrophysics
ROGER D. BLANDFORD, California Institute of Technology
MARGARET E. BURBIDGE, University of California, La Jolla
MARC DAVIS, University of California, Berkeley
DOUGLAS EARDLEY, University of California, Santa Barbara
JAMES E. GUNN, Princeton University
PAUL HOROWITZ, Harvard University
EUGENE LEVY, University of Arizona
CHRISTOPHER F. McKEE, University of California, Berkeley
PHILIP C. MYERS, Harvard-Smithsonian Center for Astrophysics
JEREMIAH OSTRIKER, Princeton University Observatory
VERA C. RUBIN, Carnegie Institution of Washington
IRWIN SHAPIRO, Harvard-Smithsonian Center for Astrophysics
MICHAEL S. TURNER, Fermi National Accelerator Laboratory
EDWARD WRIGHT, University of California, Los Angeles

The paper prepared by the Science Opportunities Panel appears in an abbreviated and adapted form as Chapter 2 of the full survey report, *The Decade of Discovery in Astronomy and Astrophysics* (National Academy Press, Washington, D.C., 1991).